T0406144

Handbook of Model-Based Systems Engineering

Azad M. Madni • Norman Augustine
Editors-in-Chief

Michael Sievers
Associate Editor

Handbook of Model-Based Systems Engineering

Volume 2

With 645 Figures and 61 Tables

Editors-in-Chief
Azad M. Madni
University of Southern California
Los Angeles, CA, USA

Norman Augustine
Lockheed Martin (United States)
Bethesda, MD, USA

Associate Editor
Michael Sievers
University of Southern California
Los Angeles, CA, USA

ISBN 978-3-030-93581-8   ISBN 978-3-030-93582-5 (eBook)
https://doi.org/10.1007/978-3-030-93582-5

© Springer Nature Switzerland AG 2023
This work is subject to copyright. All rights are reserved by the Publisher, whether the whole or part of the material is concerned, specifically the rights of translation, reprinting, reuse of illustrations, recitation, broadcasting, reproduction on microfilms or in any other physical way, and transmission or information storage and retrieval, electronic adaptation, computer software, or by similar or dissimilar methodology now known or hereafter developed.
The use of general descriptive names, registered names, trademarks, service marks, etc. in this publication does not imply, even in the absence of a specific statement, that such names are exempt from the relevant protective laws and regulations and therefore free for general use.
The publisher, the authors, and the editors are safe to assume that the advice and information in this book are believed to be true and accurate at the date of publication. Neither the publisher nor the authors or the editors give a warranty, expressed or implied, with respect to the material contained herein or for any errors or omissions that may have been made. The publisher remains neutral with regard to jurisdictional claims in published maps and institutional affiliations.

This Springer imprint is published by the registered company Springer Nature Switzerland AG.
The registered company address is: Gewerbestrasse 11, 6330 Cham, Switzerland

Foreword

As an engineering test pilot, astronaut, and especially during my tenure as the 12th NASA Administrator, it became quite apparent to me that a critical missing link in many of our projects and programs was a thorough, practical understanding of systems engineering and its critical importance in design and development of complex systems. Although complex systems have existed for many years – Apollo, the electric grid, the air traffic control system, the Internet – they could generally be designed and operated using well-established engineering practices. However, as systems have become increasingly complex and intertwined, these prior approaches, such as using interface documents to control the connectivity of individually designed components, are no longer sufficient. For example, they may not adequately identify and control the unintended consequences that can suddenly arise when one system element that has incorrectly been believed to be isolated from other system components actually proves to impact the functioning of these other components. Add to this, the unpredictability of human interactions that can take place in modern systems and the possibilities and uncertainties far exceed the ability of humans to process.

Enter Model-Based Systems Engineering (MBSE). Therein a construct is created of an entire system that enables the study of interactions throughout the system and discloses potential outcomes. It permits designers, operators, and maintainers of the system to work in concert from a common representation of the system. Yet, despite its generally recognized benefits, MBSE, when improperly scoped and applied, can, and on occasion has, led to significant inefficiencies and even system failures.

In response to such concerns, the Editors-in-Chief of this handbook have assembled a significant and timely collection of contributions from leading MBSE practitioners across the globe. The stated intent of the handbook is to clarify what MBSE is, where it stands today, and how it can be employed as a cost-effective systems engineering methodology with software tools to aid in modeling, analysis, and decision support. While there are several books and publications that focus on MBSE mechanics and languages, this is the first volume to focus on the methodological and practicable aspects of conducting systems engineering within an MBSE rubric. The handbook contains case studies that reflect the collective wisdom and experience of internationally known experts – and their lessons learned.

As is widely recognized, many of today's systems are far more complex, and their operational environments much more challenging, than ever before. Contrast the spacecraft flown by NASA 10–15 years ago with those flown today, with the latter consisting of multiple interacting computers, instruments, software, and mechanisms that must survive decades in often extremely harsh space environments. Comparing the complexity of the Mariner spacecraft of the 1960s with the James Webb Space Telescope of today is like comparing the computing power and memory capacity of a 1960's IBM 360/91 mainframe computer (once the world's most powerful computer) with a modern desktop computer that operates an order of magnitude faster and can readily accommodate more than twice the main memory. The self-driving vehicle industry has a similar tale to tell. There are as many as 500 million lines of code operating in a self-driving automobile. Even a modern, human-driven vehicle executes as many as 100 million lines of software code. In the same vein, cloud systems can have thousands of clients, servers, and data stores. Streaming services, the Internet of Things, distributed human-cyber physical systems, and fuel-efficient aircraft are continually posing challenges that press the limits of human perception and cognition. Today MBSE is increasingly recognized as essential for meeting modern engineering demands, staying ahead of the competition, and assuring the evolution and survival of enterprises.

This handbook spans topics that range from foundational MBSE concepts to real-world experiences. The sections are independent and can be read in an arbitrary order. Within each section, each individual chapter begins with a domain orientation or tutorial, followed by a discussion of practical "how-to" examples. Readers can peruse the chapters with synergy in content in a sequence of their choosing because the chapters are cross-referenced to related chapters. This feature is especially useful for practitioners and students working on MBSE projects who are also interested in foundational concepts, applications, and lessons learned from prior work. In addition, the handbook indicates how MBSE relates to other disciplines, such as digital engineering and social media. One chapter discusses transdisciplinary systems engineering that has expanded the frontiers of MBSE and provides powerful new approaches to problem solving.

In conclusion, this handbook is a working guide into the many forms, uses, and means of pursuing MBSE projects. It inspires the reader to delve into the exciting and rapidly advancing world of MBSE. Readers will learn to apply MBSE to real-world problems with unique informational and resource constraints, as well as to systems at different scales. It is a must-read for practitioners and students alike.

Major General USMC (Ret)
NASA Astronaut
12th NASA Administrator

Charles Bolden Jr.

Preface

We are systems engineers first and foremost. We are also storytellers. And we have been involved with systems engineering for more than five decades. Over that period, we have seen spectacular successes and failures that can be attributed to proper use of systems engineering and its misuse and have learned valuable lessons from them. In its early years, systems engineering was somewhat an orphan, with universities largely guiding and tracking engineers into a single practice, such as electrical engineering, chemical engineering, or mechanical engineering. Yet, most real-world problems that are encountered today include transdisciplinary aspects – that is, "systems engineering." We have seen systems engineering evolve over the past few decades to where we see transformative advances in the field paced by Model-Based Systems Engineering (MBSE), digital twin technology, AI and machine learning, ontologies, and formal methods to prove model correctness.

So how do these different advances come together? Is MBSE the rubric to bring these advances together? How can we capitalize on these advances in the work we are doing in our respective and highly varied organizations? These are but a few of the questions that we aim to address in this handbook, which is a "living document" that allows contributors to add new findings and insights that can advance the practice of systems engineering as it continues to evolve.

We, the Editors-in-Chief, have been working in systems engineering research, practice, and education for decades. Norm Augustine served as chairman and CEO of Lockheed Martin Corp., one of the largest aerospace firms in the world. He led the company to new heights with his innovative blend of systems engineering, systems management, and leadership skills. Previously, he was a faculty member at Princeton University, served as the Acting Secretary of the Army, and chaired the Advisory Committee on the Future of the US Space Program, known as the Augustine Committee. He was the Chairman of the National Academy of Engineering, of the Defense Science Board, and former President of American Institute of Aeronautics and Astronautics. A member of the National Academy of Engineering and National Academy of Science, he led numerous committees such as Rising Above the Gathering Storm, Revisited: Rapidly Approaching Category 5. He has received honorary degrees from 35 universities, is the recipient of the prestigious National Medal of Technology and Innovation, and has been recognized with numerous other international awards and honors. He is the author of *Augustine's Laws*, a highly acclaimed book that sets forth 52 laws that cover engineering and management in an entertaining and informative fashion.

Azad Madni is University Professor and holder of the Northrop Grumman Foundation Fred O'Green Chair in Engineering at the University of Southern California. He is the Executive Director of USC's flagship Systems Architecting and Engineering Program, founding director of the Distributed Autonomy and Intelligent Systems Laboratory, and faculty affiliate of the Ginsburg Institute of Medical and Biomedical Therapeutics. He is the founder and CEO of Intelligent Systems Technology, Inc., a successful hi-tech company specializing in model-based methods for complex systems engineering, education, and training. He pioneered transdisciplinary systems engineering as a next-generation systems engineering discipline and a means to address complex sociotechnical problems that appear intractable when viewed solely through an engineering lens. He also wrote an award-winning book *Transdisciplinary Systems Engineering: Exploiting Convergence in a Hyperconnected World* and is the co-author of *Tradeoff Decisions in System Design*. A member of the National Academy of Engineering and Life Fellow of IEEE, he received the NAE's Gordon Prize for Innovation in Engineering and Technology Education, and the IEEE Simon Ramo Medal in 2023, for exceptional achievement in systems engineering and systems science. Previously, he served as the Executive VP and CTO of a public company and head of the modeling and simulation group on NASA's Space Shuttle Program at Rockwell International, where he pioneered a model-based testing approach which saved substantial sums in physical testing of the Shuttle navigation system. An elected fellow of ten professional societies, he has served as Principal Investigator on 97 R&D projects totaling over $100M.

Contributors to this book are MBSE practitioners selected from experts in the field from several different countries. As MBSE gathered steam in the engineering community worldwide, several questions arose – along with some misconceptions about MBSE. This recognition spurred interest in developing a handbook that would take a broad and deep look at the use of models in doing serious systems engineering. Accordingly, we invited top practitioners to contribute their understanding of MBSE and address areas that cover fundamentals of MBSE as well as how MBSE was successfully applied to real-world problems. At the time, there was a fragmented body of knowledge in MBSE and a pressing need to consolidate MBSE concepts, methods, theories, and practices with real-world examples. This handbook attempts to consolidate many of these contributions by mixing and matching various areas and emphasizing what it takes to derive real value from MBSE. The audience for this book is practitioners, students, and educators seeking a coherent presentation of MBSE concepts, methodologies, assumptions, and illustrative examples, so as to add to the successes of the field and eliminate the failures that have sometimes come from a lack of attention to it.

Los Angeles, USA   Azad M. Madni
Bethesda, USA   Norman Augustine
July 2023   Editors-in-Chief

Acknowledgment

We would like to acknowledge Shatad Purohit, who is pursuing his doctorate under Professor Madni's guidance, for his responsible management of the handbook project and timely interactions with the contributors.

Contents

Volume 1

Part I   Introduction **1**

1   **Introduction to the Handbook** 3
Azad M. Madni and Norman Augustine

Part II   MBSE Foundations **13**

2   **Semantics, Metamodels, and Ontologies** 15
Michael Sievers

3   **MBSE Methodologies** 47
Jeff A. Estefan and Tim Weilkiens

4   **SysML State of the Art** 87
B. Bagdatli, S. Cimtalay, T. Fields, E. Garcia, and R. Peak

5   **Role of Decision Analysis in MBSE** 119
Gregory S. Parnell, Nicholas J. Shallcross, Eric A. Specking,
Edward A. Pohl, and Matt Phillips

6   **Pattern-Based Methods and MBSE** 151
William D. Schindel

7   **Overarching Process for Systems Engineering and Design** 195
A. Terry Bahill and Azad M. Madni

8   **Problem Framing: Identifying the Right Models for the Job** 257
James N. Martin

Part III   Technical and Management Aspects of MBSE **287**

9   **Model-Based System Architecting and Decision-Making** 289
Yaroslav Menshenin, Yaniv Mordecai, Edward F. Crawley, and
Bruce G. Cameron

| 10 | Adoption of MBSE in an Organization | 331 |

10 Adoption of MBSE in an Organization 331
Tim Weilkiens

11 Model-Based Requirements 349
Alejandro Salado

**12 Modeling Hardware and Software Integration by an Advanced
Digital Twin for Cyber-physical Systems: Applied to the
Automotive Domain** 379
S. Kriebel, M. Markthaler, C. Granrath, J. Richenhagen, and
B. Rumpe

13 Integrating Heterogenous Models 417
Michael J. Pennock

**14 Improving System Architecture Decisions by Integrating
Human System Integration Extensions into Model-Based
Systems Engineering** 441
D. W. Orellana

15 Model-Based Human Systems Integration 471
Guy André Boy

16 Model-Based Hardware-Software Integration 501
Joe Cesena

Part IV Quality Attributes Tradeoffs in MBSE **525**

**17 Exploiting Digital Twins in MBSE to Enhance System
Modeling and Life Cycle Coverage** 527
Azad M. Madni, S. Purohit, and C. C. Madni

**18 Model-Based Mission Assurance/Model-Based Reliability,
Availability, Maintainability, and Safety (RAMS)** 549
Luca Boggero, Marco Fioriti, Giuseppa Donelli, and
Pier Davide Ciampa

19 MBSE in Architecture Design Space Exploration 589
J. H. Bussemaker and Pier Davide Ciampa

Part V Digital Engineering and MBSE **631**

**20 Digital Twin: Key Enabler and Complement to Model-Based
Systems Engineering** 633
Azad M. Madni and C. C. Madni

**21 Developing Industry 4 Systems with OPM ISO 19450
Augmented with MAXIM** 655
D. Dori

Contents   xiii

22 MBSE Testbed for Unmanned Vehicles 675
A. M. Madni and D. Erwin

**23 Transitioning from Observation to Patterns: A Real-World
Example** 705
S. Russell, B. Kruse, R. Cloutier, and D. Verma

Volume 2

Part VI MBSE for System Acquisition and Management **723**

24 MBSE for Acquisition 725
R. A. Noguchi and R. J. Minnichelli

25 Managing Model-Based Systems Engineering Efforts 753
Mark L. McKelvin

26 MBSE Methods for Inheritance and Design Reuse 783
A. E. Trujillo and A. M. Madni

27 Model Interoperability 815
Tim Weilkiens

28 A Reuse Framework for Mode-Based Systems Engineering 833
Gan Wang

29 MBSE Mission Assurance 861
J. S. Fant and R. G. Pettit

**30 Conceptual Design Support by MBSE: Established
Best Practices** 895
S. Shoshany-Tavory, E. Peleg, and A. Zonnenshain

Part VII Case Studies **923**

31 Ontological Metamodeling and Analysis Using openCAESAR ... 925
D. A. Wagner, M. Chodas, M. Elaasar, J. S. Jenkins, and
N. Rouquette

32 MBSE Validation and Verification 955
Karen Gundy-Burlet

33 MBSE for System-of-Systems 987
Daniel DeLaurentis, Ali Raz, and Cesare Guariniello

34 NSOSA: A Case Study in Early Phase Architecting 1017
M. W. Maier

35 Cybersecurity Systems Modeling: An Automotive System Case Study ... 1045
Mark L. McKelvin

36 Assistive Technologies for Disabled and Older Adults ... 1079
William B. Rouse and Dennis K. McBride

37 Multi-model-Based Decision Support in Pandemic Management ... 1105
A. M. Madni, Norman Augustine, C. C. Madni, and Michael Sievers

38 Semantic Modeling for Power Management Using CAESAR ... 1135
D. A. Wagner, M. Chodas, M. Elaasar, J. S. Jenkins, and N. Rouquette

39 Modeling Trust and Reputation in Multiagent Systems ... 1153
Michael Sievers

40 Modeling and Simulation Through the Metamodeling Perspective: The Case of the Discrete Event System Specification ... 1189
María J. Blas and Silvio Gonnet

Part VIII   Future Outlook ... 1229

41 Exploiting Transdisciplinarity in MBSE to Enhance Stakeholder Participation and Increase System Life Cycle Coverage ... 1231
Azad M. Madni

42 Toward an Engineering 3.0 ... 1253
Norman Augustine

43 Category Theory ... 1259
S. Breiner, E. Subrahmanian, and R. D. Sriram

44 Perspectives on SE, MBSE, and Digital Engineering: Road to a Digital Enterprise ... 1301
H. Stoewer

Index ... 1339

About the Editors-in-Chief

Azad M. Madni is a University Professor of Astronautics, Aerospace, and Mechanical Engineering in the University of Southern California's Viterbi School of Engineering. The University Professor designation honors the university's most accomplished, multidisciplinary faculty, who have significant achievements in multiple technical disciplines. He is the holder of the Northrop Grumman Fred O'Green Chair in Engineering and the Executive Director of USC's Systems Architecting and Engineering Program. He is the Founding Director of USC's Distributed Autonomy and Intelligent Systems Laboratory. He is also a Professor in USC's Keck School of Medicine and Rossier School of Education. He is a faculty affiliate of the Keck School's Ginsberg Institute for Biomedical Therapeutics. He is the founder and CEO of Intelligent Systems Technology, Inc., a hi-tech company specializing in transdisciplinary model-based approaches for tackling complex sociotechnical systems problems. He is a member of the Phi Kappa Phi honor society and Omega Alpha Association, an international systems engineering honor society.

Prof. Madni is a member of the National Academy of Engineering and the recipient of the prestigious 2023 NAE Gordon Prize for Innovation in Engineering and Technology Education. A Life Fellow of IEEE, he is also the recipient of the 2023 IEEE Simo Ramo Medal. He is an Honorary Member of ASME and a Life Fellow/Fellow of AAAS, AIAA, INCOSE, IISE, AIMBE, IETE, AAIA, SDPS, and the Washington Academy of Sciences. He is the recipient of approximately 80 prestigious international and national awards from 11 different professional societies. These include the 2019 IEEE AESS Pioneer Award and the 2011 INCOSE Pioneer Award. He has 400+ publications comprising authored and edited books, book chapters, journal articles, peer-reviewed conference publications, and research reports. He has given more than 75 keynotes and invited talks in international conferences and workshops. He is a member of two NAE Lifetime Giving Societies: the Marie Curie Society and the Albert Einstein Society.

He pioneered the field of transdisciplinary systems engineering and wrote an award-winning book, *Transdisciplinary Systems Engineering: Exploiting Convergence in a Hyperconnected World* (Springer, 2018), which presented the founding principles of transdisciplinary systems engineering. He is also the creator of TRASEE™, a new engineering education paradigm based on transdisciplinary systems engineering principles. He is also the co-author of *Tradeoff Decisions in System Design* (Springer, 2016), Co-Editor-in-Chief of the Springer series Systems Engineering Research, and co-author of *3 Ds of Deep Learning – Design, Development, and Deployment* (Springer, 2023).

He transformed USC's Systems Architecting and Engineering Program based on TRASEE and provided a blueprint for other graduate engineering programs to follow. Under his leadership, the program has graduated 3200+ students and is recognized as a top graduate engineering program in the country. He has served as Principal Investigator on 97 R&D contracts and grants totaling more than $100M.

Previously, he was the Executive Vice President for R&D and the Chief Technology Officer of Perceptronics Inc., a simulation-based training and AI company that went public in 1982. Prior to that, as a lead simulation engineer at Rockwell International on NASA's Space Shuttle Program, he led the development of a model-based testing approach that generated substantial savings in navigation system performance testing for the Shuttle Program.

He received his Ph.D., M.S., and B.S. in Engineering from UCLA with a major in Engineering Systems and minors in Computer Methodology and AI. He is also a graduate of AEA/Stanford Executive Institute.

About the Editors-in-Chief

Norman Augustine attended Princeton University where he graduated with a BSE in Aeronautical Engineering, magna cum laude, and an MSE. He was elected to Phi Beta Kappa, Tau Beta Pi, and Sigma Xi.

After graduating he joined Douglas Aircraft where he worked as a Research Engineer, Program Manager, and Chief Engineer, after which he served in the Office of the Secretary of Defense as Assistant Director of Defense Research and Engineering. He then joined LTV Missiles and Space Company, serving as Vice President, Advanced Programs and Marketing. He returned to the government as Assistant Secretary of the Army for R&D and then Under Secretary of the Army and Acting Secretary of the Army. Joining Martin Marietta as Vice President of Technical Operations, he was later elected as CEO and chairman, having previously been President and COO. He served as President of Lockheed Martin Corporation upon the formation of that company and became CEO later the same year. He retired as Chairman and CEO of Lockheed Martin and became a Lecturer with the Rank of Professor on the faculty of Princeton University.

Mr. Augustine was Chairman of the Council of the National Academy of Engineering, the Aerospace Industries Association, the Defense Science Board, and former President of the American Institute of Aeronautics and Astronautics. He is a former member of the Board of Directors of ConocoPhillips, Black & Decker, Proctor & Gamble, and Lockheed Martin and served as a Regent of the University System of Maryland (12 institutions), is a Trustee Emeritus of Johns Hopkins, and a former member of the Board of Trustees of Princeton University and MIT. He has been a member of advisory boards to the Departments of Homeland Security, Energy, Defense, Commerce, Transportation, and Health and Human Services, as well as NASA, Congress, and the White House. He served for 16 years on the President's Council of Advisors on Science and Technology under both Republican and Democratic presidents. He is a member of the American Philosophical Society, the National Academy of Sciences, and the Council on Foreign Relations and is a Fellow of the National Academy of Arts and Sciences and the Explorers Club.

Mr. Augustine has been presented the National Medal of Technology by the President of the United States and received the Joint Chiefs of Staff Distinguished Public Service Award. He has five times received the Department of Defense's highest civilian decoration, the Distinguished Service Medal. He authored *The Defense Revolution*, *Shakespeare in Charge*, *Augustine's Laws*, *Augustine's Travels*, and *The Way I See It*. He holds honorary degrees from 35 universities, is a Distinguished Scholar of the University of Maryland Baltimore, and was selected by Who's Who in America and the Library of Congress as one of "Fifty Great Americans" on the occasion of Who's Who's fiftieth anniversary. He has delivered over 1,500 speeches and lectures and since retiring has served on 59 pro bono committees and commissions of which he chaired or co-chaired 43. He has traveled to 130 countries and stood on both the North and South Poles of the earth.

About the Associate Editor

Michael Sievers
Jet Propulsion Laboratory
Pasadena, CA, USA

University of Southern California
Los Angeles, CA, USA

Michael Sievers is a Senior System Engineer at the California Institute of Technology, Jet Propulsion Laboratory (JPL), and an Adjunct Lecturer in the systems architecting and engineering department at the University of Southern California (USC). He earned a bachelor's degree in electrical engineering and masters and Ph.D. degrees in computer science, all from the University of California, Los Angeles.

Dr. Sievers's graduate studies investigated Very Large Scale Integrated Circuit failure mechanisms, defect testing, and fault-tolerance. In this work, he developed a programmable logic structure and one of the earliest design automation tools used for building a self-checking Hamming Code generator and checker.

He was also a part-time academic at JPL while doing his graduate studies to develop means for automating the control of JPL's large radio antennas that were part of its Deep Space Network. He also designed a self-checking computer module that was part of a larger fault-tolerant computing project.

After completing his academic work, Dr. Sievers focused on developing fault-tolerant spacecraft command and data-handling subsystems for several Earth science experiments and US government applications. He later joined a young company developing special-purpose, high-performance computers for computational biology applications.

On returning to JPL, Dr. Sievers performs research and development in high-performance computing, adaptive optics control, system fault-tolerance and resilience, model-based systems engineering, system reputation and trust, and ground enterprise architectures. His work has earned him numerous NASA and JPL awards. At USC, he teaches classes in system architecture, model-based systems engineering, and system resilience.

Dr. Sievers is an INCOSE Fellow, AIAA Associate Fellow, and IEEE Senior Member. He is the Associate Editor of the *IEEE Open Journal of Systems Engineering* and has written more than 70 publications in refereed journals and conferences.

Section Editors

Norman Augustine
Lockheed Martin Corporation
Bethesda, MD, USA

Barry Boehm
University of Southern California
Los Angeles, USA

Joseph D'Ambrosio
General Motors
Detroit, USA

D. Erwin
University of Southern California
Los Angeles, CA, USA

Jeff A. Estefan
NASA/Jet Propulsion Laboratory
Pasadena, CA, USA

Azad M. Madni
University of Southern California
Los Angeles, CA, USA

R. J. Minnichelli
The Aerospace Corporation
CA, USA

Section Editors

R. A. Noguchi
The Aerospace Corporation
CA, USA

Alejandro Salado
The University of Arizona
Tucson, AZ, USA

M. Sievers
University of Southern California
Los Angeles, CA, USA

Contributors

Norman Augustine Lockheed Martin Corporation, Bethesda, MD, USA
Advisory Services, New York, NY, USA

B. Bagdatli Georgia Institute of Technology, School of Aerospace Engineering, Atlanta, GA, USA

A. Terry Bahill Systems and Industrial Engineering, University of Arizona, Tucson, AZ, USA

María J. Blas Instituto de Desarrollo y Diseño INGAR (UTN-CONICET), Santa Fe, Argentina

Luca Boggero German Aerospace Center (DLR), Hamburg, Germany

Guy André Boy CentraleSupélec, Paris Saclay University, Gif-sur-Yvette, France
ESTIA Institute of Technology, Bidart, France

S. Breiner Information Technology Lab, National Institute of Standards and Technology, Gaithersburg, MD, USA

J. H. Bussemaker Institute of System Architectures in Aeronautics, MDO group, German Aerospace Center (DLR), Hamburg, Germany

Bruce G. Cameron Massachusetts Institute of Technology, Cambridge, MA, USA

Joe Cesena Lockheed Martin, Sunnyvale, CA, USA

M. Chodas Jet Propulsion Laboratory, California Institute of Technology, Pasadena, CA, USA

Pier Davide Ciampa Institute of System Architectures in Aeronautics, MDO Group, German Aerospace Center (DLR), Hamburg, Germany

S. Cimtalay Georgia Institute of Technology, School of Aerospace Engineering, Atlanta, GA, USA

Norman Augustine is Retired.

R. Cloutier University of South Alabama, Mobile, AL, USA

Edward F. Crawley Massachusetts Institute of Technology, Cambridge, MA, USA

Daniel DeLaurentis Purdue University, West Lafayette, IN, USA

Giuseppa Donelli German Aerospace Center (DLR), Hamburg, Germany

D. Dori Technion, Israel Institute of Technology, Haifa, Israel

M. Elaasar Jet Propulsion Laboratory, California Institute of Technology, Pasadena, CA, USA

D. Erwin University of Southern California, Los Angeles, CA, USA

Jeff A. Estefan NASA/Jet Propulsion Laboratory, Pasadena, CA, USA

J. S. Fant The Aerospace Corporation, Chantilly, VA, USA

T. Fields Georgia Institute of Technology, School of Aerospace Engineering, Atlanta, GA, USA

Marco Fioriti Politecnico di Torino, Turin, Italy

E. Garcia Georgia Institute of Technology, School of Aerospace Engineering, Atlanta, GA, USA

Silvio Gonnet Instituto de Desarrollo y Diseño INGAR (UTN-CONICET), Santa Fe, Argentina

C. Granrath FEV.io GmbH, Aachen, Germany

Mechatronics in Mobile Propulsion, RWTH Aachen University, Aachen, Germany

Cesare Guariniello Purdue University, West Lafayette, IN, USA

Karen Gundy-Burlet Crown Consulting Inc., NASA-Ames Research Center, Moffett Field, CA, USA

J. S. Jenkins Jet Propulsion Laboratory, California Institute of Technology, Pasadena, CA, USA

S. Kriebel FEV.io GmbH, Aachen, Germany

BMW Group, Munich, Germany

B. Kruse e:fs TechHub GmbH, Gaimersheim, Germany

Azad M. Madni Systems Architecting and Engineering, Astronautical Engineering Department, University of Southern California, Los Angeles, CA, USA

Intelligent Systems Technology, Inc., Los Angeles, CA, USA

C. C. Madni Intelligent Systems Technology, Inc., Los Angeles, CA, USA

M. W. Maier The Aerospace Corporation, Hill AFB, Ogden, UT, USA

M. Markthaler BMW Group, Munich, Germany

Software Engineering, RWTH Aachen University, Aachen, Germany

James N. Martin The Aerospace Corporation, Chantilly, VA, USA

Dennis K. McBride Intelligent Systems Division, Hume Center for National Security and Technology, Virginia Tech National Security Institute, Blacksburg, VA, USA

Mark L. McKelvin University of Southern California, Los Angeles, CA, USA

Yaroslav Menshenin Skolkovo Institute of Science and Technology, Moscow, Russia

R. J. Minnichelli The Aerospace Corporation, El Segundo, CA, USA

Yaniv Mordecai Massachusetts Institute of Technology, Cambridge, MA, USA

R. A. Noguchi The Aerospace Corporation, El Segundo, CA, USA

D. W. Orellana ManTech International Corporation, Los Angeles, CA, USA

Gregory S. Parnell Department of Industrial Engineering, University of Arkansas, Fayetteville, AR, USA

R. Peak Georgia Institute of Technology, School of Aerospace Engineering, Atlanta, GA, USA

E. Peleg Metaphor Vision Ltd, Kefar-Saba, Israel

Michael J. Pennock The MITRE Corporation, McLean, VA, USA

R. G. Pettit George Mason University, Fairfax, VA, USA

Matt Phillips System Design and Analytics Laboratory, Department of Industrial Engineering, University of Arkansas, Fayetteville, AR, USA

Edward A. Pohl Department of Industrial Engineering, University of Arkansas, Fayetteville, AR, USA

S. Purohit University of Southern California, Los Angeles, CA, USA

Ali Raz Systems Engineering and Operations Research, George Mason University, Fairfax, VA, USA

J. Richenhagen FEV.io GmbH, Aachen, Germany

N. Rouquette Jet Propulsion Laboratory, California Institute of Technology, Pasadena, CA, USA

William B. Rouse McCourt School of Public Policy, Georgetown University, Washington, DC, USA

B. Rumpe Software Engineering, RWTH Aachen University, Aachen, Germany

S. Russell Johnson Space Center, NASA, Houston, TX, USA

Alejandro Salado The University of Arizona, Tucson, AZ, USA

William D. Schindel ICTT System Sciences, Terre Haute, IN, USA

Nicholas J. Shallcross System Design and Analytics Laboratory, Department of Industrial Engineering, University of Arkansas, Fayetteville, AR, USA

S. Shoshany-Tavory Technion – Israel Institute of Technology, Haifa, Israel

Michael Sievers NASA/Jet Propulsion Laboratory, California Institute of Technology, Pasadena, CA, USA

University of Southern California, Los Angeles, CA, USA

Eric A. Specking System Design and Analytics Laboratory, Department of Industrial Engineering, University of Arkansas, Fayetteville, AR, USA

R. D. Sriram Information Technology Lab, National Institute of Standards and Technology, Gaithersburg, MD, USA

H. Stoewer TU Delft, Delft, The Netherlands

E. Subrahmanian Information Technology Lab, National Institute of Standards and Technology, Gaithersburg, MD, USA

Engineering Research Accelerator/Engineering and Public Policy, Carnegie Mellon University, Pittsburgh, PA, USA

A. E. Trujillo Massachusetts Institute of Technology, Cambridge, MA, USA

D. Verma Stevens Institute of Technology/SERC, Hoboken, NJ, USA

D. A. Wagner Jet Propulsion Laboratory, California Institute of Technology, Pasadena, CA, USA

Gan Wang Dassault Systèmes, Herndon, VA, USA

Tim Weilkiens oose Innovative Informatik eG, Hamburg, Germany

A. Zonnenshain The Gordon Center for Systems Engineering, Technion – Israel Institute of Technology, Haifa, Israel

Part VI

MBSE for System Acquisition and Management

MBSE for Acquisition

24

R. A. Noguchi and R. J. Minnichelli

Contents

Introduction	726
Introducing Model-Based Systems Engineering	726
The Unique Nature of Acquisition	729
The Acquisition Life Cycle	730
MBSE Needs of the Acquisition Environment	735
The Digital Engineering (DE) Construct and Its Implications for MBSE in Acquisition	736
MBSE State of the Art and Best Practices for Acquisition	738
MBSE for Enterprise Gap Analysis	739
MBSE for Materiel Solution Analysis	743
MBSE for Contracting and Procurement	746
MBSE for the Remainder of the Acquisition Life Cycle	748
Chapter Summary	749
References	750

Abstract

Model-Based Systems Engineering (MBSE) is a rapidly advancing practice of systems engineering that employs descriptive system models as the foundation for knowledge capture and sharing within collaborative systems engineering teams throughout the system life cycle. Just as system developers are taking advantage of MBSE to improve the efficiency and effectiveness of system development processes, so also can acquisition organizations exploit MBSE to improve implementation of acquisition processes, enhance their engagement with suppliers responsible for system development, and facilitate the execution of full life cycle responsibilities for operating and sustaining the acquired systems. This chapter explores the MBSE problem space for acquisition. While some of the language and terms used are typical of US defense

R. A. Noguchi (✉) · R. J. Minnichelli
The Aerospace Corporation, El Segundo, CA, USA
e-mail: ryan.a.noguchi@aero.org; robert.j.minnichelli@aero.org

© Springer Nature Switzerland AG 2023
A. M. Madni et al. (eds.), *Handbook of Model-Based Systems Engineering*,
https://doi.org/10.1007/978-3-030-93582-5_44

acquisition processes, the approach is applicable and valuable for any acquisition effort. It contrasts the unique characteristics of MBSE implementation by acquisition organizations with the more traditional MBSE implementation practices currently used by system developers.

Keywords

MBSE · Model-based systems engineering · Acquisition · System architecting

Introduction

Introducing Model-Based Systems Engineering

Model-Based Systems Engineering (MBSE) is a modern paradigm for improving the practice of systems engineering (SE). It employs integrated or federated sets of digital system models to serve as the authoritative source of truth and the focal point for knowledge management, technical communication, and data interchange.

What distinguishes MBSE from the traditional implementation of SE is that progressively developing requirements baseline, functional architecture, and physical architecture is structured around the incremental refinement of an integrated set of increasingly detailed digital system models [1, 2]. These models offer the structure for knowledge capture and the means for information exchange about a system over its lifetime. These models replace the previously large number of discrete and disconnected documents that typically become increasingly inconsistent as each independently evolves over the system's life cycle. The disciplined use of models in lieu of documents and other static artifacts reduces the likelihood of misinterpretation and misunderstanding of system requirements, architecture, design, and inter-relationships.

In MBSE, these integrated digital system models are the foundation of a more effective SE process [2, 3]. A simplified model of the pre-MBSE SE life cycle processes is shown in Fig. 1. This depiction focuses on the inputs and outputs of those life cycle processes, which are typically captured in a large variety of documents.

Fig. 1 Systems engineering life cycle process "Vee" model (pre-MBSE)

The SE "Vee" life cycle development approach has been well-established over the past few decades. Several minor variations have been published since then. In the Vee model, SE activities that take place within each life cycle phase are intended to incrementally refine the system solution from requirements to architecture to design to realized product. Each phase takes inputs from the previous phase and creates outputs supplied to the subsequent phase. In a pre-MBSE environment, these inputs and outputs generally took the form of static artifacts such as documents. Typically, many documents were created during the development process. A few of these are identified in Fig. 1. It was difficult to keep these documents synchronized and harmonized. Furthermore, integrating these documents into a coherent body of knowledge occurred ad hoc, often resulting in incomplete and inconsistent documents.

Not surprisingly, inconsistencies among these documents, typically created by different authors at a different pace, increase in number and severity over time. While many inconsistencies arise between documents, it is also common for them to arise within a single document. Discovering and rectifying these inconsistencies is often difficult. As a result, these inconsistencies can persist, resulting in problems that manifest themselves in system architecture, design, or implementation. The later these problems are discovered, the more expensive and difficult it is to fix them, if they can be fixed at all. One intent of MBSE is to significantly reduce the risk of inconsistencies arising and propagating before discovery and mitigation.

An updated version of the SE life cycle process "Vee" model enabled by MBSE is shown in Fig. 2. This depiction is intended to illustrate the logical flow between SE life cycle processes and is not intended to illustrate a temporal waterfall development process flow.

As noted earlier, MBSE replaces a large collection of documents (static artifacts) by an integrated or federated set of digital system models that capture the content of the documents in a form that is easier to keep synchronized and harmonized. Inconsistencies are rarer, as they are more transparent and easier to discover and correct due to the connected nature of knowledge captured in the models. However, the limitations of current tools make inconsistency checking still a partly manual

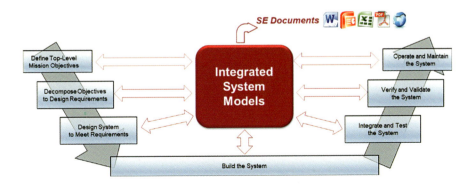

Fig. 2 Systems engineering life cycle process "Vee" model, MBSE

process, and rigorous configuration management and modeling discipline are needed to reduce the introduction of inconsistencies. As the architecture, design, and implementation of the system matures, the digital system models are concurrently refined to ensure their currency and relevance. Thus, when specific documents are required, they can be produced by transforming appropriate model content into the required artifacts in the proper formats. Importantly, these integrated digital system models link knowledge across all technical disciplines and across life cycle phases, reducing the risk of miscommunication and misinterpretation.

A further refinement of the SE "Vee" for MBSE appears in Fig. 3 [4]. Note that MBSE is applicable to any SE process model and development approach. The grey boxes represent SE life cycle processes, while the red boxes represent analysis and testing activities that are performed within a virtual or digital environment. The integrated digital system models serve as sources of authoritative inputs to the analyses and virtual testing activities and can also capture the results of the analyses and virtual tests for traceability. These system models enable execution of SE processes, while enhancing concurrency of engineering across disciplines and life cycle phases.

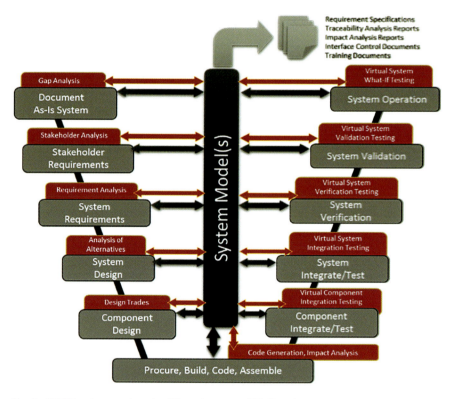

Fig. 3 MBSE systems engineering life cycle process "Vee" mode

It is important to note that while models can support the integration, testing, verification, validation, and other processes on the right leg of the Vee, those processes are still performed on the physical system. Modeling and simulation activities utilize the digital system models to provide inputs and enable much earlier and more frequent analysis to be performed to reduce the risk that design issues will manifest later in the life cycle when they are much more difficult to correct.

Furthermore, MBSE and DE facilitate using the "right" and most up-to-date data (authoritative for its purpose) within the analyses based on physical system and component experience. And more than software (code) can be represented by these models; the hardware, software, firmware, and the environment can all be captured in these digital system models.

The Unique Nature of Acquisition

In the canonical SE process, [2, 3] system developers are responsible for executing the entire set of SE life cycle technical processes, starting from a defined set of requirements and/or eliciting needs from appropriate stakeholders. The vast majority of SE and MBSE applications in the literature approach SE and MBSE processes from this traditional perspective. However, this perspective does not accurately reflect real-world constraints.

In most real-world system development problems, the system developer rarely has this prominent role for a substantial portion of the early stages of the system life cycle. In these situations, the system acquirer (i.e., an acquisition organization) performs the early SE life cycle activities that inform its decision to either purchase an existing off-the-shelf product or commission the development and production of a new product. The acquirer typically also has fiduciary responsibilities to ensure proper value (i.e., payback for the price). To this end, the acquirer often provides oversight over the development process and validation of the delivered product. The combination of context and activities performed by the acquirer is called *acquisition*.

In the acquisition context, the overall system development process is more complicated than in the traditional system development context because (a) the early life cycle activities are typically longer and more involved; (b) the system development has a greater number of stakeholders performing activities in parallel; and (c) there are defined boundaries between the acquirer and the developer that constrain their interactions and information exchanges. In this environment, SE implementation becomes more constrained, and so does MBSE implementation.

The *acquirer* is often a government agency as these agencies are often precluded by statutory or practical reasons from being the end-to-end developer and producer of large systems, and their missions often require the development of new systems to meet unprecedented requirements that cannot be satisfied by existing off-the-shelf systems. These government agencies have the resources to commission the nonre-curring engineering development work required to satisfy these requirements. However, the acquirer may also be a commercial entity or other non-governmental agency. Any acquiring agency can utilize many of the same MBSE processes, methods, and tools to improve their acquisition activities.

In many cases, systems being developed are so complicated that multiple levels of acquisition are required. An example of such a system is a military fighter aircraft. In this case, the government (e.g., the Air Force in the USA) is the primary acquirer. The Air Force may contract with a prime contractor to serve as the primary supplier and developer of this aircraft system. The prime contractor, in turn, may contract out the development of major components – such as jet engines, avionics equipment, mission equipment, etc. – to subcontractors. As a result of this arrangement, a prime contractor or subcontractor may simultaneously serve in both acquirer and supplier roles.

Due to the substantial cost of acquisitions, acquirers are usually expected to perform significant work to justify large expenditures, track and manage the contracted effort, verify acceptability of the delivered product, and furnish the product's operators and end-users with requisite operational and logistics support. For each such responsibility, SE can potentially play a significant role to facilitate and enable their execution. As important, MBSE can improve the effectiveness and efficiency of implementing SE for both the supplier and the acquirer.

The Acquisition Life Cycle

An acquiring agency rarely possesses or receives a mature set of requirements that are sufficient and ready for use in a competitive contracting action with one or more system developers. Developing these requirements *is* the job of the acquiring agency. Considerable time from the initial identification of an emergent need to the availability of a delivered product or capability to satisfy that need is spent through a variety of activities. These activities are intended to scope and validate the need, obtain political and financial support for the acquisition, develop contractual requirements, and perform contracting actions.

In most cases, an extensive set of documented processes and regulations need to be followed to ensure that all stakeholders and decision-makers develop the requisite confidence in the need to expend the resources to pursue pre-acquisition activities, much less engage in the much larger expenditures required for the acquisition, development, deployment, and operations of the new system, product or capability. As a result, many acquiring agencies have developed and documented somewhat repeatable acquisition processes comprising stages separated by decision milestones. These acquisition processes, which typically vary significantly between agencies, often follow a similar structure due to their common goal of maturing the body of knowledge needed to proceed through successive stages of increasing investment and commitment of resources. The acquisition process described here for illustrative purposes is loosely based on the U.S. Department of Defense's Defense Acquisition System (DAS) [5, 6]. However, it shares many features with the processes employed by other government agencies and commercial entities.

Enterprise Gap Assessment

The purpose of enterprise gap assessment is to determine whether a gap exists between the enterprise's existing capabilities – reflected in its current set of assets, and its current organization and other non-materiel factors – and the new context in which the enterprise must operate. If the forcing function is a new threat, can the enterprise counter that threat with its existing materiel and non-materiel solutions? Or must it acquire a new materiel solution to meet that need? Enterprise gap assessment makes this determination and presents the findings for acquisition decision makers to authorize the next step, materiel solution analysis.

An acquiring agency is typically part of or associated with an organization that is responsible for operating an enterprise comprising a wide variety of existing systems and capabilities to achieve its assigned mission. Invariably, there will be a few unsatisfied needs associated with the existing assets, perhaps due to deficiencies in their development that led to incomplete satisfaction of requirements or of the operational needs that drove their acquisition. Often, in such cases, there may be a budget to support incremental upgrades of those existing assets. Much of the time spent and the need for an acquisition is driven by external forcing functions, such as an emerging threat, an emerging new technology, the recognition of a new need, or a change in the environment or political landscape.

When such a forcing function presents itself, the acquiring agency needs to decide whether a new system acquisition – or an upgrade to an existing system – is warranted. In many cases, a non-materiel solution – one that does not require a significant investment in a new acquired system – is preferable since non-materiel solutions are often quicker and cheaper to implement. Examples of non-materiel solutions include changes in policy, doctrine, organization, or logistics to make better use of existing systems by changing how they are used and/or supported.

Materiel Solution Analysis

Once a determination is made that non-materiel solutions alone will not suffice and that a new materiel solution is needed to close the identified gaps between the existing capability and the needed capability, materiel solution analysis is performed. Materiel solution analysis focuses on identifying the types of system solutions that should be acquired. These early SE life cycle activities focus on refining the understanding of the problem space concurrently with the potential solution space, utilizing methods and processes referred to as system architecting [7, 8]. These studies provide a better understanding of the complex relationships among requirements, cost, and capabilities, facilitating a more informed decision regarding the best tradeoffs among these factors to balance the needs of the many stakeholders and interests involved in a major system acquisition program.

Stakeholder needs and desires typically begin with somewhat ill-defined and often conflicting needs. These needs (which are part of the problem space) need to be iteratively refined concurrently with the exploration of potential solutions to ensure that important decisions regarding system requirements and investments are based on what is feasible within technical, policy, and financial constraints. It is important to note that the initial recognition of the existence of a significant capability gap is often made from the perspectives of a limited set of stakeholders. In this materiel solution analysis phase, the scope of the assessment is typically broadened to include other potential stakeholders that also benefit from a new capability created to close the identified gap.

The most extended activity within this phase is the *analysis of alternatives*, or AoA. The AoA is a trade study examining the invariably large trade space of alternative solutions to determine the feasibility of closing the identified capability gaps and understanding the tradeoffs among the various class of solution concepts. Decision makers can then use the results of the AoA to determine how best to allocate their limited budgets and identify options most likely to succeed.

The AoA is primarily conducted at the architectural level of abstraction to reduce the risk of neglecting interesting regions in the trade space. However, the AoA typically needs to progress into an examination of conceptual designs of specific point solution concepts to facilitate the analysis of their performance, risks, and cost. Understanding the tradeoffs among performance, risk, and cost is critical for achievement of this maturity stage. An inability to demonstrate technical feasibility or cost or schedule executability is typically a showstopper in achieving stakeholder commitment to proceed to the next phase of the acquisition life cycle.

Critical technology maturation efforts that are needed to enable new system concept functionality or performance are identified, and recommendations are made for research and development (R&D) efforts to raise the Technology Readiness Level (TRL) [9] of specific technologies to accomplish their "on-ramping" into system acquisitions.

Once one or more families of solution concepts have been determined as suitable to meet the need and deemed to be technically feasible and affordable, acquisition decision-makers can authorize undertaking the next step-technology maturation and risk reduction or contracting.

Contracting and Procurement

The key distinction between the acquisition context and a traditional system development context is that the acquirer and supplier are distinct organizations. The first two life cycle phases – enterprise gap assessment and materiel solution analysis – are performed by the acquirer, while subsequent life cycle phases involve both the acquirer and one or more suppliers. The suppliers' activities are authorized through *contracting* or *procurement*, often on a competitive basis. In some cases, the adequate rationale can be provided to enter a *sole source* arrangement, such as when there is only one feasible supplier for a given product or system.

Before an acquirer and a supplier can enter into a contract to develop a system or produce a product, they must agree on the scope of that effort. The procurement process is straightforward if the product being procured is available as an off-the-shelf item from the supplier's product catalog. The acquirer identifies the products desired, the supplier provides a quotation, and the two negotiate the price and terms until a mutually agreeable position is established. Generally, little SE activity is needed in this circumstance.

However, often a situation arises where the acquirer's desired system does not yet exist and is not a simple variant of an existing product. In this case, substantial SE activities must occur on both sides of the agreement. The acquirer must define requirements for the desired system and represent them as formally controlled artifacts. The technical specifications for the system – functional, performance, and non-functional – are captured in one or more technical requirements specifications, often referred to as a Technical Requirements Document (TRD). The work requirements are captured in a Statement of Work (SOW) or a Performance Work Statement (PWS) – the latter providing more detail about the measurable expected outcomes and less about the process needed to achieve them. Requirements for data to be delivered at various times during the development program are expected. In the U.S. DoD context, these are identified in a Contract Data Requirements List (CDRL), and those deliverables' required content and format are captured in Data Item Descriptions (DID).

This set of technical, work and deliverable requirements are combined with other contractual requirements and requirements on the suppliers' proposals to form the body of artifacts that the acquirer releases to the supplier community in a formal Request for Proposal (RFP) or Request for Tender (RFT). Government agencies and commercial firms often have substantial regulations, processes, and oversight procedures that must be followed to ensure these contractual actions are undertaken while safeguarding proprietary information.

Suppliers answer the RFP or RFT with their proposals that respond to the proposal requirements and provide required information about their proposed product solution, work, and deliverables. Suppliers may need to perform some systems engineering activities to support the proposal development process, although typical RFP timeframes often preclude the use of a more disciplined, time-consuming approach.

The acquirer engages in a formal process to determine which supplier(s) to select and identifies points of negotiation. This *source selection* process can be particularly challenging when the proposed solution concepts are substantially different and cannot be readily compared. At the other extreme, the process can be swift if the product is a commodity for which multiple qualified suppliers exist, and the primary discriminating factor is price. In any event, this phase of activity concludes with the identification of one or more "winners" and the finalization of negotiations to enable the winning suppliers to begin work on the next phase of development.

Technology Maturation and Risk Reduction

For an acquisition program that requires significant development effort, rather than purchasing an already-developed off-the-shelf product, it is often desirable to expend some effort to retire the largest technical risks through a focused effort referred to as the technology maturation and risk reduction phase. Technology risks are retired through dedicated research and development activities. Development risks are retired through preliminary design activities, prototypes, and demonstrations.

Tradeoff analyses are often performed to refine understanding the tradeoffs between key performance parameters, system attributes, and system architectural or design features. Various planning activities are performed to prepare for full-scale development, including developing SE plans, test and evaluation plans, program protection plans, cost analyses, and risk assessments. This planning and tradeoff analysis often includes assessing opportunities to reuse designs from previously developed systems to reduce the cost or risk of developing the new system [10].

The developer's effort within this phase is often competitively procured, so the contracting activity (discussed in the next section) must also be performed. Many acquisition programs have used this phase as an opportunity to simultaneously retire technical risk among several system development suppliers to provide the acquiring agency the information needed to *down-select* to a small number of suppliers – often just one – to complete full-scale development.

Development

In the development phase, the supplier is responsible for most of the work associated with system development. The SE activities associated with the supplier role are generally congruent with the canonical case in which there are no separate acquirer and supplier roles.

The acquirer's role is to monitor the development effort and oversee the supplier's activity consistent with the terms of the contract. The extent of the acquirer's responsibility to participate in governance and engineering decisions pertaining to the development phase varies with projects. For example, if the system is particularly complex and expensive, the acquirer typically has a greater responsibility to play an active role in oversight of the development effort. In these cases, both the supplier and the acquirer often undertake substantial SE efforts to manage that complexity, while the acquirer seeks SE information from the supplier to inform its governance, requirement verification, and product assurance processes.

In many cases, one or both parties introduce changes to the project's requirements. Both parties assess the impacts of those changes, and any potential impacts on the contractual terms and the contract value are re-negotiated. SE analyses may be used by either party to perform impact assessment.

Production and Deployment

Once the product's design is finalized, prototypes have been developed, and both acquirer and supplier are satisfied with the system's level of maturity, it is common practice to enter a new phase of the system life cycle: production and deployment. Suppliers often rely on product lifecycle management (PLM) systems to manage the large body of information pertaining to the production and testing of physical products. During this phase, the supplier manages the production process and the inevitable changes that occur during the production run to account for factors such as lessons learned part obsolescence, and process changes. In this phase, SE activities focus on understanding the impacts of these changes on the system's ability to satisfy its requirements. The acquirer also needs to understand the implications of these changes on the system's ability to satisfy its requirements and the acquirer's ability to operate and sustain the system. Accordingly, the acquirer needs to execute these SE activities also.

Operations and Support

The system continues to evolve during the operations and support phase. SE activities are performed to manage the execution of the support system that has been established to sustain the continued operation of the system. These SE activities are often performed by the acquiring agency since it is usually the organization responsible for operating the system after the acquirer has taken delivery of the physical assets from the supplier.

The acquirer uses SE to perform impact assessments. Impact assessments are performed reactively when undesirable events occur, as well as proactively to prevent undesirable events from occurring. Impact assessment is also used to inform decision-making about product upgrades, process improvements, and logistics changes. Frequently, operation of the system reveals gaps between the actual capabilities provided by the system as operated, and the real-world environment that the system experiences. These gaps may result from SE escapes in the identification of the gaps in the early life cycle phase, or in the closing of those gaps in subsequent phases or could simply have resulted from the evolving context in which the system operates. When these gaps are discovered, they often trigger the acquisition life cycle to begin anew, to seek new solution concepts – materiel or non-materiel – to close the gaps.

MBSE Needs of the Acquisition Environment

MBSE adoption by system developers (suppliers) has grown significantly over the past few years, often motivated by the drive for improved efficiency from the commercial elements of their businesses. Interest in adopting MBSE on the acquirer side of the acquirer/supplier interface has also grown – although generally less quickly – motivated not only by a desire for improved efficiency but also by the need for a better approach for managing the growing complexity of the acquirers' mission enterprises. Seeing the benefits that MBSE and better integration of data across multiple technical and non-technical disciplines have brought to commercial industry and other domains, many acquiring agencies are now motivated to seek those benefits to improve program execution and enterprise management.

MBSE can bring several benefits to acquirer enterprises, including:

- Strengthened ability to architect enterprise-wide and cross-enterprise solutions by integrating knowledge and insights across the enterprise's portfolio, to better inform acquirer decisions
- Improved collective understanding of existing system capabilities, requirements, composition, functionality, behavior, interdependencies, performance, and resilience to inform acquirer decisions regarding non-materiel solutions using these existing systems or upgrades or replacements (materiel solutions) to augment or supplant existing systems

- Superior organization of technical information across system life cycle phases that reduces the impact of information loss that typically occurs at phase transitions during system's life cycle
- Clear communication across contractual and organizational interfaces that mitigates the risk of propagating misconceptions or misinterpretations that can potentially lead to SE escapes
- Improved efficiency in evaluating architectural options, particularly with substantially larger solution trade spaces that may include a broader set of upgrade and supplementation options
- Superior traceability and more efficient transition from early concept studies and capability-based assessments through all subsequent life cycle activities
- Ability to perform rapid, comprehensive impact assessment crossing architectural layers and organizational stovepipes, to better understand the dependencies and implications of acquirer decisions

Section "MBSE State of the Art and Best Practices for Acquisition" describes some of the benefits that MBSE can provide to the acquisition domain as well as within each of these system acquisition life cycle phases.

The Digital Engineering (DE) Construct and Its Implications for MBSE in Acquisition

The emerging paradigm of Digital Engineering (DE) [11] continues to gain momentum in both the acquisition and the supplier communities. DE facilitates the adoption and growth of MBSE and enables the community to take advantage of the efficiency and effectiveness benefits of improved interoperability and integration of models and data. The DE ecosystem, depicted in Fig. 4 [12], includes a wide variety of models, simulations, and data spanning many technical and non-technical disciplines, such as program management, financial management, and contracting.

Digital system models provide the foundation for MBSE processes by serving as the authoritative models and identifying the authoritative data needed and used for SE knowledge management and information sharing within the DE-enabled enterprise. The models, simulations, and data across the enterprise can – if maintained to keep pace with the evolving body of knowledge – provide the foundation for an ongoing and sustainable capability for comprehensive systems analysis in any stage of their life cycle. This "digital twin" [13–15] evolves to keep pace with the evolving "physical twin" that it represents as it grows from the conceptual design level of detail through detailed design, implementation, delivery, fielding, and as the deployed system ages.

At every level in the enterprise, decisions can be informed by the cumulative knowledge of impacts and dependencies enabled by the pervasive use of integrated MBSE tools. The hierarchical structure of models, which mirrors the hierarchical structure of organizations, is depicted in Fig. 5 [12]. The acquirer's program management offices interact directly with their prime contractors, who interact with their subcontractors. The acquirer's program management offices also need to

24 MBSE for Acquisition

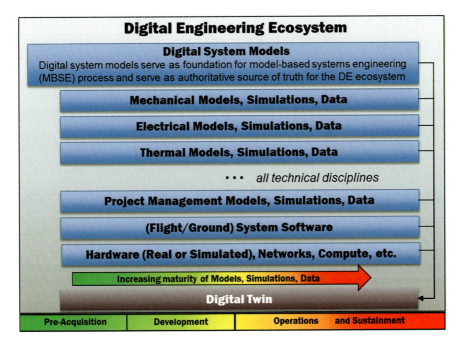

Fig. 4 How digital system models fit into a broader digital engineering landscape

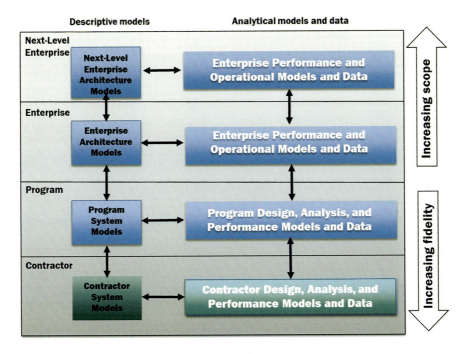

Fig. 5 How digital models interact across an organizational hierarchy

receive requirements from and provide information to the enterprises to which they belong, who in turn also have similar responsibilities to the enterprises to which they belong.

Digital system models at every hierarchical level in the enterprise are loosely coupled federated models in adjacent layers. This model federation approach provides the required balance between the need for individual layers to develop models suitable to answer their questions and the frequently competing market for these models to also contribute to answering questions by other stakeholders. These digital system models are akin to the constituent systems within a collaborative system of systems, which are each designed to achieve focused purposes for their immediate stakeholders, but which also enable the achievement of broader purposes for a wider range of stakeholders through participation in the model federation [16].

The development contractor – the supplier – employs an increasingly sophisticated set of models, including design models (e.g., CAD, electrical layout, etc.) to capture the evolving design and serve as data sources for analysis models (e.g., CFD, thermal, etc.) to characterize the design, and performance models that quantify the evolving design's ability to meet functional and non-functional requirements. On the acquirer's side of the interface, a separate set of models and simulations are employed to aid in the execution of their functional responsibilities, including analysis models to address higher-level questions such as enterprise-level requirements or operational employment questions, and mission assurance models that are used to support the independent verification and validation of suppliers' analyses to reduce risk in critical areas. Programmatic models – including cost, schedule, resource loading, and other factors – are also used to inform decision-making in the program management context.

Digital system models at each level provide inputs to digital system models at adjacent levels, and enable greater automation of analytical, design, and other models. This automated integration of descriptive and analytical models enables smoother execution of analytical and design work with reduced risk of miscommunication. Sharing of models between layers reduces duplication of work while enabling rapid assimilation of evolving data at each layer. Synchronization of models across multiple layers enables the enforcement of single authoritative source of truth. Network effects [12, 17] magnify the value of MBSE within an enterprise when adoption is more universal, and implementation is more interoperable across the enterprise. Acquirers are increasingly adopting the broader scope of DE to supplement the focused scope of MBSE, and to magnify the value that MBSE to the acquirer enterprise through the interconnectedness made possible by DE.

MBSE State of the Art and Best Practices for Acquisition

Much of the published research work in MBSE has focused on the application of model-based methods to the development process, without accounting for additional complexities and nuances associated with system acquisition. This simplified approach makes it easier to understand how to use these new modeling paradigms without encumbering the practitioner with significant additional complexity. A prime example of this is the Object-Oriented Systems Engineering Method (OOSEM) [1, 18], which is a top-down MBSE approach that originated in industry to support the development of large-scale information systems. Since then, several other MBSE methodologies have been developed [19]. These methodologies mostly focus on supporting the traditional system development process.

Fortunately, several of these methods can also be adapted to the acquisition context, and some researchers – often with funding from government agencies – are increasingly exploring the application of MBSE to the acquisition domain to address their unique needs. In the following sections, the use of MBSE at each phase of the acquisition process will be described. Within each phase, the answers to different sets of questions are sought to inform different decisions. As a result, the required content of models to inform those decisions will differ by phase and will be described.

MBSE for Enterprise Gap Analysis

Acquisition organizations are continuing to grow in complexity, particularly in terms of the organization structure, constituent sub-organizations, business processes, and systems that enable and support the enterprise. The principles and practices of SE that have been successfully applied to the development and deployment of systems within each of those enterprises can also be adapted to provide similar benefits for dealing with complexity at the enterprise level. Just as MBSE can be applied to improve the implementation of SE for individual systems, it can also be applied to the enterprise, which in some ways can be treated as a more complex set of interdependent, collaborating, continuously evolving, intelligent system-of-systems.

Today, the growing complexity of enterprises and systems has led to the creation of vast, lossy knowledge stovepipes, connected by narrow, unreliable, lossy conduits for information exchange. Information developed within one part of the enterprise, in one functional discipline, at one phase of the system lifecycle, is often trapped within that information island, unusable by anyone else, or if usable at least not in a reliable and repeatable way. Every enterprise generates, assimilates, disseminates, and archives an enormous amount of information. When this information is duplicated in multiple locations, the biggest problem created is not inefficiency, but rather the propagation of inconsistency, as updates and corrections to information items are difficult to maintain and propagate to their respective destinations. Also, sharing of information between organizations and projects is fraught with challenges, some technical, but mostly cultural or political impediments to sharing.

Enabled by the interconnection of these "knowledge islands" through the integration of their digital models, enterprise application of MBSE can enhance enterprise decision-making, and enable superior architecting of enterprise-wide solutions, by better integrating knowledge across the enterprise and its constituent programs. MBSE methods can also enable managers to quickly and comprehensively identify interdependencies within the enterprise and with external entities, if developed with this use identified. MBSE can also more efficiently trace ripple effects of changes or problems across the enterprise, enabling more effective proactive impact assessment. Enterprise models generally rely more heavily on dependencies and traceability relationships than system models do, largely to facilitate the answering of typical enterprise management and enterprise systems engineering questions.

MBSE can improve the quality of enterprise decision-making by increasing insights across stovepipes through integration of models and data at multiple levels of the enterprise. MBSE performed at the individual system level can be leveraged to reduce duplication of effort at the enterprise level, while MBSE performed at the enterprise level can be leveraged to increase standardization and integration at the individual system level.

Just as acquirers can use MBSE to improve their ability to perform enterprise gap analyses, suppliers can use MBSE in similar ways to apply SE methods to the management of their own internal enterprises. MBSE can improve the identification of gaps between the supplier's existing product line and capabilities to perform architecting, design, manufacturing, integration, testing, and other functions pertinent to the development of systems, and to prepare for the next acquisition for which they intend to compete successfully to become a supplier.

MBSE for enterprise gap analysis activities has much in common with enterprise architecture efforts. Both are focused on achieving a better understanding of the enterprise and its drivers and constraints, the prerequisites to informed decision-making pertaining to the sustainment or evolution of the enterprise. Many of the methods employed in enterprise gap analysis are a hybrid of system-level MBSE methods and methods employed in modeling enterprise architectures. The adaptation of SE principles to the scale and context of the enterprise – referred to as enterprise systems engineering (ESE) [20, 21] – is greatly aided by the adoption of model-based techniques to manage complexity and improve the quality of communication and knowledge management within the enterprise. These benefits, in turn, better inform decisions regarding the evolution and/or sustainment of the enterprise [22].

Enterprise-level application of MBSE can often benefit from the use of dedicated enterprise modeling languages, in lieu of the more general-purpose modeling languages that are frequently used for modeling systems. An example of such an enterprise modeling language is the Unified Architecture Framework (UAF) Modeling Language (UAFML) [23], which extends the standard Systems Modeling Language (SysML) [24] for modeling systems with the modeling constructs that are needed to address ESE concerns and problems. Some of these modeling constructs include built-in support for the multiple layers of abstraction that characterize enterprise descriptions, which typically include:

1. *Strategic-level issues*, such as overarching goals and objectives, capabilities – the ability to achieve a desired effect or outcome [25] – policy, doctrine, and strategy, including the gaps between the desired state and the current state
2. *Operational-level issues*, including the organizations, stakeholders, and organizational roles within the enterprise, and the high-level missions that the enterprise

is responsible for performing and the high-level functions that the enterprise is responsible for implementing

3. *Service-level issues*, which abstract solution-level implementations (software, hardware, and cyber-physical systems) to their basic implementation-independent specification of required interfaces and functions, which can then be provided by the acquisition of either dedicated systems or by contracted services
4. *Resource-level issues*, which describe the solution-level implementations and the exchanges of resources (data, electrical power, etc.) between software and/or hardware components
5. *Project-level issues*, including both internal projects and system or service acquisition projects, which result in the manifestation or provisioning of the services and/or solution-level implementations

Both enterprise management and ESE require insights at each level of abstraction and an understanding of dependencies and interactions between these levels. In some cases, the service layer is omitted, particularly when an acquiring agency has the resources or charter to develop or acquire systems at the solution level rather than at the service level. However, enterprises increasingly need to transform themselves to enable a more seamless evolution between organic legacy systems to a more open architecture comprising a larger variety of contracted services, more granularly defined system components shared by multiple users across the enterprise, and competitively procured system components to serve as a hedge against technology maturation risk. As a result, developing these service specifications and interfaces is becoming increasingly important to enterprise planning.

Enterprise-level architecture modeling to support ESE activities is a specialized application of MBSE that can exploit many of the same modeling methods, techniques, and tools, adding rigor and discipline to traditional enterprise architecture practices. Since the practice of SE is largely driven by requirements – i.e., their development, specification, allocation to system architecture and components, and their verification – MBSE practice has evolved to add rigor and discipline to the treatment of requirements.

Establishing enterprise-level requirements is an important component of ESE. These requirements represent the enduring capabilities that the enterprise needs to provide to execute its mission, and which drive the development, acquisition, and sustainment of the systems comprising the enterprise. Unlike system-level requirements, enterprise-level requirements are seldom satisfied by a single system. In reality, numerous systems are required to varying degrees. A resilient enterprise generally avoids relying on structures that lead to potentially single points of failure when providing capabilities that are essential to its mission.

When implementing MBSE at the enterprise level, the capabilities that are required to achieve the desired outcomes of the enterprise are defined in the enterprise architecture model. These capabilities represent the foundational needs of the enterprise, which drive the acquisition, development, operation, and sustainment of systems and services. The enterprise's constituent elements – systems, services, people, and processes – exist to enable the exercise of these capabilities. Enterprise architecture often focuses on one of two perspectives: the enterprise constituents that directly execute the mission and the enterprise constituents that enable execution. Using a military analogy, the former are the combatants and their weapon systems, while the latter are the frequently much larger set of individuals and systems that are responsible for acquiring systems, managing their logistics, providing for the needs of the people, etc. In each case, enterprise architecture can inform decisions about those systems, services, people, and processes. The traditional use of enterprise architecture has typically focused on an organization's information technology and associated business processes, while frequently blurring the line between mission execution and mission enablement, particularly when the organization's mission is centered around the use of information technology.

These enterprise capabilities are not requirements, per se; rather, they serve as the basis for the development of enterprise requirements. Enterprise capabilities can be incrementally achieved, with separate sets of requirements associated with each increment to reflect the evolution of each capability to achieve higher performance, greater robustness, lower cost, or other improvements. As the environment or other contexts in which the enterprise operates change, existing levels of capability achievement may transition from being entirely satisfactory to increasingly unsatisfactory. These gaps between the as-is enterprise capability and the needed capability can be defined in the enterprise architecture model and serve as drivers for subsequent activities to identify alternative solution concepts to close those gaps. Many of these potential solution concepts could potentially provide improvements to multiple capability gaps. Furthermore, the enterprise architecture model can aid in the identification of such synergistic opportunities. The enterprise architecture model also captures the current state of the enterprise's systems and services. These existing systems and services afford opportunities for synergy that can reduce the need to acquire and/or develop new systems and services by reusing existing systems and services in new configurations, known as systems of systems [16].

The enterprise architecture model typically includes information about the enterprise's systems and services at a low level of fidelity. It is the responsibility of lower-level organizations to maintain that knowledge within models or other artifacts under their control, often at a higher level of fidelity than is needed to inform enterprise decisions. However, the higher-fidelity system and service models can be connected to the enterprise architecture model to enable knowledge to traverse the boundary between organizations. This connection of models at various levels within the enterprise, depicted in Fig. 5, lies at the heart of digital engineering and can significantly increase the value of MBSE adoption by the enterprise, by enabling much more significant reuse of models and knowledge within the enterprise. This federation of models and reuse of knowledge requires achieving interoperability between models.

That interoperability can be greatly aided by relying on modeling language standards to reduce sources of variation between the modeling methods used by the different modeling teams. Even if the same modeling language is not used at every level of the enterprise, using compatible languages across the interfaces – such as UAFML for enterprise-level models and SysML for system-level models – can substantially facilitate interoperability. Since UAFML is an extension of SysML, they are able to coexist and interoperate much more readily than if substantially different modeling languages were used. However, interoperability between models is not assured even if the same modeling language is used, as the semantics of the modeling constructs are not so strictly and precisely defined that they cannot be misinterpreted. As a result, to be successful, model federation requires good communication between model builders, and ideally the establishment of mutually agreeable standards at the semantic level – capturing the meaning of modeling constructs and elements used – not just at the syntactic level.

When capability gaps are identified, and validated by stakeholders, the next set of significant acquisition decisions revolves around the determination of subsequent courses of action. In this next phase of the system life cycle (i.e., materiel solution analysis), MBSE can also play a significant role.

MBSE for Materiel Solution Analysis

Acquirers can use MBSE to improve their ability to perform materiel solution analysis activities. MBSE can potentially enable faster iteration of solution concepts and architectural trades, by exploiting reuse of previously analyzed options and improving the management of the complexity associated with increasingly large trade spaces [10, 26].

MBSE brings together the work products of various technical disciplines and integrates them into a coherent body of knowledge. This is an extension of the concept of concurrent engineering, spanning across a broad swath of technical disciplines, product organizations, and life cycle processes. Disconnects in the assumptions made by one of these entities can be more readily discovered when their evolving designs or plans are integrated with the rest of the body of knowledge. This enables the acquirer to accelerate the discovery of defects in the requirements or architecture before they can propagate for too long in the acquisition life cycle. This connected body of knowledge facilitates rigorous examination of evolving requirements, architecture, and design for disconnects, and provides greater confidence in closing in on an acceptable solution prior to progressing in the system life cycle.

Materiel solution analysis activities heavily utilize system-architecting methods for analyzing the problem space and comparing alternative options within the solution space [27, 28]. Here, the SE process is less mechanical in nature, as expert judgment and stakeholder elicitation play a significant role in shaping the description of the problems to be addressed and in constraining the solution space to enable decisions to be made to acquire a system or service. Nevertheless, MBSE can add significant value during this phase, by helping to manage complexity among the many competing objectives and solution concepts. During this concept development phase, the practice of MBSE is occasionally referred to by the moniker Model-Based Conceptual Design (MBCD) [29].

Early in the system life cycle, system concepts are developed at a relatively abstract, low-fidelity level of detail, to support assessments of the viability, feasibility, and affordability of the concept. Conceptual design activities are often performed in a concurrent engineering environment [30], enabling multiple technical disciplines to contribute their specialized expertise and design and analysis tools to facilitate rapid synthesis of system point designs. These design and analysis tools are connected in an integrated digital environment that enables them to share and exchange data, enabling multiple technical specialists to concurrently and coherently contribute to conceptual design of the same point design, without any inconsistencies or discrepancies among disciplines. In the absence of such a digitally connected environment, miscommunication of key design parameters can result in physically infeasible designs due to incompatibility between interfacing disciplines. Within such a concurrent engineering environment, MBSE methods can be used [31] to capture the evolving description of the system and its alternative conceptual designs to inform downstream SE activities.

MBSE methods can also more effectively connect descriptive and analytical models in the concept design process [32] to provide a robust, traceable network of knowledge spanning multiple technical disciplines. This knowledge can serve as the foundation for more detailed requirement development activities later in the life cycle as the system concept matures to the contracting and procurement phase. Models of these emerging system concepts can also be integrated with enterprise architecture models, resulting in a traceable network of knowledge that can enable stakeholders to understand how they fit into the broader enterprise context, allowing not only the closure of specific capability gaps but also enabling other opportunities. This traceable network is often referred to as the Digital Thread [32].

The Digital Thread provides traceability within an integrated set of models and datasets, thereby allowing the ability to answer more complex questions than can be answered by explicitly identified point-to-point relationships within a single model. For example, an enterprise stakeholder may want to understand which systems and services within an enterprise are needed to provide one of its key capabilities. Model elements representing these systems and services may not be directly connected to the model element representing the capability. In many cases, the connection will be indirect, e.g., the capability may depend on the existence of other capabilities that are performed by systems that are dependent on the availability of a service that is dependent on some infrastructure. That infrastructure element may have no direct connection to any capability in the model, and no one in the organization may be aware of that "fifth order" dependency. The federated set of models, each of which represents the collective knowledge residing within part of the organization, is able to capture the much larger scope of the collective knowledge across the enterprise, enabling decisions to be made with a broader understanding of their implications and impacts by a wider set of stakeholders. This Digital Thread – connecting information via the data representing that information throughout a single model or a large federation of models – is one of the key benefits of the MBSE paradigm, enabling the answering of more complex questions than ever before.

The analysis of alternatives is performed to aid acquisition decision-making by exploring the art of the possible and examining multiple alternative solution concepts to better understand the solution trade space. These solution concepts are examined to a level of detail sufficient to enable comparison of their effectiveness at achieving operational needs, their feasibility of implementation, and their estimated cost. Excessively detailed design at this stage of the life cycle is counterproductive in that it can expend considerable resources and time, while unduly constraining the resulting solution implementation, and inhibiting the selection of potentially, superior design options by the supplier. Therefore, conceptual designs are developed only to the extent needed to inform the acquisition decision, and no further.

MBSE methods can aid the analysis of alternatives in several ways:

1. Improve the definition of alternative solution concepts to be explored during the analysis of alternatives [33, 34]
2. Improve the automation of concurrent engineering workflows, to enable more rapid closure on conceptual designs [35]
3. Improve the integration of alternative solution definitions, captured in descriptive models, to the analytical models that will be used to quantify their performance and other important characteristics [36]
4. Improve the documentation of alternative solution concepts, improving the quality of information transfer from the analysis of alternative teams to downstream activities, such as requirement development [31]

Descriptive system models of alternative solution concepts can be created to comprehensively capture the key architectural features associated with the concepts. These models can include traceability to capability needs "drivers," and to design heritage from legacy systems. In a mature MBSE environment, in which the existing enterprise comprising systems have been already modeled, models of existing systems can be quickly modified to reflect potential upgrades, significantly reducing the time needed to flesh out the design concept.

These descriptive models can be connected to analysis models to enable multi-disciplinary design optimization, or MDO [35, 36]. In MDO, key design parameters are varied by a computer algorithm to map out the solution tradespace. The descriptive model can represent the starting point for exploration of alternatives, or excursions, with the connection to analysis models enabling automated evaluation of excursions. The result of such an MBSE-enabled MDO study results in a more thorough investigation of the solution tradespace than is possible in the same timeframe with a more manual process.

Even if MDO is not performed, the ability to connect analysis to design within a digital environment can enable automation of the concurrent engineering workflow. This enabling of seamless exchange of information between analysts and designers can improve both speed and quality of the analysis of alternatives by reducing the risk of miscommunication and reducing the timelines needed to assess each alternative solution concept [36].

The assessment of alternative solution concepts can also be performed not only on specific materiel solutions, but also on alternative operating concepts. These operational concepts can include various combinations of existing assets and new materiel solutions, interacting in different ways to achieve the required mission. In the practice of mission engineering [37, 38], the mission itself is treated as a system, and individual systems are viewed as components of that system. Applying MBSE to the mission-engineering problem [39] can improve the analysis of the performance and effectiveness of various system configurations, particularly by connecting or aggregating behavior models of system assets to assemble behavior models of the end-to-end mission. This can aid the assessment of alternative solution concepts for their suitability to enable the execution of one or more missions and provide insight to improve the quality of alternative solution concepts explored during the analysis of alternatives study.

After the analysis of alternatives is concluded and stakeholders have decided to proceed with the acquisition of interest, the acquiring agency needs to develop a comprehensive set of requirements to capture the needs in a form suitable for contracting with the supplier. This requirements development process can significantly benefit from the alternative solution concept(s) selected for acquisition being thoroughly documented. The descriptive system models resulting from the AoA can serve as a starting point for the requirement development process, enabling complete traceability from requirements to system architecture in the model. This is particularly the case when behavior models are created to support the AoA, because otherwise they need to be developed after the study to support system functional requirements' development is complete. If the behavior models are available during the study, they can be used as part of the analysis and assessment process. Ideally, descriptive system models both "enter and exit" the AoA process, enabling superior execution of the AoA, and effective execution of subsequent phases of work leading to the contracting and procurement activity.

MBSE for Contracting and Procurement

MBSE for contracting and procurement draws heavily upon the requirement development and functional architecting domains. In this phase, MBSE can be used to improve the acquirer's process of developing and representing system requirements, which are then provided to potential offerors as a statement of the acquirer's needs that need to be satisfied by proposed solutions, and which subsequently become contractually binding on the winning offeror [32, 34].

Transitioning from conceptual designs and high-level architectural concepts of these early life cycle phases to a degree of specificity that will ensure the delivered product satisfies the stakeholder needs that justified the acquisition is a significant challenge that is made even more challenging by the needs by both acquirer and supplier to carefully limit the information they exchange across the acquirer-supplier interface [34]. This interface is potentially a source of misinterpretations and communication disconnects that can result in failure of the delivered product to satisfy acquirer expectations.

For systems being acquired, MBSE can be used to capture an unambiguous and relatively complete specification of acquirer needs. Specifically, digital system models can capture requirements in a manner that reduces risk of misinterpretation, incompleteness, and inconsistency and can facilitate linking of the requirements to the evolving reference architecture that represents the acquirer's expectations regarding the solution space. Where the acquirer is a government agency, this reference architecture is often referred to as a Government Reference Architecture (GRA). This GRA is a baseline architecture created during the formative stages of a program. It serves as a reference for early life cycle activities such as system architecting, conceptual design, analyses of alternatives, cost estimation, and system requirements development. The GRA is increasingly being developed as a digital system model that can serve as the foundation for the acquirer's implementation of MBSE for the program and can also feed enterprise-level MBSE implementation.

The GRA can be adapted from the descriptive system models developed during conceptual design, either during or following the AoA. Typically, the architectural concept described in the AoA is only rudimentary, with many details left unspecified and several assumptions made about applicable constraints and dependencies. Post-AoA, the acquirer needs to elaborate the conceptual architecture model with these additional details.

Technical requirements for the system can be developed and represented in a digital requirements model that replaces the traditional technical requirements document (TRD). This requirements model can contain the same textual requirements contained in the traditional TRD but can also include additional information that improves expressiveness and reduces ambiguity in requirement specification. The requirement model can potentially contain a rich network of traceability relationships between requirements, or of allocation relationships between requirements and components of the system architecture. Key performance parameters, constituting important performance requirements for the system, can be elaborated in the model, with traceability between requirements, performance parameters, and system components that exhibit those properties.

The requirement model can also elaborate textual functional requirements with a rich set of behavioral models that more accurately and less ambiguously reflect the required behavior of the system. These behavioral models, created in disciplined fashion, can be executed within the modeling environment to verify both the accuracy and completeness of the functional requirements. These behavioral models, which are a critical part of the GRA, establish the foundation for the system's functional requirements.

In addition to the technical requirements, other acquisition requirements can also be captured in the form of a model. Work requirements, traditionally expressed in documents such as the Statement of Work (SOW) or Performance Work Statement (PWS), can be developed in descriptive system models. In these models, model elements representing work requirements can be traced to higher-level requirements or standards. Elements of the work breakdown structure can be allocated to elements of the system's product breakdown structure.

Furthermore, the acquirer often requires data and/or documents from its supplier, pertaining to the design of the system and/or its construction. The requirements for these data and/or documents are articulated in formal specifications, such as DIDs. If the supplier is using MBSE, many of these data products will be developed in model form. If the acquirer is also using MBSE, it will want to receive these data products in model form. Specifying the requirements for these models will be critical if the acquirer wants to receive models suitable for MBSE needs. These model requirements can be developed in model form by exploiting model-based methods to elicit and develop the requirement, such as the $(MBSE)^2$ approach [40]. (This is referred to by the moniker "MBSE squared" since one uses MBSE methods and tools to develop the requirements for models to support MBSE.)

If system acquisition requirements and the GRA are provided in the Request for Proposal (RFP) or Request for Tender (RFT) as digital models, they can be incorporated by the bidding suppliers into their own MBSE environments. Once this is done, they can be used to develop more traceable model-based proposals for the acquirer in the form of models rather than documents. And, of course, models can be more readily evaluated for completeness in addressing the solicitation. Traceability of a supplier's proposed design to the requirements and architectural vision captured in the GRA and elsewhere in a model-based RFP can be readily assessed using system modeling tools, much more comprehensively than can be done using pre-MBSE methods. The GRA model can continue to serve as a basis for integration of the acquirer's and supplier's MBSE efforts, thereby improving communication and ensuring that both the acquirer and the supplier have a shared and consistent understanding of acquisition needs.

One challenge posed by the acquisition environment is the introduction of semantic disconnects between the models being developed by different organizations. If the acquirer and supplier use different modeling languages, use different modeling constructs to represent the same concept, or use similar modeling constructs to represent different concepts, the acquirer and supplier may not be able to properly interpret each other's models. One benefit of the GRA model approach is that it serves as a concrete declaration of the acquirer's preferred modeling style and the semantics (meaning) of key terms and concepts. Thus, the GRA model represents a de facto standard that can improve the interoperability of models being shared between the acquirer and supplier.

MBSE for the Remainder of the Acquisition Life Cycle

Just as MBSE can improve suppliers' ability to develop complex systems more efficiently and effectively, so also can MBSE improve the acquirer's ability to more efficiently and effectively manage the development program. Regular inspection of the suppliers' digital system models, aided by automated querying and reporting capabilities, can enable a more proactive application of the acquirer's oversight role, providing early warning of potential disconnects. Using MBSE can also enable more effective verification and product assurance planning and execution; in many cases, enabling these activities to be performed early and often during the design and early integration phases, thereby reducing the risk that disconnects are not discovered too late in the formal verification process.

MBSE can also provide enduring value to the system users and maintainers long after the system has transitioned to operations. The digital system models can continue to serve as the primary source of knowledge about the system, facilitating and maintaining the distinction between "as-designed," "as-built," and "as-operated" configurations of the system. These models would also enable "what if" exercises and impact analyses – both proactive and reactive – and inform the development of tactics, techniques, and procedures (TTP) for the system.

Chapter Summary

Model-Based Systems Engineering (MBSE) is a rapidly expanding practice of SE, within which descriptive system models serve as the foundation for knowledge capture and information exchange when executing the SE process. Just as system developers are exploiting MBSE to improve the efficiency and effectiveness of the system development process, so also can acquiring agencies use MBSE to improve implementation of acquisition processes, engagement with suppliers responsible for system development, and execution of full life cycle responsibilities for operating and sustaining acquired systems.

This chapter described the problem space for MBSE in acquisition and contrasted the unique characteristics of MBSE implementation by acquiring organizations with the traditional MBSE implementation practices used by system developers. The separation of responsibilities between the acquirer and the supplier drives the need for distinct acquirer-specific MBSE activities that facilitate the evaluation of enterprise capabilities to assess gaps, identify potential materiel and non-materiel solutions to close those gaps, and develop the requirements that will ultimately be conveyed to the supplier who will be responsible for developing the new system.

Best practices associated with the use of MBSE to satisfy the unique character of acquirer needs were described for each main phase of the system life cycle that occurs prior to full-scale development and operations. In particular, the establishment of a federation of models across multiple levels of the enterprise hierarchy enables those models to contribute to enhanced decision-making at each of those levels. Models can facilitate enterprise gap analysis and analyses of alternatives, improving the decision-making processes that motivate and inform system acquisitions. Models can also improve the quality of the communication of technical requirements to the supplier, particularly through the Government Reference

Architecture. Advances in MBSE are improving both the development of systems and the execution of the acquirer process that identify the systems to be developed.

References

1. Friedenthal S, Moore A, Steiner R (2015) A Practical Guide to SysML: The Systems Modeling Language, 3rd edition, Morgan Kaufmann.
2. INCOSE (2015) Systems Engineering Handbook: A Guide for System Life Cycle Processes and Activities, 4th edition, INCOSE, San Diego.
3. ISO (2015) ISO/IEC/IEEE Standard 15288-2015: Systems Engineering—System Life Cycle Processes. International Standards Organization (ISO).
4. Noguchi RA (2019) A Roadmap for Advancing the State of the Practice of Model Based Systems Engineering for Government Acquisition. INCOSE International Symposium.
5. U.S. Department of Defense (2022) Systems Engineering Guidebook. Washington, DC, USA.
6. U.S. Department of Defense (2022) Operation of the Adaptive Acquisition System, Department of Defense Instruction 5000.02, Change 1, Washington, DC.
7. Maier, M. and Rechtin, E. (2009) The Art of System Architecting, 3rd edition, CRC Press.
8. ISO (2022) ISO/IEC/IEEE Standard 42020-2022: Software, Systems and Enterprise—Architecture Processes, International Standards Organization (ISO).
9. Heder M (2017) From NASA to EU: The Evolution of the TRL Scale in Public Sector Innovation, The Innovation Journal: The Public Sector Innovation Journal, Volume 22 (2).
10. Trujillo, A. and Madni, A.M. Evaluating Value Proposition of Design Reuse, 2021 AIAA SciTech, Nashville, Tennessee, Jan 11–15, 2021.
11. U.S. Department of Defense (2018) U.S. Department of Defense Digital Engineering Strategy. Office of the Deputy Assistant Secretary of Defense for Systems Engineering, Washington, D.C.
12. Noguchi RA (2019) Recommended Best Practices based on MBSE Pilot Projects. INCOSE International Symposium.
13. Madni, A.M., Madni, C.C., and Lucero, D.S. Leveraging Digital Twin Technology in Model-Based Systems Engineering, MDPI *Systems,* special issue on *"Model-Based Systems Engineering,"* Feb 2019
14. Zweber J, Kolonay R, Kobryn P, Tuegel E (2017) Digital Thread and Twin for Systems Engineering: Pre-MDD Through TMRR, AIAA SciTech Forum.
15. Hause M (2019) The Digital Twin Throughout the SE Lifecycle. INCOSE International Symposium.
16. Maier MW (1998) Architecting Principles for Systems-of-Systems. Systems Engineering, Vol. 1, Issue 4, Wiley.
17. Shapiro, C and Varian, HR (1998) Information Rules: A Strategic Guide to the Network Economy, Harvard Business Review Press.
18. Friedenthal S, Izumi L, Meilich A (2007) Object-Oriented Systems Engineering Method (OOSEM) applied to Joint Force Projection (JFP), a Lockheed Martin Integrating Concept (LMIC), INCOSE International Symposium.
19. Esteban JA (2007) Survey of Model-Based Systems Engineering (MBSE) Methodologies. INCOSE MBSE Focus Group 25.8, Pasadena.
20. Martin JN (2010) An Enterprise Systems Engineering Framework. INCOSE International Symposium.
21. Martin JN, Minnichelli RJ (2020) Strategy for Implementing an Enterprise Systems Engineering Capability. INCOSE International Symposium.
22. Martin JN, Noguchi RA, Minnichelli RJ, Wheaton MJ (2020) Implementing Enterprise Systems Engineering Enabled by the Digital Engineering Approach. AIAA ASCEND Conference.

24 MBSE for Acquisition

23. Object Management Group (2022) Unified Architecture Framework Modeling Language (UAFML) Version 1.2.
24. Object Management Group (2007) OMG System Modeling Language (OMG SysML) V1.0.
25. U.S. Department of Defense (2009) DoD architecture framework (DODAF), version 2.0. Washington, DC.
26. Trujillo, A. and Madni, A.M. An MBSE Approach Supporting Technical Inheritance and Design Reuse Decisions, AIAA ASCEND Conference, Designed to Accelerate Our Off-World Future, Nov 16–18, 2020.
27. Maier MW, Rechtin E (2009) The Art of System Architecting, 3rd edition, CRC Press.
28. Min IA, Noguchi RA (2016) The architecture design and evaluation process: A decision support framework for conducting and evaluating architecture studies. IEEE Aerospace Conference.
29. Flanigan D, Robinson K (2019) Exploring the Test and Evaluation Space Using Model Based Conceptual Design (MBCD) Techniques.
30. Aguilar JA, Dawdy AB, Law GW (2014) The Aerospace Corporation's Concept Design Center. INCOSE International Symposium.
31. Stevens, R (2019) Concept Design Using Model Based Systems Engineering. IEEE Aerospace Conference.
32. Stevens, R (2020) Weaving a Digital Thread into Concept Design. IEEE Aerospace Conference.
33. Do Q, Cook S, Lay M (2014) An Investigation of MBSE Practices Across the Contractual Boundary. Procedia Computer Science 28.
34. Do Q, Cook S, Fossnes T, Haskins C (2014) The Use of Models across the Contractual Boundary: Past, Present and Future. INCOSE International Symposium.
35. Sudyam A, Pyles, J (2020) Lockheed Martin Conceptual Design Modeling in the Dassault Systemes 3DEXPERIENCE Platform. AIAA SciTech Forum.
36. LaSorda M, Borky J, Sega R (2018) Model-Based Architecture Optimization for Major Acquisition Analysis of Alternatives. IEEE Aerospace Conference.
37. U.S. Department of Defense (2020) Mission Engineering Guidebook. Washington, DC, USA.
38. Hernandez A, Karimova T, Nelson D (2017) Mission Engineering and Analysis Innovations in the Military Decision Making Process. Proceedings of the American Society for Engineering Management.
39. Beery P, Paulo E (2019) Application of Model-Based Systems Engineering Concepts to Support Mission Engineering. Systems Journal, Volume 7, No. 44.
40. Noguchi RA, Martin JN, Wheaton, MJ (2020) $(MBSE)^2$: Applying MBSE to Architect, Implement, and Operate the MBSE System. AIAA ASCEND Conference.

Mr. Ryan Noguchi is a Principal Engineer in the Architecture & Design Subdivision at The Aerospace Corporation. He is responsible for developing and applying disciplined system architecting methods, processes, tools, and models for a diverse set of government agencies and internal R&D projects. He has led multiple system modeling pilot projects and advises Government enterprise and system acquisition program customers in standing up MBSE capabilities. He co-leads Aerospace's Model-Based Engineering Community of Interest, which has brought together hundreds of Aerospace personnel to share lessons learned and to collaborate on advancing the state of the practice of model-based engineering within Aerospace and with its customers. He serves as Aerospace's Representative to the INCOSE Corporate Advisory Board and participates in numerous collaborations in the Digital Engineering area with INCOSE, AIAA, and the Systems Engineering Research Center. He has extensive experience modeling in the use of Unified Modeling Language (UML), Systems Modeling Language (SysML), Unified Profile for DoDAF and MoDAF (UPDM), and Unified Architecture Framework (UAF) Modeling Language (UAFML) and is an Object Management Group (OMG) Certified Systems Modeling Professional (OCSMP) SysML Model Builder – Advanced. He has a B.S. in Mechanical and Aerospace Engineering from Princeton University and a M.S. in Mechanical Engineering from the University of California, Berkeley.

Dr. Robert J. Minnichelli is a Principal Engineer for the Systems Engineering Division at The Aerospace Corporation. He has provided system engineering analysis, programmatic support, and innovative technical capability development for over three decades, and he established and co-led a Corporate Strategic Initiative on Enterprise Systems Engineering (ESE) and Digital Engineering (DE) for 4 years. He is also a part-time lecturer in the USC Systems Architecting and Engineering Program. Dr. Minnichelli received his SB Degree in Mechanical Engineering from MIT, and his MS and PhD degrees in Electrical Engineering at UC Berkeley, all with theses in the control systems area. Bob is a co-chair of the IEEE Model Based Systems Engineering (MBSE) Technical Committee. For the IEEE Aerospace Conference, he serves as both a co-chair of the Best Paper Committee and vice-chair of the Paper Review Committee. He has been an active member of INCOSE for over 25 years. He was recently honored with the 2020 Distinguished Engineering Achievement Award from The Engineers' Council for "leadership in System-of-Systems Engineering, ESE, and MBSE for space and launch systems architecture."

Managing Model-Based Systems Engineering Efforts

25

Mark L. McKelvin

Contents

Introduction	754
Project Management	754
Project Management for MBSE Projects	757
State-of-the-Art	763
Best Practice Approach	765
Initiation and Scope Definition	766
Planning	767
Measurement	769
Execution	771
Review and Evaluation	772
Closure	773
Management Tools	773
Illustrative Example	774
Chapter Summary	778
Cross-References	778
References	778

Abstract

Despite advances and substantial investments in applying model-based systems engineering (MBSE) across the systems engineering community, approaches to managing a MBSE project are still in their infancy. In the literature, the application of MBSE projects suggests that project management remains a key challenge. Consequently, little work in the literature addresses managing a MBSE project effectively using a principled approach based on fundamental processes and activities in project management. This chapter addresses the management of MBSE projects and discusses the characteristics that make managing a MBSE

M. L. McKelvin (✉)
University of Southern California, Los Angeles, CA, USA
e-mail: mckelvin@usc.edu

© Springer Nature Switzerland AG 2023
A. M. Madni et al. (eds.), *Handbook of Model-Based Systems Engineering*,
https://doi.org/10.1007/978-3-030-93582-5_45

project challenging. An example from the literature demonstrates the application of project management processes to implement a MBSE pathfinder effort. The example illustrates the benefits and impacts of using fundamental project management processes and activities to a MBSE project.

Keywords

Project management · Model-based systems engineering

Introduction

Managing model-based systems engineering (MBSE) efforts focuses on how to improve the likelihood of success for developing, maintaining, and using model artifacts to support a systems engineering project. This chapter emphasizes the application of fundamental project management principles towards the use of MBSE to support systems engineering of an engineered system. In this context, a MBSE project represents the effort to conduct systems engineering using a model-centric approach. Similar to how project management principles are applied to a project for an engineered system, the development, maintenance, and use of model artifacts may also apply fundamental project management principles to increase the likelihood that the development of model artifacts are scoped properly to be used by and provide support to systems engineering of an engineered system. An MBSE project incorporates processes, methods, and software application tools to enable systems engineering for the engineered system. MBSE provides an integrated means for representing, analyzing, reviewing, and contributing to a systems engineering project for the engineered system.

Project management is the application of process knowledge, skills, tools, deliverables, and techniques to project activities that ensure that a project meets its stated goals and requirements [1]. Project management includes planning the activities, measuring the progress, allocating resources, and completing a project within an identified set of constraints, such as time, cost, and schedule. The construction of a building, the development of software for an automotive system, the effort to establish sales in a new market, and the application of MBSE to organize and execute systems engineering on an engineered system are all examples of projects. MBSE represents a paradigm shift in people, processes, and technology from document-centric to model-centric systems engineering, where information models are used to formalize systems engineering practice. This does not imply that documents are not used, rather, it implies the use of explicit modeling formalisms in an environment that enables machine-readable, digital products.

Project Management

A project is a set of agreed tasks with a defined time for completion, a limited budget, and a well-defined set of objectives. A project is a temporary endeavor undertaken to create a specified product, service, or result. The temporary nature of a project is due to bounding the execution time with a beginning and ending; therefore, it has a defined scope and finite resources. The characteristic of limiting the time of project execution distinguishes project management from ongoing processes in general management. The activities of a project are designed to accomplish a discrete goal. A large endeavor may be divided into iterative phases where each phase is considered a project since it manages a finite set of tasks within a duration of time and resources to complete a set of goals within a given phase. This strategy is common in methodologies that emphasize development cycles and iterative phases. Project management includes planning, organizing, directing, and controlling an organization's resources to achieve specific goals defined for a specific project [1]. The Project Management Institute (PMI) adds that project management is the application of knowledge, skills, tools, and techniques to project activities to meet or exceed a project's stakeholder needs and expectations [2]. The importance of disciplined project management is generally recognized to be a relevant factor in the success of projects, according to multiple empirical studies [3–5]. A model-based approach enables a structured and disciplined approach that could leverage computer-based verification, validation, and automation to enhance a systems engineering project. Organizations turn to project management to deliver consistent results, reduce costs, increase efficiencies, and improve customer and stakeholder satisfaction [6].

Project management generally advances a project from identifying a problem to the point where a solution is fully implemented and shared with the customer. The steps of the phases may differ by the methodology employed, but the phases are the same. For example, a project that employs Six Sigma frameworks for development projects will typically employ a sequence of *define*, *measure*, *analyze*, *improve*, and *control* [8]. On the other hand, agile development projects tend to follow phases that include *concept*, *inception*, *iteration*, *release*, *production*, and *retirement* [9], and the phase are repeated cyclically until the project is completed. While the specific project management phases might differ by methodology, the phases are intended to achieve the same goals:

- A challenge, obstacle, or need is discovered and further defined.
- Concepts for a solution are discussed and studied.
- The project is planned and the plan is executed.
- The results are monitored to ensure the project efforts are successful.
- The project ends and the results are passed off to customers or business partners.

The PMI identifies the five phases as follows: *initiation*; *planning*; *execution*; *performance and monitoring*; and *close*. A summary of each phase follows.

Conception and Initiation During this phase, a problem is discovered or a need is defined. A set of high-level solutions may be discussed with business leadership to align to organizational goals and understand whether a specific project, if enacted, would provide enough value and return on investment against the resources expended. A feasibility study may also be conducted to determine if the project would work as stated. This is where the expected goals and timelines for the project are reviewed along with existing information about budgets and resources for similar projects in the past are used to assess the ability to execute the project. As an output of this phase, a report is prepared for business leaders and stakeholders with a recommendation on the value and profitability of a project. The report will enable stakeholders to determine whether or not they want the project to proceed.

Planning This phase consists of forming a team, with the project manager leading and guiding the project effort and developing a project plan. The plan serves as an ongoing guide to keep the team on task during the rest of the project's duration. Successful projects typically include at least the following in the project plan:

- Documented specific, measurable, attainable, relevant goals, and clearly identifying a deadline.
- Budgets include where and how the team acquires funding, labor, and other resources for project requirements.
- A detailed definition of the outcome of the project.
- A task schedule with milestones, including task dependencies and when each should be completed.
- Assignments for each task to specific people on the team or resources the team has access to.

The key output of this phase is a documented plan that guides the project's execution and remaining phases.

Execution The team enacts the project plan. The role of the project manager is to lead the team and provide support to the team as needed. The project manager is usually responsible for checking in with each team member to ensure all deliverables are completed on time and facilitating communication between team members. This phase represents the bulk of the effort on a project where the work is performed. The project plan may be updated appropriately, and the project manager updates the stakeholders on project progress. The output of this phase includes the work product and documentation that describes the work, challenges, issues, gaps, risks, and decisions.

Performance and Monitoring The work of the project team and the outcome from the execution must be monitored and controlled. During the execution phase, the project manager monitors budgets and resource use, keeping the team within the project plan's scope. Testing may be employed in this phase to measure the work products and make necessary changes to control and improve the outcomes. In addition, adjustments to schedules and resources are made to keep the project on track. To monitor progress, key performance indicators are developed and tracked. Key outputs of this phase include updates to work products or updates to schedules and resources.

Close Once the project is complete, the goal is to formally close the project by turning over the project outputs to the customer. Outputs may be a service or product.

In addition, the project manager review the project results to evaluate successes and failures on the project. This phase of project management helps a team identify things that went well and areas of improvement. The project manager will need to prepare a final budget and final project report. Finally, it is good practice to collect all project documents and deliverables, and store them in a central location.

Project Management for MBSE Projects

Across industries, MBSE is gaining popularity and use as a systems engineering methodology that exploits the use of domain-specific models and modern technology in a computer-aided environment as a means to capture and exchange information in contrast to the primary use of traditional documents. The complexity of modern, large-scale engineered systems demand the ability to understand the interaction of components with one another and their operational environment. A model-based approach that includes a computer-aided modeling and analysis environment provides advantages to systems engineering. For example, in traditional systems engineering where documents are the primary artifacts for capturing the system design from different perspectives, a MBSE approach provides a single source of information for the system from which various stakeholder views are created using the same set of model elements. In a model-based approach, models are used to provide blueprints to write code, provide structure, rigor, and consistency in systems engineering artifacts. In addition, MBSE is used to improve the understanding of the system that results from increasing interactions among system components.

Despite the benefits of MBSE, adopting, planning, executing, and maintaining a large-scale project of an engineering system using a model-based approach is challenging. Systems engineering on projects for the engineered system are not likely to meet cost and target delivery dates when the a model-based approach is poorly executed. Project management principles can be applied to improve the likelihood of success to infusing and executing a model-based approach and technologies on projects for engineered systems. Managing a MBSE project applies project management activities to ensure the modeling artifacts address organization needs, and the model artifacts of the engineered system are delivered to the benefit of customers and stakeholders. This section summarizes some of the key challenges from experiences and lessons learned in deploying MBSE to various levels of scale and scope across the systems engineering community. These experiences provide valuable insight into how managing MBSE projects could be enhanced, and they raise the question, "what characteristics of MBSE projects make them challenging?" This section explores some key characteristics and factors regarding MBSE projects.

Challenges and Lessons Learned Despite the potential benefits, deploying a MBSE project is nontrivial and challenging. The challenges extend beyond the ability to develop models. Gaining adoption and value-added from a model-based approach starts at the beginning where the value of the model must be assess with respect to its purpose and goals of the project for developing an engineered system, including scoping the MBSE effort for the systems engineering of the engineered system to developing a modeling plan that supports the systems engineering approach. Experiences in the field illustrate the importance of properly staffing systems engineering projects that employ a model-based approach with the right people, skills, and processes to ensure collaboration between subject matter experts and modeling experts to ensure systems engineering success of the engineered system. An ineffective application of MBSE puts systems development and acquisition at risk of not meeting cost and time-to-market constraints or operational capability targets [17]. In reported experiences, while adopting MBSE, a set of challenges and lessons learned provide insight into how MBSE can be applied more effectively within systems engineering [10–16].

One of the more common challenges in MBSE projects across industry [18, 19], government [12, 14, 20, 21], and the systems engineering community [10, 22, 23], is the lack of matching appropriate information technology infrastructure and tools to the systems engineering user needs. For example, one of the needs on technology infrastructure is the ability to integrate commercially available tools so that data may be exchanged between tool users seamlessly for the typical user that does not have the ability to quickly develop and sustain custom tool integrations. Additional challenges on the information technology infrastructure include:

- Enabling read and write access for multiple, simultaneous users.
- Providing mechanisms for managing model changes, checking in and retrieving versioned controlled models.
- Mitigating security risks such as the unintentional loss or exposure of sensitive, confidential, or protected data.
- Providing means for assigning access roles to individuals for viewing or editing the model with only the amount of access needed.
- Providing means for stakeholders that lack modeling expertise to generate and disseminate model produced artifacts.
- Providing means for reviewers to efficiently provide tracked feedback to subject matter experts and modeling experts in a common digital environment.

The use of MBSE involves more than just deciding "the best tool" and learning a new modeling language in a new tool [10]. It is a transition into "how" SE is conducted using MBSE. The transition from document-centric to model-based involves a combination of people, processes, and an infrastructure to be effective and sustainable. The implications of the transition impacts a project manager in multiple ways, such as the need to provide oversight responsibilities to the model-based efforts and the integration of model-based activities within the system engineering processes, and the need to measure progress, and evaluate risk of the model-based activities and information technology infrastructure.

MBSE projects must be aligned to business objectives and have a clear purpose. Moreover, the alignment of a MBSE project with stakeholder and customer needs must be clearly articulated to maintain project scope and show value. As evident in experiences with MBSE projects, leadership support is critical to driving successful outcomes and influencing the broader application of MBSE within an organization.

This implies the need for strong technical leadership and "buy-in" from management. The transition to using MBSE as an approach to SE throughout an organization must endure resistance to cultural change, and this is observed particularly within organizations where MBSE is not yet in widespread use. MBSE requires a skilled workforce that is knowledgeable about the domain of interest, can leverage modern technology appropriately to achieve desired effects, and can communicate effectively to stakeholders and customers on the value of applying MBSE.

Importance of Project Management in MBSE Projects Challenges and lessons learned from deploying MBSE projects signify the importance of project management's role to enable the successful execution of MBSE projects [25]. A recent survey of MBSE state-of-practice indicates the need to support managing a successful MBSE project within an organization [26]. The survey further suggests that hurdles to implementing MBSE projects are addressed through management, such as providing resources to execute projects and providing resources to guide model development. The survey concludes that when organizations did not address management issues, they failed to establish and sustain the long-term capability to enable repeat success on subsequent MBSE projects. The study found that the lack of an organizational structure to support the execution of MBSE projects is a common inhibitor to project success across industry, academia, and government organizations. Across management levels, there is a failure to understand the conditions required to successfully manage an MBSE effort. Results from the study further indicate that successful organizations have addressed the adoption of MBSE by establishing and maintaining clear communications with leadership and ensuring leadership is committed to the success of their MBSE project.

A study by Bone and Cloutier states that "culture and general resistance to change" are the main inhibitors of MBSE adoption and thus require an emphasis on managing organizational and cultural changes effectively to be successful [27]. Similarly, Chami and Bruel [24] found in their survey that resistance to change is a common challenge to the successful deployment of MBSE projects, frequently stemming from the lack of awareness on the benefits of MBSE. As a consequence, organizations are reluctant to initiate MBSE projects. To mitigate, the authors propose that organizations provide resources to adequately define the purpose, scope, and strategy to deploy a MBSE project. To do so requires sponsorship from executive leadership and upfront financial investment. Since organizations are unique, the initiation of a MBSE project should capture organizational culture because the strategy to deploy MBSE depends on the organizational context in which it is deployed.

Essential to a successful MBSE project is a consideration for developing and executing an overall plan that spans people, processes, and information technology. As with any organizational change, the plan must be approached strategically to grow this capability and learn from experiences gained. Therefore, executing a MBSE project is a journey of change that requires learning, strategy, perseverance, and continuous improvement. It is commonly accepted that beginning a process improvement activity requires first identifying the problem to be solved. A plan is then developed that implements problem solutions. When an organization faces pressures and constraints such as time, quality, and cost, transitioning to a model-based culture can be daunting. Choosing the proper first steps to manage MBSE projects is key to a successful transition. The first steps include proper planning of the MBSE effort, and this typically includes defining the purpose of the models used to support the systems engineering effort for the engineered system, select an appropriate methodology, plan the work, including identifying the necessary staff, and developing metrics to measure progress and success of the MBSE effort.

Characteristics of MBSE Projects Significant growth in MBSE projects has led to demands on traditional project managers to manage an area that requires an understanding of the specific characteristics that make MBSE projects challenging to manage. As one example, MBSE employs different types of models and viewpoints [7] to support system engineering activities, such as data models that describe the data and relationships among data items, architecture models, simulation models, and analytical models to name a few. Consequently, MBSE projects leverage information technology and software applications to manipulate and share information within the context of systems engineering, a domain that is naturally human-centered [28]. MBSE projects leverage information technology to enable system engineers to manage the design and development of systems across the system lifecycle. Hence, an ability to manage the use and infuse technology effectively is critical to the project managers. Thus, the need for project managers to acquire a staff with the right set of skills and abilities to exercise system engineering principles and domain expertise. Let us take a look at some of the specific characteristics of MBSE projects.

1. *Intangible outcome.* Unlike an engineering project where an object being built consists of physical matter, size dimensions, and weight, a model is an abstract representation of a physical thing. A model is a digital product that is not tangible and is delivered and experienced through software and information technology. MBSE is based on logical, not physical work. It consists of ideas, designs, instructions, and formulas, as opposed to concrete and bricks, for example. The logical nature of model artifacts in MBSE projects results in a high degree of interactions that are not always clearly understood. This complexity drives the need for increased communication and collaboration between team members to understand the impacts of project dependencies. Often, the customer has limited visibility into the work on an MBSE project and the amount of effort put into developing models and digital products. This leads to customer dissatisfaction since it is more difficult to assess the value of the product while it is under development.

   As a project manager, there are several options that may help. One option is to actively use prototypes that can be periodically presented to customers throughout product development. Prototypes enable customers to visualize how the model product will look and demonstrate that the behavior is as expected within the scope of the model. Another option is to manage expectations for the final product. That is, communicate with the customer and stakeholders regarding what is expected so they are not left anticipating hopelessly. By sharing a clear project activity plan and providing periodic status reports with the customer and stakeholders leads to increased confidence in the final product. Since communication and collaboration within the team is highly critical, the project manager should establish a culture that communicates effectively within the team. This could include scheduling team meetings and one-on-one meetings with individual team members.

2. *Flexibility.* In MBSE projects, the models and technology that enable modeling are particularly subject to change. Models are malleable and therefore can be changed by anyone at any time and for any reason. There is no unique way to execute a MBSE project since it must be adapted for a specific context and purpose. It is an ideal medium to respond to changing market conditions, customer needs, and organizational changes. Developing useful models for MBSE projects is a creative endeavor that looks beyond conventional techniques to meet customer requirements and develop innovative solutions to problems. The project team needs the space and time to think, solve problems, and find new solutions. But, technical creativity is countered by modeling discipline. A lack of discipline results in models that are not reusable, challenging to maintain, and less comprehensible to different stakeholders. So, a project manager must know when and how to manage the balance between technical creativity and discipline. In addition, the information technology market is changing quickly under the pressure of volatile customer demands. Technology changes, such as new software applications and tools, are occurring at a rapid pace along with new standards appearing constantly. As a result, the IT market is flooded with a plethora of options to choose from. This flexibility in the choice of IT and software applications results in difficult decisions for the project manager.

   Project managers on MBSE projects must adapt to changes because the current technology and deployment practices will likely become irrelevant very soon. These conditions are great for adopting incremental development approaches, automation, and agility within an MBSE project, whereas an iterative approach to managing MBSE projects focuses on continuous releases and incorporating customer feedback with every iteration. Communication and collaboration within the MBSE project team and externally with stakeholders are critical to be able to keep pace with changes and ensure pieces of the product are interfaced appropriately. This is much tougher with larger teams as it increases the number of interfaces and perspectives to manage. Smaller teams that are highly specialized in skillsets foster better communication, collaboration, and agility. Moreover, these teams must be equipped with the resources and authority to get the job done; therefore, they are able to adapt to change more effectively.

3. *High degree of uncertainty.* Uncertainty in project management describes a condition in which the outcome of the project is unknown. Projects tend to have some level of uncertainty; however, the level of uncertainty in MBSE projects is relatively higher than in non-MBSE projects due to uncertainties of customer requirements, changing technology, and the dynamics of the project team. Prior to project execution, it is very difficult to accurately define the detailed requirements for a MBSE project. Requirements are not always known upfront,

and the customer often lacks understanding about what is needed and feasible. These conditions lead to poorly defined requirements and improperly scoped projects. After a project's execution has started, there is an increased understanding of the project's characteristics that leads to changing requirements and updating project scope. MBSE projects push the state-of-the-art in methods and adoption of technology. But problems arise when the underlying technology changes or no longer works as expected. It might be difficult to use new software applications or the underlying infrastructure is inaccessible or unavailable. This leads to cost overruns, time spent trying to fix the enabling technology, and schedule impacts. Interactions between teams are often dynamic and unpredictable. For example, a valued member of the team may change jobs or is no longer available. Different personalities on the team may negatively impact team cohesiveness. This increases uncertainty because human behavior is often unpredictable.

To address uncertainty, project managers should have a risk management plan in place that also contains some contingency reserves in the budget and schedule. Uncertainties could be translated into risks that could be properly managed and mitigated to avoid becoming major issues. The project manager must align the team to a common understanding of the problem that is being addressed, a clear set of goals, purpose, and expected outcomes. Like any other project, these must be defined and agreed to with the customer from the beginning so that there is a clear roadmap to success and a well-defined set of outcomes that can be measured to indicate project success. To address evolving requirements, the project manager should adopt approaches that include incremental improvements or iterations, like what is found in agile approaches.

4. *Project cost factors.* Managing the cost for a MBSE project is attributed to several factors that influence a high degree of uncertainty. As one of the key constraints on a project, poorly managing MBSE project costs lead to negative effects, such as an increasing risk of budget overrun and loss of reputation. These conditions risk loss of repeat clients, loss of new business, and failed projects. So, it is important to identify the factors that have a significant impact on the cost of MBSE projects. These factors include the cost to apply a MBSE process, infrastructure cost, training cost, and costs related to the development, integration, management, and curation of models [29]. The cost of a MBSE project varies with the choice of process, such as use of the object-oriented systems engineering method (OOSEM) [30], a systems development method that combines object-oriented concepts with traditional systems engineering practices, and the MBSE Grid framework [31], a framework that specifies different work products that express the architecture of a system from the perspective of specific concerns. Infrastructure is the information system that includes the technology stack for the software applications, licenses, data storage, networks, and computing hardware, along with the people and processes to manage the technology stack. The infrastructure decision could vary with the environment in which it operates. For example, choosing software applications a la carte or deciding on a single vendor can greatly impact the project costs. It depends largely on how the applications are intended to be used and the budget to acquire necessary

applications. Moreover, deciding on a cloud-based or on-premise infrastructure can significantly impact the cost of infrastructure deployment, as illustrated in a recent whitepaper published by Amazon Web Services (AWS) [32] on MBSE deployment strategies to cloud-based infrastructures. The cost of developing models constitutes a large fraction of the overall cost of a MBSE project. Model development includes the model's scope, purpose, model level of abstraction, the degree to which the model is federated or centralized, maintaining model consistency, verification, increased number of diverse stakeholders, model management, and curation. Each characteristic has its own set of costs that further compounds the challenge of predicting the model development cost for a MBSE project. Training is a category that involves teaching and coaching individuals on software applications, modeling languages, and the use of MBSE on system engineering activities. Training costs are influenced by human factors on MBSE projects, such as the level of organization resistance to change and the level needed to overcome the learning curve. Consequently, each MBSE project is unique in what it sets out to achieve and in the parameters that characterize its implementation. Additionally, since MBSE relies heavily on human intellectual work, human resources are the greatest assets. The labor needed dominates the project costs, unlike projects with tangible outcomes such as industrial construction projects where material dominates the cost.

The ability to adapt to changes, monitor costs, and emphasize people management can help project managers reduce the risk of not meeting cost targets. Developing a risk management plan that allows for adapting to unexpected issues enables better control of risks that may lead to increased costs. It also helps to provide margin in cost estimation for addressing unknown circumstances. Understanding of technology trends and how to leverage technology effectively is necessary to adapt to changes in the infrastructure. Invest in experienced and skilled human resources, including recruiting, training, and retaining capable team members. The experience helps develop more realistic budgets from experts that are familiar with the efforts of previous or similar projects. Comparing actual expenses against planned expenses regularly provides valuable insight into how well resources are being expended. If the costs start to run higher than expected, work with the project team to find ways to reduce costs. Regular communications with the development team and stakeholders help identify potential problems or issues that may result in increased costs due to changes in stakeholder priorities and misalignment with the project scope. To address a multitude of stakeholders, it is important to know what is most valuable for each stakeholder and then try to negotiate to achieve consensus.

State-of-the-Art

While lessons learned on MBSE projects highlight the importance of project management to project success, little work in the literature addresses how to manage and execute a MBSE project based on these principles. This section presents examples of project management principles applied to enable structured and systematic management of MBSE projects.

Hallqvist and Larsson [10] apply a systematic approach to introducing MBSE into an organization, after their initial approach met less than desirable outcomes. They reported that the shortcomings in the initial approach included the lack of a clear vision for the model development effort, varied expectations on project goals, change of scope, and lack of resource planning to scale the effort across the organization. As a result, they applied a systematic approach that eventually led to project success.

Unlike the previous attempt to introduce MBSE into the organization, management involvement increased, and the work to develop a business case for introducing MBSE was established to initiate the project. Management played a critical role in gathering stakeholders to align on the project's scope and to make decisions in the face of varying perspectives. In addition, the team developed a business case for MBSE that includes benchmarks from other organizations' introduction of MBSE, roadmaps, vision, and concept analysis for introducing MBSE.

The roadmap from the business case is used to identify the project plan and prepare for project execution. The project plan identifies all the processes expected to be affected by the introduction of models, including the decision on the modeling environment. The modeling environment includes software applications, such as the modeling tools and the infrastructure to enable their use. The plan arranges each process into smaller incremental steps to define work tasks. For each incremental step in the project plan, a risk analysis is performed to create a risk management plan that identifies the risk, its consequence, cost, probability of occurrence, and mitigation. In addition, the project plan includes a cost for training. Consequently, the project scope is redefined within the context of the project plan. A stakeholder analysis is performed to identify user needs for the modeling environment to decide on the infrastructure and software applications. Upon completion, the team proceeded to evaluate their resulting project concept and decided to move forward on project execution. This effort demonstrates how project management successfully introduces MBSE into an organization's system engineering.

Chami et al. [33] introduce a three-phase approach to manage execution of a MBSE project. Definition is the first phase, where the scope and purpose of the project are defined. This phase results in a project development and deployment plan. The phase starts by identifying small initiatives that may be expanded to cover different areas of the product under development. An analysis is performed to determine if the project will be accepted. If the project is accepted, then the initiative's scope, goals, and expected benefits are identified. The phase continues with a team of skilled personnel, external consultants, and project managers to align the team with the project goals and establish a common base of MBSE knowledge. A problem statement is defined, along with challenges, to provide motivation for the project. The problem statement informs a business case, including a clear cost of investment and an estimate of expected benefits. The risk management approach and means for measuring project quality during execution are considered. Once completed, this phase delivers a project plan for the development and deployment of MBSE.

Development of the MBSE approach is the second phase. The approach includes model-related artifacts such as the modeling ontology, model libraries, modeling guidelines, model validation, and customizations for software applications if needed. This phase also includes developing and delivering training materials on model development.

The third phase is deployment. This project execution phase emphasizes implementing a system model based on the project plan and the MBSE approach identified in the second phase. Deployment includes providing a level of governance that is necessary to monitor the model development status, measure the model quality, ensure participation of all stakeholders, and react quickly to prevent any risk of delay when changes or issues occur. Observations from these efforts lead to the conclusion that a structured approach that covers phases of project management is needed to deploy, manage, and sustain successful MBSE projects.

Best Practice Approach

A best practice approach to managing MBSE projects follows the practice of managing software engineering projects, as provided in the Guide to the Software Engineering Body of Knowledge (SWEBOK Guide) [36]. The characteristics of managing MBSE projects are closely related to the managing software intensive applications, products, and services. While there are differences in the development of system engineering models using computer-aided engineering software applications as compared to software development [34, 35], the project management for MBSE and software projects are similar. For example, similar to MBSE projects, management of software projects considers the underlying information technology, including hardware, software, policies, processes, data, facilities, and organizational management of the equipment. Managing a MBSE project must address stakeholder alignment, understanding, changing requirements, a high degree of novelty and complexity, and the rapid change of the underlying information technology. This section will summarize these practices in the context of MBSE projects.

The phases of project management as applied to MBSE efforts are summarized in Fig. 1. Since a project is a time-bound endeavor, the initiation and scope definition phase mark the start of a MBSE project. During the initiation and definition phase, a project manager establishes the purpose for doing a MBSE project and justifies what business value it will deliver upon completion, resulting in stakeholder buy-in. This phase includes a definition of the project scope and project constraints. The next phase is the planning phase. In the planning phase, a detailed set of goals and a project roadmap are identified and documented. During the planning phase, a set of processes are established that will be used to guide the project execution. The measurement phase establishes the process for monitoring and controlling the project within its scope and defined parameters to help keep the project within cost, schedule, and alignment with stakeholder expectations. The next phase, project execution, is where the processes developed during the planning and measurement phases are executed. In the execution phase, models are developed. The review and

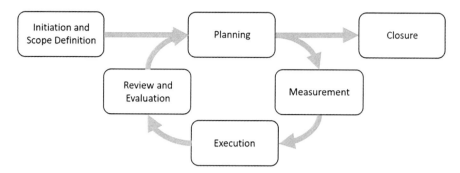

Fig. 1 Simplified model of project management phases

evaluation phase assesses the project against goals and determines how well the project met its stakeholders' needs. If the project follows an iterative process flow, then another iteration of planning, measurement, execution, review, and evaluation is performed. When the project ends, a debrief with stakeholders and within the project management team occurs during the closure phase to end the project. The flow is intended for iterative and non-iterative project lifecycle processes.

Initiation and Scope Definition

Activities within the initiation and scope definition focus on determining project requirements, assessing project feasibility, and establishing a process to review project requirements and scope. As the first phase of the project management lifecycle, an organization decides if the project is needed and how beneficial it will be. A business case and feasibility study are developed and used to evaluate the project expectations. This phase ends with developing a process to review requirements and scope throughout the project lifecycle.

Determination and Negotiation of Requirements This activity aims to determine project requirements and negotiate them with stakeholders. Stakeholders are clients, model users, and other individuals or organizations who have a stake in using the model or its development. The process of finding out what are the requirements for an intended MBSE project is to start by communicating with the stakeholders. The requirements elicitation is followed by a requirements analysis to determine whether the stated requirements are clear, complete, unique, concise, valid, consistent, unambiguous, and align with stakeholders' expectations. After validating that the requirements are aligned with stakeholder needs, they must be documented.

Feasibility Analysis A feasibility analysis is an assessment of the practicality of a proposed project that is used to determine the viability of a project. The evaluation is accomplished by analyzing technical, economic, legal, operational, and scheduling feasibility factors. For example, is there a need to create the technology to achieve what is proposed for the project? Are the people, tools, and resources available and necessary? From an operational perspective, how will the project impact daily processes in the organization and how will the project be maintained? The feasibility analysis develops alternative approaches and selects the best approach considering technology constraints, resources, finances, social, and political considerations. In addition, personnel roles must be determined along with the needed knowledge, skills, abilities, facilities, and infrastructure. One approach to assess feasibility of a MBSE project is to conduct a pilot study and identify lessons learned.

Process for the Review and Revision of Model Requirements MBSE projects are subject to frequent requirement changes. Developing a process for reviewing and revising requirements placed on the model development establishes a means by which requirements are reviewed and revised throughout the project lifecycle. As the project progresses, the project requirements and initial project scope are revisited at predetermined points in the review and revision process. If changes are accepted, then the impact of those changes should be considered. The review and revision process form the basis for evaluation of success during closure of an incremental cycle or the entire MBSE project.

Planning

Project planning is an essential phase in the management of a MBSE project. The project planning is performed before the project's execution, resulting in a plan to facilitate execution. Project planning typically consists of the following activities:

- Determining project deliverables
- Estimating project size, effort, time, and cost
- Resource allocation
- Risk management
- Quality management
- Plan management

Each of the activities within the project planning phase are elaborated in this section.

Process Planning The process planning activity identifies the model development process to be used throughout the execution of the MBSE project. Lifecycle development models span a continuum from predictive to adaptive process models. Predictive lifecycle models are characterized by developing detailed project requirements and planning that minimizes iteration through model development. In contrast, adaptive lifecycle process models are characterized by iterative development cycles designed to accommodate emergent requirements and iterative adjustments to plans. Well-known lifecycle models for MBSE projects include the waterfall, incremental, spiral [37], and various forms of agile model development [38]. As a part of planning, relevant software applications, including applications that can automate process tasks, are selected, planned for, and acquired. While many applications are chosen based primarily on technical considerations such as model construction features and usability, some of the projects are related to project management. For example, applications for project scheduling, requirements management, model maintenance, confirmation management, and quality management are also acquired.

Determine Deliverables The work products of each project activity, such as model design reports and test reports, are identified and characterized. The model deliverables are developed for specific purposes, such as communication of system architecture using the model, communication of system architecture using the model, or performing specific analyses. Determining the deliverables for a MBSE project provides an opportunity to evaluate the ability to reuse model elements, or libraries of models, from previous projects. Procurement of software applications to support model development is planned. And, application suppliers are selected, such as vendor's modeling tools.

Effort, Schedule, and Cost Estimation This activity results in estimates of the project. It relies on estimation techniques to determine the effort, schedule, and cost of a MBSE project. The effort required for a project can be determined using estimation models based on historical parametric sizing methods and other relevant methods such as expert judgment and analogy [39]. The expected schedule of tasks with projected start times, durations, and end times for more predictive MBSE projects is typically produced during planning. For MBSE projects whose requirements and scope change, an overall estimate of effort and schedule is typically developed from the initial understanding of the requirements or constraints on overall effort. The schedule may be specified and used to determine an initial estimate of the number of iterative cycles and estimates of effort and other resources allocated to each cycle.

Resource Allocation Equipment, facilities, and people are allocated to the identified project tasks. Resource allocation is based on, and constrained by, the availability of resources and their optimal use, as well as by issues that are related to personnel, such as individual and team productivity, team dynamics, and team structures. In MBSE projects, workloads are managed to ensure the best utilization of human resources. Then, if needed, people are reassigned based on current resource availability and project timelines.

Risk Management Risk is characterized by the likelihood of an event that will result in a negative impact plus a characterization of the negative impact on a MBSE project. Risk management seeks to identify risk factors and analyze the probability and potential impact of each risk factor related to the execution of a MBSE project. Risk management also emphasizes the prioritization of risk factors. Developing risk mitigation strategies minimizes the negative impact if a risk factor becomes a problem for the project. Risk assessment methods such as expert judgment, historical data, decision trees, and process simulations are used to identify and evaluate risk factors in MBSE projects. If the risk is unacceptable, then the project may be abandoned. Risks related to modeling, such as the tendency to add unnecessary features, can influence project risk management. Particular attention should be given to managing risks related to model quality, such as safety and security. Risk management is an ongoing activity that also occurs at the beginning of a project and periodic intervals throughout the project's lifecycle.

Quality Management Quality requirements for a MBSE project should be identified along with the quality of the associated model development products. Stakeholder needs and expectations guide the development of quality thresholds used to determine acceptable measurements for each requirement on the project. Quality assurance, improvement, verification, and validation activities are specified for the model product during the quality planning. Technical reviews, inspections, and demonstrations of completed functionality are examples of activities that support quality management.

Plan Management Plans and processes selected for model development are systematically monitored, reviewed, reported, and, when appropriate, revised. Plans associated with supporting processes such as documentation, configuration management, and problem resolution are also managed. Reporting, monitoring, and controlling a project should fit within the lifecycle process model that is chosen for the project. The plans also include the various artifacts that are used to manage the project.

Measurement

The importance of measurement is widely recognized to improve project management and practices [40]. Measurement for a MBSE project enables monitoring project progression and ensures alignment to project goals and purpose. A first step to measuring a MBSE project is identifying metrics to manage the development of the models. For example, a key measure in the management of the MBSE effort is how well the model satisfies its purpose. Metrics are a set of quantified attributes to help measure quality of a model product. Metrics also help measure the model development team's performance, efficiency, and productivity. The activities of the measurement process include establishing a commitment to measurements, planning the measurement process, executing the measurement process, and evaluating measures. A summary of the activities and tasks necessary to implement a software measurement process is provided in the IEEE 15939:2008 standard [41], and it is also applied to MBSE projects.

Establish and Sustain Measurement Commitment Measurement requirements establish the key metrics by which the project is measured. Measurement requirements are guided by organizational objectives and include identifying the scope of measurement, team commitment to measurement, and assigning resources such as adequate funding, training, software applications, and support to conduct the process. The organization's commitment to measurement is an essential factor for success, as evidenced by assignment of resources, such as allocating responsibility for tasks in the measurement process and providing adequate funding, training, and acquisition of software applications needed to conduct the process. The commitment of management and staff must be formally established, communicated, and staffed.

Plan the Measurement Process A measurement plan specifies what primitive metrics should be collected and what metrics should be computed during project execution to monitor progress. For a MBSE project, the plan relies on organizational context that includes the organization's constraints on the measurement process. The organizational context is stated in organizational processes, application domains, technology, organization interfaces, and organizational structure. Information needs must also be identified. Information needs are derived from the business, organizational, or the model product under development and are based on the organizational unit's goals, constraints, risks, and problems. In the planning of the measurement process, a set of measures are identified and selected with clear traces to the information needs. Measures are intended to be selected based on priorities of the information needs and other criteria such as cost of collection, degree of process disruption during collection, ease of obtaining accurate, consistent data, and ease of analysis and reporting. Resources for measurement tasks should be reviewed and approved by the appropriate stakeholders include data collection procedures, storage, analysis, reporting procedures, evaluation criteria, schedules, and responsibilities. The measurement planning also needs to include evaluating available technologies to support the measurement process. This includes evaluating available supporting technologies, selecting the most appropriate technologies, acquiring those technologies, and deploying those technologies.

Perform the Measurement Process The execution of the measurement process integrates the measurement tasks with the MBSE project. Data collection involves the collection, verification, and storage of measurement data throughout execution of the MBSE effort. Collection can sometimes be automated using software-based management tools to analyze and develop reports, such as velocity reports that collects how much effort is required to complete a model or part of a model. The data is collected and displayed in a chart that reflects the amount of the model a modeler completes within a specified time frame. Data may be aggregated, transformed, or recorded as part of the analysis process, using a degree of rigor that is appropriate to the nature of the information that is needed. Results of this analysis might include indicators such as graphs, quantitative measures, or other indications that result in recommendations and conclusions that are documented and reported to the stakeholders. Examples for using measurements to manage and improve project performance are documented in Ref. [42].

Execution

Project execution involves the implementation of plans and processes. Throughout the execution of the MBSE project, there is a focus on adhering to the selected development lifecycle process, with an expectation that adherence will lead to the successful satisfaction of stakeholder requirements and achievement of the project's objectives. Fundamental to project execution is ongoing monitoring, control, and reporting on metrics identified in the measurement process

Implementation of Plans Project activities are undertaken in accordance with the project plan. Resources, such as personnel, technology, and funding, are utilized, and modeling products are generated. Modeling products resulting from the project execution include models, diagrams, views, and test cases for the model, such as testing that model composition rules are adhered to. An example outline of topics for a MBSE plan is as follows:

- **Model Development**: Developing and improving current models sufficiently for system milestone reviews, updating the model, review and test the model for validity, implementing the model according to standards and common modeling patterns, and establishing a model baseline.
- **User Manual Development**: Emphasizes the development of user guides, navigation menus, and rules for constructing and composing models. This area includes developing guides on how to install, access, and use the modeling tools.
- **Training Plan Development**: Focuses on developing a plan to educate, instruct, and demonstrate how to apply the developed models and tools for stakeholders at different skill levels and roles.
- **Documenting Lessons Learned and Recommendations for Improvement**: Defines tasks that lead to documenting lessons learned from applying MBSE on the engineered system and consider the tasks for the next model development phase or targeted engineered system. This topic area also includes lessons learned for infusing new technology and assessing tools.

Model Acquisition and Supplier Contract Management Model acquisition and supplier contract management is concerned with acquiring models from a supplier. The supplier could be another organization that is contracted to deliver models and other model artifacts to a client. For example, in government organizations, models are commonly developed by contractor organizations who provide modeling services. The federation of models is an approach to consider given that each model in the federation of models is independently developed and maintained, but must be integrated together. Acquisition contracts come in a variety of types, such as fixed price, time, and materials [43]. The acquisition process involves the selection of appropriate contracts for acquiring model-based development services. Agreements typically specify the scope of work, deliverables, intellectual property rights, and what the acquirer is paying for. Once agreements are in place, the project can comply with the agreements' terms, and therefore, the agreements should be managed.

Implementation of the Measurement Process During the execution of the MBSE effort, the measurement process is executed throughout. A project manager uses the measures that are collected as identified in the measurement plan to continuously monitor and control outcomes of the MBSE effort. In the case of multiple models, the project manager would roll up the metrics from each federated model to obtain an aggregated results of the quality and performance of the MBSE effort.

Monitor Process The project plan is assessed continuously at predefined intervals. The outputs and completion criteria for each task in the project plan are assessed. The model deliverables are evaluated in terms of the required characteristics as identified in the planning process where the model's purpose is defined. In addition to the project plan, the effort, schedule, and costs are computed at each interval. The project's risk profile and model quality are expected to be revisited. The monitor process also analyzes project measures. When measurement thresholds have been exceeded, such outcomes should be reported to the project manager via manual communications such as a meeting or automated by means of project management software applications that enable automatic alerts when measurement thresholds are exceeded. This enables monitoring of the MBSE project to avoid cost, schedule, and scope overruns.

Control Process Outcomes of the project monitoring process provide the basis for decisions on making changes to the project as necessary. Changes may take the form of corrective action, such model reviews on specific parts of a model to ensure quality. It may consist of changing the direction of the model development if it is trending towards going out of the planned scope or abandoning the project if necessary. Decisions must be recorded and communicated to all relevant parties, and plans are revisited and revised as necessary.

Reporting Project progress is reported at specified time points to the project manager and stakeholders that have been previously agreed upon in the project planning. The reporting process informs stakeholders on the health of the project.

Review and Evaluation

Overall progress towards achieving the stated objectives of the MBSE project is reviewed and evaluated at prescribed times and as needed. Project requirements are reviewed, and assessments on effectiveness of the process, the personnel involved, and the tools and methods are conducted. Review and evaluation are analogous to system verification, where processes and procedures are implemented to check that a product meets its design specification. For a MBSE project, validation is accomplished by checking that modeling products are implemented according to guidelines or prescribed rules that govern their development or composition.

Determining Satisfaction of Requirements A principal goal of a modeling product is to achieve stakeholder satisfaction by meeting the requirements and goals of the model. In system engineering, this step is validation. Validation checks that the desired product meets the operational and end-user needs. Checking how well project requirements are being met requires that progress towards those goals are periodically assessed. Variances from the project's requirements are identified and mitigation actions are taken to correct the project's trajectory.

Reviewing and Evaluating Performance Personnel performance reviews can reveal interesting insights into possible personnel problems, such as team dynamics, which can negatively impact a model-based project. A variety of methods, tools, and practices are employed. They must be evaluated for their effectiveness, and the process used by the project should be systematically and periodically assessed. Changes are managed and made if appropriate.

Closure

Closure describes a moment in the MBSE project where all plans and processes are executed to completion concerning the project's end time or once a major project phase is reached. At this point, the project manager assesses how well the MBSE project met its goals and purpose. Closure requires identifying the criteria that defines when the project is expected to end or for a major project phase to come to an end. During the closure, the criteria for a phase, iteration, or end of a project due to completion is assessed. Once closure is established, processes for archiving, performing retrospectives, documenting lessons learned, and identifying process improvements are performed. As an example, in an agile lifecycle process, closure for an iteration could include a period where the project team gathers to provide a "retrospective" that reflects on the completed iteration and to plan for the next iteration unless the current iteration marks the completion of the project.

Management Tools

Project management tools are instruments to carry out project management functions. Tools are often implemented by software applications to enhance their use. For a MBSE project, tools are used to manage a project by providing visibility and control of the management processes. Some tools are implemented in software applications that require users to manipulate manually, while others enable more automation of the management functions. There is a recent trend towards using integrated software application suites of tools that are used throughout the project to plan, collect, monitor, control, and report on project and model product information. The categories of tools are based on their purpose. The purposes covered in this section include project planning and tracking, risk management, communications, and measurement.

Project planning and tracking tools estimate the project's effort, cost, and schedules. Planning tools often include scheduling tools that analyze the tasks within a work breakdown structure, their estimated durations, their precedence relationships, and the resources assigned to each task to produce a schedule. Tracking tools track project milestones, meetings, iteration cycles, product demonstrations, and action items. An example of tools for project planning is the Gantt chart [44]. It represents the project schedule concerning time periods. It is a horizontal bar chart with bars that represent the activities and schedule for the project activities. Gantt charts are embedded in numerous standard software applications, such as Microsoft Project.

Risk management tools track risk identification, estimation, and monitoring. These tools might use methods such as simulation or decision tree analysis to analyze effects on cost and schedule by computing estimates of the probabilities of risk events. Monte Carlo simulations compute probability distributions of effort, schedule, and risk by algorithmically combining multiple input probability distributions. The probability and impact matrix are a project management tool that combines the probability and impact scores of individual project risks and then ranks them in terms of severity.

Communication tools provide timely and consistent information that is relevant to stakeholders that are involved in a project. These tools include email notifications and message broadcasts to team members and stakeholders. Communication tools are also used to capture and communicate meeting minutes and provide charts that show task progress, such as backlogs. A modern tool for managing communication along with project issue status is an issue tracking system such as Atlassian Jira Software, provides the ability to track issues, perform resource allocation, time accounting, priority management, and oversight workflow, in addition to providing a means to communicate across large or distributed teams [45].

Measurement tools support activities that are relevant to supporting the measurement process. They are used to gather, analyze, and report project measurement data, such as project performance indicators. In model-based projects, as it is in software development projects, you cannot effectively correct or improve what you cannot identify and measure. Measurement tools are used to help identify and measure project activities despite whether the product is a model or software. As an example, Selvy et al. [46] leverage the use of test management for Jira tool to report on test execution and status within a MBSE project.

Illustrative Example

This section presents a summary of a case study from the National Aeronautics and Space Administration (NASA) to illustrate the application of project management principles towards the planning and executing a MBSE project [47]. The case study is a MBSE pathfinder effort conducted by the NASA Engineering and Safety Center in 2016. The pathfinder provides insight into the impact of project management on managing a MBSE effort.

Background The NASA Engineering and Safety Center established a pathfinder in 2016 to achieve three key goals. The first goal was to apply MBSE to complex NASA missions. This goal focuses on understanding how MBSE could be applied to complex NASA missions and to produce examples that can be shared as lessons learned. The second goal was to align MBSE efforts across NASA. This second goal focused on the technical, cultural, and management challenges of aligning MBSE efforts across the many NASA centers. Finally, the pathfinder was intended to capture issues and opportunities for informing next steps towards the enterprise-wide application of MBSE.

Initiation and Scope The pathfinder was initiated as a result of prior agency activities at multiple centers. Workshops hosted by the Jet Propulsion Laboratory and the Goddard Space Flight Center provided examples of applying MBSE that showed feasibility of a model-based approach. The pathfinder solicited nominations for teams from across NASA centers to pursue the three goals of the pathfinder. The pathfinder organizes participants from across NASA centers into virtual teams. The personnel from across NASA are from diverse geographical locations, technical backgrounds, experience in aerospace, and MBSE experience to challenge the implementation of MBSE and maximize lessons learned in this distributed work culture.

The pathfinder focused on learning, creativity, and innovation in system modeling of an existing system or concept rather than learning how to do system modeling while at the same time defining a new system. The teams are formed to focus on different levels of system architectures that range from campaign and mission to subsystems with the intent to develop a system model for each topic area. The teams looked at different parts of a project lifecycle and used the system model to generate technical products for different program milestone reviews within each topic area.

Project Planning Planning for the MBSE pathfinder began in 2015 to outline the top-level objectives, schedule, resources, training, and topic areas. Each team on the pathfinder was tasked to develop a work plan. The work plan includes detailed objectives, milestones, schedules, modeling goals, and deliverables. Human resources allocated to the pathfinder included participants having certain sets of desired qualities and skills, such as experience or ability to learn system modeling, capable of working in a fast-paced and diverse environment, ability to learn and communicate, and the ability to implement SE innovatively. Experienced team members served as the management team to provide leadership, implementation, guidance, and technical advice in an oversight capacity.

Several meetings were planned in preparation for the project and additional meetings were set to monitor progress. A kick-off meeting covered the goals and objectives of the work and how it related to advancing SE application at NASA. The kick-off also covered the approach to the pathfinder and schedule of events. The plan included face-to-face meetings over multiple sessions to provide a checkpoint on how the teams were doing, address issues, and prepare for upcoming work. Several sessions were scheduled to promote team building, preparing and reviewing of the work plans, and of these sessions, there are featured sessions where the participants focus on specific challenge topics such as tools, project management, model verification and validation, reviewing stakeholder information, and detailed modeling techniques. Each team has its schedule of activities that included management of the modeling effort for the specific topic area and meetings to check on progress and address issues.

Project Measurement Metrics were collected by each of the teams on time spent working on the pathfinder. The collected hours included time spent in meetings and with the technical advisors. Tracking the hours spent provided trending information that led to re-scoping parts of the effort. In particular, the data that was collected indicated that the team was not progressing as expected on completing project tasks. This indicator identified inefficiencies in the modeling efforts and frustration among the team. As a result, efforts were de-scoped to help mitigate less than desirable performance.

Project Execution Execution of the pathfinder involved developing models to cover the scope of the target systems within the topic areas, as planned. The authors of the report indicated that the teams planned and executed different modeling approaches. For example, one team planned to use a traditional project schedule. However, the team switched to an agile approach to focus on essential work items and accommodate team availability changes. Some teams produced more than just a systems model of the target system within their topic area. For instance, one team created a model to track cost, schedule, risk, and technical performance of the system of interest, in addition to the systems architecture model. And other teams performed parametric analysis to capture quantitative values of a sensitivity analysis.

Review and Evaluation Periodic meetings were conducted to review, evaluate, monitor, and control the effort. Monthly meetings took place to share progress and to obtain assistance on any technical or management issues encountered between meetings. The management team held weekly meetings to monitor progress across teams. These meetings allowed the management team to address cultural and technical issues as they were encountered. The authors report using several face-to-face meetings to share experiences with model construction, reuse, verification, and validation techniques. The sessions not only covered model development topics but also included discussions with team leads and organization leaders on MBSE project management issues. Technical guidance was provided by instructors who were also retained as technical advisors throughout the pathfinder to provide an external review perspective.

Closure The authors indicate that the pathfinder achieved its goals. The four teams produced a final report, presentation, and system models as deliverable artifacts. These artifacts are supplemented by a summary of benefits to each focus area. The artifacts enable assessment of the pathfinder on how well the teams achieved the goals of the pathfinder. Each of the topic area teams are successful at modeling their respective system-of-interest from selected NASA missions. The pathfinder resulted in significant lessons learned that can be applied across NASA. A strong, collaborative community of MBSE practitioners is established. The participants have become trained, experienced resources for the agency. The pathfinder increased the awareness of MBSE and a cadre of NASA personnel that can apply MBSE on their daily job duties.

The pathfinder exposes several key factors that contributed to the effort's success. For one, the teams communicated effectively in a virtual environment, especially when issues arose. Second, management support was provided at multiple levels of NASA. Management involvement in periodic meetings empowered teams and provided appropriate oversight to keep the pathfinder on schedule. Third, the success of the pathfinder is enabled by focusing on quick, real-world experiences that provide more immediate benefit and value to the participating organizations as opposed to waiting for years to see value and results. Finally, there is the active involvement of both experienced and expert personnel as leaders and team members, which enabled resolution of technical issues and provided training and guidance when needed.

Management Tools Although the pathfinder reports the use of commercial software tools for modeling, there is no clear indication of the kinds of management tools used in the effort. Despite the lack of reported management tools, a team in a topic area described creating a program model that is used for project management and measuring the return on investment for using a MBSE approach in this topic area. This program model also allows the team to track the cost, schedule, risk, and technical performance before and after the MBSE implementation of the system-of-interest in the topic area, and thereby determining the return on investment.

Summary The pathfinder provided a good example of how management engagement and the execution of fundamental project management techniques can lead to success of an MBSE effort. At the time of the pathfinder, it represented an innovative approach to building modeling and management capability across an enterprise. It provided lessons learned that can be shared with other organizations on how to engage management and implement MBSE in a way that can lead to a long-term capability. The effort exercises different aspects of SE including system architecting, analysis, and the use of MBSE to conduct technical milestone reviews. The key critical aspects of project management included the ability to plan the effort across multiple centers, each with its own culture and level of experience with the implementation and management of MBSE efforts. The planning clearly identified project objectives and the scope of the models that are to be created. Prior experiences have shown difficulty in meeting the objectives and goals of an MBSE project when the scope is not clearly defined. By providing this scope and ensuring that all teams understood the scope of their topic areas, the management team can monitor and control the project. The management team was able to identify issues and make corrective actions. It is also noted that there were experts to help guide and contribute to resolving technical issues. These resolutions enabled the teams to continue progress.

Chapter Summary

This chapter focuses on the management of MBSE efforts using project management principles. Experiences and lessons learned that are shared from MBSE case studies and pilots indicate that some of the more critical challenges with MBSE adoption and execution are in the area of project management, such as ensuring a modeling plan is identified upfront to clearly articulate project goals and scope. Other factors that pose challenges to a successful adoption and sustainment of a MBSE culture that enables meeting organization objectives include management engagement, organizational culture, and training on new methods and tools. These challenges can be mitigated by executing project management processes and activities that are used across software development projects more effectively. As MBSE adoption and execution increases across an organization, there is a greater need to be more deliberate in managing the adoption and execution of MBSE across an organization. A set of project management processes and activities are described as a best practice approach towards the management of MBSE projects. To illustrate the use of project management processes and activities, an example of a MBSE pathfinder is summarized. The results of the pathfinder demonstrate the impact of performing critical project management activities to enable successful execution of a MBSE effort that spans across multiple organizations.

Cross-References

▶ Adoption of MBSE in an Organization
▶ MBSE Methodologies
▶ SysML State of the Art

References

1. H. Kerzner, Project management: A systems approach to planning, scheduling, and controlling, 8th ed. Hoboken, NJ: John Wiley & Sons, 2003.
2. Project Management Institute. "A Guide to the Project Management Body of Knowledge: PMBOK (R) Guide." Project Management Institute, 2013.
3. H. Kerzner, Harold. Applied project management: best practices on implementation. Wiley, 2000.
4. R. G. Cooper. Winning at new products. Reading, MA: Addison-Wesley, 1986.
5. J. K. Pinto and D. P. Slevin, "Critical factors in successful project implementation." IEEE transactions on engineering management, vol. 1, 1987, pp. 22–27.

6. Project Management Institute, "The Value of Project Management," 2010.
7. Y. T. Lee, "Information modeling from design to implementation" National Institute of Standards and Technology, 1999.
8. F. T. Anbari, "Six sigma method and its applications in project management," Paper presented at Project Management Institute Annual Seminars & Symposium, San Antonio, TX. Newtown Square, PA: Project Management Institute, 2002.
9. G. Chin, "Agile project management." AMACOM, New York, 2004.
10. J. Hallqvist and J. Larsson, "Introducing MBSE by using systems engineering principles," in INCOSE International Symposium, vol. 26, no. 1, Wiley Online Library, pp. 512–525, 2016.
11. J. D'Ambrosio and G. Soremekun, "Systems engineering challenges and MBSE opportunities for automotive system design." In 2017 IEEE International Conference on Systems, Man, and Cybernetics (SMC), pp. 2075–2080. IEEE, 2017.
12. T. Bayer, "Is MBSE helping? Measuring value on Europa Clipper." In 2018 IEEE Aerospace Conference, pp. 1–13. IEEE, 2018.
13. R. A. Noguchi, "Recommended Best Practices based on MBSE Pilot Projects." In INCOSE International Symposium, vol. 29, no. 1, pp. 753–770. 2019.
14. L. Wang, M. Izygon, S. Okon, H. Wagner, and L. Garner, "Effort to accelerate MBSE adoption and usage at JSC." In AIAA SPACE 2016, p. 5542. 2016.
15. J. Suryadevara and S. Tiwari, "Adopting MBSE in Construction Equipment Industry: An Experience Report." In 2018 25th Asia-Pacific Software Engineering Conference (APSEC), pp. 512–521. IEEE, 2018.
16. K. J. Weiland and J. Holladay, "Model-Based Systems Engineering Pathfinder: Informing the Next Steps." In INCOSE International Symposium, vol. 27, no. 1, pp. 1594–1608. 2017.
17. J. Hutchinson, J. Whittle, and M. Rouncefield, "Model-driven engineering practices in industry: Social, organizational and managerial factors that lead to success or failure," Science of Computer Programming, special issue on Success Stories in Model Driven Engineering, vol. 89, pp. 144 – 161, 2014.
18. E. Asan, O. Albrecht, and S. Bilgen, "Handling complexity in system of systems projects–lessons learned from MBSE efforts in border security projects," Complex Systems Design & Management. Springer, Cham: 2014, pp. 281–299.
19. S. Bonnet, J. L. Voirin, V. Normand, and D. Exertier, "Implementing the mbse cultural change: Organization, coaching and lessons learned." In INCOSE International Symposium, vol. 25, no. 1, pp. 508–523. 2015.
20. K. G. Young, "Defense Space Application of MBSE-Closing the Culture Chasms," AIAA SPACE 2015 Conference and Exposition, 2015.
21. J. P. Hale, P. Zimmerman, G. Kukkala, J. Guerrero, P. Kobryn, B. Puchek, M. Bisconti, C. Baldwin, and M. Mulpuri, "Digital model-based engineering: expectations, prerequisites, and challenges of infusion," No. M-1435, 2017.
22. H. Eisenmann, "MBSE Has a Good Start; Requires More Work for Sufficient Support of Systems Engineering Activities through Models," INSIGHT, vol. 18, pp. 14–18, 2015.
23. M. Jackson, M. Wilkerson, and J. F. Castet, "Exposing hidden parts of the SE process: MBSE patterns and tools for tracking and traceability," In Proc. IEEE Aerospace Conference, 2016.
24. M. Chami and J. M. Bruel. "A survey on MBSE adoption challenges." In INCOSE EMEA Sector Systems Engineering Conference (INCOSE EMEASEC 2018), November 2018.
25. T. Huldt and I. Stenius, "State-of-practice survey of model-based systems engineering." Systems Engineering 22, no. 2 (2019): 134–145.
26. R. Cloutier and M. Bone, "MBSE Survey," Presented January 2015 INCOSE IW Los Angeles, CA, 2015.
27. M. Bone and R. Cloutier, "The current state of model based systems engineering: Results from the OMG™ SysML request for information 2009." In Proceedings of the 8th conference on systems engineering research. 2010.
28. S. Y. Kim, D. Wagner, and A. Jimenez, "Challenges in applying model-based systems engineering: human-centered design perspective," 2019.

29. A. M. Madni and S. Purohit, "Economic Analysis of Model-Based Systems Engineering" Systems, vol. 7, no. 1, 2019.
30. R. Karban, A. G. Crawford, G. Trancho, M. Zamparelli, S. Herzig, I. Gomes, M. Piette, and E. Brower, "The OpenSE Cookbook: a practical, recipe based collection of patterns, procedures, and best practices for executable systems engineering for the Thirty Meter Telescope," In Modeling, Systems Engineering, and Project Management for Astronomy VIII, vol. 10705, pp. 319–341, 2018.
31. A. Morkevicius, A. Aleksandraviciene, D. Mazeika, L. Bisikirskiene, and Z. Strolia, "MBSE Grid: A Simplified SysML-Based Approach for Modeling Complex Systems," In INCOSE International Symposium, vol. 27, no. 1, pp. 136–150, 2017.
32. B. Gozluklu, "Model Based Systems Engineering (MBSE) on AWS: From migration to innovation," [White paper], 2021. Retrieved from Amazon Web Services: https://docs.aws.amazon.com/whitepapers/latest/model-based-systems-engineering/model-based-systems-engineering.html?did=wp_card&trk=wp_card.
33. M. Chami, A. Aleksandraviciene, A. Morkevicius, and J. Bruel, "Towards Solving MBSE Adoption Challenges: The D3 MBSE Adoption Toolbox," INCOSE International Symposium, vol. 28, pp. 1463–1477, 2018.
34. D. L. Kuhn, "Selecting and effectively using a computer aided software engineering tool," No. WSRC-RP-89-483; CONF-891192-7. Westinghouse Savannah River Co., Aiken, SC, United States, 1989.
35. M. Jaskolka, V. Pantelic, J. Jaskolka, A. Schaap, L. Patcas, M. Lawford, and A. Wassyng, "Software Engineering for Model-Based Development by Domain Experts," 2016.
36. P. Bourque and R. E. Fairley, Guide to the software engineering body of knowledge (SWEBOK (R)): Version 3.0. IEEE Computer Society Press, 2014.
37. J. A. Estefan, Survey of Model-Based Systems Engineering (MBSE) Methodologies. INCOSE MBSE Initiative, May 23, 2008.
38. N. B. Ruparelia, "Software development lifecycle models." ACM SIGSOFT Software Engineering Notes 35, no. 3 (2010): 8–13.
39. M. Nasir, "A survey of software estimation techniques and project planning practices." In Seventh ACIS International Conference on Software Engineering, Artificial Intelligence, Networking, and Parallel/Distributed Computing (SNPD'06), pp. 305–310. IEEE, 2006.
40. H. Kerzner, Project management metrics, KPIs, and dashboards: a guide to measuring and monitoring project performance. John Wiley & Sons, 2017.
41. S. P. Berczuk and B. Appleton, Software Configuration Management Patterns: Effective Teamwork, Practical Integration, Addison Wesley Professional, 2003.
42. W. A. Florac and A. D. Carleton, "Practical software measurement: Measuring for process management and improvement," Carnegie-Mellon Univ Pittsburgh, PA, Software Engineering Institute, 1997.
43. A. Gopal and K. Sivaramakrishnan, "Research note—On vendor preferences for contract types in offshore software projects: The case of fixed price vs. time and materials contracts." Information Systems Research 19, no. 2 (2008): 202–220.
44. J. W. Herrmann, "A history of decision-making tools for production scheduling," In Multidisciplinary Conference on Scheduling: Theory and Applications, pp. 18–21, 2005.
45. D. Bertram, "The social nature of issue tracking in software engineering." Calgary, Alberta, Canada, 2009.
46. B. M. Selvy, A. Roberts, M. Reuter, C. C. Claver, G. Comoretto, T. Jenness, W. O'Mullane, A. Serio, R. Bovill, J. Sebag, and S. Thomas, "V&V planning and execution in an integrated model-based engineering environment using MagicDraw, Syndeia, and Jira." In Modeling, Systems Engineering, and Project Management for Astronomy VIII (Vol. 10705, p. 107050U). International Society for Optics and Photonics, July 2018.
47. K. J. Weiland and J. Holladay, "Model-Based Systems Engineering Pathfinder: Informing the Next Steps." In INCOSE International Symposium, vol. 27, no. 1, pp. 1594–1608. 2017.

Dr. Mark L. McKelvin, Jr., Lecturer, University of Southern California. Dr. McKelvin is a lecturer in the System Architecting and Engineering graduate program at the University of Southern California, Viterbi School of Engineering where he teaches courses in model-based systems engineering and systems engineering theory and practice. He is also a principal engineer at the Aerospace Corporation. In this role, he serves as the technical authority and the Aerospace team lead for the digital transformation and implementation of enterprise systems engineering across the National Space Systems customer base. Prior to joining the Aerospace Corporation, he led the development of model-based engineering technology and techniques for space system development at the National Aeronautics Space Administration Jet Propulsion Laboratory as a software systems engineer and fault protection engineer on major flight systems. He is a senior member of the American Institute of Aeronautics and Astronautics, and he serves on the Board of Directors for the International Council on Systems Engineering, Los Angeles Chapter. He earned a Ph.D. in Electrical Engineering and Computer Sciences from the University of California, Berkeley, and a Bachelor of Science in Electrical Engineering from Clark Atlanta University.

MBSE Methods for Inheritance and Design Reuse

26

A. E. Trujillo and A. M. Madni

Contents

Introduction	784
Design Reuse in Complex Engineering Systems	785
Taxonomy of Reuse in the Space Industry	785
Reuse Trends in the Space Industry	792
MBSE as a Facilitator of Improved Design Reuse Practices	794
Hierarchy of MBSE Methods for Reuse	794
Conditions Conducive to MBSE Adoption and Reuse	795
Best Practice Methodology for MBSE Design Reuse	796
Design Reuse Logical Process	797
MBSE Implementation of Reuse Process	801
Methodology Demonstration on Sample Problem	805
Chapter Summary	813
References	813

Abstract

This chapter explores how the MBSE paradigm may be leveraged to facilitate the reuse of designs and models and, more broadly, enable an organization to leverage (i.e., reuse) results from past work on future engineering projects. While facilitating reuse and achieving improved efficiency on engineering projects is a promising motivator for advancing MBSE methods, its potential has yet to be fully realized. Differences in rationales for MBSE, levels of adoption, means of deployment, and a lack of structured methodologies for exploring reuse scenarios have emerged as key roadblocks in fulfilling this promise of MBSE. This chapter will address and expand on these points and present a generalized

---

A. E. Trujillo (✉)
Massachusetts Institute of Technology, Cambridge, MA, USA
e-mail: alextr1994@gmail.com

A. M. Madni
University of Southern California, Los Angeles, CA, USA
e-mail: azad.madni@usc.edu

© Springer Nature Switzerland AG 2023
A. M. Madni et al. (eds.), *Handbook of Model-Based Systems Engineering*,
https://doi.org/10.1007/978-3-030-93582-5_47

method for realizing design reuse using MBSE principles. While the discussion is generally applicable across industries, the development of complex engineering systems in the space industry is specifically addressed.

Keywords

MBSE methodology · Design reuse · Inheritance · Logical process · MBSE implementation · SysML Profile · Space industry case studies

Introduction

Design reuse is a common practice that is integral to the architecting of complex engineering systems. This realization follows from the engineer's reluctance to "reinvent the wheel" and their reliance on previously successful methods and processes. Such practices not only reduce technical risk by leveraging proven artifacts, but they may also reduce project cost and improve system development cycle times. Design reuse may allow engineering teams to direct their energy on the novel aspects of the system of interest that require fresh thinking or "from-scratch" effort. Thus, it is entirely possible that a culture of efficient and appropriate design reuse may lead to more significant innovation and technology advancement. However, the effectiveness of design reuse is impacted by two interrelated factors. First, traditional document-centric methods for conducting, reporting, and archiving systems engineering (SE) products lead to "knowledge and investment [losses] between projects...increasing cost and risk" [7]. Even when projects do employ more "digital" methods for documenting design and decisions, sharing across projects is often hampered by various factors including security concerns, tool limitations, etc. Such commonly lost knowledge includes data critical to the reuse of processes and activities such as design trades, decisions capture, and rationale; design parameters and configurations; testing procedures and metrics; verification and validation; etc. Second, most engineering projects have limited structured protocols for fully assessing the feasibility and value proposition of design reuse scenarios before their adoption. This limitation often leads to misapplication of design reuse, causing either (a) only partial realization of potential reuse benefits or (b) entirely negating reuse benefits, or even becoming detrimental to the mission of interest by trying to force reuse where it may not be prudent.

As SE is transformed by MBSE, engineers and other decision-makers are afforded new opportunities to enhance design reuse practices in today's digital environments [9]. By carrying out SE processes in digitally integrated and centrally coordinated fashion, MBSE makes necessary information more readily accessible for a systematic and comprehensive assessment of reuse scenarios. This capability takes the form of a central system model that contains the most up-to-date contextual and design information for the mission of interest. MBSE can facilitate consistent archiving and retrieval of design information for proven, heritage systems in digital design libraries or element catalogs outside the mission of interest. Thus, the necessary ingredients for improving current practices in design reuse are present within the paradigm of MBSE. However, just as a recipe is needed for properly combining ingredients to make a particular dish, so too are a set of procedures, functions, and relationships needed in the form of a *methodology* to implement design reuse within MBSE.

This chapter explores exploiting MBSE to answer design reuse questions. To this end, we first investigate the current state of design reuse practices in engineering systems – with a specific focus on the space industry – to understand which practices should be retained in MBSE-based approaches and which practices need to be improved. We then explore more deeply the promising aspects of MBSE related to the reuse question. We demonstrate how design reuse and MBSE best practices can be combined within a design reuse methodology and illustrate the application of the methodology on a sample problem from the space domain.

Design Reuse in Complex Engineering Systems

To understand effective deployment of MBSE for facilitating design reuse, we must first understand how design reuse is currently achieved within traditional SE methods. This section reviews the current state of the practice and major trends related to reuse among civil and commercial players in the space industry. While many sectors (e.g., automotive, consumer electronics) extensively employ reuse, the space industry employs a diverse repertoire of reuse practices ranging from opportunistic heritage efforts to planned product platforms and platform families. Additionally, the characteristics of space missions (e.g., large-scale development, emplacement and operational costs, highly complex/multidisciplinary systems, low-risk attitude and culture, multifaceted scenarios, and inaccessibility of assets once deployed) make design reuse decisions highly consequential over the long haul. Finally, several aerospace entities have been early adopters of MBSE, with several having demonstrated significant use. This progress allows exploration of reuse-related MBSE practices already pursued in an established industry.

Taxonomy of Reuse in the Space Industry

Defining a consistent and canonical set of reuse types is necessary to developing the concept of MBSE-enabled design reuse. A review of the literature in the aerospace industry reveals various prevalent types.

Hardware Reuse

The first type of reuse is the physical reuse of hardware (e.g., piece part, assembly, subsystem, or system instances that unique part numbers may distinguish). Reuse of this type requires the physical recovery of the asset, refurbishment, reintegration, and, ultimately, relaunch. While ubiquitous in terrestrial industries, the usual inaccessibility of space assets and the difficulty of recovering space launch elements make the reuse of this type challenging and infrequent. While hardware reuse may not seem immediately relevant to MBSE, prudent redeployment of recovered hardware is only possible by understanding the hardware's context in the new missions as compared to its native mission. MBSE descriptive models of these elements with a sufficient level of detail may facilitate structural and behavioral analyses necessary for a hardware reuse decision. In this sense, libraries of MBSE models of flight-proven components may be of substantial value to new missions considering their reuse.

In the space industry, hardware reuse has almost exclusively been demonstrated for launch vehicle stages and orbiters/spaceplanes. We also note that while countless reusable or partially reusable launch vehicles have been proposed, few have ever flown, or, more importantly, been re-flown successfully. These include SpaceX's Falcon 9 orbital launch vehicle, Blue Origin's New Shepard suborbital system, and NASA's Space Shuttle program. The often-cited (but not always realized) benefits of space access hardware reuse include cost and schedule savings in materials, manufacturing, and integration. Additional benefits include lessons learned during inspection and refurbishment of recovered hardware to improve reusability properties or identify flaws exposed during hardware operation.

Beyond space access hardware, a few examples of other hardware reuse exist in the industry. These are typically in payloads that fly aboard the previously mentioned recoverable space access vehicles or crewed stations. A notable example is the Tethered Space Satellite mission, which first flew aboard STS-46 (Atlantis) in 1992 [11]. This satellite was deployed from the Shuttle bay and deployed several hundred meters of tether before becoming stuck and being retrieved by Atlantis. Four years later, the same satellite hardware was re-flown on STS-75 (Columbia), where it deployed 19.7 km of tether before it snapped and became lost in space.

A related type of hardware reuse occurs when hardware is recovered and then repurposed or reconfigured to carry out a) a new function in similar mission conditions or b) a similar function in new mission conditions [15]. In either case, some amount of processing is typically required between recovery and redeployment. Repurposing seeks to leverage operational flexibilities of already emplaced assets. This type also sees limited though growing usage in the space industry. Current or past examples of repurposing/reconfiguring space hardware include the transition of the Kepler Space Telescope from its original mission to the K2 mission with relaxed imaging constraints and new satellite operations procedures and the repurposing of the Lunar Module's Descent Propulsion system into an Earth return abort engine after the Apollo 13 failure [6].

Design Knowledge Reuse

Design knowledge reuse involves redeployment of knowledge generated and proven on previously flown missions. Indeed, hardware reuse may be viewed as a special type of design knowledge reuse, where physical entities are part of that design knowledge. However, the "knowledge" that is reused need not be confined to design parameters and configurations. Knowledge can also include requirements and specifications, test and qualification protocols, analysis algorithms, and operational procedures. As we will see in the next subsection, these artifacts may be captured in MBSE system models and libraries of models which can be queried, interrogated, and ultimately redeployed on new missions – particularly within the new mission's MBSE system model.

A typical example of design knowledge reuse is through flight heritage or legacy designs. Barley defines heritage systems as "hardware, software, and procedures with previous flight history that are reused for a new mission that enable a mission capability or reduce overall mission cost, schedule, or risk" [2]. This type of reuse is ubiquitous in the space industry. A general trend identified is that heritage designs are heavily relied on for high-risk missions. This is perhaps most clearly evident in the design of the Mars 2020 (*Perseverance*) rover which, apart from some accommodations for new instrumentation, is unchanged from its predecessor, the Mars Science Laboratory (*Curiosity*). Conversely, in missions where designers may accept higher risk, such as with low cost CubeSats or technology demonstration missions (e.g., the Mars Cube One [MarCO] CubeSats, that flew alongside the Mars InSight lander mission), new technologies may be deployed more liberally – nevertheless, even CubeSat missions tend to leverage Commercial Off-the-Shelf (COTS) elements wherever possible to keep costs down.

Redeployment of heritage systems is rarely a simple "copy-and-paste" effort. Barley explains that the *inheritance* effort should involve:

- *Understanding differences in mission context* – mission requirements and objectives, environments, operational procedures, and concept of operations (CONOPs), system of systems context
- *Assessing engineering effort to address differences* – design and integration changes, system impacts and emergent behaviors, updating interfaces/documentation, and requalification

The *NASA Systems Engineering Handbook* stresses the importance of understanding the new mission context: During technology readiness level (TRL) assessment, "if the architecture and the environment have changed, then the TRL drops to 5–at least initially." (NASA's TRL scale is a commonly used metric for assessing a technology's level of maturity. It ranges from 1–9. TRL 5 represents component validation in a relevant environment, TRL 6 represents system model or prototype demonstrated in a relevant environment, TRL 7 represents a prototype demonstrated in a space environment, and TRL 9 represents "flight proven" on a mission with successful operations.) Further assessment of the magnitude of the difference may bring the TRL back up to 6 and 7. Still, the mission designer should be aware that heritage technology can rarely be "dropped in" to a new mission with an immediate TRL of 9. Thus, the effort required to incorporate the heritage technology should be considered in cost, risk, and schedule estimates.

More intentional, or "planned," design reuse efforts revolve around commonality and modularity realized in product families or platforms. The implementation of these concepts in the industry lives roughly on a spectrum between common busses and modular platforms. Examples are shown in Fig. 1. Common busses are used in

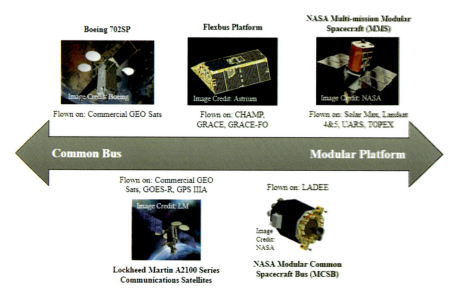

Fig. 1 Planned reuse examples on the commonality-modularity spectrum

missions with similar conditions and requirements. (A spacecraft bus is defined as the elements that provide the infrastructure and support functions necessary for the payload(s) to carry out their missions – this includes structural, power, thermal, communications, propulsion/maneuvering, and other functions.) Modular platforms work in a "plug-and-play" fashion where a set of modules with varying functions are configured into variants that each meet the needs of a unique mission; the set of variants (i.e., the family) spans a range of mission types. In these types of design reuse, firms will incur more upfront costs or performance penalties to secure the option of reuse in the future. For common busses, this entails ensuring that the bus will accommodate payloads that might evolve via upgrades; for modular platforms, this includes accepting penalties in mass that arise from standardization and modularization as well as penalties from designing toward a larger performance envelope than the initial mission.

On the "modular" side, examples like NASA's Multimission Modular Spacecraft (MMS) of the 1980s and 1990s and, more recently, NASA's Modular Common Spacecraft Bus demonstrate how modular busses are used for substantially different mission types [3, 17]. MMS family members include the Low Earth Orbit (LEO), sun-pointing Solar Max mission, and the Geostationary Earth Orbit (GEO) Landsat Earth-imaging satellite. Common bus examples include GEO commercial communications product lines from firms like Boeing and Lockheed Martin. Toward the middle of the spectrum are examples like German firm Astrium's Flexbus system, which housed common systems for various mission types but with significant modifications for novel payloads like the gravimetric experiments on GRACE and GRACE-FO [14]. Satellites and orbital systems are not the only ones leveraging this type of design reuse; concepts exist for large-scale reuse of high-level architectural elements in human exploration missions to the Moon and Mars campaigns.

Work Effort Reuse

We now distinguish between the reuse of designs and the work products and efforts that encode and describe those designs. We define *reusable work efforts* as those value-adding aspects of the systems architecting processes – primarily descriptive and analytical models – that can be adapted and redeployed across multiple missions through standardization and digitization. The prevalence of this type of reuse has increased significantly with MBSE adoption in the space industry. The first example of MBSE-enabled reuse is through the use of metamodels; if models are defined as an abstraction (or simplification) intended to answer specific questions about a system of interest, then a metamodel can be defined as the specification for that abstraction. (Refer to ▶ Chap. 2, "Semantics, Metamodels, and Ontologies," in this text for a more comprehensive discussion of metamodels and related constructs. We will revisit the concepts in later sections of this chapter.) It captures the relevant concepts, relationships between them, and rules that a model must be composed of and adhere to. Generation of metamodels and patterns are crucial and nontrivial processes for implementing MBSE concepts during architecting of complex systems. It follows that reuse of these artifacts across multiple projects can not only improve SE timelines but also promotes consistency across modeled systems, further improving design reuse potential.

Models themselves can also be reused. A common practice is cataloging models in libraries that can be imported into and queried by the larger system model. Models in these repositories, typically of low-level components or assemblies, can be dropped into appropriate areas of the system model without much additional work; this assumes that the models are prepared with a guiding metamodel that ensures consistency across levels of the system model hierarchy. Checks can be run against its design parameters (as captured in the model) that ensure it meets mission requirements or constraints.

Spangelo and Waseem demonstrate how such a cataloging effort can work in the broader application of MBSE to university-run space missions in Cubesats and scalable remote sensing satellites, respectively [16, 21]. In industry, NASA has identified a "Library of Reusable NASA Related System Models" as a key technology for supporting its exploration objectives. In the 2015 Technology Roadmap, NASA predicts "tremendous impacts on life cycle costs" from "an asset library [that] can provide the flexibility and expressiveness required to define complex systems quickly and effectively through the reuse of common entities across multiple spacecraft projects" [10]. An example of a functionally decomposed, metamodel-prescribed component catalog for Space Habitat design efforts is shown in Fig. 2. Note how each component is consistently tagged with both standard and specialized parameters, allowing for bottoms-up mass/power rollups as well as unique discipline-specific analyses. It should be noted that Fig. 2 is not a metamodel itself, but rather adheres to a modeling and archiving convention established for this effort in a separate, underlying metamodel (not shown).

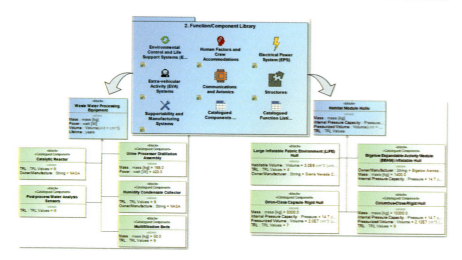

Fig. 2 Sample of a metamodel-prescribed component library for Space Habitat elements developed by this author as a part of an MBSE-enabled reuse demonstration

The third form of work effort reuse, a model-centric reuse framework known as composable design methodology, has been developed and implemented at Lockheed Martin Space Systems [8, 12]. Composable design is a concept "focusing on composing new systems from known components, designs, product lines, and reference architectures as opposed to focusing on 'blank sheet' designs based on requirements decomposition alone." The method is meant for product lines where valid predefined variants are impractical due to the wide and uncertain variety of mission types that may arise in the future. The methodology defines an abstracted structural model that acts as a "filled-in" template for each instance of a variant. A variability model captures the architect's decisions when populating the template, while a component capabilities model acts as the repository of allowable reused elements. The methodology has been used for managing the continued evolution of Lockheed's A2100 line of satellites. Fischer conducts a similar coherent implementation of metamodel and model reuse with their virtual satellite approach applied to the German DLR's S2TEP project [4].

Reuse Taxonomy

A generalized classification of reuse types into a taxonomy can be created by asking the question "What exactly is being reused?" To arrive systematically at an answer to this question, we can draw inspiration from past attempts at developing the taxonomies of technology. Van Wyk developed a widely adopted taxonomy that classifies technologies along two dimensions: (1) the underlying or fundamental function delivered and (2) an operand on which that function acts [22]. Van Wyk identified three canonical functions – transforming, transporting, and storing – and three basic operands – matter, energy, and information. No single operand/process pair is sufficient for mapping reuse types onto the original technology taxonomy matrix. Analysis of these functions reveals the complex nature of any reuse effort and introduces difficulty in the reuse classification problem. A modified version of this taxonomy maintains the operands but collapses all functions into a generic "reuse" function, as shown in Fig. 3. Hardware reuse is generalized as a reuse of matter, referred to as *Physical Article reuse*; design knowledge reuse is generalized as a reuse of information, referred to as *Codified Design Knowledge reuse*; and work effort reuse is generalized as a reuse of energy, referred to as *Design Efforts reuse*.

We now expand the three types of reuse with lower-level (higher granularity) classifications; Table 1 depicts this complete taxonomy that captures the nuances in reuse types gathered from the literature survey. In Physical Article reuse, we distinguish between recovery-dependent and in situ reuse. "Recovery-dependent" refers to Earth recovery; that is, the physical article must be returned to Earth, where it is inspected, refurbished, and integrated for future use. Recovery-dependent reuse contrasts with "in situ" reuse which covers salvaging, recycling, repurposing, or reconfiguring at or near the asset's operating environment. Within Codified Design Knowledge reuse, the unplanned reuse of legacy articles is divided into two types. "Standard Article" reuse covers those elements that are ubiquitous across space systems, and which can claim heritage from a wide set of past missions, or that rely on basic physical principles that do not require flight heritage; these can include the use of passive radiators, simple heaters, common materials, etc. Conversely,

	Matter	Energy	Information
Reuse	Physical Article	Design Efforts	Design Knowledge

Fig. 3 Modified technology taxonomy for reuse classification

Table 1 Full taxonomy for classification of reuse by type as identified in literature and industry study

Physical Article reuse	Codified Design Knowledge reuse[a]	Design Effort reuse
Recovery based	Legacy	Modeling schemes
Article reuse	Standard article	Metamodels and patterns
	Unique heritage	Stereotypes and profiles
In situ	Product families	Models
Article reuse	Common bus	Element libraries
Article repurpose	Modular platform	System models
		Discipline models (descriptive or analytic)

[a]Design knowledge includes: parameters, configurations, testing protocols and certifications, analysis algorithms, operational procedures, etc.

"Unique Heritage" reuse covers elements for which a unique and thorough inheritance process must be conducted to ensure its reusability across missions. Another category of Codified Design Knowledge reuse covers the product family types discussed previously.

Lastly, Design Effort reuse focuses on the models and modeling practices that engineers use to generate, manipulate, and represent knowledge. Model reuse includes MBSE-centric concepts such as digital design libraries and system models and traditional descriptive and analytical engineering models that leverage discipline-specific tools and environments. The increasing digitization of SE processes as prescribed by the MBSE paradigm means that the concepts of design knowledge and design effort reuse (via models and modeling practices) are best considered in tandem. This understanding is expanded in section "Hierarchy of MBSE Methods for Reuse" where we consider the *hierarchy of reuse practices* enabled by MBSE.

Reuse Trends in the Space Industry

In addition to enabling the classification of reuse types in the space industry, the literature and industry survey also revealed some broader insights into design reuse practices that must be considered when deploying MBSE methods. These are discussed next.

Planned and Unplanned Reuse

An important distinction that needs to be made about design reuse scenarios is whether reuse is planned or unplanned. Planned reuse occurs when a design is developed from the outset with the explicit intent of being reused on a subsequent product or mission. This type of reuse is implemented in the product platforms and product lines discussed in the previous section (e.g., product line of COTS CubeSat subsystems). Unplanned reuse involves redeployment of designs not originally developed with the explicit intent of being reused in the future. This type of reuse is more common in novel science or exploration missions. This difference in intent from the initial design of an element regarding future reuse has consequences on how reuse decisions for that element are made. In planned reuse the element was likely designed to satisfy a range of requirements across the constituent platform products. In unplanned reuse, the candidate element was designed solely for its native mission, requiring more careful consideration and supporting analyses prior to redeployment of the design on future missions.

The focus of some recent work in MBSE and design reuse has been toward developing methodologies for decision support in unplanned design reuse [18–20]. Such approaches had been lacking in the reuse literature, to date, where the focus was predominantly on planned reuse methods. The value of improving unplanned reuse methods is reinforced by findings from an industry survey (Fig. 4a). Most respondents indicated that unplanned reuse efforts typically do not fully realize potential benefits as measured in programmatic impacts

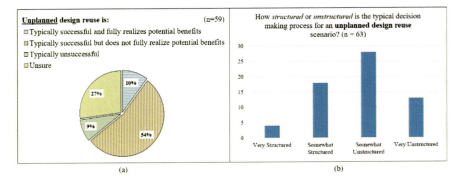

Fig. 4 Results of two questions related to the prevalence and success of unplanned design reuse. (Figures taken from a survey of industry practitioners conducted by the authors. Respondents sourced from industry, academia, and government agencies in work areas related to decision-making and architecture)

(e.g., cost and schedule). The lack of extensive literature and such findings from SE practitioners may suggest improvements can be made to current unplanned reuse methods; alternatively, this can be seen as a rejection by practitioners of the value proposition of unplanned reuse. Nevertheless, the existence of examples of successful unplanned reuse demands additional exploration.

Structured and Unstructured Methods

A typical, though not necessary, corollary to unplanned reuse is that the methods undertaken to redeploy the design or element on the new mission of interest (MoI) are rather unstructured. Respondents reflect this in the industry survey (Fig. 4b), where 65% indicated that unplanned reuse methods were either "somewhat" or "very" unstructured. Such methods are characterized by a lack of systematic, standardized, or comprehensive procedures and analyses that lead to a detrimental impact on the reuse scenario outcome. Since reuse efforts of any kind are rarely "copy and paste," such analyses are essential for (a) assessing the technical feasibility of reusing an element on the MoI (i.e., "can the element satisfy the required functionality and performance in the MoI?"), (b) enumerating the type and degree of rework or modification needed to adapt the element from its native mission to the MoI, and (c) determining the amount of break-even work effort beyond which from-scratch design would be preferable. Having a standardized set of such procedures, generalized for a wide set of mission types and applicable across diverse reuse scenarios, ensures that reuse benefits can be maximized and that decision-makers have a structured process to make logically consistent decisions. Such a generalized proceure does not replace the unique, in-depth, and rigorous technical analyses required to make reuse decisions – but rather places them in a larger methodological framework. As we propose in this chapter, this more structured framework is most effectively implemented in an MBSE environment which can leverage model patterns, libraries, and authoritative sources of information in "living" system models.

MBSE as a Facilitator of Improved Design Reuse Practices

As was discussed in the previous section, MBSE methods have seen substantial adoption in the space industry in recent years. Schoberl identifies six capabilities enabled by MBSE that have encouraged such trends during the design and SE process (and may do so in other industries):

- Reduce development cost and cycle time by improving problem understanding
- Improve collaborative, interdisciplinary engineering by establishing a shared system understanding
- Facilitate communication by using a single and up-to-date source of information
- Reuse architecture and engineering work results by documenting previous solutions
- Improve risk assessment, safety considerations, and quality assurance through transparency and traceability
- Improve later life cycle activities through easily accessible documentation

Per an industry study conducted by Schoberl, "Reuse architecture and engineering work results..." is ranked among the top 3 of these capabilities [13]. Industry practitioners recognize the potential of digitization and centralization of engineering data for improving the retention and redeployment of previously generated knowledge and models on new design efforts.

Hierarchy of MBSE Methods for Reuse

There are various mechanisms within (and related to) the MBSE paradigm for facilitating reuse across an engineering endeavor. These may be understood within a hierarchy encompassing data practices at the system design level, intermediate engineering practices, and business practices at the enterprise level (Fig. 5).

At the enterprise level, a prudent strategy is to establish standards by which all relevant organizations and engineering teams can understand and share design information using consistent terminology. Central to this strategy is to develop a domain ontology – a formal naming and definition of concepts, data, properties, and relationships relevant to the system or enterprise of interest – alongside a central

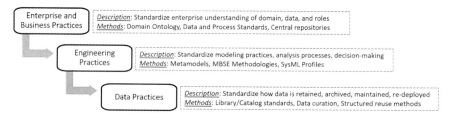

Fig. 5 Hierarchy of MBSE mechanisms for facilitating reuse

mechanism for the exchange of data across the suite of engineering tools used by the enterprise. Specific to MBSE, this data should be stored in a manner that establishes an authoritative source of truth (ASOT) for each information item and prevents duplication or falling out of date.

The next hierarchy level relates to the practices of teams carrying out engineering analyses for a MoI or system of interest (SoI). Here, reuse is facilitated by standardization of modeling/analysis processes such that common efforts are pursued in parallel across various development projects. Standardizations can occur via the use of MBSE metamodels that specify how methodologies are implemented and how modeling constructs are used. With SysML, extensions to the canonical constructs may be defined specifically for conducting the domain analysis (e.g., reuse). These extensions, described in *Profiles*, may be applied to modeling efforts on various projects to ensure consistency across them.

The last level of Fig. 5 defines data-level practices for facilitating reuse. Essentially, MBSE efforts at this level are concerned with ensuring that data (all aspects of design information) are retained, archived, maintained, and made available for future use in a consistent and standard fashion. These data-level practices are implemented via standards for model libraries, or element catalogs. Critical to this effort is understanding the data needs required of the analyses that reuse methodologies will carry out. In this way, the data may be curated appropriately for the most efficient reuse in the future.

How this hierarchy is implemented in an organization may vary. For instance, a common approach is a federated, interoperable modeling strategy. Here, various organizations within an enterprise may focus on different aspects of the modeling/digital effort. The key is commonality at the interfaces of these different efforts such that they may each be integrated into the larger digital framework. Because of such a federation, there need not be a centralized data exchange mechanism as long as there is a common exchange mechanism.

Conditions Conducive to MBSE Adoption and Reuse

Along with enterprise-wide MBSE practices beneficial to reuse efforts, it is also essential to understand the sociotechnical environment that enables these practices. Unrelated to reuse efforts, a common criticism of MBSE languages (SysML, especially) is the high barrier to entry in terms of training and understanding models that quickly become very complex. Organizational investment for improved training regimens for system modelers and engineers is required particularly for reuse efforts, where they must work with their own models and third-party models built by others. As discussed in the previous section, such training, implemented organization wide, would naturally lead to more consistent application and standardization of modeling practices. (It should be noted that there may be work-arounds to requiring substantial MBSE knowledge of all team members, including custom model input/output tools for subject matter experts to perform their analyses in environments with which they are familiar while maintaining a connection with the governing system model.)

Foundational to these sociotechnical efforts of creating environments favorable for MBSE deployment (and specific to MBSE-based reuse methods) is organizational buy-in – that is, the leadership and management must understand the value of MBSE and encourage and potentially specify the scope of its usage.

Lastly, we consider organizational conditions specifically related to reuse efforts. Agresti devised the 4A method for evaluating the reusability of a given software element by a sequential set of conditions [1]. These conditions are also applicable beyond the regime of software; they are:

1. Availability – "the reusable artifact must be available; that is, it must not have been destroyed, erased, or lost."
2. Awareness – "the developer must know of the existence of the reusable artifact."
3. Accessibility – "a transfer mechanism must exist to get the reusable artifact to where it is needed."
4. Acceptability – "the reused artifact must be acceptable to the developer for use in the new project and its environment."

Availability and accessibility relate to the approach adopted for archival and maintenance of design information and data. Awareness relates to the extent of communication and openness across teams and projects. At the same time, acceptability ultimately depends on the technical analyses necessary for determining the feasibility and work effort required for adopting a given reuse scenario. Each of these essential conditions for reuse focuses on a different aspect of an organization's cultural and technical approach to reuse. Indeed, these conditions are each in some way addressed by MBSE methods.

The procurement and acquisition process levies some constraints on the efficacy of MBSE methods. For instance, many of the components on a spacecraft are purchased and integrated by the spacecraft prime contractor. Creating a library is a substantial effort because vendors largely do not have MBSE representations of their products or if they do, those representations are considered proprietary. In the relatively rare case that vendors do have them and make them available, those models will use whatever styles/guidelines might be in place at the vendor and likely not readily work within an integrator's model. Of course, if a vendor uses a given prime's standards then it is unlikely that the vendor's model will easily plug into another prime's model. Ultimately, it will become the responsibility of a prime to develop the metamodel for procured or furnished components that matches their modeling guidelines.

Best Practice Methodology for MBSE Design Reuse

In previous sections, we focused on (a) the state of the practice of design reuse (specifically in the space industry) and (b) aspects of MBSE that may be leveraged to codify and improve these practices. In this section, we focus on the MBSE-centric design reuse methodology. This methodology is intended to provide a more rigorous approach to assessing potential unplanned design reuse scenarios.

Design Reuse Logical Process

Before exploring MBSE constructs and mechanisms to tackle the question of unplanned design reuse, we first specify what such a methodology is intended to accomplish. It follows that the first step is an implementation-agnostic generalized logical process of all necessary activities. These steps are derived from systems engineering theory and discussions with subject matter experts (SME). This logical process is depicted by SysML activity diagrams, a convenient and familiar way to visualize the process that guides eventual implementation (Fig. 6).

The logical process is structured as a sequence and hierarchy of activities – each of which can be carried out by different members of the development team, from mission planner to modeler, and to subsystem SME. The hierarchy spans three levels, with each lower level providing greater detail for carrying out higher-level steps of the logical process. The highest level of the hierarchy is shown in Fig. 6. The design reuse process can be decomposed into three central activities: (1) defining the context of the reuse decision in the new MoI, (2) conducting a technical inheritance process for assessing reuse feasibility, and (3) assessing the "value" of the reuse decision in terms of impacts on programmatic factors, such as cost and schedule, for the MoI. The overall input to the process is an MoI description including objectives, requirements/constraints, and architecture and design decisions made to date for the MoI. Assuming the organization has deployed MBSE for its design efforts, this information is kept as a system model for the MoI that details structural and behavioral features of the architecture. In a more comprehensive digital environment, the MBSE model may integrate with other modeling environments (e.g., subsystem-specific modeling tools), providing accessibility to higher levels of design detail. This methodology can then be readily consumed by the reuse methodology. If the MoI description is not kept in a system model – the likely alternative being in static documentation (e.g., MS Office, central file management systems, etc.) – then this information needs to be digitized and transformed into a system model.

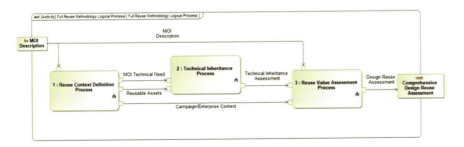

Fig. 6 Highest level of the generalized, logical process for assessing design reuse scenarios

Step 1: Reuse Context Definition Process

The first step of the process defines the context of the given reuse scenario. This context is necessary for making informed and well-supported reuse decisions. Defining context entails describing aspects related to the (1) larger campaign or enterprise within which the MoI resides, (2) the technical need, and (3) the reusable assets at the organization's disposal (Fig. 7). First, the MoI must be placed within the larger campaign or enterprise undertaken by the organization; prudent and sustainable architecture decisions, especially those relating to reuse, are best made with the campaign frame in mind. The campaign/enterprise context – which includes vision and horizon goals, objectives, constituent missions and systems, and constraints (engineering and otherwise) – is a key output of this step.

Next, the MoI technical need is defined; this is the functional need within the MoI functional decomposition for which reuse is being considered. For instance, on a robotic rover mission the instrument manipulation and deployment function may be the technical need or "reuse slot" within the MoI architecture for which flight-proven robotic arm designs from past rovers may be considered for reuse. Step 1.3 consists of (1) identification of functional need, (2) specification of the level of decomposition within the full MoI architecture (e.g., system, subsystem, assembly, component, etc.), and (3) enumeration of relevant requirements and constraints recorded in the MoI requirements documentation. These data become inputs to the technical analyses performed in Step 2. Lastly, Steps 1.4 and 1.5 concern the identification of organizational databases, design libraries, and past mission models from which reusable assets may be drawn. Those asset records relevant to the MoI and for

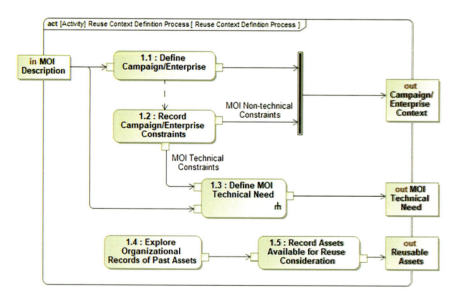

Fig. 7 Zoom in of Step 1 of the generalized logical process for a structured unplanned reuse methodology

which design information is readily accessible are collected. Steps 1.4 and 1.5 concern construction and maintenance of design libraries/archives, which has been discussed briefly in section "Taxonomy of Reuse in the Space Industry" (*Work Effort Reuse*) and are crucial steps for the preservation of organizational knowledge; these steps may be considered to reside outside of the flow of the logical process as the asset archives are incrementally populated.

Step 2: Technical Inheritance Process

Technical inheritance is defined as the process by which the feasibility of adapting legacy elements for use on new missions is assessed, including how it addresses the new objectives, environment, constraints, and context; this process also generates an estimate of required rework efforts. Figure 8 shows a more detailed view of this process (Step 2 in Fig. 6). The technical inheritance process inputs the MoI technical need and reusable assets compilation generated in Step 1. Four steps then follow a candidate element from initial down-select (Step 2.1) through a check of compatibility with the MoI (Step 2.2) and an estimate of rework required to adapt the part for reuse (Step 2.3) and finally to a technical inheritance assessment recommendation for each reuse scenario considered (Step 2.4) – the technical findings will then proceed to programmatic analyses that are performed in Step 3.

Technical inheritance begins with initial filtering of reusable assets available to the organization (Step 2.1). This filtering may be carried out differently for assets archived either in dedicated component libraries or within the system models of their native missions. In either case, mission or catalog metadata is parsed to identify elements with similar functionality to that required by the MoI. The design information for potentially viable elements is extracted (or more precisely, a pointer is created in the reuse analysis model to that information in its native models). These are labeled as reuse candidate elements (RCEs). A set of RCEs could contain one or several elements depending on the circumstances of the MoI design effort (e.g., if one predecessor mission is preconstrained as the source for reuse elements).

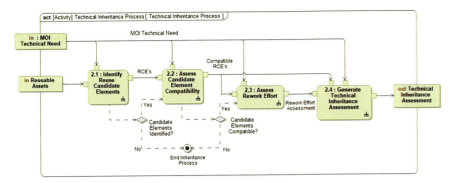

Fig. 8 Zoom in of Step 2 of the generalized logical process for a structured unplanned reuse methodology

Step 2.2 proceeds with a more refined analysis of the compatibility of the RCEs to the MoI technical need. This is done by first identifying pairs of requirements and interfaces across the native mission of the RCE, labeled reuse candidate mission (RCM), and the MoI. Requirements and interfaces are selected as key indicators of the potential of an element design to be redeployed on a new mission. Requirements are paired through a comparison of the type of requirement (e.g., functional, performance, and environmental constraints) and the text of the requirement. The figure of merit specified by each requirement in the pair is then compared to identify a "requirement gap" – the difference between what the RCE can satisfy and what the MoI requires, a difference which must be bridged by rework/adaptation. A similar process is conducted for interfaces specified at the boundary of the RCE design and the MoI slot. A final judgement is made by the designer on the basis of these gaps and the RCE is deemed: "Compatible – No Modifications," "Compatible – Minor Modifications," "Compatible – Major Modifications," and "Not Compatible."

In Step 2.3, the rework effort needed to adapt those RCE's deemed compatible with modifications must be determined. For each requirement or interface pair, designers and SMEs are tasked with enumerating all rework actions needed to bridge the gap identified previously. This process also requires relevant discipline analyses to assess feasibility of the action, as well to finalize a preliminary work breakdown structure (WBS) and to allocate an "effort" measure that will be utilized in Step 3. Lastly, impacts on testing and qualification of the reused element (i.e., "Will this rework campaign invalidate the certifications on the element, requiring retesting/re-qualification?") must be accessed. The data generated in this step – rework action enumeration, WBS, effort magnitude, and support discipline analyses – are compiled into a full rework effort assessment. In Step 2.4, the user combines this data with information generated on previous steps to make a final technical recommendation.

Step 3: Reuse Value Assessment Process

The technical analysis in Step 2 is supplemented with a valuation process that attempts to estimate the programmatic (e.g., cost and schedule) impacts of the reuse scenarios considered and the implications of reuse on the broader campaign of the MoI (see Fig. 9). For this task, the Constructive Systems Engineering Cost Model 2.0 (COSYSMO 2.0) was identified as a promising cost modeling tool to incorporate into the reuse methodology [5]. The cost estimating parameters COSYSMO employs are requirements, interfaces, algorithms, and operational scenarios – the first two parameters are those identified in Step 2.2 of this methodology. The latter two are also performed in parallel: algorithms account for unique mathematical algorithms required in the design, which may be related to the supporting discipline analyses in Step 2.3, and operational scenarios account for unique use cases of the architecture, which may be related to major performance or functional modifications made to the RCE. These four size drivers are then modulated in the governing COSYSMO 2.0 formula by a reuse designation that reduces the effort estimate based on the degree of reuse. In the determination of the reuse designation, reuse risks should be considered including: availability of components and uncertainties regarding identified rework actions.

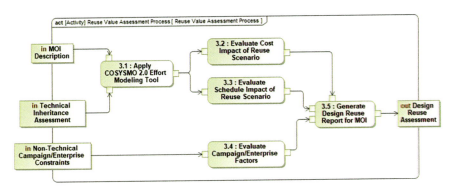

Fig. 9 Zoom in of Step 3 of the generalized logical process for a structured unplanned reuse methodology

In the preceding technical analyses of Step 2, parameters are defined and populated for each requirement or interface pair, rework action, or supporting analysis to record its size driver type, reuse category, and a "difficulty" parameter that are also specified as required by COSYSMO. The tool's output is an estimate of SE effort in person-months; this metric can then be used to draw inferences about cost and schedule impacts. The same analysis can then be run for scenarios where reuse is not employed, and all design efforts for the MoI technical need are done "from scratch." Comparing the two outcomes provides a baseline "value" for the reuse scenario, or conversely, a break-even point beyond which design reuse should not be pursued. These analyses are carried out for each RCE considered and are consolidated into the major output that is delivered to decision-makers as a design reuse recommendation at the conclusion of the methodology.

MBSE Implementation of Reuse Process

This section describes one key approach to implementing the generalized design reuse process described above within an MBSE environment. As mentioned previously, the nonuniqueness of MBSE solutions to a given modeling problem implies that there may be other plausible approaches. However, for the sake of consistency and reusability of the method itself, it is recommended that organizations implementing design reuse methodologies should adopt a single process.

SysML as the Basis for a Canonical MBSE Environment
The first question in specifying MBSE implementation of the logical process developed previously is selection of a modeling language. While many highly capable MBSE-compatible modeling languages exist – such as Modelica, MathWorks System Composer, and Object Process Methodology – SysML has become established as the de facto MBSE language throughout the industry. Its diverse set of diagram types allow for a rich graphical encoding of various aspects

and types of system information. The extension mechanism (SysML itself being an extension of the unified modeling language, UML) allows for users to tailor the language to suit the needs of a given modeling effort or discipline/industry. Additionally, various tool vendors reinforce SysML suites with capabilities to (1) create custom tables and profiles, (2) integrate discipline-specific tools into the system model, and (3) execute user-defined code for unique analyses within the system model itself. While powerful, SysML extensions and vendor-specific capabilities could lead to incompatibilities across tools and organizations within an enterprise. This leads to a tension between leveraging these features and ensuring accessibility of the model to all relevant parties.

Additionally, SysML has some fundamental limitations. First, SysML solutions are not always unique; there are multiple ways to model the same descriptive or behavioral feature of an architecture, allowing for modeling inconsistencies across modelers, organizations, or standards. Nonunique representaions may be seen as beneficial in the flexibility it provides system modelers to model as best they see fit; however, to establish consistent modeling practices, this flexibility may not be desirable. Secondly, the rich modeling enabled by SysML also leads to models becoming complex rather quickly and often intelligible only to the modeler(s). This complexity also results in a modeling language that is often described as too difficult to learn, leading to high barriers to widespread adoption at organizations and the industry. It is expected that the upcoming second edition of SysML will alleviate some of the issues discussed here. Nevertheless, SysML remains the most capable and widely used MBSE-compatible language, and, as such, is selected for implementation of the reuse methodology.

Key Decisions in MBSE Implementation

Implementing the design reuse methodology, as with implementing any unique or specialized capability within SysML, requires the definition of custom constructs and relationships (i.e., modeling elements and links/interfaces across those elements). There are multiple mechanisms within SysML by which these can be specified, each with benefits and drawbacks.

Perhaps the most intuitive approach is via SysML's generalization and specialization relationships. In this approach, a more generalized or abstract block may be specialized by a subordinate block; the lower-level block inherits all attributes of the parent block and can either adopt the default values and properties of the parent or and/or define entirely new ones (see Fig. 10a). This approach is common for discipline-specific modeling practices and in defining taxonomies of elements; applied to the reuse methodology, this approach would entail the creation of a fully generalized implementation of the logical process that can then be specialized for the given MoI. While a metamodel of this kind is capable of handling the complex modeling requirements of the methodology, it is a time-consuming and nontrivial task to propagate changes from the generic blocks to the specialized blocks once the latter have been defined, making it a rather rigid approach. (In most modeling tools, once a child block "redefines" a property of its parent

26 MBSE Methods for Inheritance and Design Reuse

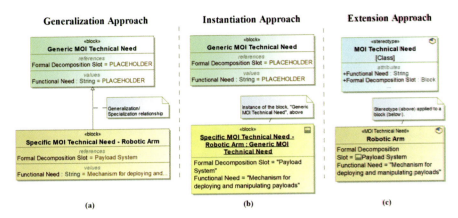

Fig. 10 Illustration of options for defining unique model elements required by the methodology

block, it is no longer updated if the parent block makes any modifications to this property.)

The simplest approach is via SysML's instance specification (or instantiation) mechanisms. Like class-instance mechanism of object-oriented programming, an instance of a user-defined SysML block inherits all of the attributes and relationships of that block (see Fig. 10b). The instance can only behave and interact as defined in its classifying block. Thus, this approach ensures traceability of (and adherence to) the implementation scheme defined in the metamodel; however, this same characteristic also makes it difficult to define custom behaviors, properties, or relationships when applying the methodology as will inevitably be required in real-world, complex design reuse case studies.

The last alternative is to employ SysML's extension mechanism mentioned in the previous subsection. The design reuse methodology may employ an extension of SysML with reuse-related model elements defined as stereotypes. Stereotypes create new model elements that are derived from existing SysML elements (e.g., "block") and are customized with specific attributes and tagged values, as shown in Fig. 10c. Stereotyped elements, the set of which is known as a Profile, then become the primitive building blocks of the modeling effort. These stereotypes may also be applied to existing modeling elements, yielding a solution to leverage existing models with minimal modification. This approach results in a flexible solution that is ideal across organization at various stages of MBSE adoption. We adopt this extension approach for the development of the metamodel guiding MBSE implementation of the reuse methodology – though we recognize that some combination of all three methods may be appropriate, depending on an organization's existing modeling standards, practices, and environments.

Reuse Profile

The SysML Reuse Profile – the container for all reuse-specific stereotypes – developed for the design reuse methodology is presented in Fig. 11. The stereotypes, shown around the perimeter of the diagram in blue, have connectors leading to the

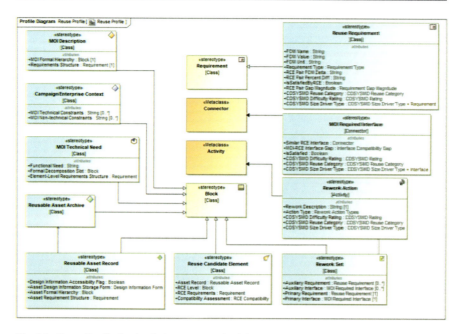

Fig. 11 Reuse Profile for the design reuse methodology

middle column of canonical SysML elements. The "attributes" in each of the reuse stereotypes indicate the special properties (or "tags") that have been defined for it and that become available to populate when the stereotype is applied to a model element. The name of the attribute is followed after the colon by the value type of which that attribute is constrained to accept, and, in some instances, the allowable multiplicity of entries for that attribute (indicated by "[]") and the default value the attribute takes when created (indicated in red text).

The first three stereotypes in the top left corner of the diagram specify model elements related to Steps 1.1–1.3 of the logical process. The MoI description stereotype contains attributes pointing to model artifacts containing (1) the highest-level element of the architectural decomposition of the MoI and (2) the requirements records of the MoI. These are presumed to have been created by the organization before implementation of this methodology. The campaign/enterprise context stereotype contains attributes for the higher-level constraints imposed on the MoI; additionally, since this stereotype is a specialized form of "block," it can also include part properties for the other constituent missions of the campaign for additional context. The MoI technical need stereotype has attributes for a) describing the functional need via text, b) identifying the "need slot" in the larger MoI architecture decomposition, and c) pointing to the MoI requirements relevant to the need slot (and necessary for the technical inheritance in Step 2 of the process).

The blocks in the lower left corner of Fig. 11 specify stereotypes related to the reusable assets available to the organization. The left-most two blocks, reusable asset archive and reusable asset record, are used within the organization's design/model repositories to record asset information. As mentioned in the discussion for Steps 1.4 and 1.5, construction and maintenance of these repositories are tasks outside the scope of this methodology; nevertheless, asset metadata of the types listed as attributes in the reusable asset record (including pointers to design information such as requirements and architectural decompositions) should be maintained. The reuse candidate element stereotype is used directly within the methodology and imports the data kept in the asset record listed as its first attribute. A key attribute populated after Step 2.2 of the process is the compatibility designation, which determines whether the RCE continues for more detailed supporting analyses and rework assessments.

The two stereotypes in the top right corner show the specialized reuse requirements and interfaces central to the technical inheritance analysis. The reuse requirement stereotype contains attributes recording the requirement type as well as the figure of merit (FOM) specified by the requirement. It also contains designations for whether the RCE requirement paired to it (see Step 2.2) fully satisfies it or whether some gap exists. Lastly, the stereotype contains attributes necessary for the use of the COSYSMO 2.0 cost/effort modeling tool. Similar attributes are kept for the MoI required interface stereotype. The rework set stereotype consolidates all rework actions necessary for bridging a given requirement or interface pair's gap. The rework action stereotype describes each action and also assigns parameters needed by the COSYSMO 2.0 tool.

Methodology: Summary and Visualization

The concepts developed for both the generalized process and the MBSE implementation of the process may be defined and represented as part of a reuse ontology. As discussed previously, an ontology defines the concepts and the relationships among those concepts related to a particular domain or type of analysis. For instance, Fig. 12 represents the relationship among terms in the conceptual development and taxonomy of reuse of section "Taxonomy of Reuse in the Space Industry" (*Reuse Taxonomy*). The key relationships relevant to the development of methodology presented in the previous sections are highlighted in gray. Likewise, we may descend one level into the ontology to illustrate the links among constructs of the MBSE implementation, as shown in the metamodel in Fig. 13. Here we see the central importance of comparing the MoI technical need and the requirements and interfaces satisfied by the RCE. We also see the complementary nature of the technical and programmatic analyses to deliver a final reuse feasibility and value assessment.

Methodology Demonstration on Sample Problem

An exemplar problem was created to demonstrate implementation of the design reuse methodology in an MBSE environment; it is formulated to provide a simple yet comprehensive illustration of the methodology's capabilities and features. A representative set of MoI objectives, constraints, and requirements is defined as well assumptions of design decisions locked in prior to reuse consideration. Notional

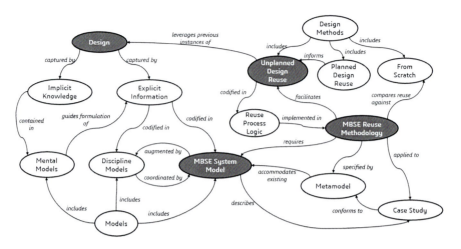

Fig. 12 High-level conceptual ontology of reuse terms

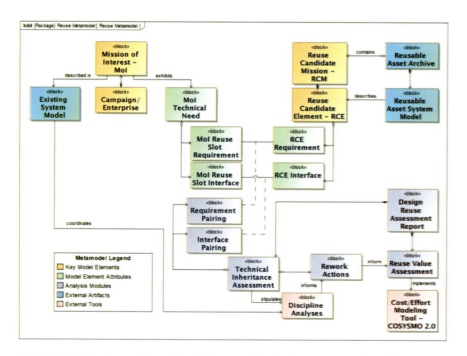

Fig. 13 MBSE implementation schematic (metamodel) of relevant terms and model constructs. (Dashed lines indicate that the "Requirement Pairing" and "Interface Pairing" analyses are carried out on the union of MoI and RCE requirement/interface elements, which are contained in the models for the MoI and RCE. The ultimate output of the methodology, the Design Reuse Assessment Report, consists of both technical and value/cost assessments. Technical inheritance assessments identify rework actions which, in turn, inform the reuse value assessment)

model elements and design information (e.g., requirements, interfaces, and design parameters) for an RCE from a real-world NASA mission are simulated from publicly available information.

Problem Selection and Description

Robotic exploration and novel science missions were identified previously as regimes of space missions where design reuse is most often unplanned but still very common. It follows that a demonstration problem from this area would be most appropriate given the focus of the methodology on unplanned reuse. Additionally, selection of a "toy" problem is constrained by the need for publicly available design data, so private or commercial ventures were not considered. The mission of interest for this problem is a robotic, unmanned multipurpose lunar rover supporting (1) surface science via remote and in situ experiments and (2) surface station buildup via payload deployment and interaction with surface assets.

The hypothetical technical need for the MoI is the experiment/payload deployment and manipulation function, typically satisfied by a robotic arm mechanism. The focus of modeling efforts to describe the MoI are on those aspects related to this technical need slot; these aspects are depicted in Fig. 14, where elements internal to the robotic arm system are shown in the dashed box – these elements and the robotic arm system as a whole must be compatible with (and interface with) the MoI elements outside of the dashed box. Specifications for internal elements are captured primarily in functional, performance, and constraint requirements, while interface specifications capture the required RCE interactions with the external elements.

Several past missions have demonstrated robotic arm systems that may be viable candidates for reuse on the lunar rover. These include the family of NASA Mars rovers including the Mars Exploration Rovers (Spirit and Opportunity), Mars Science Laboratory (Curiosity), Mars InSight, Phoenix, and others. Application of the methodology with just one of these missions as the RCM is sufficient for demonstration purposes – the Curiosity rover is selected due to substantial publicly available design information. In the absence of accessible system models for the MSL mission, a similar set of design information to the MoI description was encoded into a hypothetical RCM system model, including requirements, interfaces, and a detailed architectural decomposition.

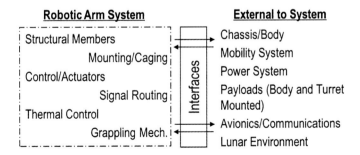

Fig. 14 Lunar rover robotic arm system schematic

Methodology Walkthrough and Sample Output

The diagram in Fig. 15 is known as the methodology dashboard and depicts the primary model elements of the design reuse methodology; it also links to additional diagrams and tables that implement the descriptive and analytical aspects of the methodology. All model elements on the dashboard (that is, elements that are not diagram icons, table icons, or standard connectors) are stereotyped to the constructs encoded in the Reuse Profile in Fig. 11; the stereotype applied to each element is indicated inside the "« »." The «Reusable Asset Archive» and «Reusable Asset Record» hierarchy on the right-hand side demonstrates how the RCE is drawn from the organization's available assets. In this case, design information related to the robotic arm is drawn from the «Reusable Asset Record» containing the MSL mission's models and is populated into the «Reuse Candidate Element» block – the construct that will be carried through the design reuse assessment. Note that the MSL record will never be altered by the methodology, it will only populate the fields in the RCE; this ensures that the activities of the design reuse methodology do not back-propagate into the archive and modify the MSL "pencils down" design.

For both the MoI and the RCM, a high-level formal decomposition is included within which the technical need slot (on the MoI side) and the RCE (on the RCM side) can be specified. Interfaces for the need slot and RCE are specified as well. These diagrams are not essential to depicting implementation of the methodology

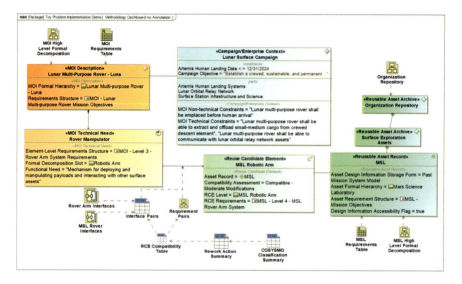

Fig. 15 MBSE design reuse methodology dashboard. (Closed-diamond end arrow type represent a composition relationship [where the arrow end is "a part" of the owning end]; circular end arrow types [with a cross] represent containment relationships in the SysML model. Other lines represent simple directed or nondirected associations in the model [where there is visibility to attributes of elements contributing to the relationship]. Icons in top right of each stereotyped block represent a visual aide defined in the Reuse Profile for more easily distinguishing each stereotype in the reuse methodology)

and are excluded from this paper – they are standard products of architecting efforts and take the form of block definition diagrams and internal block diagrams, respectively. Additionally, a requirements decomposition is imported for the MoI and RCM. In this case, requirements relevant to the robotic arm are reflected in Level 3 of the MoI hierarchy and equivalently in Level 4 of the MSL mission. Requirements (shown in Table 2) span the range of design considerations illustrated in the schematic on Fig. 14 and include: work space reach requirements, structural and loading requirements, thermal and power requirements, and others. MSL requirements were derived from known capabilities of the system where possible and inferred where public information was not available.

Pairing of the MoI-RCE requirements (per Step 2.2) occurs via a "Trace" relationship running from the requirement in the MoI to its counterpart in the MSL. (In UML, "Trace" represents a generic relationship whose use is recommended against use when employing SysML "satisfy," "derive," and "verify" relationships on requirements. However, due to the unique character of the requirements pairing approach, the more generic "Trace" is used to avoid the implication that a requirement on one mission is explicitly linked to a requirement on another mission.) The attributes for each MoI requirement (stereotyped by «reuse requirement») are then populated to capture the pair gap information; in this example, attributes are populated manually though in future iterations this may be automated by leveraging SysML's parametric capabilities. A similar pairing exercise is conducted for the MoI-RCE interfaces, though rather than a "Trace" relationship, «MoI required interface» contains a "similar RCE interface" attribute that points to the counterpart interface in the RCE. The RCE compatibility table shown in the dashboard consolidates the results of this pairing activity and summarizes the set of requirement and interface pairs as well as the gap magnitude designations determined for each.

The user now implements the rework assessment specified in Step 2.3 by constructing a «rework set» for each interface or requirement pair with a nonzero gap. Per the Profile, each set is tagged to a primary requirement or interface and may have auxiliary ones if relevant requirements or interfaces are impacted; «rework actions» are enumerated within each set with input from SMEs and discipline engineering teams. The rework campaign for the given RCE is compiled in the rework action summary table whose icon is shown in the dashboard and whose content is shown in Fig. 16. It enumerates all rework sets and actions for each of the requirement and interface pair gaps. Rework action types include modifying functionality (e.g., "Bypass Motor Control Assembly"), modifying dimensions (e.g., "Extend Upper Arm Tube"), modifying interfaces (e.g., "Modify Radiation Countermeasures"), or developing entirely new functionalities (e.g., "Develop/Demonstrate Autonomous Algorithm") as well as additional technical analyses assessing impacts of these actions on requirement/interface satisfaction (e.g., "Conduct Workspace Reach Analysis").

Each of the rework actions and the requirements and interfaces to which they are linked have also been tagged with parameters needed for the COSYSMO 2.0 effort modeling analysis in Step 3. Each of COSYSMO's size driver parameters are linked to a model element in the reuse analysis: MoI-RCE requirement pairs to the

Table 2 Representative set of requirements defined for the robotic arms on the MoI and the RCM. (This representative set of requirements for the MoI and the reuse candidate robotic arms is derived from publicly available sources [for the latter] and suitable modifications to these for transition from a Martian to Lunar environments [for the former])

MOI – Lunar Multipurpose Rover	Requirement area	RCM – Mars Science Laboratory
The robotic arm shall maneuver in situ and deployable payloads/end effectors in its workspace with 10 mm absolute positioning accuracy	Absolute positron	The robotic arm shall maneuver in situ and deployable payloads/end effectors in its workspace with 20 mm absolute positioning accuracy
The robotic arm shall maneuver in situ and deployable payloads/end effectors in its workspace with 5° orientation accuracy with respect to target surface normal	Absolute orientation	The robotic arm shall maneuver in situ and deployable payloads/end effectors in its workspace with 10° orientation accuracy with respect to target surface normal
The robotic arm shall support a payload of 60 kg at the end of the arm	Turret payload mass	The robotic arm shall support a payload of 30 kg at the end of the arm
The robotic arm shall have a maximum mass of 150 kg (not including the turret payload)	Arm mass	The robotic arm shall have a maximum mass of 65 kg (not including the turret payload)
The robotic arm shall provide a preload of 200 N for a turret instrument onto a surface target	Turret preload	The robotic arm shall provide a preload of 240 N for a turret instrument onto a surface target
The robotic arm shall consume less than 50 W of power at any time while carrying out its functions	Power budget	The robotic arm shall consume less than 30 W of power at any time while carrying out its functions
The robotic arm shall survive 15 g's of acceleration during launch and landing, while fully stowed	Launch/ Landing loads	The robotic arm shall survive 20 g's of acceleration during launch and landing, while fully stowed
The robotic arm shall survive 5 g's of acceleration during launch and landing, while fully stowed	Driving load	The robotic arm shall survive 6 g's of acceleration during launch and landing, while partially stowed
The robotic arm shall maintain all hardware within their respective operating temperatures, with the limited lower bound being -55 °C, a delta of 100 °C from shaded ambient of -155 °C	Temperature control – cold case	The robotic arm shall maintain all hardware within their respective operating temperatures, with the limiting lower bound being -55 °C, a delta of 58 °C from abient of -113 °C
The robotic arm shall maintain all hardware within their respective operating temperatures, with the limiting upper bound being 50 °C, a delta of 75 from sunlit ambient of 125 °C	Temperature control – hot case	*No similar requirement*
The robotic arm workspace shall be a cylinder whose center extends up to 0.75 m from the rover front panel	Workspace – extent	The robotic arm workspace shall be a cylinder whose center extends up to 1.1 m from the rover front panel
The robotic arm workspace shall be a cylinder of radius 400 mm	Workspace – radius	The robotic arm workspace shall be a cylinder of radius 400 mm

(continued)

26 MBSE Methods for Inheritance and Design Reuse

Table 2 (continued)

MOI – Lunar Multipurpose Rover	Requirement area	RCM – Mars Science Laboratory
The robotic arm workspace shall be a cylinder of height 1.25 m	Workspace – cylinder height	The robotic arm workspace shall be a cylinder of height 1 m
The robotic arm shall deliver surface samples to stationary main rover body instruments	Sample delivery	The robotic arm shall deliver surface samples to stationary main rover body instruments
The robotic arm electronics shall be radiation hardened for the expected dosage at the lunar south pole	Surface radiation hardening	The robotic arm shall facilitate in sample filtering via rotation with respect to the local vertical gravity vector
The robotic arm shall conduct a high-level commanded operation autonomously	Autonomous operations	*No similar requirement*
The robotic arm shall interface (translate/rotate, hold/release) with surface assets and onboard deployable payloads via a grappling mechanism	Grappling	*No similar requirement*
The robotic arm shall maintain elbow and wrist control despite the failure of their primary actuators	Fault tolerance	*No similar requirement*

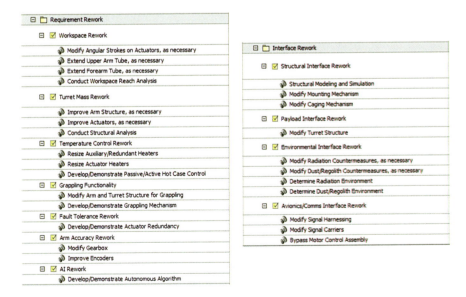

Fig. 16 SysML custom tables enumerating all rework actions deemed necessary for bridging requirement and interface gaps

"requirements" driver, MoI-RCE interface pairs to the "interfaces" driver, and rework actions to either "algorithms" or "operational scenarios" based on the type of rework. For instance, "Conduct Workspace Reach Analysis" is deemed an "algorithm" as it mainly relies on manipulation of governing equations of kinematics while "Develop/Demonstrate Autonomous Algorithms" is deemed an "operational scenario" as it introduces an entirely new functionality and set of test cases that the robotic arm solution must satisfy. Model elements are also assigned a "Reuse Category," which assesses the degree of modifications required, and a "Difficulty Rating." For instance, the workspace reach analysis may determine that the upper tube of the robotic arm needs only a minimal length increase. In contrast, the lower tube may require a substantial reduction in length – in both cases, the difficulty rating is deemed "easy." At the same time, the reuse category for the former ("Adopted") conveys fewer modifications than the reuse category for the latter ("Modified").

A major component of the methodology's final output is determining the relative value of the reuse scenario(s) compared to a nonreuse or "from-scratch" effort in terms of estimated cost, schedule, or work effort. For the lunar rover robotic arm example, a set of 18 requirements, 11 interfaces, 21 algorithms, and 6 operational scenarios were input into the COSYSMO 2.0 tool for the "with reuse case." For the "from-scratch" case, the 18 requirements and 11 interfaces remain although there "difficulty" rating was increased to account for lack of preceding designs from which to draw design aspects; the operational scenarios were examined to determine whether they were still required as part of a from-scratch effort (i.e., that they are not required as a result of changes to the RCE design) – it was determined that all operational scenarios remained necessary, albeit with increased difficulty for the same reason given previously. In most cases, a similar situation was encountered for algorithms though the "modify" designations in the rework action descriptions were altered to "design from scratch," similarly increasing the difficulty of the algorithm drivers.

The result of this comparison is as follows: the robotic arm systems engineering effort was estimated at 144.5 person-months for the "with reuse" case and 167.1 person-months for the "from-scratch" case. This is a 14% improvement in effort in the reuse case. The "effort" metric of person-months can subsequently be converted to estimates for either staffing (a potential proxy for cost) or project schedule. As an example, assuming a 25 person team working on the robotic arm design, this would result in a 6.7 month development effort in a from-scratch approach versus a 5.8 month effort with reuse. Schedule savings of up to a month are highly desirable by engineering managers and systems engineers. Conversely, planning for a 6 month effort, the from-scratch approach would require a team of 28 full time engineers (FTE) while the reuse case would require 24 – budgeting $200 K for each FTE, resulting in project savings of up to $800 K. It should be noted that these estimates only account for effort related to systems engineering activities and do not include materials, manufacturing, and operations efforts. Future work must relate the SE work effort to full project costs/schedules via estimating relationships and empirical data – it is expected that reuse effects will be compounded in these other effort areas.

Chapter Summary

In this chapter, we have discussed a powerful use case for the paradigm of MBSE, namely more effective reuse of designs in complex engineering systems. The increased digitization and centralization of engineering data promoted by MBSE allows decision-makers to employ systematic methods for assessing these reuse scenarios. The key aspects of design reuse that must be addressed are (1) the technical feasibility of a given reuse candidate element to be redeployed on a new mission given the differences in context (e.g., objectives, requirements, environment, CONOPs, etc.) and (2) the value proposition of feasible reuse scenarios in terms of cost and schedule impacts. A structured design reuse methodology based in the principles of MBSE was demonstrated. While the specific methods for implementing the methodology in an MBSE environment may vary from organization to organization, the generalized implementation-agnostic process can be widely adopted to lend systematic rigor to reuse decision-making efforts. Challenges remain in addressing issues related to the accessibility and availability of engineering data within and across organizations. Nevertheless, the increasing adoption of MBSE throughout the space industry (and beyond) provides a promising outlook for more effective and sustainable designs for the next generation of complex engineering systems, leveraging where prudent the reuse of previously proven design elements.

References

1. Agresti, William W. 2011. "Software Reuse: Developers' Experiences and Perceptions." *Journal of Software Engineering and Applications* 04 (01): 48–58. https://doi.org/10.4236/jsea.2011.41006.
2. Barley, Bryan, Allen Bacskay, and Marilyn E. Newhouse. 2010. "Heritage and Advanced Technology Systems Engineering Lessons Learned from NASA Deep Space Missions." In *AIAA SPACE Conference and Exposition 2010*, 0–10. Anaheim, CA.
3. Falkenhayn. 1988. Multimission modular spacecraft (MMS). AIAA Space Programs and Technologies Conference, 1988.
4. P. M. Fischer, D. Lüdtke, C. Lange, F. C. Roshani, F. Dannemann, and A. Gerndt. 2017. Implementing model-based system engineering for the whole lifecycle of a spacecraft. CEAS Space Journal, 9(3):351–365
5. Fortune, Jared, Ricardo Valerdi, Barry W Boehm, and F Stan Settles. 2009. "Estimating Systems Engineering Reuse." *Systems Engineering* 2009 (April).
6. Haas, M., Barclay, T., Batalha, N., Bryson, S., Caldwell, D., Campbell, J. C., ... Klaus, T. (2014). *The Kepler Mission on Two Reaction Wheels is K2*. American Astronomical Society.
7. INCOSE. 2014. "Systems Engineering Vision 2025 - A World in Motion." https://doi.org/10.1126/science.327.5970.1183-d.
8. Kaiser, Michael J., and Christopher Oster. 2015. "Managing a Satellite Product Line Utilizing Composable Architecture Modeling." *AIAA SPACE 2015 Conference and Exposition*, 1–10. https://doi.org/10.2514/6.2015-4436.
9. Madni, Azad M., and Michael Sievers. 2018. "Model-Based Systems Engineering: Motivation, Current Status, and Research Opportunities." *Systems Engineering* 21 (3): 172–90. https://doi.org/10.1002/sys.21438.
10. NASA. 2015. "NASA Technology Roadmaps TA 11: Modeling, Simulation, Information Technology, and Processing."

11. O'Connor, B., & Stevens, J. (2016). *Tethered Space Satellite-1 (TSS-1): Wound About a Bolt.* NASA.
12. Oster, Christopher, Michael Kaiser, Jonathan Kruse, Jon Wade, and Rob Cloutier. 2016. "Applying Composable Architectures to the Design and Development of a Product Line of Complex Systems." *Systems Engineering* 19 (6). https://doi.org/10.1002/sys.
13. Schoberl, Maximilian. 2019. "A Guideline for the Implementation of Model- Based Systems Engineering." Technical University of Munich.
14. Settelmeyer, M. Lampen, R. Hartmann, and G. Lippner. 1996. Flexbus - an attractive technical solution for small missions. Acta Astronautica, 43(11–12):607–613.
15. Siddiqi, A., de Weck, O., & Iagnemma, K. (2006). Reconfigurability in Planetary Surface Vehicles: Modeling Approaches and Case Study. *Journal of the British Interplanetary Society*, 450–460.
16. Spangelo, Sara C., David Kaslow, Chris Delp, Bjorn Cole, Louise Anderson, Elyse Fosse, Brett Sam Gilbert, Leo Hartman, Theodore Kahn, and James Cutler. 2012. "Applying Model Based Systems Engineering (MBSE) to a Standard CubeSat." In *IEEE Aerospace Conference Proceedings*, 1–20. Big Sky, MT: IEEE. https://doi.org/10.1109/AERO.2012.6187339.
17. S. Tietz, J. Bell, and B. Hine. 2009. Multi-Mission Suitability of the NASA Ames Modular Common Bus. 23th Annual AIAA/USU Conference on Small Satellites, (650).
18. Trujillo, Alejandro E., and Olivier L. de Weck. 2019. "Towards a Comprehensive Reuse Strategy for Space Campaigns." *70th International Astronautical Congress*.
19. Trujillo, Alejandro E, and Azad M. Madni. 2020a. "Assessing Required Rework in a Design Reuse Scenario." In *2020 IEEE INTERNATIONAL CONFERENCE ON SYSTEMS, MAN, AND CYBERNETICS (SMC)*, Submitted for Publication. Virtual.
20. Trujillo, Alejandro E, and Azad M Madni. 2020b. "Exploration of MBSE Methods for Inheritance and Design Reuse in Space Missions." In *Conference on Systems Engineering Research*, 1–8.
21. Waseem, Muhammad, and Muhammad Usman Sadiq. 2018. "Application of Model-Based Systems Engineering in Small Satellite Conceptual Design-A SysML Approach." *IEEE Aerospace and Electronic Systems Magazine* 33 (4): 24–34. https://doi.org/10.1109/MAES.2017.180230.
22. van Wyk, Rias J. 2002. "Technology: A Fundamental Structure?" *Knowledge, Technology, & Policy* 15 (3): 14–35.

Alejandro E. Trujillo is a PhD candidate in the Engineering Systems Lab in the Department of Aeronautics and Astronautics at MIT. He received an S.M. in aerospace engineering from the same department in 2018 and a B.S. in aerospace engineering from Georgia Tech in 2015. His research focuses on architecting methods and strategies for campaigns of space missions, with a focus on how MBSE may facilitate and improve on design reuse. He has experience throughout the industry including at NASA MSFC, SpaceX, and The Aerospace Corporation.

Azad M. Madni is a professor of astronautical engineering and executive director of USC's Systems Architecting and Engineering Program in the Viterbi School of Engineering. He is also the director of the Distributed Autonomy and Intelligent Systems Laboratory. He is the chair and cofounder of the IEEE SMC's award-winning Technical Committee for Model-Based Systems Engineering. He has served as general chair of the Conference on Systems Engineering Research since 2008. He is an AAAS Fellow, AIAA Fellow, IEEE Life Fellow, IETE Life Fellow, INCOSE Life Fellow, SDPS Life Fellow, and WAS Life Fellow, and has received several prestigious awards and honors in systems science and engineering including the IEEE SMC Norbert Wiener Outstanding Research Award, NDIA's Ferguson Award for Excellence in Systems Engineering, IEEE AESS Pioneer Award, and INCOSE Pioneer Award. His research has been sponsored by major government agencies including DARPA, NSF, DHS S&T, DoD-SERC, NASA, MDA, ONR, AFOSR, AFRL, ARL, RDECOM, DOE, and NIST, and several aerospace and automotive companies including Boeing, Northrop Grumman, Raytheon, and General Motors.

Model Interoperability

27

Tim Weilkiens

Contents

Introduction 816
MBSE Interoperability Concepts 816
   Scenario #0: No Interoperability 821
   Scenario #1: Connected Information Between Two Repositories 822
   Scenario #2: Interoperability by Link Management 828
   Scenario #3: Central Repository 829
   Scenario #4: Interoperability by Using Services 830
Conclusion 830
Cross-References 831
References 831

Abstract

The interoperability of engineering tools is one of the tremendous current challenges of MBSE. Logically linked systems engineering information must also be physically linked in machine-readable models, and tools must "understand" each other when they collaborate. On a purely conceptual level, which is independent of concrete modeling tools, there is no interoperability. Interoperability comes with the distribution of engineering information to different tools and models. Therefore, the discussions about model interoperability are very much dominated by tools. This chapter focuses on the concepts of tool-independent interoperability.

Keywords

Interoperability · MBSE · Modeling tools

T. Weilkiens (✉)
oose Innovative Informatik eG, Hamburg, Germany
e-mail: tim.weilkiens@oose.de

© Springer Nature Switzerland AG 2023
A. M. Madni et al. (eds.), *Handbook of Model-Based Systems Engineering*,
https://doi.org/10.1007/978-3-030-93582-5_49

Introduction

Typically, an MBSE environment contains more than one model and modeling tool. There are even more models and tools when considering the whole engineering scope, including mechanical, software, electrical, and other disciplines. In addition, the scope can be further extended beyond engineering, covering manufacturing, maintenance, logistics, financial, human resources, and other disciplines that are part of the overall system development project.

All these domains have information about the subject of interest. Conceptually, everything is covered by a single logical model, and all the information is connected, traceable, and consistent. In an ideal MBSE environment, the information can be depicted in stakeholder-oriented views and used by machines for analysis, transformation, verification, simulation, or other purposes.

In reality, information is distributed over several models and documents. The separation is due to the implementation in different organizational units and the associated technical implementations in different modeling tools and other applications.

The tools must be connected and work together to regain the links between information from the single logical model, leading to interoperability.

MBSE Interoperability Concepts

In the MBSE context, interoperability is defined here as follows:

Interoperability is the ability of MBSE tools and model repositories to work together and exchange information in such a way that users are unaware or very unaware of the technical connection between information stored in separate repositories.

Figure 1 illustrates the difference between tool and repository and where the interoperability comes into play.

The MBSE tool provides functions to view and edit the model, which is stored in a repository. The tool/repository connection means that the tool can access the information stored in the repository. The repository does not stand for the

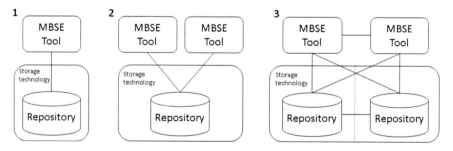

Fig. 1 MBSE tools and repositories

technology, such as a particular database, but the model information stored in a given modeling language in a storage technology. Therefore, more than one repository of the same or different modeling languages can be stored in a single storage technology like a database.

The scenario on the left side in Fig. 1 does not need any interoperability. There is no exchange of information between different tools or repositories.

The figure in the middle shows two tools working on the same repository. All connections are stored in the repository and are accessible by both tools. If the tools want to collaborate directly, they need a technical solution to do so. However, since both tools access the same repository, it is very likely not an interoperability case.

The scenario on the right side in Fig. 1 shows two tools, each with its own repository. Logically related information distributed across both repositories requires interoperability. There are many options for interoperability in this scenario: a tool could directly access the other repository, access the information in the other repository by communicating with the other tool, or the repositories provide each other access to the other repository. In addition, the two repositories could be part of the same storage technology. This enables other technical options for implementing interoperability but does not make it obsolete.

According to Niklas Luhmann, interpersonal communication consists of information, messaging, and understanding [2]. It is pretty similar to model collaboration. For messaging, they need syntactic interoperability and for understanding semantic interoperability.

Syntactic interoperability enables the exchange of information by defining the syntax and structure of the data, for example, by using the Extensible Markup Language (XML). XML defines the data enclosed by tags. Requirements data could be, for example, presented as follows:

```
<requirement>
   <id>REQ42</id>
   <text>The system shall...</text>
</requirement>
<requirement>
   ...
```

If this requirements data is transmitted from a requirements management tool to another MBSE tool, the MBSE tool can read it if it can access XML data. Additionally, it must also understand the meaning of the data, which must conform to the sender's understanding. Enabling a consistent usage of the data by the involved models and tools is called semantic interoperability.

The tags *requirement*, *id*, and *text* in the XML format give meta-information to the data. But without a common vocabulary, the models involved have no common understanding of the data. A generic XML editor could read the requirements data and present it adequately but does not know that these are requirements. Thus, the requirements management tool and the XML editor are syntactically interoperable but do not have semantic interoperability.

The Requirements Interchange Format (ReqIF) [4] is an XML-based format that gives semantics to the tags and thus provides a vocabulary and semantic interoperability.

Another example is the Systems Modeling Language (SysML) [8]. The language is specified and managed by the Object Management Group (OMG). SysML is also an ISO standard [1]. The SysML specification defines the semantic of model elements like block, requirement, or state machine. It is the common vocabulary for semantic interoperability. The XML Metadata Interchange (XMI) [3] is an XML format that can be used to exchange SysML models (syntactic interoperability).

When a SysML tool receives the above XML requirements data and can read it (syntactic interoperability), it could be adequately understood because the language SysML knows the concept of a requirement. However, the understanding of a requirement of the sending requirements management tool could be different from the understanding of the receiving SysML tool.

The tools must agree on a common vocabulary to be semantically interoperable. For example, the requirements management tool semantic can be added to SysML using its extension mechanism of stereotypes. The tools could also use an adapter that translates the information. It can translate the syntax and semantic.

The specification Distributed Ontology, Model, and Specification Language (DOL) [6] provides, besides others, semantic translation from one modeling language to another. According to [6], "DOL is a language for distributed knowledge representation, system specification and model-driven development across multiple" ontologies, specifications, and models.

A foundational concept of MBSE interoperability is the MBSE model itself. It is defined here as a model representing systems and their environments based on a modeling language covering the systems engineering domain concepts.

One purpose of this definition is that MBSE models must be understandable to machines. "Understandable" means that machines can act according to the particular semantics of a model element. For example, a document generator identifies requirements, stakeholders, and relationships in a model and creates appropriate tables in a document. Thus, in the following, an MBSE model is a description created in MBSE modeling tools and stored in a machine-readable format in a repository. The basic structure of the format is a graph, which means a set of nodes and edges.

A text document is not an MBSE model. It is based on a kind of modeling language that does not cover systems engineering concepts but text document concepts. A machine can only identify headlines, paragraphs, indices, and so on, but not requirements, stakeholders, or architecture elements.

Table 1 shows the layers which are used in this chapter for describing interoperability. The mental layer represents concepts in a logical model without any

Table 1 Mental and manifestation layers

Mental layer	A single logical model containing all relevant systems engineering information of the subject of interest	
Manifestation layer	MBSE models	Model languages, tools, and repositories

manifestation (except the brains of the engineers). The manifestation is on the next layer with models describing the information from the logical model in defined modeling languages in modeling tools stored in repositories. It can be any modeling language that fits the definition of MBSE models given above. In principle, others are also possible but not considered in this chapter. A common MBSE modeling language is the SysML. Other possible MBSE modeling languages or tools are, for example, MATLAB, Ptolemy, OPM, or Capella, to name a few.

The following simplified example illustrates the layers: A logical model on the mental layer defines requirements, a system architecture, and relationships specifying which architecture elements satisfy which requirements. Two models in two different model repositories and tools on the manifestation layer manifest the logical model: A proprietary requirements management tool stores the requirements with a tool-specific modeling language. A SysML modeling tool specifies the system architecture and satisfaction relationships. The relationships connect elements in two different tools and repositories, which require interoperability capabilities. How to achieve this is described below.

A systems engineer thinks on the mental layer but has to implement the engineering information on the manifestation layer. This mapping should be as invisible as possible. Typically, the engineer is not primarily interested in figuring out the right model element for a piece of information, linking concepts across different model repositories, or using the tools. Instead, the primary focus of an engineer should be the engineering information.

In addition to related information, tools can also collaborate by calling services to retrieve information, for example, for an analysis calculation, a document generation, a simulation, and so forth. This kind of communication requires syntactic and semantic interoperability between the tools.

In a perfect engineering world with shiny happy engineers, interoperability would not be an issue. Everything is stored and managed in a single repository with a universal modeling language covering all aspects of all involved disciplines plus perfect modeling tools for this modeling language that all involved people from different disciplines can effectively use. With only one repository, there is no interoperability needed which requires at least two repositories.

The single repository is an almost impossible scenario. Still, it leads to an interoperability principle for MBSE modeling environments: The deployment of the logical engineering artifacts on model repositories should lead to high cohesive and low coupled model repositories. Cohesion and coupling depend on different aspects like structure, function, communication, methodology, and so forth.

Coupling represents the degree to which a model repository is independent of others. A model repository should have as few links to other repositories as necessary. If it has a link, it should be as weak as necessary, for example, based on standard technologies (weak coupling) instead of proprietary repository and tool version-dependent technologies (strong coupling). The strength of coupling also depends on the form. For example, information can be coupled because the data depend on each other, because functions call other functions, because a traceability relationship between them is needed, and so forth.

The satisfaction relationships between the elements of the system architecture and the requirements from the example above are a methodical coupling between repositories to achieve traceability. Considering all coupling aspects could lead to many relationships in the repositories. They exist on the mental layer, but the manifestation layer only covers the relevant nodes and edges of the mental layer logical model. Relevant is the information needed to fulfill the purpose of the modeling, which may differ for each project.

Cohesion represents the degree to which a modeling repository forms a logical unit. It is not a rigid criterion, which makes applying this principle in practice a tricky task. For example, one could argue that requirements and architecture elements of a system component together form a cohesive unit, or one could argue that they should be separated and the requirements alone are a cohesive unit because they are all requirements, as well as the architecture elements.

Figure 2 shows some abstract examples of models on the manifestation layer. All three examples depict the graph structure of two user models in two repositories on the manifestation layer with relationships based on a single aspect like methodology.

In the left figure, all nodes are linked with each other. Thus, five relationships cross the border between the repositories. The figure in the middle shows fewer relationships, and only one of them crosses the border. Finally, the figure on the right shows even fewer relationships, and none crosses the repository borders. For example, suppose node X is deployed to repository A instead of repository B. In that case, there are still five relationships crossing the repository borders in scenario 1 but one more in scenario 2 (two relationships) and two more in scenario 3 on the right side in Fig. 2 (two relationships).

It is part of the craft competency of system architects to make the best decision on how to deploy the information in the respective context. However, in practice, a dominant driver of the deployment of engineering artifacts is the availability of appropriate modeling tools. If they are not developed in-house, you have to go by the selection of available tools on the market. Often, an MBSE project has no choice because the tools already exist in the organization and must be taken. This also partly dictates the deployment, for example, that requirements must be managed in given requirements management tools.

The graphical elements shown in Fig. 3 are used in the following to illustrate concepts and their relationships.

The rectangle depicts a concept and the arrow a unidirectional relationship. The syntax 0..* is a multiplicity defining an interval of natural numbers with the symbol *

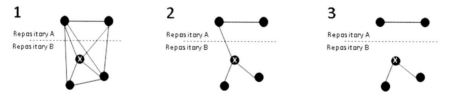

Fig. 2 Coupling of repositories

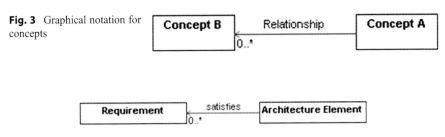

Fig. 3 Graphical notation for concepts

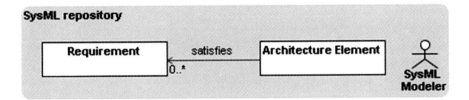

Fig. 4 Concepts in a logical model on the mental layer

Fig. 5 Single repository

for unlimited. The multiplicity specifies how many elements of "Concept B" can be involved in a "Relationship" with a single element of "Concept A." Other typical examples for multiplicities are 0..1, 1..*, or simply 1, which is an abbreviation for 1..1. This kind of description could also be described in SysML with blocks and associations.

Without loss of generality, we consider different interoperability scenarios with the above example. The following in this chapter also applies if other concepts or repositories were chosen. In our example, we have three related concepts on the mental level: requirements, architectural elements, and satisfaction relationships (Fig. 4).

A requirements management tool is used for the requirements and a SysML modeling tool for the system architecture on the manifestation layer. The separation in two different repositories with satisfaction relationships between the requirements and architecture elements leads to several interoperability scenarios discussed in the following sections.

Scenario #0: No Interoperability

The modeling language SysML covers all three concepts from Fig. 4. Therefore, we can put them in the same SysML repository on the manifestation layer, and the SysML modeler can manage the elements in a single repository (Fig. 5). Note that the notation in Fig. 5 does not depict a SysML model but shows that the concepts from the manifestation layer are inside the SysML repository. A requirements management tool is not used for the requirements in this scenario.

The sticky figure depicts the role which manages the model elements. Here, it is a SysML modeler.

Interoperability requires at least two model repositories. Therefore, scenario #0 is not an interoperability case.

Scenario #1: Connected Information Between Two Repositories

In scenario #1, the requirements and the architecture elements are stored in different repositories. Therefore, it is an interoperability scenario. The architecture elements are part of a SysML repository, and the requirements are part of a requirements management repository (Fig. 6). Each repository has its modeler role, which can be the same person. The challenging part is how to implement the satisfaction relationships, which is discussed in the following.

The satisfaction relationship leads to several interoperability use cases. For example, the SysML modeler would like to retrieve the list of satisfied requirements for a given architecture element. The query follows the unidirectional satisfaction relationship from the architecture elements to the requirements, which means a technical connection from one repository to another.

Another relationship from requirements elements to architecture elements is modeled if the other direction is also of interest, for example, if the requirements modeler would like to retrieve the list of architecture elements that implement selected requirements.

In addition to the cohesion and coupling interoperability principle presented above, another interoperability principle says that a relationship between repositories should be unidirectional. A bidirectional relationship on the mental layer would require two unidirectional relationships on the manifestation layer in scenario #1 (Fig. 7), which leads to strong coupling. It is different in scenario #2 below.

In addition to the two unidirectional relationships in Fig. 7, a constraint must be added, assuring that the correct elements are connected. For example, suppose an

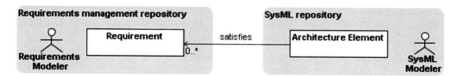

Fig. 6 Interoperability between two repositories

Fig. 7 Bidirectional relationships between concepts

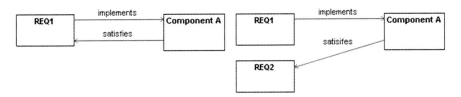

Fig. 8 Unidirectional relationships implementing a logical bidirectional relationship

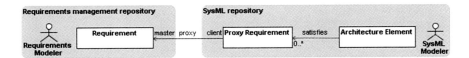

Fig. 9 Proxy element

architecture element Component A implements a concrete requirement REQ1. In that case, it must be assured that the satisfaction relationship originating from Component A points also to REQ1 as depicted on the left side of Fig. 8. Figure 7 would also allow Component A to have a satisfaction relationship to another requirement, REQ2, as depicted on the right side of Fig. 8, which is typically not intended by the modelers.

The satisfaction relationships must also be implemented in a repository on the manifestation level. How to implement this cross-repository relationship is the difference between this and the following scenarios. Scenario #1 considers only two repositories. Later, another scenario covers the case with an additional repository for the relationships.

Without loss of generality, the satisfaction relationship is stored in the SysML repository, which means the same repository as the source elements of the relationships. The following is similar if we store it in the other repository.

The endpoints of a relationship must be stored in the same repository as the relationship itself. Otherwise, the relationship would have no source or target element, which is, typically, not allowed by the modeling language.

The architecture element is the source element of the satisfaction relationship and is already in the SysML repository, but not the target requirement, which is stored in the requirements management repository. Therefore, a proxy for the requirement is created in the SysML repository that fulfills the rule that both ends of the relationship are in the same repository (Fig. 9). This resolves the issue with the satisfaction relationships between two repositories.

However, additional relationships must now link the proxy requirements with the master requirements in the other repository. It is called a proxy relationship, which specifies that the proxy (client) represents the same entity as the master.

It is a purely technical relationship that should be invisible to the modeler, as the interoperability definition also mentions that interoperability should not be apparent to the user.

The proxy element can be a complete copy of the master element, or it can be just a reference to the master element or something in between. For example, if the repositories always have a connection, a reference proxy might be sufficient since the complete data can always be retrieved on demand. On the other hand, if the repositories are only temporarily connected, a complete proxy may make more sense.

On the one hand, the proxy element can be complete, which means it contains all data like the master element. On the other hand, completeness can also refer to the presence of elements, which means a proxy is always created for each master element (complete) or only if needed (incomplete).

Finally, it depends on what is needed from the repository or the repository stakeholders and the technology used to connect.

Another interoperability principle states that only the master may be changed to update the entity. If the element is changed on both sides, conflicts can occur. The repository of the master element is specialized for this kind of element and is, therefore, the best place to change them anyway. For example, the requirements management tool should have better capabilities and domain knowledge to manage requirements than a generic SysML modeling tool.

Four general use cases access the repositories' entities: create entity, read entity, update entity, and delete entity. They are also known as the CRUD use cases. In this consideration, we remain on the more general level and call the elements entity A and entity B in different repositories and a relationship R between entity A and entity B. In total, it leads to 12 use cases (Fig. 10). The white use cases must be explicitly considered in interoperability scenarios.

Create Entity A (CA)

Creating a new entity A in the repository has no direct impact on the interoperability scenario. Maybe, the repository user would like to create a relationship to an entity B afterward, but that is another use case (CR).

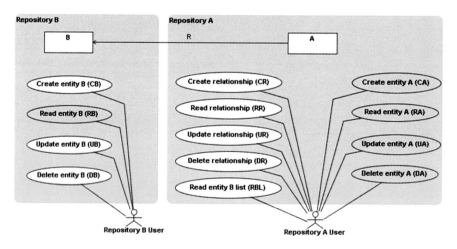

Fig. 10 CRUD use cases

Read Entity A (RA)

Reading an entity A might include reading the relationships to other elements like the relationship to entity B. That would trigger the use case (RR).

Update Entity A (UA)

Updating an entity A might include updating the relationships to entity B, if defined, and triggering the use cases CR, RR, UR, or DR.

Delete Entity A (DA)

Deleting an entity A might include deleting the relationships to entity B if defined. That would trigger the use case DR.

Create Relationship (CR)

Creating a relationship from entity A to an entity B might include retrieving the list of available B entities (see use case RBL). The proxy entity must be created first to create the relationship. Maybe it is already there because another relationship already uses it, or it is automatically created when the appropriate entity B was created (see use case CB). If it is not there, it must be created now. The exact procedure depends on the implementation of the proxy entity; see scenarios #1a–#1c.

Read Relationship (RR)

Reading a relationship is a simple standard task in repository A because it is a relationship between two entities in the same repository

Update Relationship (UR)

Updating a relationship is a simple standard task in repository A. Updating in this context means updating the properties of the relationship. If a source or target element is changed, it can be traced back to other use cases.

Delete Relationship (DR)

Deleting a relationship makes the proxy entity superfluous if it was the last connected relationship. Thus, it can be deleted, except the complete proxy approach is used (see use case CB).

Read Entity B List (RBL)

The user of repository A may be interested in getting the whole list of B entities that are a target of a relationship from entity A (see use case CR). The appropriate information must be transferred from repository A to repository B using one of the implementations described in the following scenarios #1a–#1c to display the list of B entities.

Suppose the proxies are complete representations of the master entities. In that case, the proxy entities are a complete list of all B entities, and the retrieval of the list is a simple standard task in repository A.

Create Entity B (CB)

Creating an entity B has no immediate impact on the interoperability scenario except when the complete proxy element scenario is chosen, which automatically creates the appropriate proxy entity. There could be a time offset, for example, if the synchronization does not happen on demand but only at a specified time or if only one defined baseline of repository B should be synchronized. Configuration management and working with different versions add another dimension, which is not further considered here.

Read Entity B (RB)
Reading an entity B in repository B has no impact on the interoperability scenario.

Update Entity (UB)
Updating an entity B in repository B has not a significant impact on the interoperability scenario. However, properties of the entity that should also be available in repository A must be considered depending on the implementation of the proxy element.

Delete Entity (DB)
Deleting an entity B in repository B also requires deleting the proxy entity, if available. If the proxy entity is the target of a relationship, the relationship must also be deleted, which is a standard task of the repository B tool. Otherwise, it would contain a relationship with a source, but no defined target, which is invalid.

Similar to the discussion in use case CB, the deletion of the proxy entity should happen together with the deletion of the master entity. Otherwise, there is a period during which the models are not consistent. If it is not possible to avoid the delay, the engineering process must consider the temporal inconsistency.

The following sub-scenarios cover the different concrete implementation options of the proxy entity and proxy link:

Scenario #1a: Interoperability by Import/Export

The proxy link between the requirement and its proxy element is established by exporting the requirement from the requirements management repository and importing it into the SysML repository (Fig. 11).

Only those properties of the requirements elements should be transferred that are needed in the SysML repository. Relationships between requirements elements in the requirements management repository are, typically, not imported. If these relationships and all properties of the requirement are required in the SysML repository, scenario #0 above would be a better choice because it seems to be not the right decision to separate the requirements information into another repository if it is also fully required in the SysML repository. It is very likely a violation of the cohesion principle.

However, it may be necessary to import the requirements relationships as well. Either because it requires the technology used to connect the repositories or because access to the relationships is also necessary in the importing tool. Another driver is in which repository which traceability use case is executed.

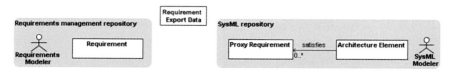

Fig. 11 Interoperability by import/export

For the import/export scenario, it is recommended that all requirements entities are imported into the SysML repository as proxy requirements, even if they have no relationships with entities in repository A. Otherwise, the SysML repository cannot easily query the list of requirements that is required, for example, in the use case CR or RBL.

If the proxies are a complete representation of the master entities, use cases CB and DB, in particular, must be considered. If new requirements are created in the requirements management repository (CB), the appropriate proxy requirements must also be created. If an existing requirement is deleted (DB), the appropriate proxy requirement must also be deleted. In both cases, the export and import of the requirements must be triggered. Importing requirements into an existing set of proxy requirements requires synchronizing the entities whereby the imported set is the master.

The import/export operation is a crucial part of this scenario. When creating or deleting a requirement in the requirements management repository until importing the change into the SysML model, the overall information is not consistent. Therefore, this period should be as short as possible, and it must be ensured that no query on the inconsistent information is made, leading to a possible invalid result.

Possible standards for the data exchange format between a requirements management and a SysML modeling tool are ReqIF or XMI. A possible common data format could also be a comma-separated file (CSV). Proprietary data formats could also be used but lead to a strong coupling between the repositories. Other modeling languages and tools require and provide other data exchange formats.

Scenario #1b: Interoperability by Linked Data

The proxy link between the requirements and their proxy elements is established by providing URL links (Fig. 12). It is still necessary to create a proxy requirement in the SysML repository to have a target for the satisfaction relationship. The proxy requirement mainly stores the URL to the appropriate requirements element, possibly identifying characteristics like the name. The URL provider of the requirements management repository enables access to the requirements by a URL. It is a technical capability that must be explicitly provided.

The proxy requirement could be a SysML requirements model element that references the URL. But it could also be just a property of the architecture element, for example, a stereotype property satisfiedRequirements:String[0..*] (Fig. 13). In the latter case, the satisfaction relationship is represented by the ownership of the property values. However, using the SysML satisfaction relationship is a more straightforward mapping to the logical satisfaction concept to the SysML language.

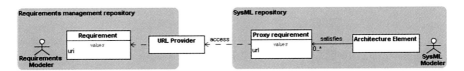

Fig. 12 Interoperability by linked data

Fig. 13 Satisfaction link implementation by stereotype property

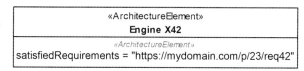

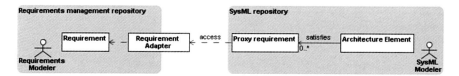

Fig. 14 Interoperability by proprietary adapters

In contrast to the import/export scenario, the complete proxy approach is inappropriate in the linked data scenario. The list of all available requirements can be retrieved on demand by accessing the URL provider.

A possible standard implementation of the linked data scenario is Open Services for Lifecycle Collaboration (OSLC) [9] (Open Services for Lifecycle Collaboration 2020). Furthermore, with the next-generation SysML v2, a SysML repository can provide an implementation of the upcoming standard SysML v2 API & Services, which will provide standard services to access the SysML model [5].

Scenario #1c: Interoperability by Proprietary Adapters

The proxy link between the requirements and their proxy elements is established by providing a proprietary adapter. The proxy requirement, respectively, the SysML repository, can access the requirements by accessing the adapter. It is similar to the linked data approach but not based on standards, and the adapter can also use other technologies than URLs (Fig. 14).

A disadvantage of this approach is the strong dependency of the SysML repository on the requirements management repository if the connection is not based on standards.

Scenario #2: Interoperability by Link Management

This scenario has no direct link between the requirements management and the SysML repository. On the manifestation layer, they are independent of each other. However, the dependency on the mental layer between the architecture element and the requirement is still there. A third entity implements the satisfaction relationship by accessing both repositories and managing the links (Fig. 15).

The link management repository must access the requirements management repository and the SysML repository. If the access is not based on standards, it depends on the specific version of the repository. It could be a challenge for the link management repository vendor to support all repositories and versions. Another challenge is the user experience of the repository modelers. The link management

27 Model Interoperability 829

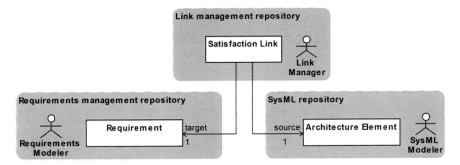

Fig. 15 Indirect interoperability by link management

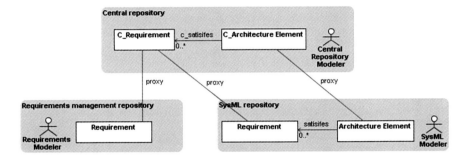

Fig. 16 Central repository

repository implements all use cases regarding the satisfaction relationship (Fig. 10: CR, RR, UR, DR, RBL). The SysML modeler must switch from the SysML tool to the link management tool to perform them, making the workflow slow. If the link management repository is provided as a plug-in in the SysML tool, it increases the problem that the link management tool strongly depends on the tool and version.

Scenario #3: Central Repository

Like the link management repository in scenario #2, a third repository is responsible for the satisfaction relationship between the SysML and the requirements management repository. But the central repository stores, in addition to the satisfaction relationship, also all the other entities from the connected repositories (Fig. 16).

The relationship between identical entities in different repositories is again the proxy relationship. But there are different options about which entity is the authoritative source. For example, in the scenario in Fig. 16, the authoritative source for the requirement could be the requirements management repository. For the architecture element and the satisfaction link, it could be the SysML repository.

If the requirements management repository can handle architecture elements, it could also retrieve the architecture element from the central repository and the satisfaction link.

The central repository could also be the authoritative source for some elements. For example, the satisfaction link could be created in the central repository, and the requirements proxy and the satisfaction link are not part of the SysML repository.

In extreme cases, the central repository could be the authoritative source for all entities, and the other repositories are only views and editors. This brings us to scenario #0, which means we would no longer have interoperability.

There are many options possible, which makes this approach flexible but, at the same time, more complex to handle.

An example of a central repository approach is the open-source project, Open Model-Based Engineering Environment (OpenMBEE). In OpenMBEE, the central repository is named Model Management System (MMS), the proxy relationships are implemented by Model Development Kits (MDK), and the MMS can be accessed by the View Editor to depict the model information in documents and to edit some model entities (see [8]).

Scenario #4: Interoperability by Using Services

Scenarios #1–#3 describe interoperability to enable cross-repository information relationships. Scenario #4 covers the case where one tool wants to access data from another tool for some reason.

The data can be accessed by one of the approaches described above in scenarios #1–#3. Additionally, the data can be provided by a tool through services, for example, an Application Programming Interface (API). These could be proprietary services or standardized services, as specified by the upcoming SysML v2 API & Services specification [5]. The usage of services includes the exchange of information and requires semantic interoperability. Syntactic interoperability is part of the technical approach of how to access the services.

Conclusion

The interoperability of engineering tools is one of the tremendous current challenges of MBSE. The logically linked systems engineering data must also be physically linked in machine-readable models. This is the only way to create the basis for mastering increasing complexity and implementing approaches such as the digital thread or the digital twin.

It will make the engineering work more effective since not so much time has to be invested in copying data back and forth, maintaining traceability matrices, or retrieving the currently valid information from the many redundant storage locations. In addition, errors will be reduced because the single-source approach of modeling always uses valid information.

On a purely mental level, which is independent of concrete modeling tools, there is no interoperability. It only comes with the distribution of the engineering information to different tools. Therefore, the discussion is very much dominated by tools. In this chapter, approaches were presented, which abstract from concrete tools.

In an MBSE environment, one should first clarify the interoperability concept and then select the appropriate tools. Of course, there is an interdependency here since the choice of available tools is limited, and the availability of the features must be taken into account in the conceptual consideration.

Another important aspect is the use of standards that reduce the dependency on individual tools and thus increase the sustainability of the MBSE environment. Noteworthy standards in this environment include ReqIF, OSLC, XMI, and the upcoming SysML v2 API & Services. In addition, sustainability is also provided by open-source tools such as OpenMBEE.

Cross-References

▶ Integrating Heterogenous Models

References

1. International Organization for Standardization. ISO/IEC 19514:2017. Information technology — Object Management Group Systems Modeling Language (OMG SysML). 2017.
2. Niklas Luhmann. Social Systems. Stanford University Press. 1996.
3. Object Management Group. XML Metadata Interchange (XMI). Version 2.5.1. formal/15-06-07.
4. Object Management Group. Requirements Interchange Format (ReqIF). Version 1.2. formal/16-07-01.
5. Object Management Group. SysML v2 API and Services RFP. ad/18-06-03.
6. Object Management Group. Distributed Ontology, Model, and Specification Language. formal/18-09-02.
7. Object Management Grou. OMG Systems Modeling Language (OMG SysML). Version 1.6. formal/19-11-01.
8. Open Model-Based Engineering Environment (OpenMBEE). https://openmbee.org. accessed October 2020.
9. Open Services for Lifecycle Collaboration (OSLC). https://open-services.net. accessed October 2020.

Tim Weilkiens is a member of the executive board of the German consulting company oose, a consultant and trainer, lecturer of master courses, publisher, book author, and active member of the OMG and INCOSE community. He has written sections of the initial SysML specification and is still active in the ongoing work on SysML v1 and the next-generation SysML v2.

He is involved in many MBSE activities, and you can meet him at several conferences about MBSE and related topics.

As a consultant, he has advised a lot of companies from different domains. The insights into their challenges are sources of his experience that he shares in his books and presentations.

Tim has written many books about modeling, including Systems Engineering with SysML (Morgan Kaufmann, 2008) and Model-Based System Architecture (Wiley, 2015). In addition, he is the editor of the pragmatic and independent MBSE methodology *SYSMOD – The Systems Modeling Toolbox – Pragmatic MBSE with SysML*.

A Reuse Framework for Mode-Based Systems Engineering

28

Gan Wang

Contents

Introduction 834
Reuse in Research and Literature 838
The Generalized Reuse Framework 839
    Reusable Resources 839
    The Reuse Process 839
    GRF Usage Scenarios 842
An Illustrative Example – Application of the Generalized Reuse Framework to Cost
Estimating and Analysis 846
    Parametric Cost Estimating for System Development 847
    COSYSMO 848
    The GRF-Based Cost Estimating Relationship 851
    The DWR and DFR Weights 853
Chapter Summary 855
Appendix A 856
    A1. DWR Detailed Weight Table 856
    A2. DFR Detailed Weight Table 857
Cross-References 857
References 858

Abstract

Reuse in system development is a prevalent phenomenon. However, how reuse is applied varies widely. The Generalized Reuse Framework is a strategic reuse model for systems engineering management in product development that addresses both investment and leverage of reuse through two interrelated and interacting processes: development with reuse (DWR) and development for reuse (DFR). This chapter summarizes the latest development of this framework by providing the taxonomic definition of DWR and DFR and analyzing the decision processes for reuse as applied to incremental development and product line

G. Wang (✉)
Dassault Systèmes, Herndon, VA, USA
e-mail: gan.wang@3ds.com

© Springer Nature Switzerland AG 2023
A. M. Madni et al. (eds.), *Handbook of Model-Based Systems Engineering*,
https://doi.org/10.1007/978-3-030-93582-5_77

engineering. It also describes how the framework is applied to the revision of the *Constructive Systems Engineering Cost Model (COSYSMO)*, a parametric cost estimating model for systems engineering. With use case scenarios, it illustrates the approach to apply the framework and to quantify the economic impact of reuse vis-à-vis investment strategies.

Keywords

Reuse · Systems engineering · System development · MBSE · Modeling · Cost estimating and analysis

Introduction

As industry embarks on an accelerated digital engineering (DE) transformation journey aiming for more efficient and effective capabilities for developing and sustaining systems, organizations seek to invest in digital infrastructure, model-based business processes, and a digitally capable workforce. At the same time, they continue to search for ways to quantify returns on investment.

Model-based systems engineering (MBSE) is considered the primary means of realizing such a transformation. The systems engineering (SE) community commonly recognizes that the model-based approach offers a unique advantage in managing the increased complexity of systems today [18] and achieving mission assurance [5]. Through the "the formalized application of modeling" throughout the system life cycle, it captures the system architecture, integrates the design source of truth, and codifies the processes and the best practices for system development and sustainment. Reuse is viewed as a key component of the SE processes.

As widely accepted in the SE community, reuse is essential and sometimes even imperative for today's system development. Reuse is fundamentally driven by economic necessities and ever-mounting pressure to deliver value and profitability to shareholders. A common rationale is cost saving through reduced work and improved quality. However, a more recent focus is the speed of delivery and time to market. In his keynote speech, Jan Bosch [3] pointed out a relatively contemporary trend of modern systems from "built to last" to "built to evolve" to adapt to a faster pace of technology change and user expectations. Reuse, especially when strategically planned, can be a fundamental enabler for the evolutionary approach to system development.

Reuse in SE happens quite frequently. However, there are different approaches to reuse in practice. It can be either opportunistic in nature focusing on "quick wins," or more strategic in implementation, with a central focus on product planning and life cycle strategy.

Reuse can be *opportunistic* in that a designer or developer spends effort searching and discovering reusable resources when the need arises inside and outside their organization and then, if successful, attempts to use the artifacts they find. For example, someone can do a Google search for freeware and lift a code segment that appears to fit a purpose. In software development, it is sometimes called "code scavenging." The outcome is almost inconstant depending upon what is available. In some cases, the reused code has to be modified. In almost all cases, its behavior has to be tested and verified. If anything fails, the developer has to debug the issues and fix the defects. This ad hoc reuse approach involves a process of discovery, assessment, modification, integration, and testing. Its main effort focuses on leveraging benefits through opportunity. For smaller efforts, this approach can be successful. But it does not scale well. The reusability is generally low and uncertainty high, especially for larger and more complex systems.

On the other side of the spectrum, reuse is *planned* and *strategic*. In this case, the developer proactively and strategically invests in reusable resources through explicit reuse processes and standards. A major focus is the up-front investment effort to make an effort easier when the actual reuse occurs later. Examples that come to mind include object-oriented software development. An object class (typically called base class) is defined to instantiate other classes of objects (called derived classes), which inherit the properties of the base class. Another strategy is to create libraries that encapsulate particular objects and functionalities that can be used in multiple applications through a set of application programming interfaces (APIs). Modern programming languages like Python and MATLAB contain large libraries of reusable objects and patterns.

A good example of strategic reuse in SE is design patterns that encapsulate the common features and actions that can be reused over and over and captured using system modeling techniques. For example, a typical data center design pattern includes data ingestion, processing, cataloging, searching, and retrieval components and management and security mechanisms. These patterns can be tailored and scaled into a physical architecture for a particular data center implementation. A pinnacle example is a service-oriented architecture (SOA), a popular software development strategy founded under the guiding principles of reuse, granularity, modularity, componentization, and interpretability. In SOA, a set of common services are developed in the so-called application service layer and offered through the enterprise service layer to the end users as reusable services. This kind of reuse focuses on reusable architectural patterns that result in more certain outcomes and more easily managed changes.

Developed through collaboration with the University of Southern California (USC) Center for Systems and Software Engineering (CSSE) and based on a series of studies [31], the *Generalized Reuse Framework* (GRF) defines two interrelated and interacting reuse processes:

- *Development with reuse* (DWR): a set of system development activities that focus on gaining benefits from utilizing or leveraging previously developed reusable artifacts, either in a planned process or an ad hoc manner.
- *Development for reuse* (DFR): a set of system development activities dedicated to developing reusable artifacts for future usages, generally in a planned manner or through an investment effort.

DFR and DWR represent the two foundational processes that bridge reuse in system development projects or, in a broader sense, for any efforts that deliver products or services. The two processes are distinguished only by intent – production and consumption of reusable resources. DWR focuses on benefits gained from using reusable resources. The basic assumption is that DWR saves labor and improves the product quality at the same time for the system that leverages these resources. DFR, on the other hand, aims to create reusable products for future usage. Acting as a producer, the DFR process feeds reusable resources into the DWR process that consumes them in its development effort. See Fig. 1.

The basic premise, however, is that DFR may incur additional upfront costs than without such a consideration to gain the benefits in DWR. But in the aggregate, it will save cost from the life cycle point of view. Table 1 contrasts the major characteristics of the two processes.

This process model addresses reuse as a central consideration for development strategy as applied to agile SE, product line engineering, and the evolution of system capability through incremental development. It assumes that a product or product line (whether it is a vehicle, aircraft, electronics, or software) is developed by a series

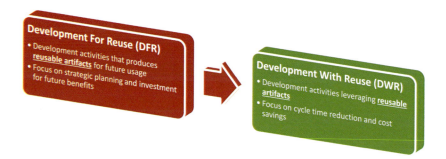

Fig. 1 The Generalize Reuse Framework consists of two interactive processes: the DFR process feeds reusable resources into the DWR process that leverages these resources

Table 1 Contrasting two reuse processes – "development with reuse" and "development for reuse"

	Development with Reuse	Development for Reuse
Role	Consumer	Producer
Purpose	Consumption of reusable resources	Production of reusable resources
Goal	Improvement of product quality Cost savings Improved speed of delivery or time to market	Investment for future benefits
Challenges	Discovery of what to reuse Decisions on how to tailor and integrate	Plans for how to reuse Design for reusability Means to verify
Reusability	If ad hoc, then generally low If planned, then generally high	Generally high

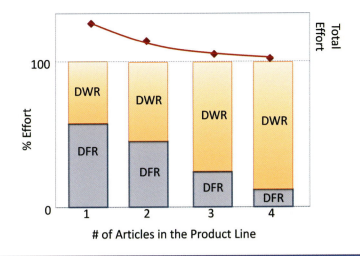

Fig. 2 A mix of DWR and DFR efforts in a project and the decreasing levels of DFR effort in the successive article releases of a project line

of projects, each of which produces several articles or baseline versions. In each article or version release, there is a mix of the DWR and the DFR contents. With careful planning, the DWR and DFR content mix changes favorably with an increasing number of articles produced or release versions deployed, while the total incurred project effort (red line) decreases, as shown in Fig. 2. As the product line matures, the percentage of the DFR content decreases while relative DWR content increases. This phenomenon can be viewed as the investment paying off over time.

This chapter provides a detailed definition of the framework, supported by a taxonomical description of reuse and several application scenarios. The goal is to provide practical guidance on planning, estimating, and managing reuse for the system development life cycle.

As an application example, the second part of this chapter describes how the GRF is applied to engineering cost estimation. Specifically, we introduce the *Constructive Systems Engineering Cost Model (COSYSMO)* [26], a parametric cost estimating model for SE effort, and describe an extension of its cost estimating relationship based on the GRF. The result is an improvement in cost realism and validity, as well as a planning and trade-off framework for development strategy and what-if analyses. Even though the discussion is centered on COSYSMO, the methodology applies to other cost estimating models.

Reuse in Research and Literature

Reuse has been widely studied from different perspectives for decades. It ranges from optimization of product line strategy by balancing commonality and differentiation [6] to the effectiveness of leveraging previously developed components and capabilities [17]. The prevailing measure is whether it would help to reduce cost and maximize profit across product families.

Product line engineering (PLE) [16] focuses on strategic reuse within a product line or product family. A particular reuse interest is in product line management [15]. One reason for focusing on product line management is that in many industries, the product sees a proliferation of incremental features [12]. A "multistream variant management" approach [11] examines reuse strategies across multiple variants of a product family through careful configuration and baseline management. The analysis highlights the complexity involved in the planned reuse of complex systems due to high levels of dependency between development streams.

Product line variability management contrasts two postures of reuse [21], in which different product variants result from deviation from a common core product versus the application (reuse) of common components in different products. It raises the rigor even higher in managing reuse for a product line. A factory approach to PLE [4] was suggested to assemble configured assets from a factory into a product. A highly commoditized view of reuse, the notion of an asset factory is a central theme of what is known as the "second-generation product line engineering" [7].

Perhaps the most noticeable efforts to date have been the focus on software reuse, examining issues range from methods and techniques [8] (Krueger 1992), design patterns [10], and processes [14, 20] to economic impacts [19, 22–24]. Modularity and scalability in product designs are considered an effective means of enabling reuse of common components and new features [25].

On the other hand, an analysis [28, 31, 32] points out that opportunistic reuse (through ad hoc search and discovery) does not generally achieve the goals and expectations for reuse. "Copy & paste engineering" often incur more cost and schedule delays due to unexpected and unplanned rework downstream in the development cycle. It is not uncommon to see those "copied and pasted" artifacts eventually being cleansed out because they ill-fit into the coherence of the architecture and limit sustainability and extensibility. Achieving desirable benefits requires careful and strategic planning and investment efforts that enable further reuse.

The Constructive Cost Model (COCOMO) [1, 2], for example, captures reuse quantitatively by characterizing source lines of code (SLOC) in terms of process and design maturity in developing new software and provides a practical approach for estimating new development with the benefit of reusable software.

Relatively more recent efforts [10, 27, 28, 29, 31, 32] examine the issue of reuse from the cost estimating point of view and establish an approach to quantify the cost and the benefit of reuse in SE. The result is an evolution of COSYSMO, a parametric cost model for SE efforts [26], with an update to its cost estimating relationship incorporating reuse in estimating the size of a system under development.

The Generalized Reuse Framework

Reusable Resources

What can be reused? A simple answer is almost everything. From an engineering perspective, however, it is the outputs from the SE processes that can be reusable, which include elements of the realized system. Then the question is, how do we express them?

We define a reusable resource as a collection of system artifacts that represent certain attributes of the system. System artifacts are physical and functional components (e.g., a piece of hardware or a software module) of a system, along with all the associated engineering and design data that specify the system at different stages of its life cycle.

A design specification can be considered as a collection of system attributes represented in requirements and logical, functional, and physical architecture descriptions. It can be functional or nonfunctional (performance) based. For example, a system attribute can be a system interface that is realized with a physical hardware connection or a software object that pulls or pushes data. Or it can be a function depicting a behavior in terms of an input and output relationship that is realized by a hardware control logic or implemented by a software algorithm.

Together, these system attributes represent what the end system is and how it functions. As a system attribute is realized through a development life cycle, a set of system artifacts are produced and, over time, culminate in the actual system built and deployed. Reversely, a system is simply an integrated collection of system artifacts developed that, together, satisfy the specified system attributes.

For definitions of the GRF process coming up next, we use the term *system attribute* to represent all the reusable resources.

The Reuse Process

Development with Reuse

The DWR consists of the following categories:

New	A system attribute that is new or unprecedented, which requires developing from scratch; or from previously defined system design or constructed product components but requiring near-complete changes in system architecture as a result of that requires developing from scratch; or from previously defined system design or constructed product components but requiring near-complete changes in system architecture due to modified or extended system functionalities.
Design Modified	A system attribute that is designed and developed by leveraging previously defined system concept, functional and logical reference architecture; or from previously designed physical architecture or constructed product components that require significant design and implementation changes or refactoring but without major changes in intended system functionality.

(continued)

Design Implemented	A system attribute that is implemented from an inherited, completed system design or a previously constructed product component that may require only limited design changes in the physical architecture to the extent that it will not impact or change the basic design, but that may require reimplementation of the component.
Adapted for Integration	A system attribute that is integrated by adapting or tailoring (through limited modification of interfaces) of previously constructed or deployed product components without changes in the core architecture, design, or physical implementation (except for those related to interface), so that the adapted element can be effectively integrated or form-fit into the new system. The change effort required is less than that of the Design Implemented category. Removing a system element from a previously developed or deployed system baseline is also included in this category.
Adopted for Integration	A system attribute that is incorporated or integrated from previously developed or deployed product components without modification, which requires complete integration, assembly, test, and checkout activities, as well as system-level V&V testing. This is also known as "black-box" reuse or simple integration.
Managed	A system attribute that is inherited from previously developed and validated product components without modification or that the integration of such an element, if required, is through significantly reduced V&V testing effort by inspection or utilizing provided test services, procedures, and equipment (so-called "plug and play"). Most of the SE effort incurred is a result of technical management.

The terminology for the category names is chosen in such a way that, colloquially, one can say a system attribute (e.g., requirement, interface, etc.) is "New," "Design Modified," "Design Implemented," and so forth.

The DWR categories capture the amount of work required to realize a system attribute in the final system by leveraging those system artifacts available for reuse at the time. From a life cycle perspective, this work typically corresponds to the life cycle stage in which relevant artifacts are available, as shown in Fig. 3. The arrows indicate the typical entry point for most of the development work. For example, "Adopted for Integration" typically commences during the integration, test, and

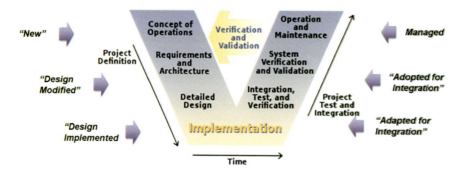

Fig. 3 Typical entry points for the work required by the DWR categories relative to the system "V" model, as a reference for different maturity level of the reusable resources

verification phase as it is possible to leverage artifacts from existing system implementation.

The concept is, in a sense, similar to that of the technical readiness level or TRL. The use of system "V" model provides a convenient reference for the maturity of the system attributes under the reuse consideration. In other words, an attribute deemed "adopted" must have sufficient maturity to be "adopted" or integrated without any modifications. However, the DWR process is not limited to waterfall development and can apply other development models, such as agile development processes.

Development for Reuse

The DFR process consists of the following categories:

No DFR	No development for reuse within the planned work scope.
Conceptualized for Reuse	This category includes a set of front-end SE activities that produce conceptual, contextual, logical, and/or functional architecture elements intended for future reuse, which must be further developed through a series of detailed design, implementation, verification, and validation testing activities to realize the final deployable product.
Designed for Reuse	This category includes a set of front-end system design activities that produce a complete system design or physical architecture elements intended for future reuse, which must be further developed through a series of implementation, integration, and verification and validation testing activities to realize the final deployable product.
Constructed for Reuse	This category includes a set of system development activities that produce a physical product or component intended for future reuse, which has been implemented and independently verified through verification testing but has not been deployed or used in an end system. These activities include required efforts at all levels of design and development, just short of final system-level integration, transition, verification, and validation testing.
Validated for Reuse	This category includes all system development activities that produce an end physical product or component intended for future reuse and operationally validated through its use in an end system.

Similarly, as in the case of DWR, Fig. 4 shows the general exit points for the DFR process relative to the system "V" model for different categories of reusable artifacts. For example, if the development activity stops after the *Detailed Design* phase, the reusable artifacts generated would be at the level of "*Designed for Reuse.*" On the other hand, if a system component has been built and united tested, it should be categorized as "*Constructed for Reuse,*" ready to be integrated into a future DWR process.

As in the case of DWR, the use of "V" model is only a reference for reuse maturity. The DFR process is not limited to waterfall development and can apply other development models, such as agile develop processes.

The DFR categories capture the work required to develop reusable artifacts at varying maturity levels. The DFR process can be a separate investment effort or occur in the same project as the DWR process. In the latter case, the same system attribute should be classified twice – once for DFR and second for DWR in an appropriate category.

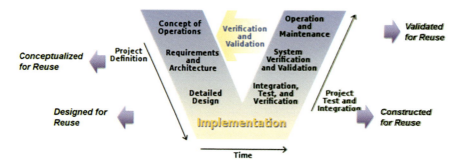

Fig. 4 The exit points of the DFR categories relative to the "System V" model, as an indication for different levels of reuse maturity

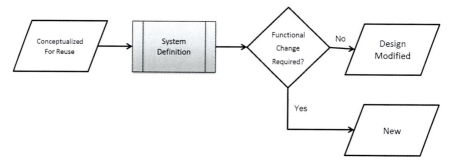

Fig. 5 Decision process for reuse of system definition

GRF Usage Scenarios

We consider four basic scenarios of how different system artifacts pass between the DWR and the DFR processes and the classification strategies for the system attributes. The decision process associated with each case can serve as a guideline for a product line manager to consider in planning the work scope and estimating the development effort, both from an investment angle and leveraging that investment.

Case 1: Reuse of System Definition

Case 1 is an elementary level of reuse. Given a system attribute from a new development effort, we examine the related system definition generated from a DFR process, as shown in Fig. 5. A fundamental question is asked: Is there a change in the functionality required in the new system? Suppose the answer is no or, in other words, the same set of requirements and functionalities can be applied to the new product configuration. In this case, the system definition can be reused from the DWR perspective to further mature into a system design. As a result, the system attribute under consideration is classified as "*Design Modified,*" which indicates the associated DWR activities required to produce a functional system based on an established reference architecture. Accordingly, the previous investment effort in question on developing the reference architecture is leveraged, and the attribute is classified as *"Conceptualized for Reuse"* for the DFR process.

Case 1 is a typical scenario in sharing common core requirements of a product line between different variants or configurations. The same scenario can also be observed in some large-scale acquisition programs where a system is defined by a systems engineering and technical analysis (SETA) team and the resulting reference architecture is handed over to an engineering manufacturing development (EMD) contract for product development.

On the other hand, if functional changes are required, such as changes in key performance requirements, then the existing system definition cannot be reused and a new system definition must be developed. As a result, the attribute is classified as *"New,"* because the DWR process must go back to the beginning of the life cycle and, in effect, start from scratch. No previous work or effort in the DFR process can be leveraged.

Case 2: Reuse of System Design

Case 2 exhibits a greater level of reuse than Case 1. A system design is developed for a particular system attribute and delivered by an investment made in a DFR process. The attribute is classified as a *"Designed for Reuse"* for the DFR process. When the same attribute is considered for reuse in a DWR process, as shown in Fig. 6, we evaluate the system design and ask the question: Are any architecture changes required? If the answer is no, we can implement the system component according to the existing design. As a result, the attribute under consideration is classified as *"Design Implemented,"* as the originally intended by the DFR effort.

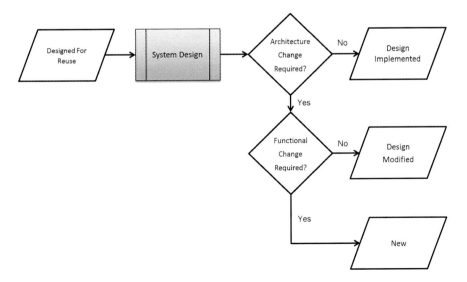

Fig. 6 Decision process for reuse of system design

Conversely, if the answer to the question of change is yes, this means we cannot reuse the design as is, and some level of design modification is required. In this case, however, we could still leverage the system definition from the DWR perspective and, therefore, ask the second question: Are there any changes in the functionality required, such as system requirements? If the answer is no, we can modify the system design based on the same reference architecture, and the category "*Design Modified*" is assigned to the attribute. If the answer is yes, indicating that we must go back and change the system definition, then the same attribute must be classified as "*New.*"

Case 2 occurs in a product family where a core component design is shared across multiple product configurations. It may happen in both EMD projects and production contracts.

Case 3: Reuse of System Implementation

In Case 3, the actual physical system component is available for integration. It may be developed by an early DFR process or provided by a vendor or subcontractor outside of the project's effort. Still, this team's responsibility is to integrate it into the final system, as shown in Fig. 7. In this case, not only is the physical component available for reuse, but also the associated development tools and relevant system tests (data and tools). The system attribute is classified as "*Constructed for Reuse*" for the investment effort in the DFR process.

In the DWR process, we first ask the question: Is this the right component for the system attribute required, and is it integration ready? If the answer is yes, the team can reuse the component and proceed directly to the integration, assembly, test, and checkout phase. As a result, the attribute under consideration is classified as "*Adopted for Integration.*"

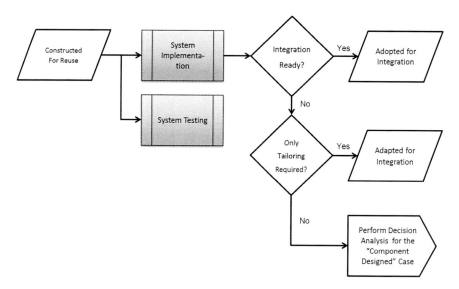

Fig. 7 Decision process for reuse of system implementation

If, on the other hand, the answer is no and the component is not suitable for reuse as is, then we must determine what level of change is required if we still wish to reuse the component. Then, a second question is asked: "Can it be tailored?" or can we perform limited modification only at the interface level to integrate the component without changing its design? If the answer is yes – the component can be simply tailored – then the attribute under consideration is classified as "*Adapted for Integration*."

This situation occurs when we have to modify a connection or a fixture so that an existing hardware component can be form-fit into the system, or when we have to write a "wrapper" or API to pass the data so that a software module or library can be leveraged.

If the answer to the second question is no and it is determined that more significant changes are required than simple tailoring, we then fall back to Case 2 and proceed with its decision process for reusing the system design.

Case 4: Reuse of Validated System Component

Case 4 assumes that a delivered and operationally validated physical system component is available. The corresponding system attribute can be classified as "*Validated for Reuse*" for the DFR process, as shown in Fig. 8.

In this case, there are two typical scenarios for assessing reuse in the DWR process:

1. The component is already part of the system under development, which is typical for incremental development in which a component from a previous release is in the current baseline under configuration control.
2. Another team (e.g., a subcontractor) provides the component as "plug 'n' play," in which case they are responsible for all effort related to platform integration, assembly, test, and checkout into the upper-tier system, which is under a separate work scope and contract arrangement.

In both of these cases, the SE effort for the prime team is not zero for the prime contractor, but at a level of technical management. Therefore, the system attribute for

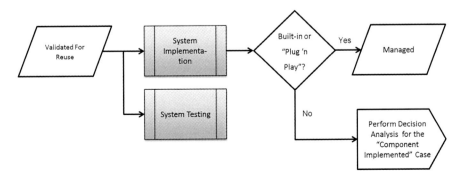

Fig. 8 Decision process for reuse of validated component

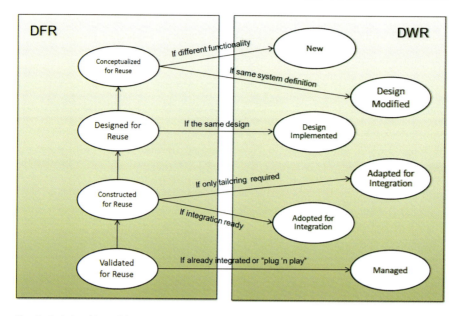

Fig. 9 Relationships of the DFR and DWR categories

this case is classified as "*Managed.*" Otherwise, we fall back to Case 3 and go through its decision process to determine its reuse as an implemented or constructed component.

Summary of the GRF Scenarios

Figure 9 summarizes the four scenarios discussed above. It shows the logical paths from the DFR space to the respective DWR categories, depending on reusability of the system attributes or the amount of changes required when applied to the DWR space.

Importantly, the question should be asked in the reversed direction. If certain reusability is desirable in the DWR process, what level of investment is required for a DFR process? This thought process is critically important for systems engineers and product managers in planning strategic reuse.

An Illustrative Example – Application of the Generalized Reuse Framework to Cost Estimating and Analysis

Cost estimating and analysis is crucial to all modern systems engineering practices. It is an integral part of system development and acquisition. As a fundamental pillar for economic analysis and business decision-making, cost estimation is critical in evaluating the merit of system architecture and provides the essential criteria for design trade space.

Parametric Cost Estimating for System Development

Parametric estimating is commonly recognized as one of the most effective cost estimating methods. It is based on statistical analysis of historical data. In essence, it uses parametric equations between cost (and effort) and one or more parameters. These parameters are derived from the characteristics of the system under estimation and may be physical, performance, operational, programmatic, or cost in nature. They are the independent variables into the equation and commonly known as *cost drivers*. The parametric equation is commonly known as *cost estimating relationship* (CER). It is based on the statistical inferences of multiple similar systems or development efforts. Its output is cost or effort required for development of the system.

Considered as one of the most trusted and reliable basis of estimate, parametric methods are often preferred by the source selection authorities for system acquisition. It is used throughout the system development and acquisition life cycle.

The system architecture is matured incrementally by engineering teams through different stages during development. One of the outcomes is a set of "technical baselines," a collection of technical design data under configuration management associated with progressing project milestones. Incrementally populated and matured by the design activities, the data in the baselines describes the system under development corresponding to its maturity at each juncture. Figure 10 shows an exemplar procurement life cycle, depicting three basic program stages (business capture, program execution, and postdelivery capability upgrade) and the typical program milestones associated with each stage.

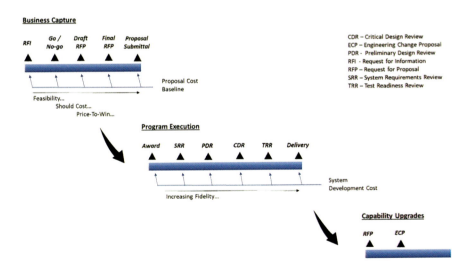

Fig. 10 Systems engineers must support multiple cost estimation activities throughout the system life cycle

Throughout the business capture stage, rough order of magnitude (ROM) estimates are developed to support an evaluation of and response to a request for information (RFI), cost-effectiveness analysis (CEA), analysis of alternatives (AoA), and should-cost analysis in support of competitiveness assessment and Price-to-Win (PTW) criteria [13]. These estimates are typically generated with limited information about the system and/or its development and must be created in a fast-turn manner to support time-to-market decisions. In these situations, parametric estimating is a preferred method, as it requires a relatively small amount of data and can be developed in a short turnaround time.

Cost baselines are developed during the bid and proposal phase to support each proposal stage relative to the draft request for proposal (RFP), final RFP, and the final proposal, including all the intermediate internal milestones and decision gates for management reviews and approval. These cost baselines support the technical solution development. They also form the foundation for the proposed contract price, typically supported by a rigorous basis of estimate. While the final pricing model typically requires a detailed estimating approach that requires a large quantity of supporting data and significant workforce to develop, parametric estimating is often preferred as a validation of the bottom-up approach.

During program execution, cost estimation is performed continuously. As system design matures and more detailed design data becomes available, cost estimates mature accordingly. High-level designs evolve into detailed designs and eventually into first articles. Gradually, realized costs replace estimated costs. At key project milestones (e.g., SRR, PDR, CDR, etc.), cost estimates are reviewed as part of the programmatic baseline and as part of the basis for milestone decisions. As the project progresses and development matures, cost estimates are revised accordingly. When engineering change proposals affect the system design, cost impacts are assessed as inputs to the decision process for approving or rejecting the change.

During postdelivery sustainment, each capability upgrade must be proposed, bid, and designed and built. As with any other design activity, candidate solutions may be subject to AoA design trades and design-to-cost (DTC) constraints. Rapid estimating methods such as parametric estimating are again needed to support the business decision process and validate bottoms-up cost estimates.

After each project, actual costs are analyzed and compared to cost estimates. Sizing drivers and cost drivers, along with project cost and effort, are revised to reflect the "as-built" system. These data calibrate or recalibrate the cost models for future estimation uses.

COSYSMO

COSYSMO, developed at the University of Southern California, is a parametric model for estimating the end-to-end SE and integration effort required in developing and deploying a system. COSYSMO defines a CER that estimates a development project's SE and integration effort using four sizing parameters, also known as *size drivers*. They are:

28 A Reuse Framework for Mode-Based Systems Engineering

- System requirements (REQ)
- System interfaces (INT)
- Critical system algorithms (ALG)
- Operational scenarios (SCN)

The nominal effort is further adjusted by 14 *effort multipliers*, also known as *cost drivers*, representing the product and project environment and complexity factors.

Mathematically, the effort, under a nominal schedule, is described as a function of weighted counts of the four size drivers as shown in Eq. (1):

$$PM_{NS} = A \cdot \left(\sum_k (w_{e,k}\Phi_{e,k} + w_{n,k}\Phi_{n,k} + w_{d,k}\Phi_{d,k}) \right)^E \cdot \prod_{j=1}^{14} EM_j \qquad (1)$$

where,

PM_{NS} = effort in Person Months (nominal schedule)
A = calibration constant derived from historical project data
k = {REQ, IF, ALG, SCN}
w_x = weight for "easy," "nominal," or "difficult" size driver
Φ_x = quantity of "k" size driver
E = represents (dis)economies of scale
EM_j = effort multiplier for the jth cost driver; the geometric product results in an overall effort adjustment factor to the nominal effort.

On an intuitive level, the weighted sum term in the equation above describes how "big" a system is and represents the "size" of the job for the development effort. We call it the "*system size*," which aggregates the effect of four size drivers into a single quantify, with a unit of measure called "*eReq*" or equivalent requirements. The size drivers are each counted at three levels of difficulty.

For a detailed description of the model, including the definitions of the size and cost drivers, as well as the quantitative aspects of its CER, please refer to [26]. The remaining discussion in this section assumes the reader is familiar with the basic model definition.

This model definition, known as COSYSMO version 1.0, does not consider reuse. As represented in Eq. (1), the CER implicitly assumes that all system developments start from scratch – not a very realistic and practical situation in today's world.

The lack of reuse semantics proved to be problematic in practical applications. It severely affects the estimating accuracy and cost realism [27]. An initial effort was to provide a formal construct for reuse similar to the DWR definition today [28] and to augment the COSYSMO model definition. Practical applications of this extended COSYSMO proved effective and significantly improved the estimating accuracy. Figure 11 compares a set of historical project data, collected and validated in COSYSMO [27], with two scatter plots showing the same set of projects captured

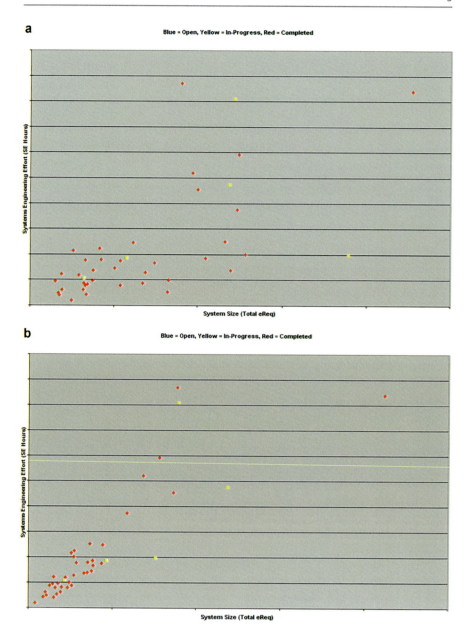

Fig. 11 Scatter plots of the same dataset representing a group of historical system development projects, before (**a**) and after (**b**) implementing a reuse extension in COSYSMO

before and after the reuse extension. Figure 11a shows the projects captured in COSYSMO version 1.0, without consideration of reuse, and there is no distinctive trend observed. Figure 11b shows the same dataset, now captured with consideration of reuse, and the data converges and displays a distinct trend.

Additional studies [29–31] expanded the model to account for the entire scope of reuse – both the investment and the leverage of reusable resources – leading to the GRF today.

The GRF-Based Cost Estimating Relationship

The Generalized Reuse Framework extends the COSYSMO model definition to account for both the efforts of DWR and DFR. Thus, better characterizes the modern system development process and, as a result, significantly improves the fidelity of the estimated cost.

The generalized reuse framework inherently implies two parallel efforts in a development project, each driven by the DWR and DFR processes. Therefore, the total project effort is the sum of the two, as below:

$$PH_{\text{Total}} = PH_{\text{DWR}} + PH_{\text{DFR}} \tag{2}$$

where,

PH_{Total} = total SE effort in person-hours for the entire project
PH_{DWR} = total SE effort in person-hours spent on the DWR process
PH_{DFR} = total SE effort in person-hours spent on the DFR process

In essence, PH_{DWR} is the total effort dedicated to developing the end system with benefit of reuse. It is less than what it would be if it started from a "clean slate." PH_{DFR} is the effort devoted to developing reusable artifacts that could be reused either within the same project or in future projects.

Thus, the total effort can be expressed as

$$PH_{\text{Total}} = A_{\text{DWR}} \cdot SS_{\text{DWR}}^{E_{\text{DWR}}} \cdot CEM_{\text{DWR}} + A_{\text{DFR}} \cdot SS_{\text{DFR}}^{E_{\text{DFR}}} \cdot CEM_{\text{DFR}} \tag{3}$$

where, SS represents the system size under development in the DWR and the DFR processes, respectively. It is with the same unit of measure of "*eReq*" or equivalent requirements, as in Eq. (1). The respective system size is expressed mathematically as

$$SS_{\text{DWR}} = \sum_{k} \left(\sum_{r} w_r \left(w_{e,k} \Phi_{e,k} + w_{n,k} \Phi_{n,k} + w_{d,k} \Phi_{d,k} \right) \right) \tag{4}$$

and

$$SS_{\text{DFR}} = \sum_{k} \left(\sum_{q} w_q \left(w_{e,k} \Psi_{e,k} + w_{n,k} \Psi_{n,k} + w_{d,k} \Psi_{d,k} \right) \right) \tag{5}$$

where,

Φ_x = quantity of "k" size driver, accounted for in the DWR process
Ψ_x = quantity of "k" size driver, accounted for in the DFR process
$k = \{REQ, IF, ALG, SCN\}$
$e = \{Easy, Nominal, and\ Difficult\}$
$r = \{New, Design\ Modified, Design\ Implemented, Adapted\ for\ Integration, Adopted$
$\quad for\ Integration, and\ Managed\}$
$q = \{No\ DFR, Conceptualized\ for\ Reuse, Designed\ for\ Reuse, Constructed\ for$
$\quad Reuse, and\ Validated\ for\ Reuse\}$
w_x = weight for "easy," "nominal," or "difficult," for the respective size driver
w_r = weight for defined DWR levels of the respective size driver
w_q = weight for defined DFR levels of the respective size driver
A_{DWR} = calibration constant for DWR, typically derived from historical project data
A_{DFR} = calibration constant for DFR, typically derived from historical project data
E_{DWR} = nonlinearity for the DWR productivity curve, representing a diseconomy of scale
E_{DFR} = nonlinearity for the DFR productivity curve, representing a diseconomy of scale
CEM_{DWR} = composite effort multiplier for DWR
CEM_{DFR} = composite effort multiplier for DFR

Putting it all together, we can express the CER for the total project effort, including both the DWR and DFR efforts, as.

$$PH_{Total} = A_{DWR} \cdot \left[\sum_k \left(\sum_r w_r (w_{e,k}\Phi_{e,k} + w_{n,k}\Phi_{n,k} + w_{d,k}\Phi_{d,k}) \right) \right]^{E_{DWR}} \cdot CEM_{DWR}$$

$$+ A_{DFR} \cdot \left[\sum_k \left(\sum_q w_q (w_{e,k}\Psi_{e,k} + w_{n,k}\Psi_{n,k} + w_{d,k}\Psi_{d,k}) \right) \right]^{E_{DFR}} \cdot CEM_{DFR}$$

$$(6)$$

This cost estimating relationship captures the total project effort, including the part for investment and the part with the benefit of reuse. It shows that a development project may contain both the DFR and DWR efforts, in different proportions. Reversely, a system attribute may purposefully be developed in both the DFR and the DWR processes. When counting the size drivers for a system or a project, we classify them in the corresponding DWR and DFR categories, respectively, to accurately account for the collective effort.

For example, a critical system algorithm may be implemented as part of a standard library and it would be classified as *Constructed for Reuse* in the DFR process. It can then be used in the same project and would be classified as *Adopted* in the DWR process, if it can be directly integrated into the end system.

As any experienced systems engineer would say, "Not all requirements are created equal!" COSYSMO extended by the GRF is a powerful proof of that statement.

The DWR and DFR Weights

The reuse weight values in Eq. (6) were obtained through a series of wide-band Delphi analyses and further validated with the dataset shown in Fig. 11. They are elaborated next.

The DWR Weights

Table 2 provides the derived values for the DWR category weights. The percentage values in the first row are the relative weights for the six DWR categories, with *New* at 100% and other categories at an incrementally lower level. Mathematically, they are the numerical values for the coefficients, w_r, in Eq. (6). At an intuitive level, they represent the partial set of SE activities required to realize a size driver (REQ, IF, ALG, or SCN) in the end system due to leveraging reuse, relative to the complete set of end-to-end activities corresponding to New. The rest of the 24 (4 × 6) decimal values in the table correspond to the individual DWR weights for each of the four size drivers, in respective rows, at the nominal level of difficulties.

A graphical representation is provided in Fig. 12 that gives an intuitive impression of the relative weight distributions for the four size drivers in each of the six reuse categories. The height of the bars corresponds to the decimal values in the table above in the respective reuse categories. The downward trend for each of the size drivers (shown in a different color) represents a decreasing level of required development effort due to the benefit of increasing levels of reuse (from *New* to *Managed*), as previously elaborated.

When resolving the weight values for the three-dimensional weighted sum term in Eq. (6), we get four sets of 18 (3 × 6) weight values, one for each of the four size drivers, or 72 values in total. This is the result of expanding the six DWR categories to consider all three levels of difficulties at the same time. A detailed table is

Table 2 The numerical weights of the six DWR categories for each of the four COSYSMO size drivers at the nominal level of difficulty

	New	Modified	Impl'ted	Adapted	Adopted	Managed
	100.00%	66.73%	56.27%	43.34%	38.80%	21.70%
System Requirements	1.00	0.67	0.56	0.43	0.39	0.22
System Interfaces	2.80	1.87	1.58	1.21	1.09	0.61
Critical Algorithms	4.10	2.74	2.31	1.78	1.59	0.89
Operational Scenarios	14.40	9.61	8.10	6.24	5.59	3.13

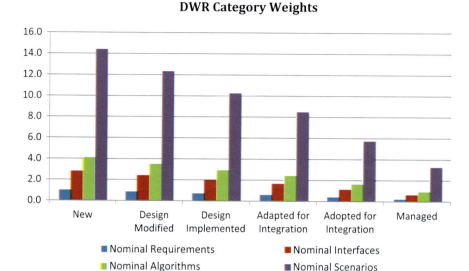

Fig. 12 The graphical representation of the six DWR category weights for each of the four COSYSMO size drivers at the nominal level of difficulty

provided in Appendix A.1 for the convenience of those readers who may wish to implement this model in an estimating tool of their own.

The DFR Weights

Similarly, the values for the DFR weights are listed in Table 3. Mathematically, these values define the numerical values for the coefficients w_q in Eq. (6). At an intuitive level, however, they indicate an increasing level of SE effort required to realize a size driver (REQ, IF, ALG, or SCN) to a higher level of reusability. In particular, the percentage values in the first row represent the derived DFR category weights, with 0% for *No DFR* and other categories at increasing levels up to 94.7% for *Validated for Reuse*.

Figure 13 represents the DFR category weights for the four size drivers (No DFR not represented). Each of the four groups represents the weights of a category for the four size drivers, each shown by different color bars. Once again, the height of the bars corresponds to the decimal values in the table above, in the respective reuse categories for all the drivers. Contrary to DWR, the upward trend in the graph, from *Conceptualized for Reuse* to *Validated for Reuse*, represents an increasing level of SE effort for all four size drivers.

After fully resolving the weight values for the three-dimensional weighted sum term in (2–2), we get four sets of 15 (3 × 5) weight values, one for each of the four

Table 3 The numerical weights of the five DFR categories for each of the four COSYSMO size drivers at the nominal level of difficulty

	Nom. weights	No DFR 0.00%	Conceptualized for Reuse 36.98%	Designed for Reuse 58.02%	Constructed for Reuse 79.15%	Validated for Reuse 94.74%
System Requirements	1.00	0.00	0.37	0.58	0.79	0.95
System Interfaces	2.80	0.00	1.04	1.62	2.22	2.65
Critical Algorithms	4.10	0.00	1.52	2.38	3.25	3.88
Operational Scenarios	14.40	0.00	5.33	8.36	11.40	13.64

size drivers, or 60 values in total. These values are the result of expanding the five DFR categories to consider all three levels of difficulties at the same time. Appendix A.2 includes a detailed table for those readers who may wish to implement this model in an estimating tool of their choice.

Chapter Summary

Reuse is a fundamental feature of SE. However, only planned reuse with a proactive product strategy likely yields economic benefits. The Generalized Reuse Framework provides an effective tool for product line managers and systems engineers to manage reuse in a system development effort. We provided the definitions of the DWR and DFR processes and showed how they can be applied to trade-offs of development approach and reuse planning in product line engineering and incremental system development.

We also described an application of this framework to cost estimating. Specifically, we describe an extension of COSYSMO model definition by applying the GRF, including the modified cost estimating relationship with a set of coefficients calibrated by the reuse weights. The extended COSYSMO more closely captures actual system development processes and significantly increases the fidelity of the cost model and the accuracy of its estimates.

With the competitive market environment and continuing push by enterprises for improved productivity and efficiency, reuse has become a primary consideration in system development. However, only when it is systematically integrated into the end-to-end MBSE process, will we be able to realize the promises of improved architecture understanding, better development decisions, increased agility and collapsed cycle time, and reduced life cycle cost for the complex systems we develop today and tomorrow.

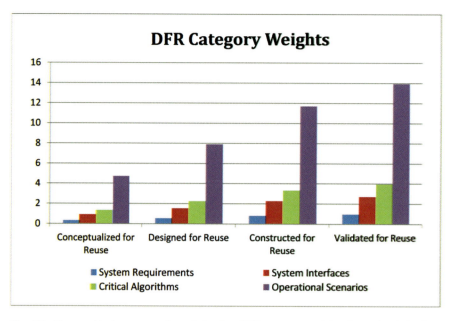

Fig. 13 The graphical representation of the five DFR category weights for each of the four COSYSMO size drivers at the nominal level of difficulty

Appendix A

A1. DWR Detailed Weight Table

The detailed DWR weight values for the 4 COSYSMO size drivers at 3 levels of difficulties and in 6 categories of reuse, or 72 values in total. These values are derived from resolving all the multiplications between the three indices (k, r, x) in the three-dimensional weighted sum term (i.e., 4 size drivers × 3 difficulties × 6 DWR categories) representing the system size in Eqs. 4 and 6.

Reuse category		*New*	*Modified*	*Adapted*	*Deleted*	*Adopted*	*Managed*
Reuse weights		**100.00%**	**66.73%**	**56.27%**	**43.34%**	**38.80%**	**21.70%**
System requirements	Easy	0.5	0.3336	0.2814	0.2167	0.1940	0.1085
	Nominal	1	0.6673	0.5627	0.4334	0.3880	0.2170
	Difficult	5	3.3364	2.8137	2.1671	1.9402	1.0852
System interfaces	Easy	1.1	0.7340	0.6190	0.4768	0.4268	0.2387
	Nominal	2.8	1.8684	1.5757	1.2136	1.0865	0.6077
	Difficult	6.3	4.2038	3.5453	2.7305	2.4447	1.3673
System critical algorithms	Easy	2.2	1.4680	1.2380	0.9535	0.8537	0.4775
	Nominal	4.1	2.7358	2.3073	1.7770	1.5910	0.8899
	Difficult	11.5	7.6737	6.4716	4.9843	4.4625	2.4959
Operational scenarios	Easy	6.2	4.1371	3.4890	2.6872	2.4059	1.3456
	Nominal	14.4	9.6088	8.1036	6.2412	5.5878	3.1254
	Difficult	30	20.0183	16.8824	13.0026	11.6413	6.5112

A2. DFR Detailed Weight Table

The detailed DFR weight values for the 4 COSYSMO size drivers at 3 levels of difficulties and in 5 categories of reuse, or 60 values in total. These values are derived from resolving all the multiplications between the three indices (k, q, x) in the three-dimensional weighted sum term (i.e., 4 size drivers $\times$ 3 difficulties $\times$ 5 DFR categories) representing the system size in Eqs. 5 and 6.

Reuse categories		No DFR	Conceptualized for reuse	Designed for reuse	Constructed for reuse	Validated for reuse
Reuse weights		**0.00%**	**36.98%**	**58.02%**	**79.15%**	**94.74%**
System requirements	Easy	0.0000	0.1849	0.2901	0.3958	0.4737
	Nominal	0.0000	0.3698	0.5802	0.7915	0.9474
	Difficult	0.0000	1.8492	2.9011	3.9577	4.7368
System interfaces	Easy	0.0000	0.4068	0.6383	0.8707	1.0421
	Nominal	0.0000	1.0356	1.6246	2.2163	2.6526
	Difficult	0.0000	2.3300	3.6554	4.9867	5.9684
System critical algorithms	Easy	0.0000	0.8137	1.2765	1.7414	2.0842
	Nominal	0.0000	1.5164	2.3789	3.2453	3.8842
	Difficult	0.0000	4.2532	6.6726	9.1028	10.8947
Operational scenarios	Easy	0.0000	2.2931	3.5974	4.9076	5.8736
	Nominal	0.0000	5.3258	8.3553	11.3982	13.6420
	Difficult	0.0000	11.0954	17.4068	23.7463	28.4208

Cross-References

▶ Digital Twin: Key Enabler and Complement to Model-Based Systems Engineering
▶ Exploiting Digital Twins in MBSE to Enhance System Modeling and Life Cycle Coverage
▶ Exploiting Transdisciplinarity in MBSE to Enhance Stakeholder Participation and Increase System Life Cycle Coverage
▶ Managing Model-Based Systems Engineering Efforts
▶ MBSE for Acquisition
▶ MBSE Methodologies
▶ MBSE Methods for Inheritance and Design Reuse
▶ MBSE Validation and Verification
▶ Model Interoperability
▶ Model-Based Requirements
▶ Model-Based System Architecting and Decision-Making
▶ Overarching Process for Systems Engineering and Design
▶ Role of Decision Analysis in MBSE

References

1. Boehm, B. W., *Software Engineering Economics*, Prentice Hall PTR, 1981.
2. Boehm, B., Abts, C., Brown, A.W., Chulani, S., Clark, B., Horowitz, E., Madachy, R., Reifer, D., and Steece, B., *Software Cost Estimation with COCOMO II.* Upper Saddle River, NJ, Prentice-Hall, 2000.
3. Bosch, J., "Examining the need for change in strategy, innovation methods and R&D practices," *Keynote, the 24th INCOSE International Symposium*, Las Vegas, NV, July 2014.
4. Clements, P.C., "Product Line Engineering Comes to the Industrial Mainstream," *Proceedings of the 25th Annual INCOSE International Symposium*, Seattle, WA, July 2015.
5. Cornford, S., and Feather, M., *Model Based Mission Assurance in a Model Based Systems Engineering (MBSE) Framework*, NASA/CR—2016–219272, Jet Propulsion Laboratory, California Institute of Technology, Pasadena, CA (US), 2016.
6. De Weck, O., Suh, E. S. and Chang, D., "Product Family and Platform Portfolio Optimization," Proceedings of the ASME Design Engineering Technical Conferences - Design Automation, 2003.
7. Flores, R., Krueger, C., Clements, P. "Mega-Scale Product Line Engineering at General Motors," *Proceedings of the 2012 Software Product Line Conference (SPLC)*, August 2012.
8. Freeman, P., "Reusable Software Engineering: Concepts and Research Directions", ITT Proceedings of the Workshop on Reusability in Programming, 1983.
9. Fortune, J. and Valerdi, R., A Framework for Systems Engineering Reuse, *Systems Engineering*, Vol. 16, No. 2, 2013.
10. Gamma, Erich; Helm, Richard; Johnson, Ralph; Vlissides, John (1995). Design Patterns: Elements of Reusable Object-Oriented Software. Addison-Wesley. ISBN 0-201-63361-2.
11. Gery, E. and Scouler, J.L., "Strategic Reuse and Product Line Engineering," *developerWorks*®, IBM Corporation, October 2014.
12. Hillhouse, B. and Ishigaki, D. T., Strategic Reuse: A Fundamental Approach for Success in E/E Engineering, IBM Rational webinar, 2011.
13. INCOSE. (2015). *Systems Engineering Handbook* (4th ed.). Hoboken: John Wiley and Sons.
14. Kim, Y. and Stohr, E. A., "Software Reuse: Survey and Research Directions", *Journal of Management Information Systems*, Vol. 14, No. 4, Spring 1998.
15. Knauber, P. Bermejo, J. Bockle, G. Sampaio do Prado Leite, J., van der Linden, F., Northrop, L., Stark, M., Weiss, D., "Quantifying Product Line Benefits", *Software Product-Family Engineering*, 4th International Workshop, PFE 2001 Bilbao, Spain, October, 2001
16. Le Put, A., *Systems Product Line Engineering Handbook.* Association Francaise d'Ingenierie Systeme, 2015.
17. Nazareth, D.L. and Rothenberger, M.A., "Assessing the Cost-effectiveness of Software Reuse: a Model for Planned Reuse", *Journal of Systems and Software*, Vol. 73, No. 2, 2004.
18. Peterson T., *Systems Engineering: Transforming Digital Transformation, NDIA 22nd Annual Systems Engineering Conference*, Tampa, FL (US), 2019.
19. Poulin, J.S. and Caruso, J.M., "A Reuse Metrics and Return on Investment Model", *Second International Workshop on Software Reusability*, 1993.
20. Redwine, S. T. and Riddle, W. E., "Software Reuse Processes," Proceedings of ACM Software Process Workshop, 1989.
21. Reiser M.O., "Managing Complex Variability in Automotive Software Product Lines with Subscoping and Configuration Links," Suedwestdeutscher Verlag fuer Hochschulschriften, June 2009.
22. Succi, G. and Baruchelli, F., Analyzing the Return of Investment of Reuse, *ACM SIGAPP Applied Computing Review*, 1996.

23. Selby, R. W., "Quantitative Studies of Software Reuse", in T.J. Biggerstaff, A.J. Perlis (eds.), Software Reusability, vol. 2, Applications and Experience, Reading, MA: Addison-Wesley, 1989.
24. Selby, R., "Enabling Reuse-Based Software Development of Large-Scale Systems." *IEEE Transactions on Software Engineering,* Vol. 31, No. 6, June 2005.
25. Simpson, T. W., Siddique, Z. and Jiao, J., Eds., 2005, *Product Platform and Product Family Design: Methods and Applications*, Springer, New York, NY.
26. Valerdi, R., *The Constructive Systems Engineering Cost Model (COSYSMO)*, PhD Dissertation, University of Southern California, May 2005.
27. Wang, G., Valerdi, R., Roedler, G., Ankrum, A., and Millar, C.., "COSYSMO Reuse Extension", Proceedings of the 18th INCOSE International Symposium, Utrecht, the Netherlands, June 2008.
28. Wang, G., Valerdi, R., Fortune, J., Reuse in Systems Engineering, *IEEE System Journal*, Vol. 4, No. 3, 2010.
29. Wang, G. and Rice, J., "Considerations for a Generalized Reuse Framework for System Development", Proceedings of the 21st INCOSE International Symposium, Denver, CO, June 2011.
30. Wang, G., Valerdi, R., Roedler, G., Pena, M., "Quantifying Systems Engineering Reuse – a Generalized Reuse Framework in COSYSMO", Proceedings of the 23rd INCOSE International Symposium, Philadelphia, PA, July 2013.
31. Wang, G., Valerdi, R., Roedler, G., Pena, M., "A Generalized Systems Engineering Reuse Framework and Its Cost Estimating Relationship," *Proceedings of the 24th INCOSE International Symposium*, Las Vegas, NV, July 2014.
32. Wang, G., "The Generalized Reuse Framework - Strategies and the Decision Process for Planned Reuse." *INCOSE International Symposium Volume 26, Issue 1*, July: 175–189. 2016.

Gan Wang is the vice president for systems engineering ecosystem at Dassault Systèmes (3DS). Gan provides leadership for 3DS' collaboration with the systems engineering communities and represents the company in industry associations and standardization bodies. Gan is an established leader in systems engineering with decades of experience in developing software-intensive systems. His areas of specialty include systems engineering processes, system and enterprise architecture, design reuse methodology, cost estimating and analysis, and multicriteria decision support methods. He is a frequent author and instructor in these subject areas. Gan is an INCOSE Fellow and ESEP. Prior to 3DS, Gan spent many years at BAE Systems where he was a Global Engineering Fellow and served as the chief engineer for its Integrated Defense Solutions business area. He also worked for many years as a software engineer and control systems engineer developing real-time geospatial data visualization applications, man-in-the-loop flight simulation, aircrew training systems, and a virtual reality-based game system.

MBSE Mission Assurance

29

J. S. Fant and R. G. Pettit

Contents

Introduction (Problem Statement, Key Concepts, Terms and Definitions)	862
Definitions	864
Challenges	865
State-of-the-Art (Review of the Literature)	866
Best Practice Approach (What's New and Different, Benefits and Payoffs)	868
Program Assurance Core MA Process Best Practices	868
Requirements Analysis and Validation Core MA Process Best Practices	872
Design Assurance Core MA Practice Best Practices	875
Manufacturing Assurance Core MA Process	880
Integration, Test, and Evaluation Core MA Process	880
Operations Readiness Assurance (ORA) Core MA Process	881
Operations, Maintenance, and Sustainment Core MA Process	882
MA Reviews and Audits	882
Illustrative Examples (A Couple to Show Uses of Approach)	883
Program Assurance Example	883
Design Assurance Example	885
IT&E Assurance Example	887
Chapter Summary	888
Cross-References	889
References	890

Abstract

Mission assurance (MA) is the disciplined application of proven scientific, engineering, quality, and program management principles toward the goal of achieving mission success (Guarro SB, Johnson-Roth GA, Tonsey WF, Mission

---

J. S. Fant
The Aerospace Corporation, Chantilly, VA, USA
e-mail: julie.s.fant@aero.org

R. G. Pettit (✉)
George Mason University, Fairfax, VA, USA
e-mail: rpettit@gmu.edu

© Springer Nature Switzerland AG 2023
A. M. Madni et al. (eds.), *Handbook of Model-Based Systems Engineering*,
https://doi.org/10.1007/978-3-030-93582-5_72

Assurance Guide Revision B, El Segundo. The Aerospace Corporation, CA, 2012). As companies evolve from traditional document-based SE approaches to model-based systems engineering (MBSE)-based approaches, the "what" MA needs to do largely remain the same. However, the "how" MA is carried out can be expanded to specifically leverage MBSE modeling artifacts. Leveraging these digital artifacts has the potential to improve the efficiency and timeliness of the MA process. This chapter examines the state of the art for MBSE MA and compares it with traditional document-based MA. The chapter also references some specific MBSE MA benefits, challenges, and analysis techniques. Finally, we demonstrate some MBSE MA analysis techniques using a case study.

Keywords

MBSE · Mission assurance · Mission success · SysML · Digital thread

Introduction (Problem Statement, Key Concepts, Terms and Definitions)

Mission assurance (MA) is "the disciplined application of proven scientific, engineering, quality, and program management principles toward the goal of achieving mission success. MA follows a general systems engineering (SE) framework and uses risk management (RM) and independent assessment as cornerstones throughout the program life cycle" [1]. Thus, a typical MA effort involves performing MA processes over the course of a system life cycle. The Mission Assurance Guide (MAG) defines eight core MA processes as shown in Fig. 1 and Table 1 [1].

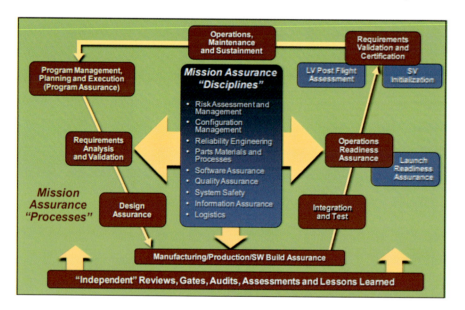

Fig. 1 SE view of core MA processes [1]

29 MBSE Mission Assurance

Table 1 Core MA processes from MAG [1]

Core MA process.	Objective
Program Assurance	Perform technical assessments of the programmatic practices (cost, schedule, performance, and risk) to ensure the program delivers the required capability for required operations within the allocated budget and schedule for overall mission success
Requirements Analysis and Validation	Independent Technical Assessments (ITA) supporting requirements validation activities
Design Assurance	Iterative set of planning, analysis, test, and inspection activities which are performed from conceptual to preliminary to detailed design stages to improve the probability that the system will meet their intended requirements through all operating conditions and throughout the design life
Manufacturing Assurance	Conduct various activities to ensure the manufacturing processes are able to produce hardware that meets the design requirements and to translate the design into a reliable, durable manufactured item using manufacturing processes that are highly repeatable and error free
Integration Test and Evaluation	Evaluate and ensure end item requirements satisfaction (e.g., functionality, performance, design/construction, interfaces, and environment) at all levels of assembly as those end items (e.g., units) form a system
Operations Readiness Assurance (ORA)	Ensure readiness planning, activation, and mission operation
Operations, Maintenance, and Sustainment	The goal is to maintain performance through continued monitoring, predictive assessment, periodic maintenance, and asset replenishment
MA Reviews and Audits	Assess the technical maturity within a program, evaluate program risks and opportunities, understand stakeholder expectations, and ensure readiness for the next phase in the overall program life cycle of events and milestones

Within each of the core MA processes, there are multiple core activities that are performed. Together these capture the "what" needs to be done for MA. The "how" the core activities are performed can vary. To define the "how," specific analysis techniques are selected to achieve the core activity. A graphical summary of relationships between these terms is shown in Fig. 2.

When MA is applied to a program, the core MA processes, core activities, and analysis techniques are scoped appropriately based on the program's risk, budget, and schedule. This plan for the MA effort is captured in a Mission Assurance Plan (MAP). The Mission Assurance Planning process is described in more detail in the Mission Assurance Guide [1].

As companies evolve from traditional document-based SE approaches to MBSE and Digital Engineering (DE)-based approaches, the MA goals and core MA processes largely remain the same. What is changing are the analysis techniques being applied, and in some cases, the core MA processes need to be expanded to specifically leverage MBSE modeling artifacts. For example, if the MBSE MA

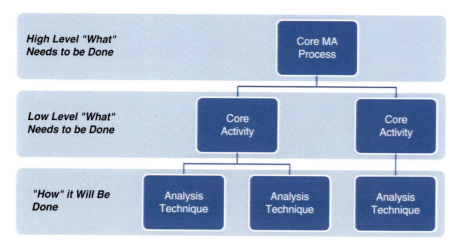

Fig. 2 Terminology hierarchy for MA

practitioners have access to the digital MBSE model, analysis techniques that leverage models rather than document-based artifacts can be employed. For the rest of this chapter, application of MA to leverage MBSE artifacts will be called MBSE MA.

MBSE MA has the potential to improve timeliness and reduce the effort required to perform MA core processes. This is because it can leverage MBSE artifacts in a digital format, which can be easily queried for relevant information or integrated with other models to provide more complete picture of the system. This is an improvement over document-based artifacts where manual reviews are required to extract relevant information. It also makes MBSE MA more conducive on agile SE approaches, like Full Scaled Agile Framework (SAFe) [2], where the SE activities are performed with each planning increment and models are an integral part of the process. The MBSE MA activities on agile programs need to be timely to keep up with the velocity of agile programs.

Definitions

Digital engineering (DE): An integrated digital approach that uses authoritative sources of system data and models as a continuum across disciplines to support life cycle activities from concept through disposal [3]

Digital thread: "refers to the communication framework that allows a connected data flow and an integrated view of a physical asset's digital data (i.e., its Digital Twin) throughout its life cycle cutting across traditionally siloed functions" [4]

Mission assurance: "the disciplined application of proven scientific, engineering, quality, and program management principles toward the goal of achieving mission success. MA follows a general systems engineering (SE) framework and uses risk management (RM) and independent assessment as cornerstones throughout the program life cycle" [1]

Model-based systems engineering (MBSE): The formalized application of modeling to support system requirements, design, analysis, and verification and validation activities beginning in the conceptual design phase and continuing throughout development and later life cycle phase [5, 6]

MBSE MA: mission assurance with an emphasis on leveraging MBSE artifacts to perform technical assessments.

Mission Assurance Plan (MAP): "Is the program specific MA plan to validate and verify the concept development, design, manufacturing, integration, test, deployment, and operations of a space system" [1]

Core MA process (CMP): "an SE process that is defined and applied to support MA goals" [1] It defines the high level "what" needs to be done.

Core activity: a specific activity that is performed to achieve a core MA process. It defines the low level "what" needs to be done to execute the core MA process [1]

Analysis techniques: the specific techniques used to achieve a core activity. They define the details for "how" the core activity will be executed.

Challenges

MBSE MA is not without challenges, and these challenges should be addressed when establishing MAP. The first challenge that MBSE MA practitioners face is currently there are no comprehensive MBSE MA guides or handbooks. So MBSE MA practitioners must rely on existing MA guides, strategies, and articles and adapt them appropriately to MBSE.

A second challenge facing MBSE MA practitioners is getting access to the model data. In some cases, it may be as simple as sharing or providing access to a MBSE model. However, if the MBSE MA analysis is being performed by an independent third party, then sharing the model may become more challenging. For instance, the MBSE model may contain proprietary information within the design or it could also be integrated with other data sources in the digital thread that contain proprietary information. In this case, if the proprietary information cannot be easily removed prior to sharing the model with an independent third party, then getting access to the model data for analysis will become more challenging. Another challenge is if the MBSE model is one component in a DE effort where the MBSE model references data stored in other data sources, such as from a requirements management tool or external simulation tool. If that data is needed for the MBSE MA analysis, it will require either an additional export of data or additional access to that data.

Another challenge facing MBSE MA practitioners is the quality of MBSE artifacts produced by the SE processes. Even if an MBSE model is developed in a MBSE tool, it is still possible the resulting model will be of poor quality and difficult to efficiently analyze. Quality issues can stem from many issues, such as lack of conformance to a style guide and improper use of a MBSE tool. A missing modeling style guide or lack of conformance to a modeling style guide and naming conventions leads to understandability issues, such as ambiguous model elements that lack definitions and descriptions making them difficult to understand and using inconsistent linkages between model element types, which makes it difficult to query and understand the model. This is also the case when using modeling languages with weak semantics (e.g., SysML) that leave the semantic meaning of modeling elements and their connections up to loose interpretation or tool-dependent interpretation. Improper use of the MBSE tool can also negatively impact model quality. For example, the power of MBSE tools relies on defining design elements in only one location and then referencing them in multiple diagrams. This enables the MBSE tool to track these references and maintain a comprehensive picture of the design elements role within the system. If an MBSE model contains multiple definition for the same design element, each of the duplicate design elements will only contain a subset of the relationships and usage within the system. Thus, comprehensive view for that design element cannot be easily created, and the model will be more difficult to analyze from a MA Design Assurance core process perspective.

Another challenge MBSE MA practitioners face in selecting MBSE analysis techniques is managing the maturity of MBSE analysis techniques. MBSE analysis techniques are evolving and are at various stages of maturity. Some techniques are more mature and may even have commercial tool support. However, others are still in the research and development phase, so tool support options are limited and the effectiveness of these techniques on large-scale programs maybe unknown. However, if they are promising enough, MBSE MA practitioners may still wish to apply them or even develop their own tool support to improve the practical application of the analysis technique.

Finally, the last challenge MBSE MA practitioners face is tool interoperability. There can be tool interoperability challenges if an independent third party for the MBSE MA effort doesn't utilize the same MBSE tool as the model developer. Currently, MBSE tools have some built-in ability to import models from other tools [7–9]. In some cases, the model can be imported without data loss, but in many others cases, the data loss from the import process will be unavoidable. This is particularly the case if a tool has allowed nonnormative models to be created that may have no comparable representation in another tool. Some vendors offer additional tool support to exchange models without or with minimal data loss [10]. However, those options may be too cost prohibitive for the model developer if the model only needs to be exported a few times for MBSE MA analysis.

State-of-the-Art (Review of the Literature)

MBSE MA is a growing field that will keep evolving to take advantage of model artifacts and digital threads provided by MBSE and DE. Currently, there are no comprehensive MBSE MA guides that provide detailed guidance on MBSE analysis techniques that can be leveraged by MBSE MA practitioners. The Mission Assurance Considerations for Model-Based Engineering for Space Systems provide a high-level guidance on leveraging modeling artifacts. However, it does not describe any specific analysis techniques that can be applied [11]. The chapter extends this work by taking a deeper dive into the MBSE MA and providing some analysis techniques that can be leveraged and in demonstrating MBSE MA on a case study.

One area of related work is existing MA guides, strategy documents, and articles which include the Mission Assurance Guide [1], Systems Engineering Guide's chapter on Mission Assurance Chapter [12], Department of Defense (DOD) Mission Assurance Strategy [13], Mission Assurance—A Key Part of Space Vehicle Launch Mission Success article [14], and many others. While these may not specifically address MBSE, they can be leveraged and adapted for MBSE MA.

Another area of related work is on individual organizational MBSE MA efforts. These efforts are tailored to the organizational needs of a company. One of these efforts is from the National Aeronautics and Space Administration (NASA) Office of Safety and Mission Assurance (SMA). NASA SMA has developed their vision and framework for MBSE MA, which they call Model-Based Mission Assurance (MBMA). Their MBMA approach has strong emphasis on harnessing and leveraging the MBSE model being developed as part of the SE process to perform a variety of safety assurance analyses earlier in the life cycle when changes to design are less costly to make. NASA SMA's MBMA vision has a strong emphasis on safety cases and assurance as graphically depicted in Fig. 3 [15, 16].

A final area of related work is on developing analysis techniques for modeling artifacts. These analysis techniques can be leveraged by MBSE MA approaches. These are discussed in more detail in their appropriate core MA process sections in section "Best Practice Approach (What's New and Different, Benefits and Payoffs)."

Fig. 3 NASA OMA's MBMA vision

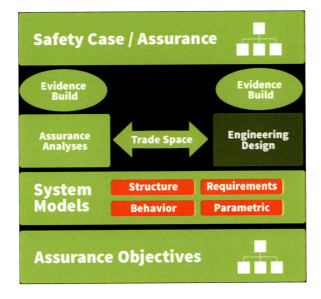

Best Practice Approach (What's New and Different, Benefits and Payoffs)

MBSE MA is an evolving discipline with significant potential to improve the way core MA processes are executed. First, by leveraging models where SE data is in digital format, MBSE MA practitioners will be able to directly query the data. This will result in a more effective and efficient core MA process execution when compared to doing manual searches through multiple document-based artifacts to find the relevant information and then put it in a format that can be analyzed.

A second benefit of MBSE MA is it can improve the ability to find authoritative data for analysis. Assuming the MBSE model is considered a single source of truth for the architecture, it should contain all the latest up-to-date information about the system architecture. Thus, any data that is pulled from the MBSE model is considered current. There will be no more wondering if the MBSA MA practitioner received the latest version of the document or requesting additional referenced documents to find the data they require.

Another advantage of MBSE MA is it can improve the way MBSE MA practitioners perform their analysis. MBSE MA practitioners can utilize automated or partially automated analysis techniques on digital MBSE artifacts. By automating or partially automating some of analysis, MBSE MA practitioners will be able to save time and be more efficient with their reviews. Additionally, as MBSE modeling techniques evolve and models integrate more data from SE and other process, the analysis techniques to evaluate the models will also evolve alongside them.

Finally, the last benefit of MBSE MA is it has the ability to improve the ability of MBSE MA practitioners to locate related data. If the program has leveraged a more comprehensive DE approach where the MBSE model is integrated with authoritative sources of data from other disciplines, such as requirements, then it will be much easier for the MBSE MA practitioners to find data across disciplines needed for their analysis. This is a significant improvement over having to coordinate, find, and review the right set of documents from a document-based SE approach.

MBSE MA spans broad spectrum of core MA process, and each core MA process will utilize MBSE differently. Therefore, in the remainder of this section, each core MA process is examined in more detail to understand what is new, different, and beneficial.

Program Assurance Core MA Process Best Practices

"Simply stated, Program Assurance objectives are cost, schedule, performance, and their associated risk" [1]. The core MA activities are sorted along three primary disciplines, program management, SE, and acquisition, and they are graphically depicted in Fig. 4. The core MA processes where MBSE will have the most impact are P1 Program Definition and P3 Plan Program, all the SE processes, and A2 RFP/Source Selection Activities. The impact MBSE has on each of these core activities is described below in more detail.

29 MBSE Mission Assurance

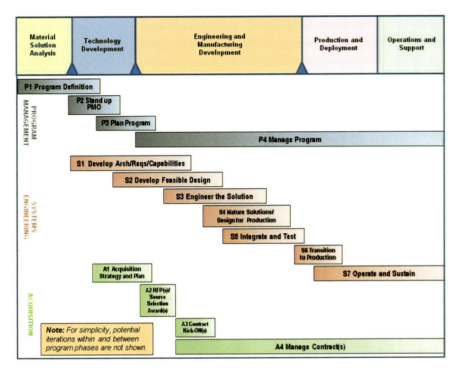

Fig. 4 Program Assurance core MA processes

The output of the P1 Program Definition and P3 Plan Program core processes are "operational needs defined in a set of high-level requirements and defining and developing the concept of operations (CONOPS), implementing configuration management control and tracking, tailoring the life cycle to best support the program's needs and outcomes, implementing a strong risk management process, beginning the planning for testing, updating the initial schedule to include more detail, and updating the initial high-level cost estimate, to create a full life-cycle cost estimate (LCCE)" [1]. On document-based acquisitions, the government typically created these artifacts as independent documents and MA practitioners manually review the documents for consistency and completeness.

On MBSE-based approaches, it is possible for some of this planning information to be included within the MBSE model. For example, the Department of Defense Architectural Framework (DoDAF) and the Unified Architectural Framework (UAF) have the ability to model and create traceability between Capability Views/ Strategic Views that capture the desired capabilities and capabilities evolution to Operational Views that capture the tasks, activities, and operational performers needed to achieve mission goals [17–18]. The traceability helps to ensure all the desired capabilities are represented and accounted for in the operational views. If traceability is defined and maintained in an MBSE model, the MBSE MA practitioners can use analysis techniques that leverage traceability matrices from these frameworks that are commercially available in MBSE tools [19]. For example, if the Operational Activities to Capabilities traceability is maintained in the MBSE model, then the MBSE MA practitioner can view this data in a traceability matrix to quickly identify any inconsistencies between the views. As shown in Fig. 5, gaps in this traceability are easy to identify when data is presented in this manner. MBSE MA practitioners will still need to manually review the correctness of the traceability, but this is easy to accomplish using the traceability matrix as well. Additionally, if the capabilities roadmap is created in the model using the UAF Strategic Roadmap, then MBSE MA practitioner can quickly inspect the diagram to identify any capabilities that are not included in the roadmap, as shown in Fig. 6.

The outputs from the P1 and P3 processes are used as inputs and evolve with the S1 Develop Architecture/Requirements/Capabilities core activity. The goal of S1 is to refine the system into a more detailed set of operational and technical requirements and to begin defining the architecture [1]. This typically starts out as a

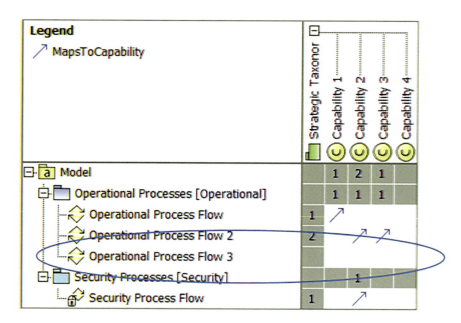

Fig. 5 Example UAF traceability between capabilities and operational activities

Fig. 6 Example UAF traceability between capabilities and strategic roadmap

government activity, where the government develops a reference architecture. Then the work transitions to the contractor to develop the detailed solution after the contract is awarded. On document-based acquisitions, the government typically creates independent document-based artifacts, and MA practitioners manually review for consistency and completeness.

During the S1 phase, it is possible for the government to use MBSE approaches to capture operational views, high level architecture views, and requirements and create traceability to requirements. Models can leverage architectural frameworks like DoDAF and UAF, which already have predefined terminology for creating operational, resource/system, and capability views. Additionally, they contain traceability relationships between the different views [17, 18]. Alternatively, SysML can be extended with custom stereotypes to capture custom operational, system design, and capability views and terminology [20, 21]. By using a MBSE model to capture this information, the traceability mechanisms in the model help ensure the model is complete and consistent. Additionally, MBSE tools can help enforce consistency within the views, which makes them easier to develop and maintain. This is illustrated in Fig. 7. The top diagram is a UAF operational connectivity diagram

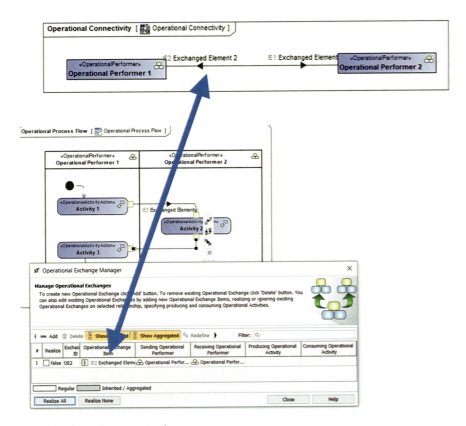

Fig. 7 UAF consistency example

that shows two operational performers and the exchanges between them. The lower figure shows an operational process flow, which show what operational activities the operational performers are doing along with the exchanges between the processes. When adding the exchanges to the diagram, this MSBE tool enables the modeler to select from the existing valid exchanges between these operational performers from a pop-up context window as shown below. The is just one example of how MBSE tools help enforce consistency between views. Additionally, requirements can be included or imported into the model and traced to the operational views. This can help with the requirements development effort.

In the S2 Develop Feasible Design and S3 Engineer the Solution core processes, the architecture is further developed and refined. The activities are aligned with the activities described in the section on Design Assurance. Thus they are discussed in more detail in section "Design Assurance Core MA Practice Best Practices."

Finally, the core activity A2 Request for Proposal (RFP)/Source Selection Award goal is "to develop the RFP and supporting Source Selection Plan and to conduct the actual source selection" [1]. On document-based approaches, the RFP is document containing the relevant information, and the source selection involves manually reviewing document-based proposals. On MBSE-based approaches, it is possible to include the model or portions of the model developed during the S1 phase as part of the RFP. It is also possible to request models as part of the RFP response. Using model artifacts as part of the RFP/Source Selection Award is still an emerging concept. A more detailed look at this topic is available in [22].

Requirements Analysis and Validation Core MA Process Best Practices

The Requirements Analysis and Validation Core MA Process involves Independent Technical Assessments (ITAs) on the SE requirements process. There are three core activities that are typically included in the Requirements Analysis and Validation Core MA Practice from the Mission Assurance Guide, which include Requirements Development, Requirements Validation, and Requirements Verification [1]. They can all benefit from new MBSE-based analysis techniques, so each of the core activities is described below in more detail.

First, the MA Requirements Development core activity involves Independent Technical Analysis (ITA) of requirements analysis, functional analysis and allocation, and synthesis [1]. On document-based SE approaches, MA practitioners primarily relies on manual top-to-bottom and bottom-to-top analysis techniques to identify orphaned, widowed, or derived requirements. These techniques are manual because the data needed to perform the ITAs on the requirements traces is typically located in multiple documents and multiple stove-piped requirements databases. For example, ITAs of top-to-bottom requirements traceability involves top-level system requirements documents such as Capabilities Development Document, CONOPS, and government or procuring agency directives and policy. MA practitioner performing the ITAs needed to manually pull the important data from these

29 MBSE Mission Assurance

documents for tracing. Then, they build traces to the lower level requirements, which are captured in documents and/or stored across multiple stove-piped requirements databases. During tracing process, orphaned, widowed, or missing derived requirements are manually identified and reported as part of the ITAs. It can be a lengthy and time-consuming process.

On a MBSE approach, MBSE MA practitioners have the potential to leverage more automated analysis techniques to identify for orphaned, widowed, or derived requirements. This is because MSBE models can be queried since MBSE modeling languages and architectural frameworks have mechanisms for capturing and tracing requirements. For instance, SysML contains requirement elements including the ≪allocate≫ relationship to create traceability into the system design and the ≪derive≫ relationship to create traceability to lower level requirements. SysML can also be extended with stereotypes to create custom categories [20]. The stereotypes can be used to define the different categories of requirements in a requirements hierarchy. The Unified Architecture Framework (UAF) and the DoD Architectural Framework (DoDAF) have mechanisms to capture and high level capabilities with multiple relationships to trace how it is realized through its various views of the system [17, 18]. Simple examples of these are graphically depicted in Fig. 8.

If the requirements are not authoritatively tracked in the MBSE tool, it is possible to import them into the MBSE model for analysis. MBSE tools and integration tools offer the ability to import requirement data directly from requirements databases [23–25] or through importing requirements in Requirements Interchange Format (ReqIF) and comma-separated value (CSV) format [26]. MBSE MA practitioners can leverage this functionality integrate data from multiple stove-piped requirements databases to create a complete picture for their review. Once the requirements

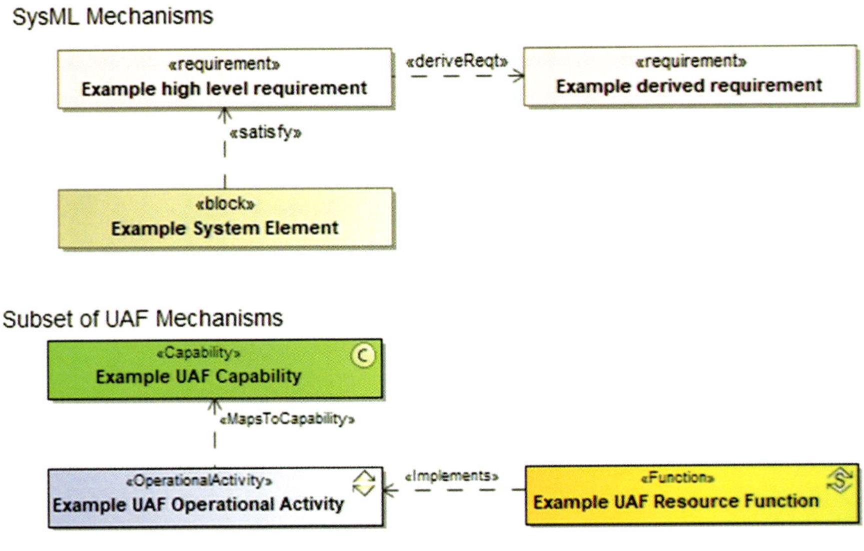

Fig. 8 Example MBSE requirements traceability mechanisms

traceability is in the MBSE model, MBSE MA practitioners can apply automated analysis techniques to query the model. Commercially available techniques include completeness queries [27], coverage metrics [28, 29], tables [30, 31], requirements matrices [32, 33], and requirements traceability visualizations [34, 35] to locate orphaned and widowed requirements. This will save the MBSE MA practitioners a significant amount of time on their ITAs and enable them to provide faster results. However, MBSE MA practitioners still need to manually validate the correctness of the requirements traces and manually identify missing derived requirements. MBSE MA practitioners can leverage the traceability views they created for their traceability reviews as well.

Second the MA Requirements Validation core activity includes evaluation of user operational scenarios and the establishment of design reference cases, evaluation of KPPs, and evaluation of architecture alternatives against operational scenarios and KPPs. The primary means to achieve this are through modeling and simulation [1]. On document-based SE approaches, the analysis techniques applied involve creating independent models and simulations to verify the operational scenarios and KPP. This commonly involved manual review of multiple documents to create the models and/or identify data for the simulations. The models and simulations are often not integrated so data from one model would need to be generated and fed into other models.

On MBSE efforts, there is the potential to integrate MBSE architectures with simulations used to analyze KPPs and architectural alternatives. SysML contains parametric diagrams and constraint blocks, to enable the capturing of constraint properties and their parameters [20]. Parametric diagrams can be used to create systems of equations that can constrain the properties of blocks and enable the computation of aggregate properties of a system. For example, in [21] orbital analysis is performed in a simulation tool. The parameters on a SysML parametric diagram are used to specify the inputs for the external simulation and parameters are also used to ingest outputs of the external simulation. This parametric diagram can be used to analyze different system alternatives for orbits. Parametric diagrams aren't the only option for enabling simulations. Some commercial MBSE tools offer simulation capabilities like executing parametric diagrams, executing activity-based behavior, and executing state-based behavior [36–39]. Additionally, some tools offer the ability to interface with mainstream mathematical analysis tools MATLAB and Simulink [36, 38]. From a MBSE MA perspective, it is possible to leverage these mechanisms. Currently, there are a variety of analysis techniques that MBSE MA practitioners could leverage, and new ones are still emerging. The analysis techniques vary from integrating model-based designs and externals simulations via parametric diagrams or MATLAB blocks to building executable models themselves [21, 40–43]. From an MBSE MA practitioner's perspective, it is possible to leverage these analysis techniques for use their evaluations of KPPs and architectural alternatives.

Finally, the requirements verification core MA activity involves establishing an engineering and program management consensus on the verification methods applied to each requirement, tracking tools, and the roles/responsibilities of

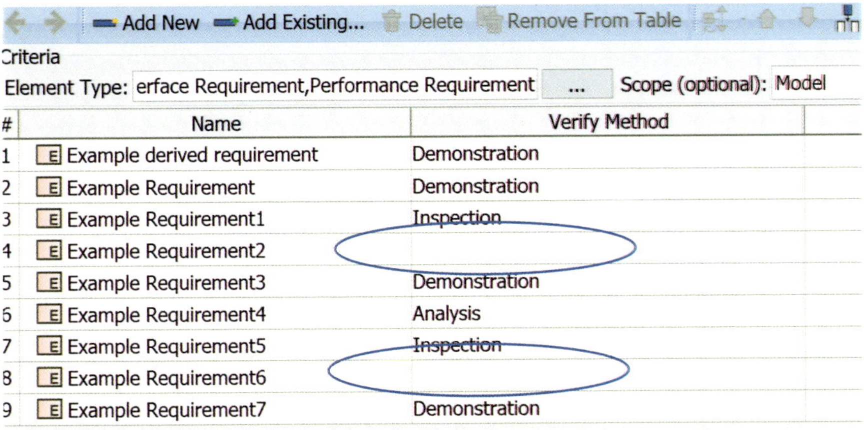

29 MBSE Mission Assurance

Fig. 9 Example query for verification method

organizations and individuals [1]. On document-based SE approaches, the analysis techniques employed are typically manual reviews of stove-piped requirements databases. Any requirements lacking a verification method need to be manually identified and tracked to ensure resolution.

On a MBSE-based approach, it is possible for all the verification method data to be integrated into a MBSE model. The SysML specification includes a verification method field on requirement elements [20]. MBSE MA practitioners can leverage automated analysis techniques on the model if this field is populated with authoritative data. If the verification method is not authoritatively tracked in the MBSE model, then MBSE MA practitioners can leverage MBSE and integration tools to integrate data from multiple stove-piped requirements databases to create a complete picture for their review. MBSE MA practitioners can leverage commercially available query functions in MBSE tools, such as table views [30, 31], to automate the analysis technique. Table views can be customized to display and analyze requirements without verification methods. A sample table view query from a MBSE tool is shown in Fig. 9, and it could be further filtered to only display the requirements lacking verification methods.

Design Assurance Core MA Practice Best Practices

The Aerospace Corporation's studies suggest that design issues account for 40 percent of on-orbit spacecraft anomalies [1]. Therefore, the goal of Design Assurance core MA process is to reduce these anomalies by finding and correcting design errors early in the life cycle when changes are less costly to make. The core activities in the Design Assurance core MA practice from the Mission Assurance Guide [1] that are impacted by MBSE are summarized in Table 2. Since the design is so strongly

Table 2 Core activities in Design Assurance core MA process

Core activity	Objective
Conduct Mission Design Analysis	Determine that the mission system is capable of operating as intended with sufficient margin to guarantee mission success
Audit the management of specifications, their communication, and design incorporation	Ensure there is a clear and auditable process of accurately translating requirements into a design
Examine design methodologies	Design methodologies should be examined to ensure design models are properly evaluated, reviewed, and documented
Control Documents	Ensure configuration management (CM) of program documentation is demonstrated
Perform Change Control	Examine the processes and include the mechanisms, whereby changes to documents are carried out
Conduct Independent Design Audits	Perform targeted independent audits at different levels of design maturity
Perform Test	Ensure verification of the implementation of design against the requirements

impacted by MBSE, all the core activities will be impacted by MBSE. These changes are described below in more detail.

First, the Conduct Mission Design core activity typically involves using analysis techniques to verify adequate mission planning for all operational conditions, to examine system level and integration requirements, to ensure the baseline reliability is preserved and requirements have been met [1]. On document-based approaches, the analysis techniques applied are largely ad hoc manual reviews of completeness and consistency across document-based artifacts, such as System Design Documents, ICDs, stove-piped requirements databases, and other design artifacts. Additionally, it may include mission analysis techniques to examine the physical constraints, such as whether or not the system has adequate weight and power.

On MBSE approaches, there is the potential to leverage automated analysis techniques to the Conduct Mission Design core activity. Ensuring all requirements have been met in the design can leverage the same requirements traceability analysis techniques as previously discussed in section "Requirements Analysis and Validation Core MA Process Best Practices." During the design phase, the traceability completeness can be quickly queried for lower level requirements and lower level design artifacts. However, this traceability will still need to be manually reviewed for correctness.

Another set of analysis techniques that can be applied during the Conduct Mission Design core activity are model-based simulations. For example, designs can also be analyzed for functional flow correctness. Commercial MBSE tools include the ability to animate and/or simulate the control flow through a model [37–39]. MBSE MA practitioners can use these analysis techniques to validate functional flows through model to ensure the system behaves as expected, as

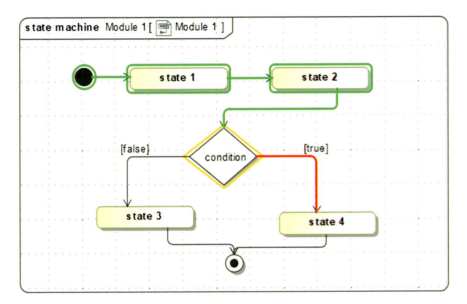

Fig. 10 Example of model animate to validate functional flow

shown in Fig. 10. Past locations are shown in green, the current location is in yellow, and the next location is in red. If the simulated path deviates from expected path, then the issue can be noted and provided as part of the assessment. Another analyses that can be leveraged are parametric diagrams and constraints to examine the design constraints. MBSA MA practitioners have the option to review outputs from model developer's analysis techniques if they are included on the MBSE approach or apply them independently to the models and augment the models with the required data needed for analysis.

Another set of analysis techniques that can be applied to the Conduct Mission Design core activity include model based non-functional analysis techniques. Model-based analysis techniques for reliability, availability, maintainability, and safety (RAMS) analysis are on the rise [16, 43–47], and approaches for cyber and resilience are starting to emerge [43, 48, 49]. MBSA MA practitioners have the option to review outputs from analysis techniques if they are included on the MBSE approach or apply them independently to the models and augment the models with the required data needed for analysis. The latter could be time-consuming depending if the data needed for the analysis is not readily available.

One other aspect of the Design Core Activity that can be improved with MBSE is communication between MBSE MA Practitioners and the Model Developers. MBSE MA practitioners can leverage SysML comments blocks and anchor them to specific design elements to capture their comments [20]. The comments can then be tracked, queried, and resolved directly in the MBSE modeling tool rather than a separate external comment resolution matrix. The SysML comments can also be customized to capture all the desired fields needed for a comment resolution matrix. One existing approach for doing this is in [50].

The next core activity is Audit the management of specifications, their communication, and design incorporation. This core activity involves ensuring an auditable process of accurately translating requirements into a design [1]. On document-based SE approaches, the analysis techniques applied are largely manual reviews for completeness of the process documentation for the program, as well as manually checking requirements and specifications during development to ensure conformance to the process.

On an MBSE-based approach, the process of checking the requirements and specifications for conformance to the process can leverage automated analysis techniques. Using commercially available features to query the model data, such as table and matrices [29–33], the requirements traceability can be easily checked to see if the process is being followed. For example, Fig. 11 shows a simplified example of a table view created to inspect the traceability. In this example, the requirements should use the ≪derive≫ relationship for any derived requirements and the ≪satisfy≫ relationship to trace it to the design. In this example, the MBSE MA practitioner can quickly see that Requirements D5, D6, D8, D9, and D10 violate this traceability. While not shown in this table, it is also possible for the analyst to create queries to check for ≪refine≫ relationships between use cases and requirements to ensure proper mapping between these artifacts. The MBSE MA practitioner can then document the issues and work to help remedy the problem.

#	Name	Satisfied By	Derived
1	R Requirement D1	Design Block 1	R 3 Requirement 2
2	R Requirement D2	Design Block 1	R 16 Requirement 3
3	R Requirement D3	Design Block 2	R 4 Requirement 1
4	R Requirement D4	Design Block 1	R 4 Requirement 1
5	R Requirement D5		R 4 Requirement 1
6	R Requirement D6		R 3 Requirement 2
7	R Requirement D7	Design Block 2	R 3 Requirement 2
8	R Requirement D8	Design Block 1	
9	R Requirement D9	Design Block 2	
10	R Requirement D10	Design Block 2	
11	R Requirement D11	Design Block 1	R 16 Requirement 3
12	R Requirement D12	Design Block 1	R 16 Requirement 3
13	R Requirement D13	Design Block 1	R 17 Requirement 5
14	R Requirement D14	Design Block 2	R 17 Requirement 5

Fig. 11 Example traceability relationship table view

The next two core MA activities are Control Documents and Perform Change Control. The Control Documents and Perform Change Control are primarily focused on ensuring that program documentation changes are controlled and managed [1]. On document-based SE approaches, the CM and change control processes are focused around managing different versions of a document and the changes that are in each version. On the MA side, the analysis technique commonly used is manual inspection of the CM and Change Control process documentation and output from these processes.

On MBSE-based programs, these core MA activities should be expanded to not only examine documentation, but models as well. Ideally, models on a program will be controlled on commercially available model management tools, and the changes made to the model are governed by a change control process. Examples of model management tools and processes are available in [51–53]. On the MBSE MA side, the analysis techniques can leverage log files from the model management tools to assess conformance to the process. For example, branching and merging functionality provided by model management tools can be analyzed to ensure the configuration management process is followed and the model baseline is being controlled. A branch-merge log for a model with all the changes occurring on the main baseline trunk rather than development branches would be an indicator that CM branching and merging processes are not being followed or are not properly established. This is graphically depicted in Fig. 12 where all changes are on the trunk and development/maintenance branches do not exist. This analysis technique could be done by manual inspection, or it has the potential to be automated by creating custom tools that directly query and analyze data from the model management tools via their development application program interfaces (APIs) [54]. Change tracking tools such as JIRA can also be used in conjunction with the modeling tools for creating and maintaining change request matrices (CRMs).

The fourth core activity is the Conduct Independent Design Audits. It includes performing targeted independent audits at different levels of design maturity throughout the life cycle [1]. On document-based SE approaches, the analysis

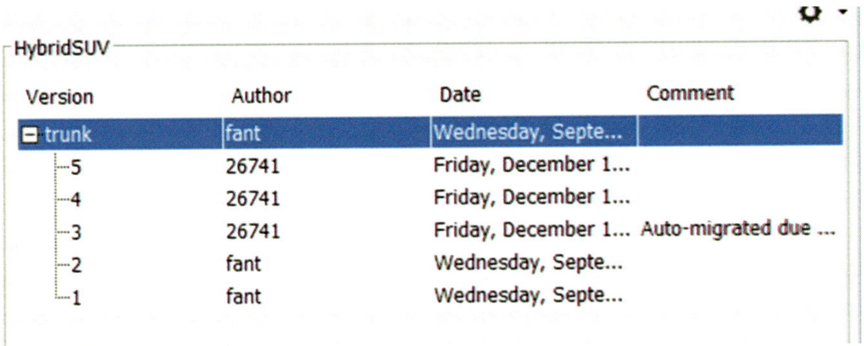

HybridSUV			
Version	Author	Date	Comment
trunk	fant	Wednesday, Septe...	
5	26741	Friday, December 1...	
4	26741	Friday, December 1...	
3	26741	Friday, December 1... Auto-migrated due ...
2	fant	Wednesday, Septe...	
1	fant	Wednesday, Septe...	

Fig. 12 Examples of branching

technique applied is usually a manual review of design artifacts, both official deliverables and unofficial deliverables.

On MBSE-based programs, the independent design audit can directly analyze the design and leverage the analysis techniques described in the Conduct Design core activity section. The audits could also be short quick assessments to monitor and track model metrics for development. MBSE metrics for managing MBSE is still an evolving area [55], and most metrics are focused on return on investment metrics. Metrics to measure the MBSE development progress are emerging, such as Model Assurance Level (MAL) Assessments for System and Software Models [56–58].

Finally, the Perform Test core activity involves the verification of the implementation of design against the requirements occurs, and it is important that design engineering is checking to ensure design intent is truly being demonstrated [1]. On document-based approaches, this is typically done through manual review of the documentation and design.

On MBSE-based approaches, there are several analysis techniques that can be applied depending on how detailed of a verification is needed. First, requirements traceability analysis techniques and simulations of functional flows previously described help to ensure the requirements are met and the system design executes as planned. Second, some MBSE tools have the ability to compute system parameters and evaluate them against the requirement. This is accomplished by leveraging SysML constraint blocks and has the tool calculate system parameter, such as total mass. The MBSE tool can also automatically evaluate parameters against the system requirement and give the verdict on whether it is satisfied or not [59]. Advanced analysis techniques are emerging such as [60].

Manufacturing Assurance Core MA Process

The objective of the Manufacturing Assurance core MA process is "to (1) to ensure the manufacturing processes are able to produce HW that meets the design requirements and (2) to translate the design into a reliable, durable manufactured item using manufacturing processes that are highly repeatable and error free" [1]. There are some proposed MBSE approaches for the manufacturing domain, such as [43, 61]. However, they are not mainstream yet. Therefore, most of the core MA activities are focused around the manufacturing materials and processes and thus will unlikely benefit from MBSE. On larger DE approaches where models are integrated with other hardware and manufacturing data, then more automated analysis techniques maybe possible. However, a full DE approach is outside the scope of this paper.

Integration, Test, and Evaluation Core MA Process

Integration, test, and evaluation (IT&E) is a "broad process whose purpose is to verify end item requirements satisfaction (e.g., functionality, performance, design/ construction, interfaces, and environment) at all levels of assembly as those end

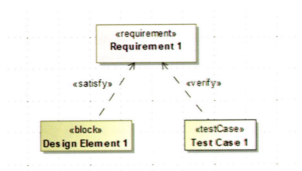

Fig. 13 Simplified test planning

items (e.g., units) form a system" [1]. The MA core activities in this area are primarily focused on reviewing and ensuring the appropriate testing, especially development, qualification, and acceptance testing, is appropriately planned, performed, witnessed, and documented. On document-based approaches, this typically involves manual reviews and/or participation in evaluating that the contractor-provided evidence of completion (EOC) satisfies the requirements and specification baseline and evaluates analysis, simulation, inspection, and test results to determine readiness to proceed to subsequent test or program activities and periodic independent review team review and assess the overall test program as well as specific test results and the resolutions of test failures and participate/attend Test Readiness Reviews (TRRs), Test Exit Reviews (TERs), and Failure Review Boards (FRB) [1].

On MBSE-based approaches, many of these activities will remain the same such as attendance and participating in TRRs, TERs, and FRBs. However, MBSE analysis techniques can be used to help in other areas. First, MBSE analysis techniques can be leveraged to help in the test planning. SysML offers the test case elements to capture the steps to perform in a specific test cast, a verifies relationship to capture traceability between test cases and requirements, and the satisfies relationship between requirements and design element [20]. A simplified example of this traceability is shown in Fig. 13. Some specific approaches to using these SysML elements are captured in [62, 63]. If these are leveraged in the MBSE model, then the MBSE MA practitioners can create traceability tables and matrices to quickly query the traceability for completeness. The correctness of the test case themselves and the traceability would remain a manual review.

Operations Readiness Assurance (ORA) Core MA Process

Operational readiness assurance (ORA) covers all the activities required to transport, receive, accept, store, handle, test, and deploy a system so that it is conducted safely and successfully. ORA involves readiness planning, activation, and mission operation, and it occurs throughout the program life cycle [1]. Some of the ORA reviews, such as review transporting and storing readiness, are outside the scope of MBSE.

However, there are some ORA reviews that are within the scope of MBSE such as review test plans for readiness. For MBSE-based ORA reviews, the same analysis techniques that were used during development can be applied. But rather than monitor them for progress, they are assessing them for completeness and readiness. These analysis techniques are discussed in previous sections. Therefore, they will not be described in detail here.

Operations, Maintenance, and Sustainment Core MA Process

The Operations, Maintenance, and Sustainment core MA process is focused on perform engineering assessments, detail changes, and impacts and addresses subsequent changes to the system during the operations phase of the program life cycle [1]. The MA activities are similar to the MA activities during development, which are described in previous sections. Rather than monitoring them for progress and completeness, they will be examined for impact analysis. For example, if a data exchange or interface is changing, traceability tables and matrices [30, 33] can be used to identity downstream impacted system elements. This is shown in Fig. 14 which contains a traceability table for UAF resource exchanges. If Data Element 1 is changing, from this traceability table, the MBSE MA practitioner can quickly see that this change impacts the interface between Software 1 and Software 2, as well as Software 1 and Resource Artifact 1. Since the analysis techniques are very similar, they will not be discussed in detail here.

However, a new core activity should be added to specifically address maintaining models. If the MBSE model contains authoritative data on an architecture or design, it needs to be updated to remain consistent with the operational system and evolve as changes to the operational system are made. If the operational system and the MBSE deviate, the MBSE model no long contains authoritative data about the operational system and thus will no longer be providing value to the system.

MA Reviews and Audits

The purpose of conducting mission assurance (MA) reviews is to assess the technical maturity within a program, evaluate program risks and opportunities, understand stakeholder expectations, and ensure readiness for the next phase in the overall program life cycle of events and milestones [1]. The MA analysis techniques to assess the system artifacts at given milestones are largely the same as the ones discussed in previous sections. For example, audit that is performed during SRR can

#	Exchange ID	Resource Exchange Item	Sending Resource	Receiving Resource	Producing Function	Consuming Function
1	RI1	DE1 Data Element 1	Software 1	Software 2	Function 1	Function 2
2	RI2	DE1 Data Element 1	Software 1	Resource Artifact 1	Function 1	Function 3
3	RI3	DE2 Data Element 2	Software 2	Resource Artifact 1	Function 2	Function 4

Fig. 14 Example impact analysis

leverage the analysis techniques discussed in "Requirements Analysis and Validation Core MA Process Best Practices." Therefore, the techniques will not be repeated in this section.

Illustrative Examples (A Couple to Show Uses of Approach)

This section will highlight some of the MBSE MA analysis techniques using the FireSat II MBSE model provided from the Architecting Spacecraft with SysML book [21, 63]. The mission objective for the FireSat II case study is to detect and monitor fires in the USA and Canada. The FireSat model is implemented in SysML, and the diagrams were built following the Mission & System Specification and Design Process described in the Architecting Spacecraft with SysML book [21]. In this chapter, we will be discussing it as if it were a government acquisition and highlight some core MA processes that can be leveraged during the acquisition life cycle. We added additional data and we purposely removed some traceability from the FireSat II MBSE model to show how it can be identified with MBSE MA analysis techniques.

Program Assurance Example

The Fire Sat II case study begins with the pre-acquisition Program Assurance core MA process. First, the MBSE MA practitioner is tasked with assuring the requirements development and the concept of operations behavior since that data is included in the FireSat II MBSE model. The FireSat II MBSE model includes Mission Object Use Case Diagrams, Mission Requirements Diagrams, Refined Mission Requirements Diagrams, Mission Element Structure, Operational Use Cases, and Mission Behavior Activity Diagrams. The program chose to also add traceability relationships between the design elements in these views as shown in Fig. 15.

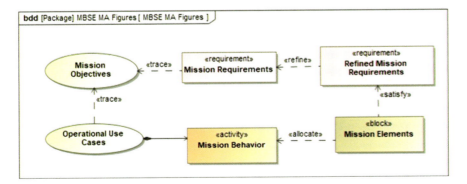

Fig. 15 MBSE artifact traceability

The MBSE MA practitioner can leverage traceability in the MBSE model to create traceability analysis techniques described in section "Program Assurance Core MA Process Best Practices." First, the MBSE Practitioner will assess the traceability of the mission requirements. So, they will create a traceability table for mission requirements showing their upstream traceability to mission objectives and downstream traceability to refined mission requirements and mission elements. A shown in Fig. 16, the MBSE MA practitioner can quickly identify gaps in the requirements traceability in the mission requirement.

This traceability table is created using the set of mission requirements as the starting point. So, it will only identify gaps related to the mission requirements, and it will not show any refined mission requirements that are not traced to mission requirements nor mission use cases not traced to mission requirements. To get that information, the MBSE MA practitioner needs to create a separate traceability table. For example, Fig. 17 shows another traceability table using the set of mission objectives as a starting point. This table contains the mission objectives upstream traceability to CONOPS operational use cases and downstream traceability to mission requirements. From this table, the MBSE MA practitioner can see that the

#	Traced From	Name	△ Refined By	Satisfied By
1	Provide Forest Fire Data in Near Real Time Detect and Monitor Forest Fires in US and Cana	Performance		
2	Detect and Monitor Forest Fires in US and Cana	Risk		
3	Detect and Monitor Forest Fires in US and Cana	Political		
4	Detect and Monitor Forest Fires in US and Cana	Development Constraints		
5	Detect and Monitor Forest Fires in US and Cana	Coverage	5.1.5 Coverage 5.1.3 Detection Time	Spacecraft
6	Provide Forest Fire Data in Near Real Time Detect and Monitor Forest Fires in US and Cana	Data Distribution	5.7 Data Distribution & Delivery	Ground Station Relay Sattelite Spacecraft
7	Detect and Monitor Forest Fires in US and Cana Provide Forest Fire Data in Near Real Time	Data Content, Form, and F	5.7 Data Distribution & Delivery	Ground Station Relay Sattelite Spacecraft
8	Detect and Monitor Forest Fires in US and Cana Provide Forest Fire Data in Near Real Time	User Equipment	5.7 Data Distribution & Delivery	Ground Station Relay Sattelite Spacecraft
9	Detect and Monitor Forest Fires in US and Cana Provide Forest Fire Data in Near Real Time	Timeliness	5.1.3 Detection Time	

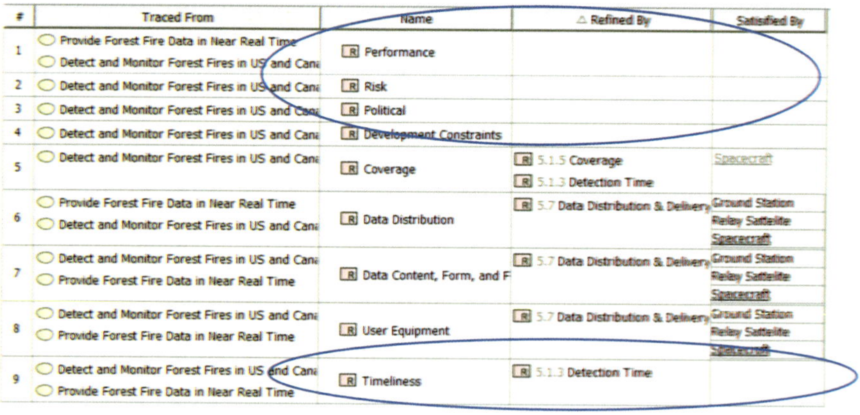

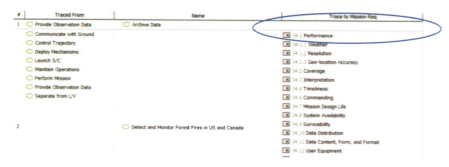

Fig. 16 Traceability table for mission requirements

#	Traced From	Name	Trace to Mission Req
1	Provide Observation Data	Archive Data	
	Communicate with Ground Control Trajectory Deploy Mechanisms Launch S/C Maintain Operations Perform Mission Provide Observation Data Separate from L/V		34.1 Performance 34.1.1 Weather 34.1.2 Resolution 34.1.3 Geo-location Accuracy 34.2 Coverage 34.3 Interpretation 34.4 Timeliness 34.6 Commanding 34.7 Mission Design Life 34.8 System Availability
2		Detect and Monitor Forest Fires in US and Canada	34.9 Survivability 34.10 Data Distribution 34.11 Data Content, Form, and Format 34.12 User Equipment

Fig. 17 Traceability table for mission objectives

Archive data mission objective does not have traceability to mission requirements. The MBSE MA practitioner creates other traceability tables as needed to identify issues with the other requirements and use cases. Once all the gaps are identified, the MBSE MA practitioner can work to get them resolved.

Design Assurance Example

After contract is awarded, the contractor establishes the system and subsystem specification and designs. The MBSE MA practitioner will follow the Design Assurance core MA process to find anomalies by finding and correcting design errors early in the life cycle when changes are less costly to make.

First, the MBSE MA practitioners can leverage the MAL analysis technique to monitor the model development, as previously described in "Design Assurance Core MA Practice Best Practices." The MBSE MA practitioner performs a MAL assessment on the model to quickly gather some metrics on the effort to decide where they should begin, shown in Fig. 18. The MBSE MA practitioner can quickly see that to date the modeling effort is focused on the requirements and structural analysis of the system. Thus, the MBSE MA practitioner starts reviewing the structural diagrams and requirements.

The MBSE MA practitioner can ensure the requirements are being met by creating traceability tables and matrices for system and subsystem specifications back to upstream requirements and downstream system design elements. They can use traceability table analysis techniques again as discussed in section "Program Assurance Example." For example, Fig. 19 shows the traceability of the payload sensor specification to its structural design elements. The MBSE MA practitioner

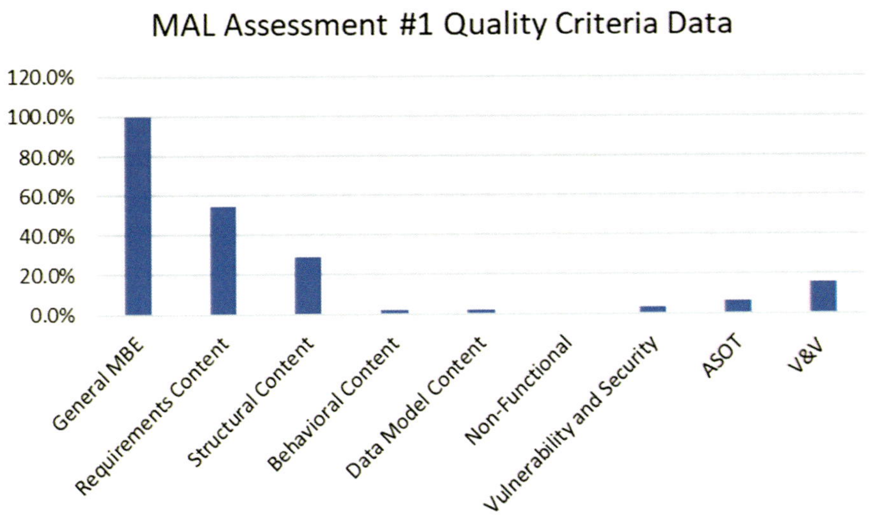

Fig. 18 MAL assessment #1 data

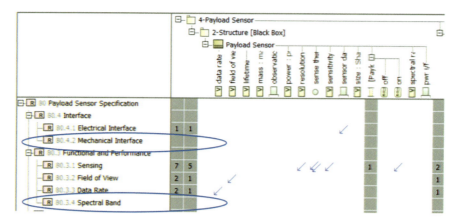

Fig. 19 Payload subsystem traceability matrix to design

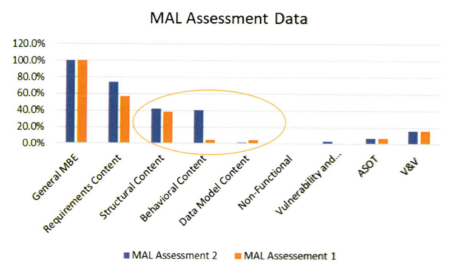

Fig. 20 MAL assessment data #2

can quickly see the mechanical interface and spectral band are not traced and they can work to get the issue resolved. They should also manually review the traceability for correctness.

Later on in the life cycle, the MBSE MA practitioner performs a second MAL assessment and compares it against the first MAL assessment to understand the progress that is being made on the model. The results of the assessment are shown in Fig. 20. The MBSE MA practitioner can quickly see that behavioral views are developed now, so they can now assess those views, as well as rechecking the previously created traceability tables and matrices to ensure the traceability issues previously identified were resolved.

29 MBSE Mission Assurance

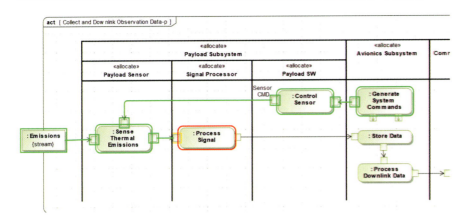

Fig. 21 Example functional flow validation with simulation

#	Scope	Date	Requirement	Covered By Design	Covered By Design Percentage	Covered By Test Cases	Covered By Test Cases Percentage
1	Data	2014.12.15 11.55:38	35		92.1053	0	0

Fig. 22 Requirement coverage metrics

Next, the MBSE MA practitioner begins to validate functional behavior or the design. To accomplish, the MBSE MA practitioner will leverage the model-based simulation of flows analysis technique described in Design Assurance Core MA Practice Best Practices in section "Design Assurance Core MA Practice Best Practices." This is graphically depicted Fig. 21, where activities that are executing are highlighted in green, the current location is highlighted in yellow, and the next location is highlighted in red. The flow can be traced to ensure it executes and expected. Any issues such as incomplete flows, activities with no exits, or invalid flows should be noted and submitted for resolution. In fact, these high level activity diagrams can essentially be used as a table of contents to more detailed sequence or state diagrams, thus allowing the model to be simulated at increasingly finer granularity as maturity progresses.

IT&E Assurance Example

The MBSE MA practitioners involved on the IT&E effort are monitoring and assessing the test planning that is occurring in the MBSE model. In the FireSat II model, the test cases are being designed in the model; they are being traced back to the requirements and are sequenced to show the order in which they will be executed [21]. During the planning process, MBSE MA practitioners can leverage analysis techniques such as coverage metrics to monitor the progress. For example, Fig. 22 shows a coverage table from the beginning of the IT&E effort. The MBSE MA

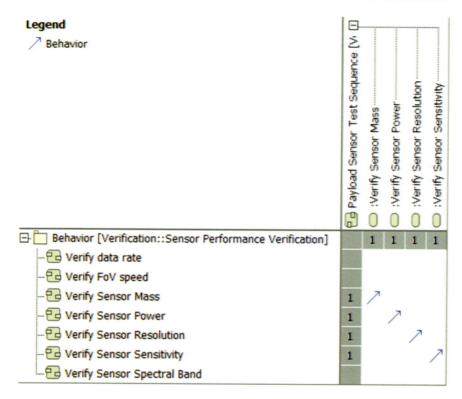

Fig. 23 Test cases in sequencing traceability table

practitioner can see that the requirements are traced to the design for 92% of the requirements and the covered by test cases hasn't started yet.

Since the FireSat II MBSE model also contains test case sequencing, the MBSE MA practitioner can also query the MBSE model to see if all the test cases are included in the test case sequence. The analysis technique chosen is a traceability matrix to show where test cases are included in the test sequences. Using the view in Fig. 23, the MBSE MA practitioner can quickly see that three of the test cases for the payload subsystem still need to be included in a test sequence. The MBSE MA can then work the issue and ensure these test cases get planned.

Chapter Summary

In summary, this chapter discussed current state of the art for MBSE MA. It contains discussions comparing and contrasting how the core MA processes and associated analysis techniques on programs utilize MBSE with traditional document-based approaches. All the MA core processes that can be positively impacted by MBSE

29 MBSE Mission Assurance

Table 3 Summary of core MA process impacted by MBSE

Core MA process	Core activity	Core activity change	New analysis technique
Program Assurance	P1 Program Definition		X
	S1 Develop Architecture/ Requirements/ Capabilities		
Requirements Analysis and Validation	Requirements Development		X
	Requirements Validation		X
	Requirements Verification		X
Design Assurance	Conduct Mission Design Analysis		X
	Audit the management of specifications, their communication, and design incorporation		X
	Control Documents and models	X	X
	Perform Change Control on models	X	X
Integration Test, and Evaluation			
Operations Readiness Assurance (ORA)	Maintain models	X	
Operations, Maintenance, and Sustainment	Maintain models	X	X
	All		X
MA Reviews and Audits	All		X

are summarized in Table 3. Furthermore, this chapter also referenced some specific MBSE analysis techniques that can be leveraged by MBSE MA practitioners and demonstrated them on a case study.

MBSE MA is evolving, so future work should include updating the chapter with new information as it develops, such as new MBSE MA guides and handbooks, MBSE- and DE-based analysis techniques, comprehensive case studies on MBSE MA, and lessons learned on MBSE MA.

Cross-References

▶ MBSE for Acquisition
▶ MBSE Validation and Verification
▶ Model-Based Mission Assurance/Model-Based Reliability, Availability, Maintainability, and Safety (RAMS)
▶ Model-Based System Architecting and Decision-Making

References

1. S. B. Guarro, G. A. Johnson-Roth and W. F. Tonsey, Mission Assurance Guide Revision B, El Segundo, CA: The Aerospace Corporation, 2012.
2. Scaled Agile, "Full SAFe Agile," 2020. [Online]. Available: https://www.scaledagile framework.com/#. [Accessed December 2020].
3. DoD, "DoD Digitial Engineering Strategy," DoD, 2018.
4. D. Tolle, "MBSE, PLM and the Digital Thread: Market Update," in *INCOSE Internaltion Workshop 2019*, 2019.
5. INCOSE, "INCOSE Systems Engineering Vision 2020," INCOSE, 2020.
6. INCOSE, "INCOSE Systems Engineering Vision 2007," INCOSE, 2007.
7. No Magic, "MagicDraw 19.0 LTR Documentation: Importing data from other UML tools and formats," [Online]. Available: https://docs.nomagic.com/display/MD190/Importing+data+from +other+UML+tools+and+formats. [Accessed December 2020].
8. IBM, "IBM Knowledge Source: Exchanging model data by using XMI," 2020. [Online]. Available: https://www.ibm.com/support/knowledgecenter/SSB2MU_8.3.0/com.ibm.rhp. importing.doc/topics/rhp_c_dm_xml_metadata_interchange_and_rhp.html. [Accessed December 2020].
9. Sparx Systems, "Enterprise Architect User Guide: XMI Import and Export," 2020. [Online]. Available: https://sparxsystems.com/enterprise_architect_user_guide/14.0/model_publishing/ importexport.html. [Accessed December 2020].
10. Dassualt Systems, "Cameo Inter-Op Product Page," [Online]. Available: https://www.nomagic. com/products/cameo-inter-op#intro.
11. M. Wheaton, "TOR-2017-01695 Mission Assurance Considerations for Model-Based Engineering for Space Systems," The Aerospace Corporation, 2017.
12. MITRE, "Systems Engineering Guide," MITRE, 2014.
13. DOD, "Department of Defense Mission Assurance Strategy," DOD, 2012.
14. E. Pawlikowski, "Mission Assurance—A Key Part of Space Vehicle Launch Mission Success," *High Frontier,* 2008.
15. J. Evans, S. Cornford and M. Feather, "Model Based Mission Assurance (MBMA): NASA's Assurance Future," in *2016 Annual Reliability and Maintainability Symposium (RAMS)*, Tucson, AZ, USA, 2016.
16. S. Cornford and M. Feather, "NASA/CR—2016–219272 Model Based Mission Assurance in a Model Based Systems Engineering (MBSE) Framework: State-of-the-Art Assessment," NASA, 2016.
17. OMG, "Unified Architecture Framework Specification Version 1.1," OMG, 2020.
18. DoD Chief Information Office, "DoD Architectural Framwork Version 2.02," DoD, 2010.
19. Dassault Systems, "UAF Plugin User's Guide," [Online]. Available: https://docs.nomagic.com/ display/UAFP190/User+Guide. [Accessed December 2020].
20. OMG, "System modeling language (SysML) Specification Version 1.2," Object Mangement Group (OMG), 2010.
21. S. Friedenthal and C. Oster, Architecting Spacecraft with SysML: A Model-based Systems Engineering Approach, CreateSpace Independent Publishing Platform, 2017.
22. A. Hoheb, "ATR-2019-01782 Model-Centric Source Selection," The Aerospace Corp, 2019.
23. IBM, "Managing requirements with Rhapsody Gateway and DOORS," [Online]. Available: Managing requirements with Rhapsody Gateway and DOORS. [Accessed December 2020].
24. Dasaullt Systems, "Introducing Cameo DataHub," [Online]. Available: https://docs.nomagic. com/display/CDH190/Introducing+Cameo+DataHub. [Accessed December 2020].
25. Intercax, "Syndeia Product Page," [Online]. Available: http://intercax.com/products/syndeia/. [Accessed December 2020].
26. OMG, "Requirements Interchange Format (ReqIF) Specification," 2020. [Online]. Available: https://www.omg.org/reqif/. [Accessed December 2020].

29 MBSE Mission Assurance

27. Dassault Systems, "Tracking Unlinked Requirements in Cameo Systems Modeler 19," [Online]. Available: https://docs.nomagic.com/display/MD190/Tracking+unlinked+require ments. [Accessed December 2020].
28. IBM, "Lesson 5: Linking requirements to Rational Rhapsody model elements," [Online]. Available: https://www.ibm.com/support/knowledgecenter/SSB2MU_8.2.0/com.ibm.rhp. doors.tutorial.doc/topics/lesson5_apply.html. [Accessed December 2020].
29. Dassault Systems, "Metrics," [Online]. Available: https://docs.nomagic.com/display/ CRMP190/Metrics. [Accessed December 2020].
30. Dasaullt Systems, "Tables," [Online]. Available: https://docs.nomagic.com/display/MD190/ Tables. [Accessed December 2020].
31. IBM, "How to set diagram elements to appear in the Table view of a Rational Rhapsody model," [Online]. Available: https://www.ibm.com/support/pages/how-set-diagram-elements-appear- table-view-rational-rhapsody-model. [Accessed December 2020].
32. Dasault Systems, "Requirements Matrices," 2020. [Online]. Available: https://docs.nomagic. com/display/SYSMLP190/Requirement+matrices. [Accessed December 2020].
33. IBM, "Creating Matrix Layouts," [Online]. Available: https://www.ibm.com/support/ knowledgecenter/SSB2MU_8.2.0/com.ibm.rhp.matrix.doc/topics/rhp_t_dm_creating_matrix_ layouts.html. [Accessed December 2020].
34. Dassault Systems, "Relation Maps," 2020. [Online]. Available: https://docs.nomagic.com/ display/MD190/Relation+Map. [Accessed December 2020].
35. M. Jackson and M. Wilkerson, "MBSE-driven Visualization of Requirements Allocation," in *IEEE Aerospace Conference,*, 2016.
36. Dassault Systems, "ParaMagic Plugin," [Online]. Available: https://www.nomagic.com/ product-addons/magicdraw-addons/paramagic-plugin. [Accessed December 2020].
37. Dassault Systems, "Cameo Simulation Tookit product page," [Online]. Available: https://www. nomagic.com/product-addons/magicdraw-addons/cameo-simulation-toolkit#features. [Accessed December 2020].
38. IBM, "Simulating use cases," [Online]. Available: https://www.ibm.com/support/ knowledgecenter/SSB2MU_8.2.0/com.ibm.rhp.sysml.doc/topics/t_simulatingusecases.html. [Accessed December 2020].
39. Sparx Systems, "Dynamic Simulations," [Online]. Available: https://sparxsystems.com/enterprise_ architect_user_guide/15.2/model_simulation/how_it_works2.html. [Accessed December 2020].
40. R. Peak, R. Burkhart, S. Friedenthal, M. Wilson, M. Bajaj and I. Kim, "Simulation-Based Design Using SysMLPart 1: A Parametrics Primer," in *INCOSE International Symposium*, 2007.
41. R. Karban, A. Crawford, G. Trancho, M. Zamparelli, S. Herzig, I. Gomes, M. Piette and E. Brower, "The OpenSE Cookbook: a practical, recipe based collection of patterns, procedures, and best practices for executable systems engineering for the Thirty Meter Telescope," in *SPIE Astronomical Telescopes + Instrumentation,*, 2018.
42. J. Hodge, K. Duncan, M. Zimmerman, R. Drupp, M. Manno, D. Barrett and A. Smith, "An integrated radar model solution for mission level performance and cost trades," in *Disruptive Technologies in Sensors and Sensor Systems;*, 2017.
43. A. Madni, B. Boehm, R. Ghanem, D. Erwin and M. Wheaton, Disciplinary Convergence in Systems Engineering Research, Springer International Publishing, 2018.
44. Y. Zhou, Z. Ren and R. Wang, "Principle and method of integrating reliability design in the MBSE proces," in *New Trends in Civil Aviation*, 2018.
45. M. Hecht, A. Chuidian, T. Tanaka and R. Raymond, "Automated Generation of FMEAs using SysML for Reliability, Safety, and Cybersecurity," in *Annual Reliability and Maintainability Symposium (RAMS)*, 2020.
46. Z. Huang, R. Hansen and Z. Huang, "Toward FMEA and MBSE Integration," in *Annual Reliability and Maintainability Symposium (RAMS)*, 2018.
47. E. Brusa, "Digital Twin: Towards the Integration Between System Design and RAMS Assess- ment Through the Model-Based Systems Engineering," *IEEE Systems Journal,* 2020.

48. A. Dwivedi, "Implementing Cyber Resilient Designs through Graph Analytics Assisted Model Based Systems Engineering," in *IEEE International Conference on Software Quality, Reliability and Security Companion (QRS-C)*, 2018.

49. A. Barreto and P. Costa, "Cyber-ARGUS – A mission assurance framework," *Journal of network and computer applications*, 2019.

50. J. Fant, "TOR-2021-00406 Leveraging built-in Rhapsody features to generate and manage model assessment comments," in *The Aerospace Corp*, 2020.

51. "Teamwork Cloud Product Page", [Online]. Available: https://www.nomagic.com/products/teamwork-cloud. [Accessed December 2020].

52. IBM, "Rhapsody Model Manager Product Page", [Online]. Available: https://www.ibm.com/support/knowledgecenter/en/SSB2MU_9.0.1/com.ibm.rational.rmm.overview.doc/topics/rhp_c_rhapsody_model_manager.html. [Accessed December 2020].

53. E. Parrott and L. Spayd, "Configuration and Data Management of the NASA Power Propulsion Element MBSE Model(s)", in *IEEE Aerospace Conference*, 2020.

54. Dassualt Systems, "Teamwork Cloud 19.0 REST APIs", [Online]. Available: https://docs.nomagic.com/display/TWCloud190SP1/REST+APIs. [Accessed December 2020].

55. OMG, "Methodology and Metrics", [Online]. Available: http://www.omgwiki.org/MBSE/doku.php?id=mbse:methodology#Metrics. [Accessed 2020 December].

56. J. Fant and R. Pettit, "Model Assurance Levels (MALs) for Managing Model-based Engineering (MBE) Development Efforts", in *7th International Conference on Model-Driven Engineering and Software Development*, 2017.

57. J. Fant, R. Pettit and D. Gayak, "A Quantitative Approach for Calculating Model Assurance Levels", in *IEEE 22nd International Symposium on Real-Time Distributed Computing (ISORC)*, 2019.

58. J. Gaskell and C. Harrison, "Improved System Engineering Technical Review's Entrance/Exit Criteria with Model Maturity Metrics", in *International Symposium on Systems Engineering (ISSE)*, 2019.

59. Dassualt Systems, "Requirements verification", 2020. [Online]. Available: https://docs.nomagic.com/display/SYSMLP190/Requirements+verification#:~:text=Systems%20Modeling%20Language%20(SysML)%20is,to%20support%20analysis%20and%20verification. [Accessed December 2020].

60. M. Rahim, A. Hammad and M. Ioualalen, "A methodology for verifying SysML requirements using activity diagrams", *Innovations in Systems and Software Engineering*, 2016.

61. O. Batarseh, L. Mcginnis and J. Lorenz, "MBSE Supports Manufacturing System Design", in *INCOSE International Symposium*, 2012.

62. B. Selvy, C. Claver and G. Angeli, "Using SysML for Verification and Validation Planning on the Large Synoptic Survey Telescope (LSST)", in *SPIE Astronomical Telescopes + Instrumentation*, 2014.

63. Friedenthal & Oster, "Architecting Spacecraft with SysML Models Page", Friedenthal & Oster, 2017. [Online]. Available: http://sysml-models.com/spacecraft/models.html.

Julie S. Fant, PhD, Senior Engineering Specialist, The Aerospace Corporation. Dr. Julie Fant has over 15 years of experience in the software engineering industry with expertise in large-scale, mission-critical software intensive systems. Dr. Fant has expertise in software and systems model-based engineering, model-based analysis, Unified Modeling Language (UML), SysML, architectural frameworks, mission assurance, and software product lines. Dr. Fant joined The Aerospace Corporation in Chantilly Virginia in 2004. Over the years, she supported mission assurance activities for both space and ground systems. Her responsibilities include performing mission assurance activities on government acquisitions, providing technical advice and guidance on model-based engineering activities, and performing cutting edge research. Additionally, she has published several papers, given presentations, served on program committees for national and international conferences.Dr. Fant received her B.S. in Integrated Science and Technology with a minor in Computer Science from the James Madison University. She earned her M.S. in Information Systems from the George Mason University (GMU) and received the Department of Information and Software Engineering's Academic Excellence Award in Information Systems. Dr. Fant completed her PhD in Information Technology with concentration in Software Engineering from GMU.

Robert G. Pettit, PhD, Professor of Practice and Director of New Graduate Programs, Department of Computer Science, George Mason University. Rob Pettit has over 30 years of experience in the software engineering industry with expertise in large-scale, mission-critical software systems. Dr. Pettit is internationally recognized in the fields of real-time embedded software systems; model-based software engineering; and in the Ada programming language. He has authored over 40 refereed papers and co-authored the textbook globally referenced as the Ada9 coding standard. Following 23 years at The Aerospace Corporation, Dr. Pettit joined the Computer Science faculty at George Mason University in Fall 2021 where he is responsible for graduate and undergraduate education in the fields of model-based software, real-time embedded software engineering, and leveraging his industry experience to prepare the next generation of software engineers. Dr. Pettit received his B.S., Computer Science/Mathematics, degree from the University of Evansville and his M.S., Software Systems Engineering, degree and Ph.D., Information Technology Engineering/ Software Engineering, from George Mason University.

Conceptual Design Support by MBSE: Established Best Practices

30

S. Shoshany-Tavory, E. Peleg, and A. Zonnenshain

Contents

Introduction	896
State-of-the-Art Survey	897
MBCD Objectives	897
Conceptual-Design Methodology and Methods	898
MBCD Implementation Issues	900
SysML for MBCD	901
MBCD Best Practices	901
Defining the Process	901
Addressing MBCD Needs and Objectives	903
Addressing MBCD Issues	912
Model Interoperability and Data Management	913
Organizational Support for MBCD	913
Solutioneering	914
Lack of "Stopping Criteria" for Modeling	914
Low Stakeholder Engagement	915
Return on Investment (ROI) Concerns	915
Requirements' Uncertainty	916
Lack of Common Taxonomy	916
Methodology Elaborated	916
Reports	918

S. Shoshany-Tavory (✉)
Technion – Israel Institute of Technology, Haifa, Israel
e-mail: sharons@technion.ac.il

E. Peleg
Metaphor Vision Ltd, Kefar-Saba, Israel
e-mail: epeleg@metaphor.co.il

A. Zonnenshain
The Gordon Center for Systems Engineering, Technion – Israel Institute of Technology, Haifa, Israel
e-mail: avigdor@sni.technion.ac.il

© Springer Nature Switzerland AG 2023
A. M. Madni et al. (eds.), *Handbook of Model-Based Systems Engineering*,
https://doi.org/10.1007/978-3-030-93582-5_84

MBCD Methodology Discussion and Future Direction 918
Chapter Summary 920
Cross-References 920
References 920

Abstract

Conceptual design of a product or a system is the preliminary phase of the system's life cycle. Innovative ideas arise, high-level decisions are made, key technologies are identified, alternatives are compared, and major components are selected. Conversely, model-based systems engineering (MBSE) targets the entire life cycle of systems' design and operation. MBSE is supported by tools for visualization, connecting requirements and structure, and promising productivity increase and better traceability. MBSE offers nonspecific support for conceptual design, and the interpretations for Model-Based-Conceptual-Design (MBCD) are still evolving. This chapter reviews current MBCD implementation trends. A comparison of these trends to existing document-based conceptual design methods produces best-practices and gap analysis. The Systems Modeling Language (SysML) is used for MBSE representation while Integrated-Conceptual-Design-Method (ICDM) represents the document-based methods for Conceptual-Design. ICDM was selected for its extensive toolset and concise process, while SysML was selected for its inherent extendibility and popularity among the MBSE community. This analysis highlights conceptual design processes and constructs, and the current capabilities of supporting tools. The chapter emphasizes MBSE features that flawlessly support conceptual design and the needed tailoring and extensions. Value-proposition is discussed on the merits of MBSE use within a conceptual design. The chapter is of interest to practitioners, methodologists, and tool vendors.

Keywords

MBCD · Conceptual design · Modeling · Creativity · Abstraction

Introduction

Systems Engineering (SE) encompasses analysis, design, and decisions concerning the entire life cycle of a system – from conception to retirement. Model-Based SE (MBSE) transforms the traditional document-centered approach into the digital-engineering era where connected models replace textual data. Such connectivity allows the model to become the "Authoritative Source of Truth" [1], in which each stakeholder and discipline have their own views. Consequently, it enables the digital engineering vision of the US Department of Defense (DoD) [2].

MBSE application can lower development and operation risks through improved collaboration and communication between all stakeholders through connected artifacts, improve complexity management by better formality and visualization, enhance knowledge capturing through standard abstraction mechanisms, and facilitate capturing a systems' behavior, and early validation support [3]. While MBSE encompasses the full life cycle development, one of the areas where knowledge is still evolving is the early phase of Conceptual Design (CD). The INCOSE SE Handbook acknowledged the importance of CD: "if the work is done properly in the early stages of the lifecycle, it is possible to avoid recalls and rework in later stages" [3]. The need for special considerations led to the establishment of the Model-Based-Conceptual-Design Working group (MBCD WG) [4] to explore MBSE support for CD. The work of the MBCD WG and the MBSE efforts of the Space Systems Working Group (SSWG) (e.g., [5]) resulted in multiple publications, offering an interpretation of MBSE for CD, i.e., MBCD, appearing in the past decade. Concurrently, numerous mature methodologies exist in the design domain – a society (https://www.designsociety.org/) that coexists with the systems engineering community, offering methods and tools to tackle the elusive CD. These methods are document-based or use specific tools to generate supporting but unrelated models.

The next section establishes through state-of-the-art review the emerging set of MBCD requirements, and those originating from design domain applications. For the latter, an existing CD framework is explored. Additionally, specific concerns arising from MBCD implementation are explored. In the following section, practitioners' needs are merged with the current MBCD literature to offer best practices and interpretations. Next, we highlight specific issues that require further extensions, and finally, future research directions are discussed.

State-of-the-Art Survey

MBCD Objectives

Full-scale development (FSD) supported by modeling follows an exhaustive structured process, where each iteration refines the requirements, design, and verification into a formal model with high fidelity. The evolving models serve as the specific iteration's "authoritative truth" where each such iteration can be assigned its own well-defined modeling goals in advance. On the other hand, CD serves exploratory purposes, where "the need to be creative, flexible and non-solution specific" drives the applicable SE process [6]. At this stage, formality and fidelity may be considered restrictive, establishing completion conditions is challenging, and uncertainty – both needs and solutions – dominates the knowledge. Thus, multiple alternative "truths" may be considered concurrently.

Additionally, the goals of the CD process may vary between different applications, depending on the mitigated risks. For example, some CD applications target better requirement understanding (e.g., prephase in [7]). In contrast, others establish a set of solutions to be tested through feasibility studies or even define the ultimate product architecture. Therefore, CD can be viewed as a set of explore, create, evaluate, and manage iterations before FSD can commence [8] or as a specific risk mitigation activity.

The MBCD WG, through its practitioners' survey and workshop [6], and Concurrent CD Proposal (CCD) [9] identify which goals to pursue through an MBCD application, to facilitate both creative thinking and rapid iterations. Key objectives found in those analyses are the following: (1) problem definition that determines the problem and the system's scope, and identifies the stakeholders; (2) identification of stakeholders' needs and requirements for development, production, utilization, support, and retirement – while focusing on concept development rather than writing a specification; (3) definition of concept of operations; (4) consideration of technological solutions alternatives and their different attributes (e.g., risk, cost); (5) designing architectural alternatives, configurations, and interfaces between systems/subsystems; (6) allocating measures of performance to design solutions, including general figures of merit (FOM) (e.g., cost, value, and reliability) or domain-specific system budgets (e.g., mass, power, and data link budget in spacecraft industry). In some cases, FOMs are referred to as design variables; (7) analysis, evaluations, and estimations of solutions, including simulations; and (8) tradespace exploration and solutions selection. The CCD further highlights the need (objective 9) for an iterative process to establish CD, where "coarse estimates of the design parameters are refined." While CD of a new product or system-of-interest (SOI) may follow sequential phases (for example, from Cocktail Napkin to Integrated Design [9]), on multiple occasions CD is exercised over an existing SOI that needs modification. Thus, MBCD for existing systems necessitates modeling (or reverse engineering) the existing solution (objective 10).

Conceptual-Design Methodology and Methods

While MBCD methodology needs to address the abovementioned goals, which arise from the SE community literature, it should also address best-practices of the design community which offers multiple tools and methods for new product development. To elicit the design community's envisioned goals, the methods and tools of the Integrated-Conceptual-Design-Method (ICDM) framework are analyzed. We chose ICDM [10] as a reference framework for our CD exploration, based on its prescriptive nature, and demonstrated capabilities. ICDM provides specifically tailored quantitative processes by integrating methods originating from the design research domain. Such processes promote and support creative thinking and innovative solutions, enabling teamwork and synergy of designers and stakeholders [11]. ICDM's support for creativity at the very early stages of a new product design [10] was demonstrated in multiple case studies [12–14].

The initial seven steps of innovation and creativity of ICDM are shown in Fig. 1. While presented as sequential, Waterfall-like steps, the process may be carried iteratively and commence at different steps. These include the following: (1) the identification of the customers and their needs, generating customer requirements; (2) translation of the voice of the customer into product definition and specification

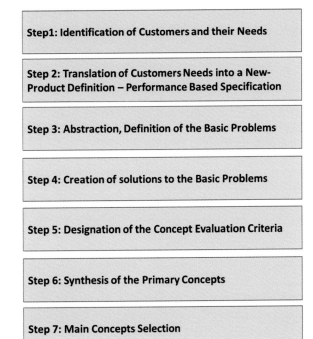

Fig. 1 ICDM initial steps, after Hari et al. [10]

that typically uses Quality Function Deployment (QFD) and House-of-Quality (HoQ) [15], which provide a structured approach for defining customer requirements and translating them into product characteristics through a planning matrix. In conjunction with Customer Satisfaction Rating (CSR) functions that express the metrics value of each target value fulfillment, it allows balancing and prioritizing needs, and required product characteristics [16]; (3) abstraction uses functional analysis tools to produce functional requirements (where customer requirements described the end-user requirements, the functional requirements describe what the system must do). The "solution-free" functions are then turned into a set of Basic-Problems (BP) to be solved by the system. For example, if the functional analysis elicited sense-fire as a required function, "how to sense fire" becomes the BP for which a solution must be found; (4) solution principles for each BP are created using Theory of Problem-Solving methods (TRIZ encompasses methodology, tool sets, a knowledge base, and model-based technology for generating innovative solutions for solving problem) [17] and its likes. To allow synthesis of solutions principles into integrated solutions, they are presented as sketches and words on Fritz Zwicky-adapted Morphologic Diagram table. The morphological table is a method for generating ideas in an analytical and systematic manner. Its outline lists BP versus their respective potential solution components; (5) evaluation criteria are then derived from the selected specification (of step 2), using rating methods such as

Analytic Hierarchy Process (AHP) [18] or Nominal Group Technique (NGT). The latter allows small groups to reach consensus on the identification of these criteria; (6) synthesis of primary integrated concepts is performed, where a concept is a feasible combination of component-solutions to all the selected BP. This process supports bottom-up creativity, using the Morphological Table that was populated in step 4. A full solution of the problem is a combination of (at least) one solution principle from each row of the morphological table. Each crafted principal concept is a carefully chosen combination of solution components that together form an integrated conceptual solution; and (7) main concepts for further elaboration are selected via multiattribute solution decisions methods such as Pugh matrix-based process or brainstorming techniques that weight and compare the conceptual designs [19].

While the ICDM methodology follows a traditional requirements-design-implementation-verification process, it highlights the following CD specificities: (1) At this stage, the customer (ICDM uses "Customer" where SE literature would use "Stakeholder") requirements may be both uncertain and tradeable and depend on the implications of the systems' design and characteristics. For example, spacecraft sensor accuracy might be renegotiated if the solution mass exceeds the launch vehicle lift capacity; (2) ICDM preaches that for CD, only a limited set of selected artifacts (i.e., needs, requirements, or BPs) are expected to influence the concept and need further elaboration. For example, energy consumption requirements in a power-plugged device may be ignored in the CD phase as they are not considered concept-driving requirements. Thus, modeling support should allow not only in-depth analysis (e.g., requirement elicitation and elaboration), but also selection and pruning of the resultant artifacts, for concept driving ones; and (3) systematic creativity needs specific support, for example, by allowing bottom-up creativity through combining alternative subsolutions into integrated solutions.

Where ICDM uses models in its different steps and methods, it does not present itself as an MBSE methodology. First, it does not attempt to holistically create a systems' model. Second, its digital interpretation is still limited – while current attempt at creating such consolidated view is underway [20], the framework lacks an encompassing tool other than spreadsheet. Therefore, in our best-practice discussion we integrate the ICDM methods into SysML-based MBCD interpretation.

MBCD Implementation Issues

While the explored literature provides requirements and interpretations for MBCD, the MBCD WG survey [6] identified a set of common issues, constraining more comprehensive and successful use. The recurring hurdles included the following: (1) model interoperability and data management; (2) gaining organizational support for MBCD; (3) "Solutioneering" – offering solutions before thorough problem understanding and design-space exploration; (4) lack of "stopping criteria" for the modeling effort; (5) lack of CD process and methodology; (6) lack of stakeholder engagement; (7) lack of MBCD best practice; (8) lack of examples and Return on

Investment (ROI) information; (9) requirements' uncertainty; and (10) lack of a common taxonomy.

SysML for MBCD

For manifestation of MBCD, we need language, methods, and tools. Though other MBSE-supporting languages exist, we use SysML as a baseline language for our MBCD interpretation. SysML is defined to specifically address system-modeling needs, providing extensive descriptive capabilities for structure and behavior, as well as some engineering analysis support through parametric diagrams. The language is supported by multiple tools, and its definition also addresses tool interoperability.

As the MBCD methodology and process are still being shaped, we harness SysML expandability in the form of metamodeling and metalanguage – specification of custom-made constructs using UML profiles, stereotypes, class inheritance, etc. [ISO/IEC 19514:2017 Information technology – Object management group systems modeling language (OMG SysML)]. We use profile-based extensions for two purposes: First, it enables us to describe best-practice methodology through a process-model. The Methodology-as-a-Model approach allows improved understandability of the suggested development process. For this interpretation, we use Use-Case (UC) as the general description of the process and Activity Diagrams (AD) as its detailed version. Second, as SysML strives for expressiveness at the price of complexity, best-practice calls for simplifying the language for specific applications. Thus, a tailored variant of the SysML is used both for Methodology-as-a-Model and for CD of the problem and system scopes. The Domain-Specific-Languages (DSL) provide such a UML standard tool. In fact, SysML is a DSL for systems engineering. DSL supports simplification of the otherwise abstract SysML through incorporating domain terminology, CD-specific symbols, diagram customization, and even users-interface tailoring of the generic tool [21]. The details of our SysML tailoring are described in the section Methodology Elaborated.

MBCD Best Practices

Defining the Process

Before delving into specific MBCD needs, the first issue to be addressed is the CD process definition. MBCD may start at different points (e.g., new product design versus improvement to an existing project) and end when alternative goals are achieved (e.g., list of alternatives for feasibility studies versus comprehensive list of customer requirements). SysML can be used to define the tailored process through description of Methodology-as-a-Model. For example, in Fig. 2, we illustrate the process steps outlined for ICDM (Fig. 1) and depict them by implementing DSL extensions.

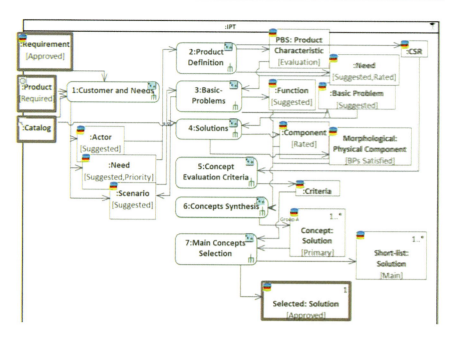

Fig. 2 ICDM Process as an Activity Diagram

We use an *Activity Diagram* (AD) (Activity modeling emphasizes the inputs, outputs, sequences, and conditions for coordinating role-based behaviors (OMG, SysML v1.6).) to express the Methodology-as-Model and include activities, methods, inputs, and outputs to describe the CD process. Knoll et al. [22] made similar use when defining their process for conceptual tradespace exploration. These capabilities serve for: (a) capturing the best practices process and make it repetitive; (b) tailor the intended CD process and communicate it to project team and stakeholders; and (c) establish a collaborative process among different teams to increase efficiency. The Methodology-as-a-Model allows clear understanding of the relations between activities and their consumed and produced artifacts.

In Fig. 2, the inputs to the process (input and outputs are highlighted by bold and brown frame and are referred to in the text using their capitalized names) are Product (not only system purpose, but also results of previous iterations) and Requirements definitions. At this point, they may be vague, or formal, depending on the business model. To these ICDM constructs, we add an additional input in the form of Catalog of reusable modeling assets. The catalog-modeling artifacts allow capturing of a new product development within its context – the surrounding organization and its assets, as well as existing legacy systems. The Catalog concept will be further elaborated in our discussion.

The different process Actions are depicted with a symbol denoting an action between human-machine. More example of available tailoring for better visualization is described in the section Methodology Elaborated. Each action creates and consumes persistent interfaces-objects (symbolized by database picture). For example, in Fig. 2, the Customer and Needs action analyzes stakeholders and customers' concerns to produce the Needs, the Actors who will interact with the system, and the preliminary set of use Scenarios that are envisioned at this stage. Each object may have additional parameters which are not shown on the diagram, such as association of Need to a specific stakeholder. In another object example, Criteria are produced by the Concept Evaluation Criteria action and consumed by the Main Concepts' Selection action which also consumes the integrated Concepts solution.

Each object may have multiple states (denoted in square brackets). For example, Needs are initially Suggested and then Prioritized. Thus, out of the integrated Concepts which are initially in Primary state, the Short-list of solutions (in Main state) and the finally Selected solutions (Approved for further elaboration) are produced. Each action is either a simple action – an action representing local functionality that can be described by text – or complex activity represented as CallBehaviorAction (CBA). Usually, at this operational level, all actions are CBAs. The more detailed actions, as implied by the "rake" symbol, are happening behind the scenes, where the objects' states change. To simplify our presentation, we did not include the detailed AD diagrams. DSL adaptation in Fig. 2 includes, for example, the specific pictograms to different types of entities or omitting of start and end symbols.

Using AD may also support role-activity definition, using "Partitions" (previously called "Swimlanes"), where each partition is typed by a specific role (which is an extension of "Actor"), representing suborganizations, people, and machines. For example, instead of generally assigning the CD process to the Integrated Project Team (IPT) as implied by Fig. 2, we can add the Marketing partition and assign it the Customer and Needs action that transfers the needs and operational scenarios to the IPT. Such presentation associates each role with its own actions, and the interfaces-objects between the roles are made explicit.

Figure 2 addresses a single CD iteration as outlined by ICDM, which is usually short; thus, no control gates are specifically depicted. In longer CD processes and between iterations, gates and reviews should be considered, for example, as proposed in the Concept Maturity Levels concept [8]. Such iteration, gates, and reviews can be illustrated on a higher-level process diagram where needed.

Addressing MBCD Needs and Objectives

The following sections outline modeling support for CD – as provided by SysML, and how specific concerns are addressed. SysML modeling constructs are italicized when used in the diagrams. In cases where SysML support for CD faces challenges or modeling gaps are recognized, we highlight MBCD-specific concerns and initial thoughts on how to close the gap. To allow coherent discussion of best-practices, the analysis groups together the MBCD objectives identified in section "MBCD Best Practices" and the ICDM framework principles into the following subsections: (1) problem definition, operational aspects, and stakeholders' needs;

(2) performance-based specification and system characteristics; (3) abstraction; (4) designing alternative solutions; and (5) selecting candidate solution(s).

Our examples draw upon the FireSat case study used by the SSWG [5]. FireSat is a fictitious space mission used in Larson and Wertz's Space Mission Analysis and Design (SMAD) textbook [23]. The goal of the FireSat mission is to detect, analyze, and monitor forest fires, and provide warnings to the Forest Service. In the provided examples, we consider extensions such as processes performed by the firefighting ground control stations.

Problem Definition, Operational Aspects, and Stakeholders' Needs

The first step in analyzing a system involves its scoping – defining the SOI and the systems and organizations interacting with it. SysML offers *Block* diagrams (BDD, IBD) that allow capturing of the system "parts" and their relations. For example, Delp et al. [5] define the FireSat SOI to interact with the external "Earth" (encompassing the atmosphere, the forest, and the fires). "Parts" of the system may include both technical systems and operating organizations. Both may reside outside the system's borders – using its services and interacting with it, or inside its borders, being part of the solution system.

Once scoping or scoping alternatives are established, the second step requires exploration of the operational aspects, needs, and requirements. These inputs may be supplied by the customer (for example, via an acquisition document). However, in many cases, the IPT must analyze the stakeholders' needs by investigating the operational scenarios and deriving requirements, roles, and processes. SysML enables such exploration through several analysis presentation tools. First, a *UC* diagram can be used to depict the required operational processes. Figure 3 illustrates

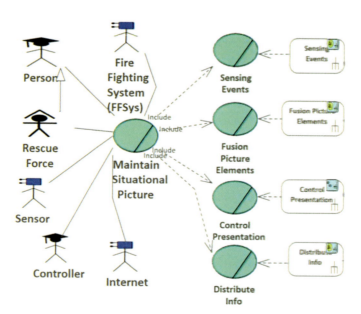

Fig. 3 High-level operational processes diagram

such an example, where the services of an imaginary forest firefighting situation-awareness picture system are described. The diagram contains high-level operational processes and the roles participating in creating and maintaining the situational picture. The picture merges information regarding all fire sites, sensing facilities, fire extinguishing forces, and health, police, and other involved parties. For maintaining the situational picture, the UC diagram illustrates high-level services required to be provided by the system, relations to their operators (*Actors*), and relationship to other operational processes extending the specific UC.

In the business concept described by Fig. 3, the Maintain Situational Picture UC is responsible to keep the situational picture, including Fire-Picture up-to-date at all times. It includes Sensing Events (supplied, for example, by FireSat) and other event sources, Fusion Fire-Picture (FP) Elements with the new events, control presentation of display parameters, such as presented and hidden layers, update rate, etc., and finally Distribute Info of FP to various recipients, each according to its distribution characteristics. All the elements in Fig. 3 are DSL-related elements. They extend of *Actors* and *USeCases* metaclasses. For example, each actor is represented by a different pictogram, such as ⩓ representing a technical device. The definitions of all these elements can be found in the section Methodology Elaborated.

Once the main UC are established, a detailed functional analysis of each operational process may commence to produce *AD* – such description identifies for each UC the participating actors (organizations or systems), the initiating *Condition*, the transferred and consumed data objects and their state, and the actions involved with the flow of the process.

An example of an AD for the forest firefighting is given in Fig. 4. It illustrates one of the firefighting UC – the process of Maintain Situational Picture. The flow commences when Fire-Fighters System (FFSys) starts and ends for any Rescue Force actor while they personally stop the system, or when the system controller shuts down the system. The system accepts event reports from various sources, such

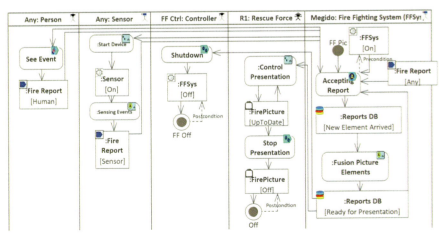

Fig. 4 Detailed Operational Process (OP) of fire picture creation as an Activity Diagram (AD)

as citizens, pilots, and automatic reports of various sensors (which may include the FireSat satellite). After the FFSys system, processes the reports and stores them in the system's database (DB) as persistent information, the unified fire picture (part of the situational picture) is built and distributed for actors upon request or periodically. This is an ongoing process that lives until the fire situation picture is no longer needed. The *ActivityFinal* ends a Flow with a *Postcondition*. The postcondition is an object in state, connected to the ActivityFinal with a ≪Postcondition≫ Dependency. Such analysis allows better understanding of the operational aspects and the customer needs or system requirements.

The above operational process (OP) uses several adjusted notations, such as manual actions (🖐), or persistent database objects (🗄). The process commences upon a true *Precondition* represented as *Object* in *State* (e.g., FFSys [On]). It terminates with a *Postcondition* as object in state, where multiple termination conditions are allowed. Each *Partition* is an instance (or object) of an *Actor*, either a human or machine actor. More details are provided in the section Methodology Elaborated.

Comment: Objects and partitions may have detailed values for each attribute of their classifier within their slots. These capabilities may be used for simulation of the process.

The resultant requirements can be documented and connected to various entities through SysML *Requirements* diagram. This connectivity can be presented using the *Traceability matrices* and on *Traceability* diagrams. Multiple types of operational elements may be defined and tagged according to the IPT/modeler's needs. Several examples are suggested in Table 1.

Table 1 Operational elements mapping to SysML

#	Operational elements	Description	MBSE element	Comments
1.	**Needs**	Customer needs of any type	Requirement	SysML Requirement
1.1	Functional	Needs for system functionality	≪Functional≫ Requirement	Stereotyped SysML Requirement
1.2	Performance	Needs for specified performance	≪Performance≫ Requirement	-"-
1.3	Safety	Defined safety needs	≪Safety≫ Requirement	-"-
2	Role	Manned or technical role	≪type≫ Actor	Types: human role, technical role (machine), and team
3	Operational Process (scenario)	A process performs by roles and executes actions to produce something	≪OP≫ UseCase ≪Manual, Computerized, Automatic≫ Activity/Action	Activity Diagram implementation for OP

≪name≫ – represents a stereotyped element using UML profile capabilities

30 Conceptual Design Support by MBSE: Established Best Practices

MBCD-specific challenges are associated with the Problem definition, operational aspects, and stakeholders' needs step, including the following:

1. In the conceptual phase, even the scoping of the SOI may still be evolving. For example, a space-vehicle may boost itself out of a parking orbit; thus, such capabilities may be included in the SOI, or alternatively the existing facilities of a Launch Vehicle (assuming such legacy is defined in advance) will be externally provided. Thus, where the line is drawn between internal and external entities is a matter of concept exploration and trade studies. While multiple scopes can be drawn and investigated in SysML – as separate UC diagrams, completing a scoping diagram may inhibit further investigation and may contribute to the noted hurdle of "Solutioneering." Therefore, the methodology should attempt to exhaust several potential scopes before defining the SOI. For example, we may choose a different SOI to the Fig. 2 alternative that includes extinguishing devices and efforts. Additionally, the supporting tool should accommodate the different configurations. The SOI decision reasonings should be documented in case such definitions need revisiting in the detailed design process.
2. It is advisable and supported by SysML to create models of the external systems, and to highlight interfaces and any functional dependencies. The reverse-engineering modeling of structure, services, and interfaces should be created at the CD onset. Such models allow better understanding of what can be expected of these entities and support SOI decisions. Such mechanism can be used for new systems designed into an existing eco-system, or for alteration of an existing legacy system. A multiproject organization may further benefit from keeping a repository of such models and sharing it among its projects through the catalog mechanism.
3. While stakeholders and subject matter experts' contributions are mandatory, specifically at this step, they may have a limited understanding of the meaning of SysML artifacts. Additionally, the technical complexity, abstraction, and the number of items of the SysML language are often confusing for nonmodelers and may lead to the noted low engagement of stakeholders. A common way to solve such complexity is using DSL extensions of SysML through UML standard profiles. These extensions allow structured tailoring of formal and organizational ontologies, eliminating language and ontologies gaps [21]. DSL can be used for defining domain-specific entities, properties, relationships, diagrams, and tools. An example of DSL is given in Fig. 5 in which a *Class* is extended by Sensor and specifically ThermalSensor, to be shared with domain experts. This diagram illustrates an example of profile definitions. The Sensor Stereotype is an extension of *Metaclass* of type *Class* with its internal properties. The Stereotype has its own Icon, Weight, and other attributes. Other Stereotypes, such as Optical and Thermal sensors, specialize (subclass through inheritance) properties of the generic Sensor Stereotype. Such Stereotypes may be placed on *component* diagram toolbox, to be directly used by the modeler (User-Interface adaptations). Stereotype attributes become Tagged-Values of the element defined by this Stereotype.

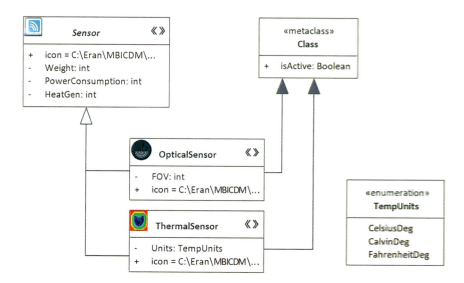

Fig. 5 DSL example of Sensor entity

4. According to ICDM best practices, only a limited set of requirements or even operational scenarios influence the CD of the system. However, due to a lack of "stopping criteria" for modeling, the modelers may find themselves elaborating the model to extensive depth, uncalled for at this stage, and as a result, drowning in the details. Thus, the CD process should include a selection step (similar to ICDM as captured in Fig. 2), in which the different customer-related artifacts are ranked and only the concept-influencing ones selected. Ranking is supported by tagging each requirement or required scenario. The remaining requirements should be kept for FSD use, or future iterations.

Performance-Based Specification and System Characteristics

System performance is defined by a set of measures typical to the specific domain and problem. For example, a FireSat has mass and detection time as two of its design-distinguishing characteristics – mass because of satellite launching sensitivity and detection time defines the speed by which a fire can be detected. These so-called figures of merit (FOM) that characterize the system design will typically include attributes of performance, cost, risk, etc. FOM can be a physical descriptor (mass in kg) or an ordinal enumerator (low, mid, and high). FOM modeling is supported in SysML through *parametric* diagrams, such as those used in the FireSat example of Delp et al. [5]. Additionally, system characteristics can also be defined as a *requirement* (for example, "5 hours maximum delay-to-detection time"). The use of FOM in MBCD is of particular importance as it allows comparison of different CDs. Each configuration or solution can be assessed either through computational tool-specific support, or by exporting the data into an external computational platform.

MBCD-specific challenges of performance-based spec include the following:

In cases, such as the FireSat, where the system domain (satellites) has been previously explored and well-known architectural patterns and best-practices exist, most of these measures are known in advanced (e.g., power, mass, and volume) and can even be associated with the different components (e.g., power consumption); therefore, only specific application FOM needs to be added (e.g., detection latency of FireSat). However, in new product CD, defining FOMs, their relative importance, and their system budgets is challenging. Additionally, FOMs may compete (e.g., optical resolution of a payload may compete with its mass). No specific SysML support addresses and represents these challenges. ICDM, on the other hand, offers specific methods for this purpose:

1. To resolve competition and rank importance, ICDM uses the HoQ and QFD. The methods help transform the voice of the customer into engineering characteristics and establish their relative importance. Such integration of requirements-characteristics association and prioritization techniques had been used in multiple studies [24–26]. For example, the "fast detection" requirement will influence both detection time and weight characteristics.
2. CSR (or similar utility function) allows the understanding of the satisfaction level that a customer associates with the different values – for example, detection time may be unsatisfactory when values are over 10 h, 50% satisfactory within 5 h, and 90% satisfactory with 2 h – meaning further improvement is marginal.

While the application of these methods may be carried outside the SysML modeling tool and only the results incorporated in the model (e.g., via tag values), it is advisable to create DSL extensions to allow seamless integration.

Abstraction

During the abstraction process, the modeler may use various functional analysis techniques to analyze the required functionality and functions expected from the system. Specifically, FAST (Functional Analysis System Technique) [27] which provides a graphical representation of how functions are linked or work together in a system for delivering the intended goods or services, or the Enhanced Functional Flow Block Diagram (EFFBD) [28] which is more flow oriented, may be used for functions elicitation. SysML supports the process through its behavior views, such as AD. Where more formal time analyses are required, sequence diagrams (SD) provide such solution [29]. The abstraction process creates a set of functions deployed by the system that are allocated to specific components.

MBCD challenges specific to abstraction are the following:

1. According to ICDM best practices, only a limited number of functions influence the CD; thus, the functional analysis process should be followed by merging (removing duplicates), sorting, and pruning (removing nonessential-for-CD functions, based on ranking functions importance) and selecting (choosing which functions will be pursued in the CD).

2. The merging-sorting-pruning process can be supported by tagging the functions' entities and associating similar functions. Tool support is recommended to allow automation of these processes, for example, through SQL add-in. By relying on element description, names, and tags, an automatic search (e.g., via Artificial Intelligence) can assist modelers in identifying similar elements that already exist in the model or the catalog – thus eliminating duplication. Additionally, tagging for "importance" can be used as a parameter for an automatic prune process.

3. To preserve model consistency and knowledge base, as well as take advantage of sunk costs, all prepruning functions should be kept in the repository. While filtered views (for example, Enterprise Architect (EA) implementation of SysML viewpoints) allow presentation of only the pruned functions, the other functions may be considered in later CD iterations, FSD, or other projects.

Designing Alternative Solutions

SysML-based MBSE offers extensive description tools that facilitate and communicate a system solution. Multiple model views allow description of solutions' structure, such as the abovementioned IDB/BDD diagrams, and traceability to requirements and functions, where the architecture and the underlying technologies can be described. Each "part" can have multiple parameters describing its characteristics, and interfaces to connect with other parts. Parts models may contain physical entities (for example, a CCD camera) or logical entities (for example, a sensor pattern). Hierarchical decomposition allows drilling down, and various tools allow filtering the views to just-needed scope. Additionally, behavioral views such as the abovementioned AD and the more detailed and temporal representation of SD can be used to annotate the dynamics of a solution. Most SysML tools provide synchronization between the different views.

MBCD challenges specific to designing solutions:

1. While the formal descriptive power of SysML is an asset in FSD, where a single authoritative truth is required, CD pursues different goals. Specifically, it requires support for solution generation, representing multiple concurrent alternatives and comparing them. Additionally, depending on the stage of the CD, it requires being sketchy rather than formal, for example allowing "Cocktail Napkin" ideas [9].

2. SysML tools and MBSE-based methods do not explicitly support the process of creating solutions, leaving this crucial step outside the modeling realms. Additionally, an overly detailed model produced too early might inhibit creativity, giving the team a false sense of completion and leading to the termination of solution-space exploration.

3. ICDM, on the other hand, offers an exhaustive and systematic creativity-supporting process for new products. Using a bottom-up innovation methodology, it guides the SE to turn the selected-for-CD functions into BPs, create and rank component solutions for each BP. A component can be a well-defined part, a

technological concept, or a logical component. For example, sensing function is turned to "how to sense" and can be accomplished via existing physical sensor, futuristic IR-based component, or left as a functional sensor only known by its performance. Each component solution can be assigned multiple FOM attributes, such as the general semiquantitative estimates of risk or performance or more specific physical features. Through the use of morphological table, the component solutions are combined into multiple integrated solutions to be used as alternative designs. This bottom-up process supports creative and systematic solutions generation – both at the component level and at the solution integration level. Such use allows solution generation, active debate, and comparison through the morphological table visual aid. While current tools offer no build-in support for this manual process, SysML can be extended via DSL to support the process visual aids. It can also be extended to automatically produce "best solution" based, for example, on FOM as tagged values and SQL add-ins for mathematics.

4. While creativity is concerned with generation of new and innovative ideas, reuse is still an important feature of systematic creativity, where existing assets may be combined to produce new designs. Reuse can encompass physical assets (e.g., specific propulsion systems) or logical assets (e.g., the general architecture of a satellite's flight system, including avionics, propulsion, and power systems). Extending the design libraries idea suggested by Kruse and Shea [30], the recommended best practice is an organizational asset catalog comprising models of multiple artifacts, both physical and logical, as well as management tools. For more detail, see future directions.

5. While SysML allow alternative presentation as unrelated models or parametric diagrams, some SysML tools offer support for tradespace exploration, where multiple solution can be generated from the abovementioned parametric models, either internally or exported to external analysis tools. For example, alternative system architectures are described as complex models representing multiple performance parameters (e.g., total number of satellite planes, orbit height) versus cost [31]. Additionally, parameters (expressed as properties and tags) may have interdependencies. Such dependencies may be managed outside the SysML model; however, the language allows associations between parts, such as "mutual exclusive," "included," etc. that enable restriction of the solution domain.

Selecting Candidate Solution(s)

Selection among candidate solutions is a typical CD activity. At this stage, our models represent several alternative solutions in the form of different models and/or configurable alternatives (such as those supported by parametric models). The selection process, by which one or several alternatives are selected as CD candidates for future work, may comprise of manual semiquantitative comparison and selection or automatic tradespace exploration. The latter method allows the selection of a Pareto optimal design. (The most efficient design that gives the acceptable performance, and stable within its scope.) Based on the predefined FOMs or design variables, multiattributes solution space exploration (e.g., [32]) can be used to create novel solutions [31]. Multiattribute tradespace evaluations are best performed using a computational search due to the large number of options. Performing an automated tradespace analysis necessitates linking an SysML tool to an external tool that performs the tradespace search [33] or using an extended tool with inherent capabilities.

MBCD-specific challenges (solutions' selection):

1. Automated tradespace analysis best fits areas where prior knowledge exists (e.g., CubeSat or FireSat examples). For new paradigms, semiquantitative approaches present a better choice as human experts, and their face-to-face negotiation should be supported [22]. Such methods are implemented as part of ICDM, for example, by using the Pugh multiattribute solution decision method [19]. The first step in its application is to establish a set of criteria by which the solutions are compared, and their relative importance (weighted criteria). Within the Pugh table, all concept solutions are listed in columns, while the rows denote criteria for comparison. One of the concepts is taken to be the datum (usually the one whose characteristics are better known), and each concept is then compared to the datum and scored accordingly. The total weighted score of each concept allows the selection of best candidate/s. The method thus strikes a balance between being quantitative and addressing knowledge gaps.
2. Such table presentation of the multiple solutions may also support solutions development – by identifying and integrating successful features from one solution into another. For example, if one forest firefighting solution relies on FireSat constellation as a sensor and another on surveillance balloons, we can create an improved solution where a local balloon would be used over hazard or active fire zones. Such integration allows achieving faster local situational awareness while lowering overall solution costs because constellation detection time requirements can be relaxed.
3. Currently, no tool supports Pugh or similar multiattribute representations or analysis. However, criteria may be represented using the SysML stereotype mechanism, DSL can be used for capturing the analysis results, and SQL scripts can automatically retrieve and compute the scores.

Addressing MBCD Issues

While the MBCD best-practices analysis of the previous section provides recommendations that address MBCD requirements, generated through the state-of-the-art survey, this section provides a concise exploration of issues raised in the MBCD WG survey [6] as hampering its successful implementation. Each subsection follows the raised issue, explores it, and offers mitigations.

Model Interoperability and Data Management

Model interoperability and data management are nonspecific to MBCD and inhibit many attempts to apply MBSE throughout a project lifecycle. The interoperability and management challenges arise from the different modeling tools used by other disciplines, the necessary two-way data mapping, and configuration management requirements. Current working solutions address model management, tool compatibility, and interfaces to domain-specific modeling tools. Specifically, models generated by SysML tools adhere to OMG's XMI standard and usually reside in a synchronized repository, where all model data and views are kept coherent. Additionally, a model from one SysML tool may be exported to XMI and then imported into another SysML supporting tool, though some specialized constructs are harder to import.

Interoperability with other tools that accept standard XMI models is also possible. However, interoperability with other specialized modeling tools presents a challenge. While some SysML tools already have transformation adopters to non-SysML tools, the translation is not always an effective two-way process. As a result, system-level modeling may be abandoned once domain-specific modeling commences. Better futuristic interoperability may be possible through contemporary standards such as the Open Services for Lifecycle Collaboration (OSLC)[4].

Until then, IPT groups need to set their goals and collaborate with other teams to define the modeling process, including which tools are used, and which models serve as the "actual truth" and at what stage. The process can be described using the Methodology-as-a-Model to establish clearer procedures.

While modeling tools provide data retentions, configuration management is a tool-dependent concept. While basic capabilities such as establishing a baseline version exist, MBCD with its multiple choices and different alternatives of different artifacts may require a tailored supportive configuration control at the different granularity. Such control allows specific model interrogation to allow comparison and merging between solutions. This feature may be achieved where database-based tools are used and through variant modeling. Such solution may comprise of additional components to existing ones [34] or different values of Tags properties of the Connectors between the "Solution" and its part components.

Organizational Support for MBCD

Regardless of the MBCD process and best practices, modeling in most organizations still requires organizational support to successfully perform CD within a given project. This support carries the form of (1) a champion (a person who represents the organizational commitment to MBSE application); (2) a modeling support team, who provide hands-on help to reduce the tool complexity and assist the less-experienced SEs, or even model their ideas; and (3) assets management. To further run an effective MBCD process, organizational best-practice identifies two additional roles: (1) a mentor for the conceptual phase who defines the CD process, its scope, tools, the required iterations, and the completion goals, in collaboration with the IPT and the supervising SE; (2) assets manager – the person responsible for managing the multiple modeling artifacts created and adapted in the CD processes that can benefit the organization even if not further used in the current project.

Solutioneering

The purpose of the initial steps of customer needs analysis, performance characteristic, and abstraction is to avoid solutioneering, by exploring the "what" and "why" before delving into the technical "how." However, modeling during CD may sometimes give a false sense of completing the required analysis process, where the modelers terminate their efforts after modeling an initial concept and refrain from exploring additional alternatives to scope, requirements, and design. The recommended remedy for solutioneering is to adhere to the ICDM process and set qualitative goals for generating alternatives at each stage, for example, maintaining at least three but no more than ten viable alternative designs at the alternative solutions' design step.

Lack of "Stopping Criteria" for Modeling

While MBSE requires formal and detailed modeling, excessive modeling at the CD phase may waste the team's time and energy on details that do not influence the high-level CD decisions. However, the stopping criteria changes with the specific domain and problem. For example, the power supply for a mission to the dark side of an asteroid requires further elaboration to establish feasibility, while a fixed-position medical device can be plugged-in to any outlet meriting little CD attention. Therefore, the modeler is advised to use common sense when elaborating the system in width (its scope) and depth (its details). The SE modeler is also advised to use tags to rank the modeling elements' relative importance. Tags can also be used to extend relations between objects, for example, adding satisfaction-level tags to *Dependency*. Additionally, the CD mentor should set other stopping mechanisms, such as who should establish which artifacts, which are needed during the current iteration, following just-in-time logic.

If used properly, the model can be interrogated and filtered for high-influencing artifacts without having to delete less meaningful artifacts. However, all artifacts are kept in the repository and may be used during repeated modeling iterations, where more details are required.

Low Stakeholder Engagement

Lack of stakeholders' engagement is sometimes associated with the technical complexity of "out-of-the-box" SysML language. Several techniques are used to enhance their engagement: (1) DSL elements are used instead of the more general language artifacts; (2) as presented, some tools allow conversion of model element to more meaningful representation for decision-makers (e.g., distinguishing between different roles using pictograms); and (3) modeled operational processes can be simulated through automatic UC generation directly from the SysML diagrams. This process allows visualization of the scenario dynamics, while using the real-world stakeholder representations. Such simulations increase stakeholders' engagement and contribute to the quality and correctness of the operational specifications.

Additionally, most tools provide reporting facilities that allow nonmodelers and review sessions access to the modeling artifacts in filtered views and document form. However, document forms may defy the purpose of modeling, as the model ownership and details are not shared.

Return on Investment (ROI) Concerns

Implementing MBSE in earlier stages of a project had been associated with improved requirements, logically leading to more successful projects [35]. Additionally, recent publications demonstrate significant ROI when considering end-to-end MBSE-supported projects [36]. However, the criteria for ROI for MBCD are more complex hence ROI is still debatable, as the MBCD WG had shown. Specifically, in this phase, the CD-induced costs are hardly the right measure. Rather creativity is a better measure of success, for example, the number of relevant suggested alternatives, the exhaustiveness of the tradespace exploration, and the time it takes for effective (but creative) convergence of an iteration are more appropriate measures than mere savings.

Additionally, it becomes apparent that a single process fails to fit all, thus the complexity level, the number of collaborating developers involved, and the CD phase should influence the decision of when and how much to MBCD. For example, the short "Cocktail Napkin" design sessions of Concept Maturity Level 1 [8] may be more appropriately served, at least initially, by collaboratively waving physical or computerized sticky notes [37], instead of drawing precise models. Thus tailoring of the MBCD process and iteration sessions are required [22]. Additionally, concerns for creativity loss by a more rigid MBCD should be addressed. This conflict stems not only from proficiency but also from the nature of the brainstorming process. Thus, it is advisable to allow initial sessions to have free format, while recording and formalizing the emerging models at the end of the session. The models can thus be elaborated and presented for approval later. More advanced sessions such as when subsystems' interfaces are addressed can carry a more formalized form, such as by using Design Structure Matrices (DSM) during the session itself [22].

Requirements' Uncertainty

Since requirements' uncertainty is an essential part of CD, the SE modeler should explore such uncertainty and use the abovementioned steps to lower it. As with solution-alternative modeling, alternative requirements can be modeled, and performance measures compared in a process aiming that gradually lowers uncertainty. Such a process needs the inputs of the stakeholders, making their engagement mandatory. Additionally, the use of CSR method to derive the actual satisfaction the stakeholders/customers associate with conformance level allows the development of a balanced set of requirements.

Lack of Common Taxonomy

The SysML provides a common metamodel that can be used to establish common grounds. However, as pointed out, its "out-of-the-box" form complexity is overwhelming. Therefore, projects and organizations are advised to create their own adapted DSL-based interpretation. These should include the domain constructs used in the specific project (e.g., a sensor instead of class). Additionally, DSL-based adaptation can also change the look and feel of the models, for example adding pictograms or coloring to enhance readability. Figures 2, 3, and 4 depict such use. For example, in Fig. 3, different appearances are assigned to different types of actors. This was implemented using stereotypes, each with a different icon and properties. See the section Methodology Elaborated for more information. One of the CD mentor missions should be establishing such taxonomy, matching it to the disciplines and the problem domain and the selected toolset.

Methodology Elaborated

A single modeling language for systems engineering is essential for improved communication; however, different projects and teams have different needs and legacy. Specifically, the ad-hoc processes of MBCD require customization to the appropriate level of simplicity, while adhering to the common language and tools, to allow transferability between tools. Customization should address several subjects:

(a) Customization of the language using extension Profiles
(b) Customization of tool behavior using automation (Add-ins)
(c) Connecting the MBSE environment with existing modeling tools to create true model interoperability
(d) Adapting the methods, and the culture of the organization

The use of Profiles enables both the customization of the language to represent the organization culture and specific customer requests, while remaining standard adhering at the same time. Visualization of standard SysML icons and properties may change; however, the diagrams, rules, and roles remain. Further Simplification is suggested to reduce toolboxes and options – for example, proposed options for dependency connectors can be limited, to the project needed taxonomy. Such customization can improve the learning curve, by lowering the required proficiencies. Additionally, the methodology flow can be defined and automatically guided by the modeling environment, preventing mistakes. Tailoring the modeling environment may also enable each type of modeler (e.g., of different disciplines) with their own modeling look-and-feel tools.

Automation offers another simplification venue, for example, by providing search facility in an available resources catalog, simplifying construction of complex elements, or creating toolset integration. Simplifying the tool is shown in our experience to be mandatory for the success of MBCD/MBSE and extending the modelers circle. The diagrams provided in this chapter were customized through Stereotyped Meta-Model elements, using UML Profile defining element types, their properties and pictograph, Diagram types and their Toolbox, element relationships, Constraints, and shape appearance via Stereotype-related scripts. The customization was implemented using Enterprise Architect (EA) Model Driven Generator (MDG) technology and Metaphor Builder.

An **Operational Process** (OP, ●) describes any organizational process – the project-specific process for extinguishing fire, or the CD development process. Each OP is executed by Roles (either human or technical entity) carried by stereotyped extensions of Actor. An OP has Activity and Actions, data elements, defined flow, constraints, and start and end points. It starts upon a specific state of an object (trigger) and end ups with something (Object in State).

OP combines specialization of Use-Case (UC) as the general OP property, and Activity as its detailed description. Roles are presented by Actors, where each actor may be associated with a different visual representation. Role (♥) – duty of a human being, Team (♠) – a group of people implementing a role within an organization, and a technical device (♥) operating a process.

An **Activity Diagram** contains Partitions representing Roles, Object-Nodes representing data elements with States (Object of a defined Class and related State of its Statemachine), Actions describing the functions executed on objects, Flow lines with Constraints, Precondition triggering the flow, and Postconditions as Objects generated by the process.

Additional elements used in the diagrams include: ObjectNode with bolded edge as input or output object (with state) – ▭ Action (either local or CallBehavior) – ▭ 4:Solutions ▭; Arrowed connector – Flow; Role in AD as Partition – ▭ SE ▭. Additionally, Domain Specific Language (DSL) is being used – for example, an operational process, such as the one described in Fig. 4, contains three types of Actions: Manual (●), Computerized (●), and Automatic (●). An action is done manually by a person, between a person and a machine, and automatically by a machine, respectively. Data objects in this example can be either Persistent (●), temporary (▭), Form (▭), or User-Interface (▭).

Functions of functional analysis are represented by UseCase (high-level) and Action (detailed). The use case is represented by OP, and Actions may carry the type of Manual, Computerized, or Automatic. One Activity may be part (CallBehaviorAction) of many detailed OP, and vice versa. However, functional hierarchy can be built from these elements in a model.

Requirements provided by SysML are extended as presented in Table 1 – for example, Basic Problems are defined as Originating Requirements (●).

Components (🔧—) are implementation elements, either logical or physical. A component is a system element with defined Interface, Services, resources, and User-Interface (UI). They are implemented as an extension of a SysML Block; Service is an extension of UseCase; Interface is built on SysML's Proxy Ports with Flow direction and Standard Port with Provided/Required interfaces. Component properties include pictogram of its type, and Tags from Meta-model's and Attributes of Stereotypes. Components are traced to Basic problems using ≪Satisfy≫ Dependency.

Solution – an integrated Solution is built of a set of components, each satisfying a BP. One component can participate in many solutions. Thus, its instance (Part) is participating in a solution and is connected by Allocate relationship. Solution is an extended SysML's Requirement.

Reports

SQL and simple Queries serve for generating reports from models. All model elements, such as Stereotype, Properties, Tags, and relations, can serve as query parameters.

MBCD Methodology Discussion and Future Direction

While CD had traditionally used various models during feasibility studies, MBCD strives toward accomplishing the holistic vision of MBSE, where the model is the truth. However, compared to the formal mainstream MBSE methodology, MBCD needs to support flexibility and creativity, in less formal situations. Specifically, it serves ad hoc teams, containing experts and disciplines who are less familiar with the MBSE process. While the MBCD WG survey highlighted the lack of methodology as inhibiting MBCD application, several authors and organizations contributed their CD methodology vision over the years. Specifically, the space domain contributed multiple insights. Early Team-X [9] publications by Jet Propulsion Laboratory (JPL), and later the Team-A approach [38], have demonstrated reduced time and cost of CD studies and applicability for architecture and trade studies. JPL further addressed embedding CD into the larger system life cycle, described by concept maturity levels [39], from Cocktail Napkin to Integrated Design. Recently, Knoll et al. [22] defined a process model for Concurrent Conceptual Design (CCD) based on a survey of best practices in the space industry [40] to include the following: (1) multidisciplinary team of experts; (2) colocated in a shared workspace; (3) producing an integrated system model; (4) using collaboration tools; and (5) following a managed design

30 Conceptual Design Support by MBSE: Established Best Practices

process. Addressing the challenges of experts' availability, integrated toolchain, and capturing engineering knowledge, the proposed process highlights requirements for time efficiency, quality of designs, and efficiently bringing diverse temporary teams to work together.

The analysis provided in this chapter supplements these practices by offering some additional insights. First, as a CD iteration may serve alternative goals and different organizations, it preaches the need to tailor the process upfront, and communicate it to the collaborating teams using Methodology-as-a-Model of the agreed process. The tailoring should also address tools selection, tool-task allocation, and interoperability guidelines. Second, using extensions mechanisms such as DSL, the modeling environment should be adapted to support the goals of establishing a taxonomy that better matches the problem domain and the experts' vocabulary. DSL extensions should also support the uncertainty of the stage, the required selection process typical to CD, and the creativity-supporting views and tools. Third, the organization support is advised to include the role of CD mentor providing guidance and best practices. Additionally, knowledge retention should be offered in the form of an organizational catalog of modeling assets to be reused. The catalog should include not only models of multiple assets (technical and physical, component-solution, functions, and requirements), but also logical patterns of assets (e.g., architecture patterns). Using the catalog, the MBCD team can efficiently mix legacy and existing elements, with new and innovative noncatalog elements. The catalog management tools should include advanced searching capabilities and creation of "new" elements according to predefined patterns. The catalog should be used with care to optimize repeated modeling efforts, while not inhibiting creativity.

Lastly, the analysis highlights the need to incorporate methods from the design domain, to seamlessly integrate with the modeling efforts (such as the QFD, CSR, Morphological Table, and Pugh Table), and provide modeling solutions for their implementation.

Future work should pursue several research directions including the following: (1) more CD case studies, exploring the aforementioned best practices, specifically examples of new products designs; (2) tools interoperability and exploration of solutions to avoid tools lock-in and allow two-way synchronization of models; and (3) advanced methods for stakeholders' engagement. The latter may include integration of Conops definition with three-dimensional graphics technology to facilitate advances in the computer-gaming industry. Such an integrated concept-engineering framework was envisaged by Cloutier et al. [41] to link scenario generation, mission analysis, and systems analysis by simulating and visualizing operational scenarios.

Another direction to be further explored is the relations between the CD phase models and the FSD MBSE. Two straightforward approaches are present – use the artifacts as a starting point to evolve or use just the essential ideas and start from scratch. We postulate that a third solution is needed, specifically regarding specialty architectures. For example, when establishing a safety architecture or a cybersecurity one [42], the expectation is that this architecture should adhere to chapter and verse, rather than just a starting point. Thus, future extensions should support these distinctions.

Chapter Summary

The chapter presents and demonstrates the methodology and best practice of model-based CD. This methodology enables the systems engineer and designer to design the concept of the system supported by the model-based systems' engineering approach. The challenges and benefits of practicing this methodology are presented. Also, future directions for MBCD in research and practice are discussed.

Cross-References

▶ MBSE for Acquisition
▶ MBSE Methodologies

References

1. A. M. Madni and M. Sievers, "Model-based systems engineering: Motivation, current status, and research opportunities," *Syst. Eng.*, vol. 21, no. 3, pp. 172–190, May 2018, https://doi.org/10.1002/sys.21438.
2. DoD - OUSD R&E, "DoD - DIGITAL ENGINEERING STRATEGY," Washington, DC, 2018. [Online]. Available: https://fas.org/man/eprint/digeng-2018.pdf.
3. INCOSE, *INCOSE Systems Engineering Handbook: A Guide for System Life Cycle Processes and Activities*, 4th ed. INCOSE, 2015.
4. K. P. Robinson, "Model-based Conceptual Design Working Group (MBCD WG) Charter," 2013. https://www.incose.org/docs/default-source/wgcharters/model-based-conceptual-design.pdf?sfvrsn=920eb2c6_6.
5. C. Delp et al., "The Challenge of Model-based Systems Engineering for Space Systems, Year 2," *INSIGHT*, vol. 12, no. 4, pp. 36–39, Dec. 2009, https://doi.org/10.1002/inst.200912436.
6. B. A. Morris, D. Harvey, K. P. Robinson, and S. C. Cook, "Issues in Conceptual Design and MBSE Successes: Insights from the Model-Based Conceptual Design Surveys," *INCOSE Int. Symp.*, vol. 26, no. 1, pp. 269–282, Jul. 2016, https://doi.org/10.1002/j.2334-5837.2016.00159.x.
7. NASA, "NASA System Engineering Handbook Revision 2," *Natl. Aeronaut. Sp. Adm.*, p. 297, 2016, [Online]. Available: https://www.nasa.gov/sites/default/files/atoms/files/nasa_systems_engineering_handbook_0.pdf.
8. R. R. Wessen, C. Borden, J. Ziemer, and J. Kwok, "Space Mission Concept Development Using Concept Maturity Levels," 2013.
9. P. Zarifian et al., "Team Xc: JPL's collaborative design team for exploring CubeSat, NanoSat, and SmallSat-based mission concepts," in *2015 IEEE Aerospace Conference*, Mar. 2015, pp. 1–10, https://doi.org/10.1109/AERO.2015.7119221.
10. A. Hari, M. P. Weiss, and A. Zonnenshain, "ICDM – An Integrated Methodology for the Conceptual Design of New Systems," 2004.
11. A. Hari and M. P. Weiss, "ICDM - AN INCLUSIVE METHOD FOR CUSTOMER DRIVEN CONCEPTUAL DESIGN," 1996.
12. J. Herscovitz and A. Hari, "Systems Engineering with ICDM - A Case Study," *INCOSE Int. Symp.*, vol. 12, no. 1, pp. 989–996, Aug. 2002, https://doi.org/10.1002/j.2334-5837.2002.tb02565.x.
13. A. Hari, D. H. Cropley, and A. Zonnenshain, "Agile System Engineering for Creative Anti-Terror Solutions," 2005.

14. M. P. Weiss and A. Hari, "Extension of the Pahl & Beitz Systematic Method for Conceptual Design of a New Product," *Procedia CIRP*, vol. 36, pp. 254–260, 2015, https://doi.org/10.1016/j.procir.2015.03.010.

15. A. Hari, J. E. Kasser, and M. P. Weiss, "How lessons learned from using QFD led to the evolution of a process for creating quality requirements for complex systems," *Systems Engineering*, vol. 10, no. 1. pp. 45–63, 2007, https://doi.org/10.1002/sys.20065.

16. A. Hari, M. P. Weiss, and A. Zonnenshain, "Design Quality Metrics Used as a Quantitative Tool for the Conceptual Design of a New product," 2001.

17. G. Caligiana, A. Liverani, D. Francia, L. Frizziero, and G. Donnici, "Integrating QFD and TRIZ for innovative design," *J. Adv. Mech. Des. Syst. Manuf.*, vol. 11, no. 2, pp. JAMDSM0015–JAMDSM0015, 2017, https://doi.org/10.1299/jamdsm.2017jamdsm0015.

18. E. H. Forman and S. I. Gass, "The Analytic Hierarchy Process—An Exposition," *Oper. Res.*, vol. 49, no. 4, pp. 469–486, Aug. 2001, https://doi.org/10.1287/opre.49.4.469.11231.

19. S. Pugh, "Concept Selection - a Method that Works," 1981.

20. S. Shoshany-Tavory, E. Peleg, A. Zonnenshain, and G. Yudilevitch, "MBSE for conceptual design: an integrative approach," *Syst. Eng.*, vol. Under Revi, 2022.

21. B. Cole *et al.*, "Domain-specific languages and diagram customization for a concurrent engineering environment," in *2013 IEEE Aerospace Conference*, Mar. 2013, pp. 1–12, https://doi.org/10.1109/AERO.2013.6497134.

22. D. Knoll, C. Fortin, and A. Golkar, "A process model for concurrent conceptual design of space systems," *Syst. Eng.*, vol. 24, no. 4, pp. 234–249, Jul. 2021, https://doi.org/10.1002/sys.21576.

23. W. J. Larson and J. R. Wertz, Eds., *Space mission analysis and design*, 3rd ed. Springer Netherlands, 1999.

24. D. N. Mavris and K. Griendling, "Relational Oriented Systems Engineering and Technology Tradeoff Analysis (ROSETTA) Environment," in *2011 6th International Conference on System of Systems Engineering*, Jun. 2011, pp. 49–54, https://doi.org/10.1109/SYSOSE.2011.5966572.

25. M. Abdelrazik, A. Elsheikh, M. Zayan, and A.-B. Elhady, "New Model-Based Systems Engineering Methodology Based on Transdisciplinary Quality System Development Lifecycle Model," *J. Eur. des Systèmes Autom.*, vol. 52, no. 5, pp. 465–476, Nov. 2019, https://doi.org/10.18280/jesa.520505.

26. J. C. Tejada, A. Toro-Ossaba, S. Muñoz Montoya, and S. Rúa, "A Systems Engineering Approach for the Design of an Omnidirectional Autonomous Guided Vehicle (AGV) Testing Prototype," *J. Robot.*, vol. 2022, pp. 1–13, 2022, https://doi.org/10.1155/2022/7712312.

27. B. John, "FAST Diagrames: The Foundation for Creating Effective Function Models," *Trizcon 2011*, pp. 1–10, 2011.

28. W. Power, A. Jeffrey, and K. Robinson, "Applying model-based system engineering to modelling and simulation requirements for weapon analysis," in *2018 IEEE Aerospace Conference*, Mar. 2018, pp. 1–16, https://doi.org/10.1109/AERO.2018.8396501.

29. T. Bayer *et al.*, "Update on the Model Based Systems Engineering on the Europa Mission Concept Study," in *2013 IEEE Aerospace Conference*, Mar. 2013, pp. 1–13, https://doi.org/10.1109/AERO.2013.6496855.

30. B. Kruse and K. Shea, "Design Library Solution Patterns in SysML for Concept Design and Simulation," *Procedia CIRP*, vol. 50, pp. 695–700, 2016, https://doi.org/10.1016/j.procir.2016.04.132.

31. R. E. Thompson, J. M. Colombi, J. Black, and B. J. Ayres, "Disaggregated Space System Concept Optimization: Model-Based Conceptual Design Methods," *Syst. Eng.*, vol. 18, no. 6, pp. 549–567, Nov. 2015, https://doi.org/10.1002/sys.21310.

32. A. M. Ross and D. E. Hastings, "The tradespace exploration paradigm," *15th Annu. Int. Symp. Int. Counc. Syst. Eng. INCOSE 2005*, vol. 2, pp. 1706–1718, 2005.

33. K. R. Duncan and R. Etienne-Cummings, "A Model-Based Systems Engineering Approach to Trade Space Exploration of Implanted Wireless Biotelemetry Communication Systems," *IEEE Syst. J.*, vol. 13, no. 2, pp. 1669–1677, Jun. 2019, https://doi.org/10.1109/JSYST.2018.2874102.

34. D. Bilic, "Managing Variability in SysML Models of Automotive Systems," 2020, [Online]. Available: https://www.diva-portal.org/smash/record.jsf?pid=diva2:1470374.

35. S. J. Saunders, "Return on Investment Using Model-Based Concept Design," *INSIGHT*, vol. 17, no. 4, pp. 23–25, Dec. 2014, https://doi.org/10.1002/inst.201417423.
36. E. B. Rogers and S. W. Mitchell, "MBSE delivers significant return on investment in evolutionary development of complex SoS," *Syst. Eng.*, vol. 24, no. 6, pp. 385–408, 2021, https://doi.org/10.1002/sys.21592.
37. P. Dalsgaard, K. Halskov, and C. N. Klokmose, "A study of a digital sticky note design environment," *Sticky Creat. Post-it Note Cogn. Comput. Des.*, no. January, pp. 155–174, 2019, https://doi.org/10.1016/B978-0-12-816566-9.00007-0.
38. J. K. Ziemer, R. R. Wessen, and P. V. Johnson, "Exploring the science trade space with the JPL Innovation Foundry A-Team," *Concurr. Eng. Res. Appl.*, vol. 26, no. 1, pp. 22–32, 2018, https://doi.org/10.1177/1063293X17740406.
39. L. S. Wheatcraft and L. Lewis, "Concept Maturity Levels," *INCOSE Int. Symp.*, vol. 28, no. 1, pp. 1592–1607, Jul. 2018, https://doi.org/10.1002/j.2334-5837.2018.00570.x.
40. D. Knoll, C. Fortin, and A. Golkar, "Review of Concurrent Engineering Design practice in the space sector: state of the art and future perspectives," in *2018 IEEE International Systems Engineering Symposium (ISSE)*, Oct. 2018, pp. 1–6, https://doi.org/10.1109/SysEng.2018.8544387.
41. R. Cloutier *et al.*, "Prototype of a Graphical CONOPS (Concept of Operations) Development Environment for Agile Systems Engineering," p. 142, 2013.
42. D. Mažeika and R. Butleris, "MBSEsec: Model-based systems engineering method for creating secure systems," *Applied Sciences (Switzerland)*, vol. 10, no. 7. 2020, https://doi.org/10.3390/app10072574.

Sharon Shoshany-Tavory is an expert in analysis and systems engineering of resilient cyber-physical systems. Her multidisciplinary capabilities were acquired during her 35+ years of industrial career as a senior manager and developer of defense and transportation systems. As an adjunct lecturer, she mentors ME students in the Technion, Israel Institute of Technology, Systems Engineering Program.

She is an avid advocate of MBSE. Her interest dates to the late 1980s, as an early user of Stalemate, continued as UML/SYSML user, and finally as a researcher of MBSE practice.

Sharon holds an undergraduate and master's degree in electrical engineering from the Technion and PhD in natural resources and environmental management in Haifa University.

Eran Peleg is the founder and CEO of Metaphor Vision Ltd., an Israeli start-up company, developing a model-based Digital-Twin and MBSE (Model-Based System Engineering) tools. Metaphor Vision is operated in the Druze village of Hurfaish. Eran's Partner is Eiad Bader and developers are all from Hurfaish.

Eran is a veteran of the IAF as a combat pilot and computer unit commander.

Eran has BA degree in computer science and economy, from the University of Bar-Ilan, Ramat Gan, Israel. Currently, Eran is involved in a few research programs of the Technion in System Engineering and MBSE and consults to several leading Israeli companies.

A. Zonnenshain is currently the Senior Research Fellow at The Gordon Center for Systems Engineering and at the Neaman Institute for National Policies Research at the Technion, Haifa, Israel. He has a PhD in systems engineering from the University of Arizona, Tucson, USA. Formerly, he held several major positions in the quality, reliability and systems engineering areas in RAFAEL and in the Prime Minister's Office.

He is an active member of the Israel Society for Quality (ISQ). He was also the Chairman of the Standardization Committee for Management & Quality in the Standardization Institute of Israel. He is a Senior Adjunct Lecturer at the Technion-Israel Institute of Technology.

He was a member of the board of directors of the University of Haifa.

He is an active member of INCOSE & INCOSE_IL (past president). He is a Fellow of INCOSE.

Part VII
Case Studies

Ontological Metamodeling and Analysis Using openCAESAR

31

D. A. Wagner, M. Chodas, M. Elaasar, J. S. Jenkins, and N. Rouquette

Contents

Introduction	926
Language	926
Abstraction	928
Automation	929
The Meaning of *Modeling*	929
The Meaning of *Metamodeling*	930
Combining Modeling Languages and Semantics	930
Ontological Metamodeling	933
State of the Art	934
Vocabularies and Ontological Reasoning	938
Ontological Modelling Language	939
Vocabulary	941
Vocabulary Bundle	941
Description	941
Description Bundle	942
OML Analysis Workflows	942
openCAESAR Tools	943
OML Workbenches	943
Analysis Services (https://langserver.org/) (https://theia-ide.org/)	944
OML Bikeshed	945
OML Merge	945
OML to OWL Adapter	946
OML Reason	946
OML Load	947
OWL SPARQL	947
OWL SHACL	947
OWL Start/Stop Fuseki	947
Tool Adapters	948

D. A. Wagner (✉) · M. Chodas · M. Elaasar · J. S. Jenkins · N. Rouquette
Jet Propulsion Laboratory, California Institute of Technology, Pasadena, CA, USA
e-mail: david.a.wagner@jpl.nasa.gov; mark.chodas@jpl.nasa.gov; maged.elaasar@jpl.nasa.gov;
j.s.jenkins@jpl.nasa.gov; nicolas.rouquette@jpl.nasa.gov

© Springer Nature Switzerland AG 2023
A. M. Madni et al. (eds.), *Handbook of Model-Based Systems Engineering*,
https://doi.org/10.1007/978-3-030-93582-5_78

Example Ontologies	948
Core Vocabularies	948
IMCE Vocabularies	949
FireSat Example	951
Chapter Summary	951
References	952

Abstract

All modeling depends on having some vocabulary in which a model can be expressed. For the purpose of this chapter, we will define *metamodeling* as the modeling of vocabularies that can then be used to build useful models. This chapter will explain how standards developed for the semantic web can be used to produce precise vocabularies (ontologies) that can then be used as the foundation, on which to build precise descriptive models.

Keywords

CAESAR · openCAESAR · OML · OWL · OWL 2 DL · Semantic modeling · Metamodeling · Ontology · Description logic

Introduction

Engineering is by nature a problem-solving activity tempered by scientific and mathematical rigor. The feature that distinguishes engineering from other problem-solving is the use of science and mathematics to develop a proper understanding of the relevant phenomena and to produce evidence that a proposed solution will be satisfactory before deploying it.

Systems engineering *is* engineering, with all that entails, but it almost exclusively deals with large-scale endeavors that cannot be accomplished without multi-disciplinary teams in effective collaboration. Moreover, systems engineers often do not design a solution per se, as an electrical engineer might. Often the role of the systems engineer is to guide the collaborative design of complex systems to successful conclusion; in this context success is rarely a point solution in final form. Especially in the early phases of the life cycle, the aim is to identify a family of solutions, each of which can be shown by analysis to exhibit behavior that achieves mission objectives under applicable constraints and with high confidence.

Effective systems engineering, therefore, hinges on collaboration and analysis. These, in turn, are enabled by language, abstractions, and automation. We discuss each below.

Language

Mature fields of engineering have long relied upon regular language and notation for fundamental concepts and properties. In electrical engineering, for example, we find established terms for:

- Elemental objects: resistor, capacitor, and inductor
- Properties of element: resistance, capacitance, and inductance
- Metrology: ohm, farad, and henry
- Aggregations: L-network, transmission line, and amplifier
- Properties of aggregations: cutoff frequency, characteristic impedance, and gain

The language of electrical engineering is more or less universal and understood worldwide across cultures and customs.

Various specialized sublanguages have been created to serve specific purposes, e.g.:

- Very High-Speed Integrated Circuit Hardware Description Language (VHDL)
- Simulation Program with Integrated Circuit Emphasis (SPICE)
- Modelica

Around each of these (and other languages) is a community that understands and improves the language; uses the language to communicate designs, analyses, etc.; and creates economic value in the form of products and services. It is clear that human creativity is unleashed by common language.

Despite considerable progress in the SysML community, there is not yet widespread consensus on a unifying language for systems engineering. Part of the challenge is that the scope of systems engineering knowledge is for all practical purposes unbounded: certainly larger than electrical, mechanical, thermal, telecom, etc.

There is a wealth of published information (e.g., [1]) with definitions of terms and box-and-line drawings that describe practices in detail, but systems engineering as a discipline lacks uniform language for the most fundamental systems engineering concepts, e.g., system breakdown structure. We have a name for it, but no consensus way to construct, describe, or exchange it.

The way out of this morass will require serious thinking about the fundamental concepts, properties, and relationships that undergird systems engineering. It will *not* be achieved by merely creating glossaries or dictionaries that attempt to expound the meaning of words. Our problem is precisely the opposite: to discover what is important for the practice, assign names to those things, and provide constraints for using and combining these things in a meaningful way.

A correlate from physics is the notion of *energy*. Physicists cannot define what energy *is*, and they do not care. What they can tell us is how to measure energy under various circumstances and that certain conservation relations involving energy appear to be true. Energy is an important concept because (and only because) these metrics and laws allow us to make predictions about the behavior of things in the world.

Abstraction

Electrical engineering has from the beginning made use of mathematical abstractions to empower analysis: calculus, differential equations, Fourier series, transfer functions, and Laplace transforms; more recently measure theory, Hilbert space, functional analysis, and probability; and even more recently abstract algebra and undoubtedly many others.

These abstractions are the bridge between real-world problems and scientific and mathematical rigor. Engineers transform real-world problems into mathematical problems, apply perhaps centuries of mathematical techniques to solve those problems, and then transform those mathematical solutions back into their corresponding real-world manifestations.

Engineering language and engineering abstractions are closely related. As an illustration, consider the question "What is a capacitor?" By *capacitor* one often means a circuit element inserted by design to evoke a desired behavior, but that just shifts the question to "What is the desired behavior?" That behavior, of course, is *capacitance*, a precise relation between voltage and current represented by the abstraction of a first-order linear differential equation, $I = C\frac{dV}{dt}$. In the most fundamental sense, a capacitor is anything that exhibits capacitance, by design or not (e.g., a finger), desired or not (e.g., parasitics).

This example illustrates the close linkage between language and abstractions that pervades engineering. The important concepts in engineering are more often characterized by behavior than structure; other examples include *truss, suspension bridge, full-wave rectifier, Butterworth filter, heat pipe*, and *Stirling engine*. Each of these terms represents an important concept because we understand the abstractions behind each and their implications for behavior.

Systems engineering as a discipline has historically not identified a set of foundational abstractions. There are exceptions in the literature, notably Wymore [2], but they appear to have little adoption in real-world practice.

Because relatedness plays such a central role in systems engineering, one could argue that the discipline's most fundamental abstraction is the *graph*. A graph is a mathematical structure used to model pairwise relations between objects. A system breakdown structure, for example, is a graph that relates parts to wholes. In this case, the graph is a particular specialization called a *directed rooted tree*. This specialization is defined by four formal graph-theoretic properties:

- **Connected**: every node is included at least once.
- **Acyclic**: there is exactly one path between any two nodes.
- **Directed**: the whole-part relation is antisymmetric (if B is a part of A, then A is not a part of B).
- **Rooted**: there is a single node that belongs to no whole.

Defining a system breakdown structure in this way immediately gives us the tools to describe breakdowns (e.g., via an edge list) and for any party exchanging such a description to immediately and unambiguously determine whether it is a well-formed system breakdown.

Another common pattern in systems engineering analysis is the binding of specific property values to roles in physical laws or other models of behavior. For example, the dynamics of a vehicle in a gravity field obey Newton's second law $F = ma$ with the gravity field playing the role of F and the vehicle mass playing the role of m. The assignments of specific values to roles are not necessarily invariant, with some assignments holding only for certain envisioned scenarios. Mathematical graphs are well-suited to describing such situations, and graph-theoretic concepts such as reachability are useful for specifying completeness or other well-formedness.

There are undoubtedly other useful abstractions. As a discipline, systems engineering would benefit from a comprehensive effort to advance a solid theoretical foundation for all its practices. Practices that do not lend themselves to such grounding should be viewed with suspicion.

Automation

It is beyond question that machine computation is important for modern systems engineering. Many interesting problems do not have closed-form mathematical solutions and require approximation well beyond the ability of humans. An important point to bear in mind regarding computing, however, is that machine computation brings an important feature beyond mere speed: reliability. If computers made errors at the same rate as humans, high speed would not be a virtue. Automation, therefore, is important for maintaining rigor. Computation (with appropriate formalisms) should be employed for any operations that humans cannot do fast enough *or* reliably enough.

Fortunately, precise language and mathematical abstractions pave the way for automation. Many mathematical abstractions have corresponding computational abstractions (e.g., symbolic and numerical computation libraries, graph theory libraries, etc.)

The Meaning of *Modeling*

The word *modeling*, as applied to systems engineering, is often self-referentially described as "systems engineering with models." That is not particularly helpful unless *model* itself is subsequently defined. Ultimately, such questions are not satisfactorily resolved by resorting to dictionaries.

We characterize (but do not define) modeling as the effective use of precise language, appropriate mathematical abstractions, and automation. Model-based systems engineering, therefore, is simply those practices applied to systems engineering. Inasmuch as precise language, mathematical abstractions, and automation are essential features of modern engineering in every mature discipline, it is difficult to defend the proposition that model-based systems engineering is separable from systems engineering.

The Meaning of *Metamodeling*

Every model is created using the constructs and relationships defined in a modeling language, which also defines rules constraining the organization of these constructs and relationships in a model. Metamodeling is the process of designing a modeling language. This process requires selecting a language with which to express the design of a modeling language, a metamodeling language. In practice, the Eclipse Modeling Framework ([3]) is a popular metamodeling language widely used for the design of many systems engineering modeling tools. However, EMF is not required for metamodeling. For example, one could use SQL, XML schema, spreadsheets, OWL, or existing modeling languages (e.g., UML, SysML, Matlab, Mathematica, Modelica) as a metamodeling language. This choice requires careful consideration about important aspects of the modeling languages that will be designed – concrete syntax, serialization, and semantics – and how these aspects will ultimately affect the resulting characteristics of these modeling languages: precision, rigor, and automation support.

In practice, selecting a metamodeling language and technology stack boils down to how much support the stack provides for simplifying the tedious but necessary aspects of designing and integrating modeling languages. Given that choosing such a stack requires an investment in training, processes, and methodology, it makes sense to reduce the number of different stacks for managing multiple modeling languages. The EMF stack is a popular choice because it provides a rich ecosystem of extensions for tackling important aspects such as specifying constraints for a modeling language (e.g., OCL, ECL), transformation across multiple languages (e.g., QVT, Viatra), and specifying the semantics of a modeling language (e.g., Xsemantics). Outside the modeling technologies mentioned above, support for metamodeling is also available in the form of domain-specific language (DSL) tooling in tools widely used in systems engineering such as mathematical modeling (e.g., Mathematica, Maple), product life cycle management (e.g., Siemens Teamcenter [4]), and computer-aided design (e.g., Dassault Systemes Catia [5]).

Combining Modeling Languages and Semantics

As explained earlier, with the absence of a unifying language for systems engineering, we need to figure out an effective strategy to support practices involving multiple modeling languages with a coherent strategy for combining their semantics. When different modeling languages have compatible semantics, then it becomes practically sensible to look for opportunities to unify models across semantically compatible modeling languages or to map them. Doing so could help reduce the problems with reconciling differences across multiple models that, at least conceptually, ought to be semantically compatible but practically may have significant differences due to the different modeling tools involved. An example of this situation would be using multiple requirements modeling tools whose modeling languages have some degree of semantic compatibility but also tool-specific differences. The challenge in practice arises from managing the tool-specific differences. Transforming/mapping information across different requirements modeling tools on the basis of a common semantics is relatively easy.

A different situation occurs when modeling languages have heterogeneous semantics. Consider, for example, dynamic simulation of a system whose semantics is temporal traces of state variables and probabilistic risk assessment (PRA) for evaluating risks associated with life cycle events of the same system. Typically, PRA focuses on assessing the risk of initiating events leading to failures in a system to assess their potential outcomes. The heterogeneity of the semantics raises questions like the following: Suppose that PRA identifies an undesirable outcome of a likely initiating event: can the dynamic simulation provide a temporal trace from this initiating event all the way to that undesirable outcome? The answer is obviously no if the dynamic simulation model covers so-called nominal behaviors to the exclusion of off-nominal behaviors corresponding to failure conditions. Even if the dynamic simulation models cover a broad range of possible behaviors, the probabilistic nature of the PRA relationship between a specific initiating event and a specific potential outcome may be difficult to replicate in a dynamic simulation model: the dynamic simulation model may be deterministic, instead of probabilistic in which case additional input conditions need to be carefully chosen such that the trace reaches the PRA's outcome; alternatively, the simulation model may be probabilistic, which means that the state variables are random variables with a probabilistic distribution. In that case, the challenge involves comparing the probabilistic distributions of the dynamic simulation model with those of the PRA model.

The point of this example is that modeling languages with heterogeneous semantics exacerbate the difficulty of transforming/mapping information among such languages: even if one can establish correspondences among the models – e.g., probabilistic state variables in a dynamic simulation with a corresponding probabilistic variable in PRA – one needs to carefully consider whether the heterogeneity of the semantics makes such correspondences practically useful, e.g., if the dynamic simulation cannot produce probabilistic traces corresponding to the PRA relationship between a given input event and a potential outcome, then the correspondence of probabilistic state variables between the models provides no practical value because the variables cannot relate behaviors of the two models.

There are two broad strategies for managing inter-model correspondences that can be used separately or in combination:

- Implicit correspondences by transformation: If the semantics of a modeling language subsumes that of another, then latter models may be produced by transformation from the former models. See [6] for a survey of model transformation tools.

- Explicit correspondences across federated models: If two modeling languages with heterogeneous semantics are used for the same systems engineering activity, there must be some common ground of information between them; otherwise the modeling languages would be completely unrelated to one another. This common ground of information provides the basis for managing inter-model relationships according to the common ground of element correspondences.
- Consider, for example, a mechanical CAD language that describes a physical artifact with geometric properties, whereas a thermal design language description involves thermal properties: given that there are typically many such artifacts in these models, it is necessary to explicitly assert which physical artifacts in the CAD and thermal models represent different views of the same artifact in the real world so that one can combine the information from both models into a unified view.

There are many strategies for managing correspondences across federated models. Query-based strategies include Answer-Set Programming [7] and GraphQL [8].

NIST contributed a new perspective on inter-model correspondences with the concept of a "digital thread" of information across models [9], which, in the context of manufacturing, further contributes to the improvement of standards for the exchange of 3D data (see https://ap242.org).

Managing inter-model correspondences is an example of a broader concern about the meaning of the logical information represented in systems engineering models. Databases provide a useful analogy for this perspective where this broader concern is typically addressed by augmenting the database schema with database integrity and validation constraints, whereas managing inter-model correspondences amounts to linking database records to each other or to external representations in some way. Besides databases, the logical paradigms available for managing this logical information fall into two broad categories:

- A property-oriented specification language where a system model is a collection of logical axioms with a focus on describing the required logical properties that a system conforming to the model must satisfy. Examples include description logics (DL) such as W3C's OWL 2 DL [10], which enjoys support for automated reasoning for computing logical consequences of asserted logical axioms in a system model.
- A model-oriented specification language where a system model specifies the states of the system and possible operations on the system state with a focus on describing the required behavioral properties that a system conforming to the model must satisfy.

    Examples of standard-based modeling languages include OMG's UML/SysML [11,12] with extensions for execution like FUML/ALF (https://www.omg.org/spec/FUML) [13]. Examples of formal-methods languages include Coq (https://coq.inria.fr/)[14], Lean (https://leanprover-community.github.io/

index.html)[15], VDM (https://en.wikipedia.org/wiki/Vienna_Development_Method)[16], and ISO/IEC 13568:2002 (https://www.iso.org/standard/21573.html) (Z formal specification) [17].

In practice, both paradigms can be combined; however, it is necessary to exercise care to ensure that the semantics of the underlying logic involved for property-oriented and model-oriented specification can be combined in a way that ensures derived logical properties are logically valid [18]. In practice, this is challenging to do with some combinations like W3C's OWL 2 DL and OMG's modeling languages because the semantics of the latter is unfortunately beyond what can be expressed in the former. There is a wealth of established theoretical computer science involving category theory for the integrating multiple modeling languages that have been codified in the OMG standard for a Distributed Ontology, Model, and Specification Language, DOL [19].

Ontological Metamodeling

Atkinson and Kühne introduced the distinction between linguistic and ontological metamodeling [20] according to the purpose of the instantiation relationship. In linguistic metamodeling, the instantiation relationship specifies that a model is an instance of a metamodel, thereby inducing a hierarchy of linguistic modeling levels. In ontological metamodeling, the instantiation relationship appears within a single level of linguistic modeling to specify a logical typing relationship between model elements representing ontological instances and classes. Consequently, ontological metamodeling depends on the choice of a linguistic metamodeling paradigm called a technological space in [21]. The OMG MOF [22] architecture is famous for introducing a four-layer linguistic metamodeling architecture as the foundation for UML and SysML. Given the complexity of linguistic metamodeling technological spaces, ontological metamodeling is clearly better suited for systems engineering because it provides support for defining new domains (as ontological metamodels) within the same technological space as long as it is adequate.

Since the original notion of ontological metamodeling in [20] lacks a criteria for an adequate linguistic metamodeling paradigm beyond the relation of instantiation, we postulate that such paradigm must support, at minimum, using W3C's OWL 2 DL augmented with W3C SWRL (https://www.w3.org/Submission/SWRL/) [23] for specifying linguistic and ontological metamodels as DL ontologies and rules and for modeling their instances as DL ontologies as well. This strategy reduces the problems of combining logical information from heterogeneous system models created with different modeling languages to a simpler problem of combining their respective ontologies. In practice, this strategy requires careful consideration to ensure that basic reasoning tasks such as checking for consistency and calculating entailments are computationally efficient. For example, using OWL 2 with the RDF-based semantics makes basic reasoning tasks such as consistency and entailment computationally undecidable; in practice, it is advisable to use a subset of

OWL 2 for which reasoning tasks become computationally tractable, such as one of the predefined OWL 2 profiles [24, 24] or a suitable subset that experimentally demonstrates adequate performance for reasoning.

In practice, ontological metamodeling involves the several activities that critically depend on computationally efficient reasoning about OWL 2 DL and SWRL rules:

- Specifying the syntax of a modeling language in the form of a DL ontology vocabulary of concepts, relations, properties, and datatypes, the core constructs of OWL 2.
- Verifying that the logical classes of the modeling language vocabulary are satisfiable, that is, that one could potentially construct a model using any of the vocabulary classes
- Asserting logical constraints using OWL 2 DL axioms and SWRL rules about what it means for a model to be logically well-formed
- Deriving closure axioms for a set of modeling languages to force vocabulary concepts that are intended to be asserted as logically distinct from each other
- Computing the logical inferences for a model about the logical axioms asserted in the model combined with those of the modeling language vocabulary
- Verifying that all asserted and inferred logical axioms are consistent – that is, that the model conforms to the constraints defined in the modeling language vocabulary

Note that this is distinct from linguistic ontology where the intent is to try to model the meaning of natural language. In ontological metamodeling, the meaning is specifically asserted to be whatever the metamodel says it is through the vocabulary and rules it expresses. Metamodelers will tend to borrow nouns and verbs from natural language to try to make a language more intuitive. This can be a slippery slope for users of the language to intuit more meaning than the vocabulary explicitly supports, but formally there is no more meaning than that given in the metamodel or by formal inferences derived from the metamodel. Inferences derived from any other natural language understanding of the terms are irrelevant. Homonyms can also create confusion in natural language where readers only have the context of usage to infer the correct meaning. Formal languages support the notion of namespaces which define formal syntactic context and permit the same term to be redefined with different semantics in different contexts. A namespace typically adds a prefix to each term that disambiguates usage. Thus, *project:Product* refers to the specific definition of a product in the context of the *project* vocabulary.

State of the Art

The OMG modeling standards have been widely adopted by vendors of systems engineering tools although with significant differences such as interoperability (e.g., support for OMG's XML Metadata Interchange, OASIS' Open Services for Lifecycle Collaboration (OSLC)) or the modeling language (e.g., OMG's Unified

Modeling Language, UML, Systems Modeling Language, SysML). These characteristics, albeit important, are largely independent of the tool's paradigm for managing logical information in models, a concern that warrants a closer look at the tool's metamodeling paradigm:

- Catia/NoMagic Cameo Systems Modeler (https://www.nomagic.com/products/cameo-systems-modeler) [25]

The breadth and scope of this commercial tool support for OMG's standards (UML and SysML) reflect the engagement of the company in the development of such standards. The underlying metamodel is based on the Eclipse Modeling Framework (EMF), whereas the logical information management reflects an emphasis for standards: e.g., OMG's Object Constraint Language (OCL) (https://www.omg.org/spec/OCL/2.4/PDF) [26] *for logical information management, OMG's SysML Extension for Physical Interaction and Signal Flow Simulation* [27]*, SysPhS for transforming and federating to/from Modelica, and OASIS' OSLC for transforming and federating information with Siemens' Teamcenter. Extending Cameo Systems Modeler for domain-specific modeling involves creating extensions of the underlying UML or SysML language. Consequently, the biases and limitations inherent in UML bear heavily on the flexibility of defining domain-specific extensions as profile extensions of UML or SysML. Additionally, although UML and SysML are intended to have a logical semantics according to the OMG standards, the logical semantics of UML suffer from intrinsic issues in UML's foundation in OMG's MOF* (See "Simplifying MOF-based metamodeling" [28] for a critique of MOF with respect to issues preventing a clean logical mapping to a normalized database schema) *and profiles* [28].

- Syndeia [29]

Syndeia focuses on managing logical information across the diversity of tool-specific systems engineering models with the concept of a digital thread, "a graph whose nodes are elements in various enterprise repositories, tools, and version control systems, and whose edges are intra-model relationships within each tool and inter-model relationships created and managed by Syndeia between the nodes." Thanks to a broad range of partnerships with systems engineering tool vendors, Syndeia provides a powerful platform for transforming and federating logical information to/from the leading systems engineering tools used in practice.

- GME: Generic Modeling Environment (https://www.isis.vanderbilt.edu/projects/GME) [30]

GME is a configurable metamodeling toolkit for defining systems engineering modeling languages in an integrated environment. GME supports transformation (via GReAT) and federation (via UDM) with other systems engineering tools in the Microsoft ecosystem thanks to its architecture based on the Microsoft Common

Object Model, COM, as a binary interface standard for data and tool integration. GME's metamodel provides the common constructs for managing logical information of domain-specific extensions in terms of folders for organizing domain-specific models and the concept of first-class object, FCO, for specifying the domain-specific contents of such models in terms of FCO's key abstractions: atoms, references, connections, and sets. Although FCO provides adequate flexibility for defining domain-specific modeling languages, the logical semantics of FCO is unique to GME and consequently limits reuse of standard [31].

- SysMLv2

As explained in the SysML v2 Request For Proposals (RFP), "SysML v1" was adopted in 2006 as a general-purpose graphical modeling language for specifying, analyzing, designing, and verifying complex systems that may include hardware, software, information, personnel, procedures, and facilities. The language provides graphical representations with a semantic foundation for modeling system requirements, behavior, structure, and constraints. The goal of SysML v2 is to address the shortcomings of SysML v1 such that SysML v2 provides a better foundation for model-based systems engineering. At this time, SysML v2 has not reached the adoption stage in OMG's standardization process.

Whereas SysML v1 was defined as a profile extension of UML, SysML v2 uses a different language architecture with the Kernel Metamodeling Language, KerML, as a simpler replacement of UML and with a library extension mechanism as the principle for defining SysML v2 itself as well as domain-specific library extensions. From a systems engineering perspective, the SysML v2 library provides fundamental constructs for modeling dependencies among elements, for distinguishing definitions of concepts from their context-specific usages, for modeling variability with explicit variation points for configuring systems by selecting compatible choices. Although it is too early to discuss the logical information aspects of SysML v2 and its precise relationship with W3C OWL's 2 DL, the overall strategy is discussed in this chapter [32]. This chapter explains the four-dimensional (4d) paradigm for modeling object behavior as changes to objects in both space (3 dimensions) and time (4th dimension). From an ontological perspective, this chapter corresponds to a particular pattern for modeling change in a 4d paradigm. Are there alternatives that warrant considerations, and what would their pros and cons be? It would be constructive to consider how the biomedical community is addressing these questions. In this chapter [33], the authors describe and evaluate five patterns, including two variants for a 4d paradigm. It is unfortunately difficult to tell whether the SysML v2 4d paradigm corresponds to either variant or is entirely a new, third variant. Recent updates to the SysML v2 standard are summarized in [51]. The language now has a formal ontological footing that focuses particularly on execution semantics with equivalent graphical and textual syntax.

- OntoUML [34]

Before SysML v2, several researchers in academia and industry explored combining conceptual modeling languages like OMG's UML and SysML with a rigorous ontological foundation to ensure that the vocabulary of the conceptual modeling languages has a precise meaning defined in a formal ontology. Thus, OntoUML identified a small subset of UML class diagrams as a suitable language for lightweight conceptual modeling according to a UML profile for a rigorous formal ontology called the Unified Foundational Ontology, UFO. OntoUML resembles SysMLv2 where the UML class diagram subset corresponds to KerML and UFO corresponds roughly to the SysML v2 library. *OntoUML is designed as an expressive language for capturing the ideal ontology of a domain resulting in an axiomatization that requires an intentional modal logic, which is more expressive than the description logic used for formalizing OWL2 as explained in this chapter* [35]. *From a practical standpoint, this means that specifying a domain in OntoUML requires a transformation to approximate the intended domain in OWL2 and SWRL as explained in this chapter* [36]. *The difference between the intended semantics of a domain specified in OntoUML and the corresponding lightweight encoding in OWL2+SWRL after transformation can be difficult for non-experts to appreciate.*

- Semantic Application Language (SADL) (https://github.com/SemanticApplicationDesignLanguage/sadl) [37]

Although developed independently of each other, SADL and OML share a common design objective: instead of expanding the richness of conceptual vocabularies with computationally expensive semantics like SysML v2 and OntoUML, the idea is to restrict the expressiveness of the modeling constructs such that the semantics of the conceptual models remains within the scope of decidable logics, such as W3C's OWL 2 DL. SADL emphasizes providing a controlled English vocabulary as a concrete syntax for SADL models that correspond to a subset of W3C's OWL 2 DL; SADL queries that can be written in English for a subset of SPARQL or in opaque strings for unrestricted SPARQL; and SADL rules for inferences. The SADL tooling environment provides several mechanisms to integrate SADL in a broader environment for model-based systems engineering: a plugin architecture for using Apache Jena or Prolog as a model repository, reasoner, rule engine, and query service.

It is also worth mentioning that, apart from the semantics, languages can provide textual, graphical, or programmatic representations or sometimes all three. These different representations should not affect the semantics of the language though they may affect the ways in which the language can be used. A programmatic interface such as EMF enhances the ability to automate transformations and build tools that natively understand the language. Graphical representations may enhance certain modes of authoring, and textual representations can enhance interoperability with a wide variety of other tools.

Vocabularies and Ontological Reasoning

The semantic web is an approach to information representation that makes the regularly human readable information available on the web also machine readable. The approach was born from the need to improve the precision of the search function on the web. In particular, the idea was to give meaning to the information expressed on the web in natural language, in order to make sense of it, and use it in response to a web search formulated also in natural language.

An analogy can be drawn between the benefit of using semantic web to describe and analyze the web information and the benefit of using it to describe systems engineering information. In the latter, the intent is to improve the precision of the systems engineering information, from being captured in documents written in natural language to being captured in documents written in a formal language that has well-defined vocabulary, syntax, and semantics. Documents of the latter kind are called ontologies.

An ontology is a directed labeled graph that is identified by a unique IRI (Internationalized Resource Identifier) and that describes a set of things by a set of propositional statements that are regarded to be true in some context. Those axioms are represented in a format called RDF (Resource Description Framework), which is a standard for representing information in the form of triples, each of which consists of a subject, a predicate, and an object. For example, "*europa:EuropaClipper rdf:type mission:Mission*" is a triple whose subject is "*europa:EuropaClipper*," predicate is "*rdf:type*," and object is "*mission:Mission*." Triples describe individuals in the world (e.g., EuropaClipper) using terms that give them meaning. These terms can represent either a) classes of the individuals (e.g., Mission) or b) properties of the individuals (e.g., *rdf:type*) and are typically defined in terminology ontologies (TBox), whereas individuals are defined in assertion ontologies (or ABox).

The Web Ontology Language (OWL), a W3C standard [38], is a common language for expressing ontologies. OWL allows a TBox ontology to define a set of terms and provide them with precise semantics that are grounded in logic. A popular class of logic that has proven to be tractable in reasoning is called description logic (DL) [39]. A profile (subset) of OWL that supports DL is called OWL DL [10, 40]. OWL has gone through a major revision and is now in version 2 (OWL 2). Therefore, a language that has good potential to be used to describe and reason on systems engineering information is called OWL 2 DL. This language is supported by off-the-shelf reasoners (e.g., Pellet [3]) that can produce logical entailments (inferred axioms) based on a set of asserted axioms by applying DL inference rules. Those built-in rules can also be augmented by a set of extra inference rules encoded in a language called the Semantic Web Rule Language (SWRL) [23] that has been integrated with OWL 2 DL and is supported by some DL reasoners (like Pellet).

The inference semantics of OWL 2 DL provide a number of advantages when used to describe and reason on systems engineering information. We mentioned that such semantics allow the generation of entailments from assertions. Those entailments are generated for both TBox and ABox ontologies. Since the TBox ontologies describe the terms of a particular domain, along with their rules, such asserted rules,

31 Ontological Metamodeling and Analysis Using openCAESAR   939

and the extra ones that can be inferred from them, should not be contradictory. DL reasoners can verify this with a feature called satisfiability analysis, which checks whether a valid system description is possible under all these rules. If no such description is possible, then the set of TBox assertions are called unsatisfiable. This is important to verify before such TBox can be used to describe systems in ABox. On the other hand, ABox assertions, and their entailments, should also be non-contradictory. DL reasoners can also verify this with a feature called consistency analysis. It allows verifying that system description, encoded in ABox statements, is logically consistent. This is again very important to know before further analysis can be performed on such system description, since a self-contradictory description cannot be a basis or for any downstream analysis.

Furthermore, the entailments produced by DL reasoners have another advantage, which is to simplify queries that are made in downstream analyses. Such queries are typically formulated using a language called SPARQL [41], which is a standard query language for RDF. A SPARQL query defines a triple pattern containing variables to match in some graph scope. The triples that are matched could have been either asserted in the original dataset or inferred by the DL reasoner based on the inference rules. Without such capability, each query must account for that inferencing. For example, consider a *base:contains* property, which has a domain of *base:Container* and a range of *base:Contained*. It can express a relation from some container element to another element contained by it. Such property is defined as *transitive*, meaning that if we have an axiom *:A base:contains :B* and another axiom *:B base:contains :C*, then a DL reasoner can infer that *:A base:contains :C*. This allows formulating a simple SPARQL query that matches the last axiom directly, as opposed to having to match the first two axioms. The utility of this advantage becomes clear when analyzing a large dataset that is expressed with a vocabulary that is rich in inference semantics.

Ontological Modelling Language

While OWL 2 DL can effectively be used to describe information about systems engineering, it can be tricky to encode information with it to achieve the desired analysis result. This is because unlike most other information description formalisms, like XSD, MOF, and ER, which use the closed-world assumption (what is not stated to be true must be false), OWL 2 DL adopts the open-world assumption (what is not stated to be true may still be true). This assumption may lead to surprising results if not carefully taken into account, for example, stating that all members of some class must have a value for an *ID* property and failing to give some member *A* of the class such a value is an error under the closed-world interpretation: the lack of an *ID* value means *A* has no *ID* value. In the open world, the interpretation is that *A might* have an *ID*; we haven't said one way or the other. To reach the equivalent closed-world interpretation, we have to additionally assert that *A* has *no ID* value. Then an OWL 2 reasoner will detect inconsistency.

Furthermore, OWL 2 DL is a very flexible and intentionally methodology-agnostic language so that it can be used in a variety of contexts. However, when used in a given context, like systems engineering, it is often more practical to specialize it and/or constrain it so that it is used in a specific methodology. Such methodology will ensure that the language is used consistently and allow making assumptions and optimizations based on that.

When we started using OWL2 DL for our systems modeling, we quickly realized that verifying logical consistency and satisfiability of systems model ontologies was insufficient to verify their compliance with respect to a systems modeling methodology unless we also verified that methodology-specific patterns were correctly encoded in these ontologies. This led us to defining methodology-specific audit rules to check systems model ontologies for violations of our methodology. While effective in finding genuine errors after the fact, these audit rules also produced many false negative and positive errors because several methodology patterns, while conceptually simple, required syntactic complexity to represent properly in OWL that, in turn, led to errors of omission and syntactic conflicts respectively. The gap between the concise intent of our methodology-specific vocabularies and the brittle minutiae of their syntactic representation in OWL2-DL led us to create a new language that could express the intended OWL2 vocabularies more concisely while retaining semantic precision as much as possible. Working across multiple systems engineering domains gave us key insights into the tension between authoring extensible domain-specific vocabularies versus composing them for analysis purposes. For example, negative axioms and disjunctions are useful for the latter but counter-intuitive to author properly in the former. Restricting class expressions to named class concepts facilitates understanding of reasoner explanations later in analysis but requires counter-intuitive discipline earlier in authoring. Acknowledging that mastering these subtleties of the OWL2-DL language requires significant experience and wisdom that few domain-specific systems engineering experts may have, we sought to create a sub-language of OWL2-DL that makes authoring reusable and composable vocabularies easier and that introduces a powerful notion of ontology management policy for generating the most error-prone, counter-intuitive disjunctive axioms that provide the most useful methodological guidance when reasoning about a system in the context of a commitment to a closed-world set of vocabularies. By defining the syntax and semantics of this sub-language to a subset of OWL2-DL extended with SWRL, this approach allows us to take advantage of the many tools available for working with OWL2-DL while simplifying the authoring and management of models and vocabularies.

We call such language the Ontological Modeling Language (OML). The language abstractions map to a set of patterns defined on a subset of OWL 2 DL, which makes describing information with OML more concise and streamlined than using OWL 2 DL directly. The sections below describe the main abstractions of OML.

Vocabulary

An OML vocabulary is an ontology for defining terms (and rules) that are used to describe information in a particular domain (with open-world semantics). The terms can include types of things in the domain, their properties, and their interrelations. In each case, OML has abstractions to define these things precisely. For example, an OML type can be an aspect (a mixin capability used as a supertype), a concept (a class of real things identified by id), a structure (a class of anonymous objects unique by value), or a scalar (a primitive type that classifies literals). An OML property can be an annotation property (has no semantics) or a feature property (has semantics). The latter can either be a scalar property (with literal values) or a structured property (with structured values). An OML relation can be a forward relation (unidirectional) or a reverse relation (unidirectional but opposite to some forward relation). Both relations have a corresponding relation entity that reifies them (defines a type of objects representing the relations) and can have its own properties and interrelations. Vocabulary can extend each other and can also import description ontologies to use them as standard libraries.

Vocabulary Bundle

An OML vocabulary bundle is an ontology that imports a set of OML vocabularies and gives them closed-world semantics. This means adding assertions that declare certain types in those vocabularies as disjoint, i.e., cannot be mixed together as types of instances (individuals). This is a very useful capability for defining methodology-based closed-world vocabularies to use for reasoning (since reasoning on open-world vocabularies often lead to unexpected results) without committing to this closedness upfront. This means the types in vocabularies are extensible by definition and can participate in different bundles driven by a given methodology. Vocabulary bundles can also extend each other, allowing the creation of methodology layers.

Description

An OML description is an ontology for describing information in a given system using some vocabulary. Descriptions can define instances classified by types from the vocabulary, along with their property values and relationships. Instances can either be named (e.g., concept instances and relation instances) or anonymous (structure instance). They can also be multi-classified by more than one type. Descriptions often describe time-insensitive information (like requirements or static configurations) or time-sensitive information (like the state of a system at a point in time). The latter can be used to describe various instances of the system state (start, end, middle, etc.) in a simulation.

Description Bundle

An OML description bundle is an ontology that imports a set of OML descriptions and gives them closed-world semantics. This means adding assertions that limit the truth to the set of already existing assertions in the bundle. For example, for a property with a lower cardinality restriction, there will be assertions on instances of the domain of this property stating that the number of values is restricted to the actual one asserted. This will allow a reasoner to detect violations of lower cardinality restrictions that are otherwise not detectable under open-world semantics. Moreover, description bundles represent self-contained datasets to reason on. They can be used to model exclusive variants in a bigger dataset since they allow including ontologies that belong to each variant.

OML Analysis Workflows

After having created ontologies, both vocabularies and system descriptions, using OML, we need to create workflows that [42, 43] (https://github.com/stardog-union/pellet) (https://github.com/opencaesar) consist of model transformations, analyses, view generators, and so forth and automate them to the extent possible in order that they may be used in a continuous integration (CI) process. That is, rather than having analyses be performed ad hoc on the models, they are integrated into an automated process triggered by model changes made through formal commits or releases in a configuration management system such as git. This not only provides a quality gate, through which authored information must pass before being released to a wider user community; it also provides feedback to both communities on completeness and other quality metrics of the product that can be measured on a continuous basis.

CI processes, or workflows, for model data will typically render a set of analytical views as primary products. These views can be simple projections of model data into discipline-specific (user-friendly) visualizations – typically as tables or visualizations of data derived through analytical reductions of the raw model data. In the simplest case, the model data for a workflow may be a single physical model (authored by a single tool but possibly by multiple people). In this case, the workflow is used to analyze (validate, reason on, solve, simulate, etc.) and publish information from that specific model.

More complex workflows can be used to merge and reconcile multiple models. Here, the process is more complicated because inconsistencies can take the form of various kinds of conflicts or disconnects between the input models that may require corrections in some or all of the source models (we prefer to use the term reconciliation here rather than synchronization because this suggests more of a negotiated process, which is more typically the case. In rare cases, where one source model can be declared correct, it may be possible to automate a transformation whereby a secondary model can be generated from a primary model by transformation (derivation), but then it isn't reconciliation anymore – it's just a transformation. Reconciliation processes typically result in views that report differences between the source models that will have to be negotiated between authors of the source models and repaired through changes to the source models (e.g., element X is found to be in model A but not in model B. Is that important? Is it actually present under a different name?).

The ability to compare and analyze models is facilitated by having them in a single consistent form, and OWL 2 DL is an ideal form for doing that due to its expressiveness and modularity. Thus, the middle part of every analysis workflow consists of transformations needed to render the OML models (vocabularies and descriptions) in OWL 2 DL, merging any model fragments into a single "model bundle" from which inferences can be generated and storing the assertions and inferences in a way that makes them accessible to a SPARQL (https://www.w3.org/TR/rdf-sparql-query/) query endpoint.

A SPARQL interface makes the model data accessible to a wide range of analysis tools that can then be used to query, reduce, and render the data into user-friendly visualizations or perform additional analysis.

openCAESAR Tools

Many useful tools have been developed in the semantic web community for working with OWL ontologies, including editors such as Protégé [42] and reasoners such as Pellet [43] (https://github.com/stardog-union/pellet). In order to better support and streamline the OML methodology described above, we developed a set of additional tools that can be used in combination with available open source or commercial tools. We packaged these tools into a common platform called openCAESAR (Computer Aided Engineering for System Architecture) and opened sourced it on GitHub (https://github.com/opencaesar). openCAESAR is best characterized as an agile multi-disciplinary ontology-based system modeling platform. In this section, we discuss some of tools of the platform that were developed to support model authoring and definition of workflows that perform useful model transformations, analyses, and reporting using user-friendly views (https://opencaesar.github.io/).

OML Workbenches

A the heart of the openCAESAR platform is the Ontological Modeling Language (OML), which is a domain-specific language that is used to describe information. OML makes describing information more concise and streamlined than using OWL 2 DL directly. OML has a concise textual syntax that lends itself to editing with ordinary text editors and that makes it more usable with text-based change control systems such as Git. OML also supports a Java-based API that is packaged and published as a maven artifact. This API is designed with the Eclipse Modeling Framework (EMF), which allows it to be used with a wide range of tools that work with EMF. For example, openCAESAR provides an extensible Eclipse-based workbench for OML called Rosetta. The workbench packages the OML API as a plugin. It also supports a context-sensitive textual editor for OML designed with EMF's XText framework [44]. Moreover, it supports model-aware domain-specific authoring viewpoints and user interfaces that are designed with EMF-compatible user-interface frameworks, such as Sirius [45] (https://www.eclipse.org/sirius/overview.html). Rosetta also supports popular code management systems, such as git, and collaboration features like comparing and merging OML models, using their textual syntax, but also the more precise abstract syntax (language abstractions).

The openCAESAR platform also provides a language server for OML that supports the standard Language Server Protocol (LSP) [46] (https://langserver.org/). This allows it to be supported by any textual editor that supports the LSP, including traditional IDEs like Eclipse and VS Code, but also emerging web-based IDEs like Theia [47] (https://theia-ide.org/). The latter also supports a graphical viewer for OML implementing its graphical notation (Figs. 1 and 2).

Analysis Services (https://langserver.org/) (https://theia-ide.org/)

To support continuous integration of information, openCAESAR provides a number of model transformation and analysis services (listed below as subsections). Such services are packaged as command line interfaces, but also as Gradle task types. Gradle is a widely used, Java-based build tool commonly used to build continuous integration workflows. Gradle is able to download analysis tools on demand from Maven repositories, where they are published with semantic versions, and orchestrate them in the build process. openCAESAR also leverages Gradle scripts to publish and download OML models as artifacts with semantic versions to/from Maven repositories. This allows OML models to declare dependencies on other

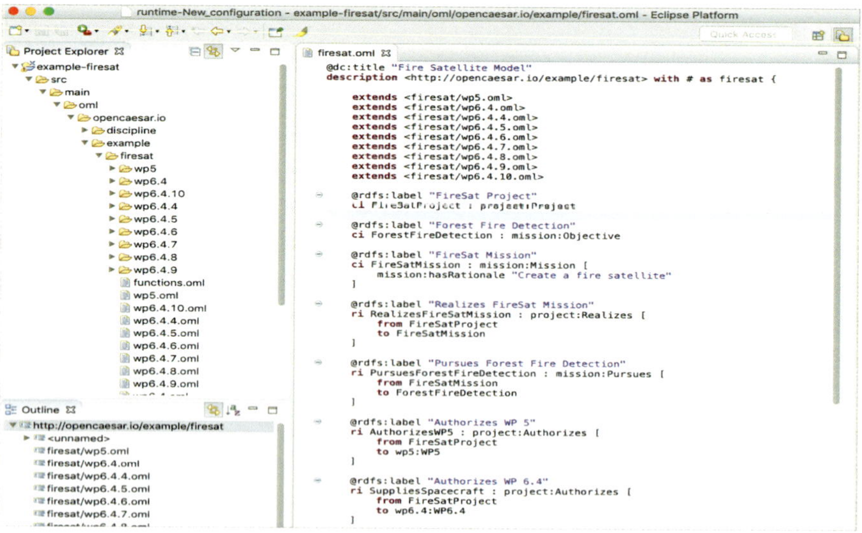

Fig. 1 The Eclipse-based Rosetta workbench

31 Ontological Metamodeling and Analysis Using openCAESAR 945

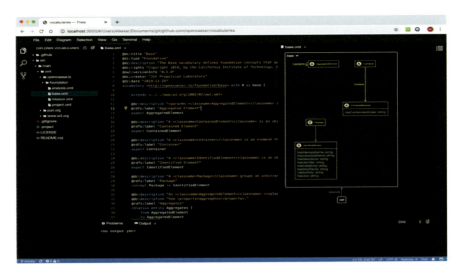

Fig. 2 Theia web browser-based workbench

models with semantic versions without having to manage them together in the same repositories. It also allows models to compose other models in the same way that large software systems are composed of a number of separately managed software libraries.

OML Bikeshed

This service generates canonical spec documents for an OML dataset (a set of related models) using Bikeshed [48], (https://tabatkins.github.io/bikeshed/) which is a spec document platform used by the W3C specifications. Every ontology in the dataset has its own Bikeshed spec that has sections corresponding to its named members. In the case of a vocabulary ontology, these would be the terms and rules, whereas in the case of a description ontology, these would be the named instances.

The Bikeshed format extends the Markdown format, which can be rendered to HTML. Therefore, the generated documents are a set of crosslinked HTML pages that can be published together as a web site. When invoked from a CI script, this service can rebuild the website corresponding to the latest commit of an OML dataset. The website can then be delivered to a web server (e.g., GitHub Pages) or an artifact repository (e.g., Bintray).

OML Merge

This service can take as input a set of OML datasets and produces as output a single OML dataset that is the union of those input datasets. When ontologies have unique

IRIs, they are simply copied over to the output dataset. When ontologies in different datasets have the same IRIs, they are checked first for equivalence and will only be copied if they are equivalent. If they are not, then the service will give an error. This is useful in order to alert the user that incompatible versions of the same ontology are being merged. This situation may occur when more than one dataset depends on different versions of a common ontology (e.g., a vocabulary). It may also occur when multiple authorities mistakenly use the same IRIs, where they are supposed to be unique.

This service is often used from a Gradle script analyzing a dataset to download and merge its OML dependencies. In this case, those dependencies are downloaded from a Maven repository using their maven coordinates and merged as a read-only dataset. Fortunately, Gradle provides ways to address the problem of multiple dependencies transitively depending on different versions of a common dependency. By default, it chooses the most recent semantic version of the common dependency. This often works well when different patch or minor versions are used together. However, when different major versions (typically incompatible) are mixed, Gradle can be configured to give errors.

OML to OWL Adapter

This adapter transforms OML models to the equivalent OWL syntax so that models can be ingested and analyzed by a wide variety of tools that understand OWL. It takes as input an OML dataset and produces as output an OWL dataset that is syntactically and semantically equivalent. Both datasets represent a set of axioms that are asserted by their authorities.

OML Reason

One of the key benefits of mapping OML to OWL 2 DL is that generic DL reasoners (like Pellet[43]) (https://github.com/stardog-union/pellet) can be used to reason on the consistency of ontology bundles. The OWL reason service enables that. When run against a vocabulary bundle, this service can perform satisfiability analysis, i.e., checking that classes in the bundle are not over-constrained such that no valid instances can be described, for example, if a class C restricts the value of an integer property to the value 0 and the value 1 at the same time. This is equivalent to making sure that software libraries compile before using them in an application. When run against a description bundle and its used vocabulary bundles, this step will find cases where the bundle violates the rules of the vocabularies (e.g., the number of property values are not in the restricted cardinality range) or when contradictions exist or can be inferred (e.g., an instance is classified by two classes A and B, but those classes are asserted to be disjoint). Any violations that are detected by this service will be logged to a file that follows the JUnit result format, which is widely recognized by CI tools, allowing benefiting from existing visualization support.

As a side effect of this process, the reasoner will also produce a set of inferences that can enrich the model and enable simplifying the SPARQL queries in a subsequent step (as opposed to building such inference in the queries themselves). For example, if component *C1 contains* component *C2, C2 contains C3*, and *contains* is transitive, the reasoner will conclude that *C1 contains C3* and add axioms to that effect. With these additional axioms in scope, queries can interrogate what is true, not merely what has been asserted.

OML Load

This service can load OWL ontologies into a graph database (specified as a URL) with a SPARQL query endpoint. Every ontology will be loaded to its own named graph in the database. The set of ontologies to load can be specified as the whole dataset or as a specific ontology and its import transitive closure. The latter enables loading different subsets of the dataset to different databases. Such subsets, often represented as description bundles, can represent variants in the datasets (e.g., alternative spacecraft configurations) that can be analyzed independently.

OWL SPARQL

This service allows a set of SPARQL queries, captured as input files, to be run as a batch on a SPARQL endpoint (specified as a URL) and the resulting data frames (tables with the selected variables as columns) persisted in output files. These frames can be requested in various formats, including XML, CSV, or JSON. Such frames can then be reduced by other analysis tools or rendered into report or document views.

OWL SHACL

This service allows a set of SHACL [49] (https://www.w3.org/TR/shacl/) rules (called shapes), captured as input files, to be run on a SHACL endpoint (specified as a URL) exposed by a triple store. SHACL is a standard language for detecting the existence of patterns (shape instances) in a dataset. A common use case is using them to validate a dataset by detecting anti-patterns (undesired patterns). The result of the detection is then serialized as output files that can be further processed.

OWL Start/Stop Fuseki

This service allows starting an instance of a headless Fuseki [50] (https://jena. apache.org/documentation/fuseki2/) server in the background and stopping it later. Fuseki is an open-source triple store that supports creating RDF/OWL databases (called datasets) and exposes SPARQL and SHACL endpoints on them. This service can be used from a Gradle script to conveniently spawn a Fuseki instance, load OWL datasets to it, and then run queries or rules on them. At the end, the Fuseki instance can be stopped. While one could also use a cloud-based triplestore to load and query data in a more public way, this method of locally deploying a private instance works better in a continuous integration environment because it avoids all contention for a shared resource.

Tool Adapters

The openCAESAR project also provides a number of transformations (called tool adapters) that can translate between OML and other data formats used by various engineering tools. So far, an adapter has been developed for the Ecore [3] (https://www.eclipse.org/modeling/emf/') format. Several adapters to some UML/SysML tools are also being developed now.

Example Ontologies

The openCAESAR project contains GitHub repositories that exemplify the use of OML for modeling different vocabularies and systems and using the openCAESAR tools to analyze them. Those repositories contain OML models and Gradle build scripts that run the analysis tools (described above) on them. Those Gradle scripts are invoked (https://travis-ci.org/) by Travis CI (https://travis-ci.org/) scripts whenever commits are pushed to the repositories. If the build is successful, the models get archived and published with semantic versions to a Maven repository (on Bintray), where other projects can declare them as read-only dependencies. The build process also generates Bikeshed HTML documentation for the models and publishes them as GitHub Pages for the repositories.

In the subsections below, we describe some of those OML example repositories that are provided by the openCAESAR project.

Core Vocabularies

The *core vocabularies* repository contains a set of building block OML vocabularies that can be incorporated in any systems engineering methodology. Some of those vocabularies (e.g., w3c, purl) represent standard ontologies that have been (partially) translated to OML. Others (e.g., metrology, iso) represent information from the literature that have been represented ontologically in OML directly. Yet, others (e.g., diagram) represent CAESAR specific vocabularies that enable certain analysis tools (Fig. 3).

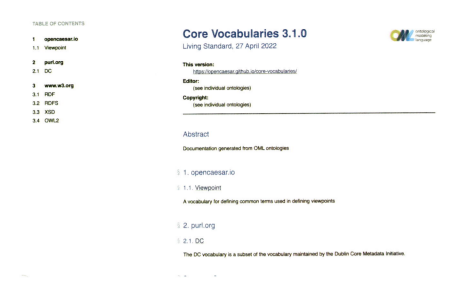

Fig. 3 Index page of openCAESAR core vocabulary documentation

IMCE Vocabularies

The *IMCE vocabularies* repository contains a bundle of OML vocabularies contributed by the IMCE (Integrated Model Centric Engineering) team at JPL and represents the foundational basis of the JPL systems engineering methodology. The bundle, named foundation, allows modeling of systems, projects, and missions and analyzing their characteristics. The *IMCE vocabularies* declare core vocabularies as a dependency in order to import them (Fig. 4).

Keep in mind that the *IMCE vocabularies* were originally developed to serve as a basis for describing the systems engineering methodology at JPL and may or may not be suitable for general-purpose use (however they definitely serve as examples).

The *IMCE vocabularies* consist mainly of the Mission and Project vocabularies. These vocabularies extend some primitive concepts from the base vocabulary such as the *contains* relation. They also import a number of other core vocabularies that define common data types or annotations. For example, the Dublin Core provides a standard set of metadata attributes that can be used to describe the terms of the vocabularies beyond their DL semantics.

The **Project** vocabulary describes the organizational aspects that govern the process within a project. The concept of a *Project* and its hierarchical allocation of authority through a work breakdown structure are defined here. JPL uses a matrix organization style where the projects are distinct from the implementing organizations who do the work. So, the vocabulary defines *organizations* populated by *persons* who can be allocated responsibility over *work packages* via *roles*. Work

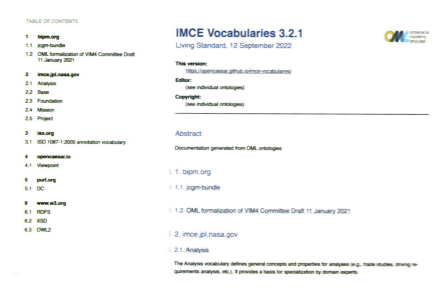

Fig. 4 Index page of openCAESAR IMCE vocabulary documentation

packages can produce **products** which, in this vocabulary, describe things such as specifications, plans, and designs which are the deliverables of systems engineering.

The *Mission* vocabulary focuses on describing a functional system that is the subject of project work. A *component* is defined as an element of a functional decomposition of a system. At JPL a *mission* is typically the root system. *Components* perform *functions* and can exist in a compositional hierarchy. To connect the functional architecture to the project, a *component* can be supplied by a *work package*. This relation defines the subject of specification, design, or implementation for the *work package*. *Components* may also present *interfaces* through which they can interact with other *components* via *junctions*. A mission may also include environments that are not supplied or engineered but can interact with *components*.

In several applications where these vocabularies have been used, we have defined discipline specializations of some of this vocabulary. So, for example, the specification used for harnessing specification defines assembly as a physical (hardware) *component* that can present *EndCircuit Interfaces* to represent abstractly the internal electrical circuits that need to be joined between assemblies.

And then, projects can further specialize concepts and relations in project libraries. For example, a project can define types of *assemblies* such as antenna or heater with distinct interfaces and properties (IMCE intends to provide a curated model library with some common types in future). This layering is important because all of the application tooling including the authoring and reporting capabilities is designed around the discipline vocabulary (the common abstractions), but the generic tooling can still understand specializations defined in a library without having to change any tooling. When developing local vocabularies from scratch or extending these examples, this is an important pattern to keep in mind.

FireSat Example

The *FireSat example* repository provides an example of using OML and the IMCE vocabularies to describe an open-source spacecraft called FireSat using the IMCE vocabularies. The model is intended to illustrate the kinds of things and relations that can be expressed along with examples of constraints, queries, and transformations to document presentations enabled using the openCAESAR tools previously described. It defines some simple discipline-specific extensions to the imported IMCE vocabularies in the src/oml/imce.jpl.nasa.gov/discipline/fse folder which are then used in the instance models to describe a project work decomposition, a basic functional architecture with allocation of architectural components to subsystems and supplier work packages, some requirements, and some electrical interfaces. This is sufficient to then demonstrate the ability of the openCAESAR tools to analyze the model for consistency, produce entailments, load that data into a triplestore database, execute several queries, and transform the model data into a general specification document and a more specifically focused document whose content is derived from the query results.

Chapter Summary

Metamodeling is essentially about extending vocabulary to specialize meaning as distinct from merely enumerating instances of some well-defined concept or relation. Modeling the world includes both of those aspects. Ontological metamodeling is merely using semantic web technologies for the purpose of defining vocabularies with which to model real-world applications. The Web Ontology Language (OWL) is a good language for both direct description modeling and metamodeling as it was designed to do both. Adding description logic semantics ensures that the language is limited to expressing decidable logical statements from which deterministic inferences can be drawn.

Practical metamodeling should focus on defining vocabulary to precisely and concisely describe the things you need to describe. This is not necessarily the same thing as trying to capture the meaning of natural language used in similar circumstances (the word ontology can also refer to the study of the meaning of language or just the description of language; here we are more specifically defining language with the specific meanings we intend).

The OML was defined by openCAESAR as a particular encapsulation of OWL 2 DL into both a programmatic API and a textual syntax that concisely enforces some best practices of OWL. But it retains all of the descriptive modeling power of OWL and compatibility with a wide range of commercial and open-source tools. openCAESAR also provides some useful transformation and analysis tools wrapped as Gradle tasks in order to support automated continuous integration processes for merging, validating, and transforming models into useful review products such as reports and documents or data products for use in other tools.

Acknowledgment Some of the work described here was performed at the Jet Propulsion Laboratory, California Institute of Technology, under a contract with the National Aeronautics and Space Administration.

References

1. "A Guide to the Systems Engineering Body of Knowledge (SEBoK)." [Online]. Available: https://www.sebokwiki.org/wiki/Guide_to_the_Systems_Engineering_Body_of_Knowledge_ (SEBoK).
2. W. Wymore, *Mathematical Theory of Systems Engineering: The Elements*. Malabar, FL: Krieger Pub Co, 1977.
3. "Eclipse Modeling Framework (EMF)." [Online]. Available: https://www.eclipse.org/modeling/emf/.
4. "Siemens Teamcenter." [Online]. Available: https://www.plm.automation.siemens.com/global/en/products/teamcenter/.
5. "Dassault Systemes Catia." [Online]. Available: https://www.3ds.com/products-services/catia/.
6. N. Kahani, M. Bagherzadeh, J. R. Cordy, and E. Al., "Survey and classification of model transformation tools," *Softw. Syst. Model.*, vol. 18, pp. 2361–2397, 2019.
7. R. Eramo, A. Pierantonio, J. R. Romero, and A. Vallecillo, "Change Management in Multi-Viewpoint System Using ASP," in *12th Enterprise Distributed Object Computing Conference Workshops*, 2008, pp. 433–440.
8. P. Stünkel, O. Bargen, A. Rutle, and Y. Lamo, "GraphQL Federation: A Model-Based Approach.," *J. Object Technol.*, vol. 19, p. 18:1, Jan. 2020.
9. T. Hedberg, Jr., J. Lubell, L. Fischer, L. Maggiano, and A. Barnard Feeney, "Testing the Digital Thread in Support of Model-Based Manufacturing and Inspection," *J. Comput. Inf. Sci. Eng.*, vol. 16, no. 2, Mar. 2016.
10. "Web Ontology Language with Description Logic." [Online]. Available: https://www.w3.org/TR/owl2-syntax/.
11. "Unified Modeling Language (UML)." [Online]. Available: https://www.uml.org/what-is-uml.htm.
12. "Systems Modeling Language (SysML)." [Online]. Available: https://www.omg.org/technology/readingroom/System-Modeling-Language.htm.
13. "Functional UML (FUML)." [Online]. Available: https://www.omg.org/spec/FUML.
14. "The CoQ Proof Assistant." [Online]. Available: https://coq.inria.fr/.
15. "Lean Prover." [Online]. Available: https://leanprover-community.github.io/index.html.
16. V. S. Alagar and K. Periyasamy, "Vienna Development Method," 2011, pp. 405–459.
17. "ISO/IEC 13568:2002." [Online]. Available: https://www.iso.org/standard/21573.html.
18. "Wikipedia: Soundness." [Online]. Available: https://en.wikipedia.org/wiki/Soundness.
19. "The Distributed Ontology, Modeling and Specification Language – DOL." [Online]. Available: http://wiki.dol-omg.org/index.php/DOL. [Accessed: 02-Dec-2021].
20. C. Atkinson and T. Kuhne, "Model-driven development: a metamodeling foundation," *IEEE Softw.*, vol. 20, no. 5, pp. 36–41, 2003.
21. I. Ivanov, J. Bézivin, and M. Aksit, "Technological Spaces: An Initial Appraisal ." pp. 1–6, Oct-2002.
22. O. M. Group, "Meta-Object Facility, version 2.5.1." [Online]. Available: https://www.omg.org/spec/MOF/2.5.1/PDF. [Accessed: 12-Feb-2021].
23. "A Semantic Web Rule Language Combining OWL and RuleML (SWRL)." [Online]. Available: https://www.w3.org/Submission/SWRL/.

24. "OWL2 Computational Properties." [Online]. Available: https://www.w3.org/TR/owl2-profiles/#Computational_Properties.
25. "Catia Cameo System Modeler." [Online]. Available: https://www.nomagic.com/products/cameo-systems-modeler%0D%0A%0D%0A.
26. "Object Constraint Language (OCL)." [Online]. Available: https://www.omg.org/spec/OCL/2.4/PDF.
27. "Simulink and Modelica Transformation Plugin." [Online]. Available: https://docs.nomagic.com/display/SMTP190SP2/19.0+LTR+SP2+Version+News. [Accessed: 02-Dec-2021].
28. B. Henderson-Sellers, O. Eriksson, C. Gonzalez-Perez, P. J. Ågerfalk, and G. Walkerden, "Software modelling languages ," *A wish list* . Institute of Electrical and Electronics Engineers (IEEE) , Piscataway, NJ , pp. 72-77 BT-Proceedings-7th International Works, 2015.
29. "Syndeia." [Online]. Available: http://intercax.com/products/syndeia/.
30. "Generic Modeling Environment." [Online]. Available: https://www.isis.vanderbilt.edu/projects/GME.
31. T. L. and Á. L. Maróti, M., T. Kecskés, Róbert Kereskényi, Brian Broll, P. Völgyesi, L. Jurácz, "Next Generation (Meta)Modeling: Web- and Cloud-based Collaborative Tool Infrastructure," in *MPM@MoDELS*, 2014.
32. C. Bock and C. Galey, "Integrating four-dimensional ontology and systems requirements modelling," *J. Eng. Des.*, vol. 30, no. 10–12, pp. 477–522, Dec. 2019.
33. P. Burek, N. Scherf, and H. Herre, "Ontology patterns for the representation of quality changes of cells in time," *J. Biomed. Semantics*, vol. 10, no. 1, p. 16, 2019.
34. "Onto UML." [Online]. Available: https://ontouml.org/.
35. G. Guizzardi, "On Ontology, ontologies, Conceptualizations, Modeling Languages, and (Me-ta) Models," in *Proceedings of the 2007 conference on Databases and Information Systems IV: Selected Papers from the Seventh International Baltic Conference DB&IS'2006*, 2006, pp. 18–39.
36. M. M. and A. G. Barcelos, P., Victor Amorim dos Santos, Freddy Brasileiro Silva, "An Automated Transformation from OntoUML to OWL and SWRL," in *ONTOBRAS (2013)*, 2013.
37. "Semantic Application Design Language." [Online]. Available: https://github.com/SemanticApplicationDesignLanguage/sadl.
38. "Web Ontology Language." [Online]. Available: https://www.w3.org/OWL/.
39. *The Description Logic Handbook: Theory, Implementation and Applications*, 2nd ed. Cambridge: Cambridge University Press, 2007.
40. B. C. Grau, I. Horrocks, B. Motik, B. Parsia, P. Patel-Schneider, and U. Sattler, "OWL 2: The next step for OWL," *J. Web Semant.*, vol. 6, no. 4, pp. 309–322, 2008.
41. "SPARQL Query Language for RDF." [Online]. Available: https://www.w3.org/TR/rdf-sparql-query/.
42. "Protege OWL Editor." [Online]. Available: https://protege.stanford.edu/.
43. E. Sirin, B. Parsia, B. Grau, ... A. K.-W. S. science, and U. 2007, "Pellet: A practical owl-dl reasoner," *J. Web Semant.*, vol. 5, no. 2, 2007.
44. "XText." [Online]. Available: https://www.eclipse.org/Xtext/. [Accessed: 12-Feb-2021].
45. "Sirius UI Framework for Eclipse." [Online]. Available: https://www.eclipse.org/sirius/overview.html.
46. "Language Server Protocol." [Online]. Available: https://microsoft.github.io/language-server-protocol/. [Accessed: 21-Feb-2021].
47. "Theia Cloud & Desktop IDE Platform." [Online]. Available: https://theia-ide.org/. [Accessed: 12-Feb-2021].
48. "Bikeshed specification generator." [Online]. Available: https://tabatkins.github.io/bikeshed/.
49. "Shapes Constraint Language (SHACL)." [Online]. Available: https://www.w3.org/TR/shacl/.
50. "Apache Jena Fuseki." [Online]. Available: https://jena.apache.org/documentation/fuseki2/.
51. "Systems Modeling Language (SysML v2) Support for Digital Engineering", [Online], Available: https://incose.onlinelibrary.wiley.com/doi/abs/10.1002/inst.12367

David A. Wagner manages both the System Modeling and Methodology group as well as product development on the CAESAR project at the Jet Propulsion Laboratory. He holds a bachelor's degree in Aerospace Engineering from the University of Cincinnati, a master's degree in Aerospace from the University of Southern California, and over 40 years of experience in systems engineering large and small projects at JPL.

Mark Chodas is a systems engineer at the Jet Propulsion Laboratory. He is a product owner for the CAESAR project and contributes to other model-based systems engineering tools at JPL. He previously was the Instrument Systems Engineer for the REXIS instrument onboard NASA's OSIRIS-REx asteroid sample return mission. He holds bachelor's, master's, and PhD degrees in Aerospace Engineering from the Massachusetts Institute of Technology.

Maged Elaasar is a senior software architect at the Jet Propulsion Laboratory (California Institute of Technology/NASA), where he technically leads the Integrated Model Centric Engineering program and the CAESAR product. Prior to that, he was a senior software architect at IBM, where he led the Rational Software Architect family of modeling tools. He holds a PhD in Electrical and Computer Engineering and MSc in Computer Science from Carleton University, (2012, 2003) and a BSc in Computer Science from American University in Cairo (1996). He has received 12 US patents and authored over 30 peer-reviewed journal and conference articles. He is a regular contributor to and leader of modeling standards at the Object Management Group like UML and SysML. Maged is also the founder of Modelware Solutions, a software consultancy and training company with international clients. He is also a lecturer in the department of Computer Science at the University of California Los Angeles. His research interests span model-based engineering, semantic web, big data analytics, and cloud computing.

Steve Jenkins is a Principal Engineer in the Systems Engineering and Formulation Division, Jet Propulsion Laboratory, California Institute of Technology. Since 2009 he has served as the Chief Engineer of JPL's Integrated Model-Centric Engineering Initiative. His interests include the integration of descriptive and analytical modeling and the application of knowledge representation and formal semantics to systems engineering. He was awarded the NASA Outstanding Leadership Medal in 1999 and was a co-recipient of the NASA Systems Engineering Award in 2012. He holds a BS in Mathematics from Millsaps College, an MS in Applied Mathematics, and PhD in Electrical Engineering from the University of California, Los Angeles.

Nicolas Rouquette is a Principal Computer Scientist at the Jet Propulsion Laboratory where his pioneering work on comprehensive code generation for JPL's Deep Space One mission paved the way for applying such techniques since then. He made key contributions to several revisions of modeling standards developed by the Object Management Group (OMG) for UML and SysML, including for UML 2.4.1 which became the first revision of UML for which formal verification of consistency was performed by mapping to OWL 2 and for SysML 1.4 which became the first revision of SysML for which comprehensive support 13 for the 3rd revision of the Vocabulary of International Metrology (VIM) is available in SysML's ISO 80000 library. He holds a PhD in Computer Science from the University of Southern California where he enjoyed guidance on applied graph algorithms from Pavel Pevzner and then a PostDoc at USC's Math department. He also holds an engineering diploma from the French ESIEE engineering school in Paris.

MBSE Validation and Verification

Case Study for LADEE

32

Karen Gundy-Burlet

Contents

Introduction 956
State of the Art 960
Best Practice Approach 961
    Software Overview 962
    Prototyping the Process 963
    Development Cycle 964
    Requirements 966
    Simulation Hardware 968
    Test Infrastructure and Traceability 969
    Model-Based Unit Testing 969
    GOTS Components 970
    Command and Telemetry Dictionary 970
    ConOps-Driven Testing Methodology 972
    Test Patterns for Verification of Level-4 Requirements 975
    Guidance, Navigation, and Control 975
    Commanding Modes and Transitions 975
    Fault Management 976
    Commands and Telemetry 976
    Testing Cadence During Development 976
    Formal Inspection 977
Illustrative Examples 977
    Training and Mission Operations 977
    Ambiguous ICDs 978
    Emergent Behavior 979
System Reboot 980
    Final Load 982
Conclusions 983
Cross-References 984
References 984

K. Gundy-Burlet (✉)
Crown Consulting Inc., NASA-Ames Research Center, Moffett Field, CA, USA
e-mail: kgundyburlet@crownci.com

© This is a U.S. Government work and not under copyright protection in the U.S.; 955
foreign copyright protection may apply 2023
A. M. Madni et al. (eds.), *Handbook of Model-Based Systems Engineering*,
https://doi.org/10.1007/978-3-030-93582-5_58

> **Abstract**
>
> The Lunar Atmosphere Dust Environment Explorer (LADEE) mission orbited the moon in order to measure the density, composition, and time variability of the lunar dust environment. The successful mission launched September 7, 2013 and was de-orbited and impacted the moon's surface on April 17, 2014. The ground-side and onboard flight software for the mission was developed using a Model-Based Software Engineering (MBSE) methodology combined with strong reuse of Government and Commercial Off-The-Shelf (G/COTS) components. Models of the spacecraft and flight software were developed in a graphical dynamics modeling package. Flight software requirements were prototyped and refined using the simulated models. After the model was shown to work as desired in the simulation framework, C-code software was automatically generated from the models. The auto-generated software was then tested in real-time Processor-in-the-Loop and Hardware-in-the-Loop test beds. Traveling Road Show test beds were used for early integration tests with payloads and other subsystems. Traditional techniques for verifying computational sciences models were used to characterize the spacecraft simulation. A lightweight set of formal methods analysis, static analysis, formal inspection, and code coverage analyses was utilized to further reduce defects in the onboard flight software artifacts. These techniques were applied early and often in the development process, iteratively increasing the capabilities of software and fidelity of vehicle models and test beds.

> **Keywords**
>
> Model-based systems engineering · Validation and verification · LADEE · Software engineering · Automated testing · Auto-code

Introduction

The Lunar Atmosphere Dust Environment Explorer (LADEE) [1] was a small explorer-class mission that was launched on September 7, 2013 and successfully de-orbited and impacted the moon's surface on April 17, 2014. LADEE was the first deep space mission launched from Wallops Flight Facility and was also the first mission to launch on the Minotaur 4. The goal of the science mission was to determine the density, composition, and variability of the lunar exosphere by identifying the sources, sinks, and surface interactions of the dust atmosphere. This was inspired in part by Astronaut Gene Cernan's on-orbit observations of light rays radiating from the horizon during lunar sunrise (Fig. 1) and confirmed by the Clementine mission [2] (Fig. 2) which took images of dust loft on the limb of the moon. This led to a theory that electrical effects associated with the sun's terminator were a significant contributor to the lofting of dust in the atmosphere. The observations highlighted the fact that the lunar surface has a poorly understood thin dust atmosphere that could affect future missions to the moon. To investigate the

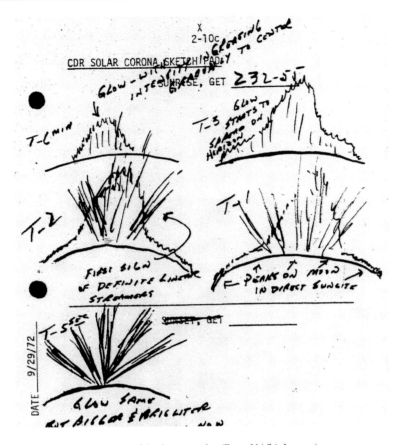

Fig. 1 Gene Cernan's drawings of the lunar sunrise (From NASA Images)

phenomena, LADEE was placed in an equatorial orbit with main-mission orbital science altitudes between 30 km and 50 km. In the extended mission, the science altitudes were as low as 2 km from the surface. Phasing burns, orbital insertion, and altitude station-keeping required precise control of the propulsion system, while the science payloads placed stringent requirements on the attitude control system. The active continuous control required to successfully perform the mission highlighted the need for careful assurance activities for the spacecraft subsystems.

To minimize fabrication and design costs, the modular common bus (MCB) spacecraft was designed with standard structural components that could be connected together to form the spacecraft. LADEE was formed with four such modules, housing the propulsion, avionics, and three instruments for observation of the moon. The instruments were a Neutral Mass Spectrometer (NMS) [3], an Ultraviolet Spectrometer (UVS) [4], and the Lunar Dust Experiment (LDEX) [5]. The mission also carried the first deep-space optical communication device, the Lunar Laser Communication Demonstration (LLCD) [6]. The total mass of the spacecraft was 383 kg, with main dimensions of 2.37 m tall × 1.85 m width and

Fig. 2 Clementine spacecraft image of moon dust corona (From NASA Images)

depth. Figure 3 presents a depiction of the spacecraft. For further details on the mission and spacecraft, please see Hine, Spremo, Turner and Caffrey [7].

The philosophy for the flight software development was complementary to the physical construction of the spacecraft, also proceeding with a low-cost rapidly prototyped product-line effort. Flight software development has an extensive history of cost overruns, schedule slips, and failures in operation. There are many drivers for these failures, including:

- Increasing complexity of missions
- Lack of implementation of best practices
- Difficulties forming a stable set of requirements
- Communications problems between stakeholders

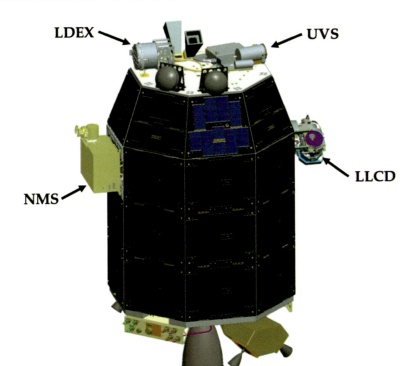

Fig. 3 LADEE spacecraft observatory

LADEE had the additional pressures of a firm fixed cost cap as well as a firm launch date driven by an extended period of solar eclipses that commenced shortly after the launch date. With these pressures on the Flight Software (FSW) development team, inspiration was sought from earlier successful rapidly prototyped small spacecraft missions, such as XSS-10 and XSS-11. The software team drew on lessons learned from these prior missions to shape the development approach. A set of requirements was enforced on the development process itself to enable the goal of an on-schedule, on-budget launch of the spacecraft. It was decided that the software should be designed to be tested from the start. Model-based approaches allowed for early rapid prototyping of requirements, algorithms, and tests, and the auto-code reduced transcription errors in the flight software. Extensive automation of the verification and validation test suite was done to enable constant regression testing of the model and software base.

The program emphasized best practices use of layered, modular software architecture with strong reuse of Government Off-The-Shelf (GOTS) and Commercial Off-The-Shelf (COTS) components. The software layers are discussed in detail in the Software Overview section. Guidance, Navigation, and Control (GN&C) engineers used model-based methods to develop models of the high-level spacecraft control functions, and software engineers then auto-code the models to the C programming language for integration with the rest of the software layers. This minimized the opportunity for communication and transcription errors between algorithm designers and qualified software developers. The model-based methodology also enhanced early prototyping of requirements, verification during early stages of development, and provided a common platform for communication between subsystems, software engineers, and stakeholders. It also enabled rapid analysis and software updates in response to errors found during validation activities in integration and test with payloads and on the spacecraft hardware.

State of the Art

Petit et al. [8] analyzed the special needs and challenges of Model-Based Engineering (MBE) for spaceflight systems and software, noting a particular need to be reliable before launch. This need is driven by difficulties and risks of uploading updates in flight, schedule pressures induced by limited launch windows, and increasing complexity of the missions. MBE provides an attractive approach because size and complexity are addressed through abstraction and analytical models developed early in the process. The paper predicted the needs for executable models that lend themselves to automated verification activities, as well as benefits from auto-generated code reducing the possibility of transcription errors by software developers. The following analysis will focus on practitioners realizing those necessary advances in the use of MBE for the particular area of Validation and Verification (V&V) of spacecraft flight software. In this chapter, Model-Based Software Engineering is denoted as MBSE while the larger area of model-based systems engineering is denoted as MBE. There is a limited set of MBSE literature, as V&V practitioners in the midst of missions rarely have time to publish journal or conference papers. Presentation only workshops such as the annual Flight Software Workshop (http://flightsoftware.jhuapl.edu/) provides the most recent view of the state of the art in practitioners efforts.

One such set of efforts incorporated model-based verification and testing for the Space Launch System [9, 10]. The effort focused on early requirements' development for fault management (FM) logic. Initially, they utilized System Modeling Language (SysML) to develop fault management diagrams but found that they could not easily be executed or simulated. Instead, they developed a State Analysis Model in Stateflow [11], which is a control logic tool developed by MathWorks to model state machines and flow charts within a Simulink model. This allowed for auto-code of software and incorporation in a Hardware-in-the-Loop facility to test functionality and obtain realistic estimates of compliance with time-response requirements for the FM system. In this case, the resulting executable code was not developed to the standards for incorporation into the final flight software: it was only used for requirements verification.

Another effort at Marshall Space Flight Center seeks to evaluate the use of MBSE coupled with automated testing for Guidance, Navigation, and Control (GN&C) Flight Software development for the Viper Lunar Lander program [12]. The group

32 MBSE Validation and Verification

was able to take advantage of upgraded MathWorks Model-Based Design (MBD) tool, illustrated in a flow diagram on the MathWorks website [13]. Approximately 80% of the NASA Procedural Regulations (NPR) 7150.2C [14] requirements are covered by the workflow integrated across the MathWorks/Simulink product line. As with LADEE, requirements were captured in Excel spreadsheets but were associated with models using the Simulink Requirements Editor. The group also utilized new MathWorks workflows for requirements testing, continuous model and flight code testing, enforced DO-178C [15], auto-code generation, static code analysis, and report generation. For the model, the group generated a Generalized Lander Simulation in Simulink (GLASS) using MathWorks Simscape for simulation of the lander. They were able to develop a highly automated system to develop, auto-code, and test, which enabled them to produce a highly disciplined process for FSW development. Several other efforts at NASA Ames Research Center adopted the LADEE FSW MBSE philosophy and parts of the software stack: BioSentinel [16], Resource Prospector [17], and the Starling Swarm Demonstration mission [18]. Of these, BioSentinel is scheduled for flight and will be flown as a secondary payload in 2021 on the Artemis-1 mission.

Best Practice Approach

When development of the onboard flight software (OFSW) for the LADEE mission started in 2008, the applicable NASA Procedural Requirements (NPR) for software development and software assurance were NPR Software Engineering Requirements 7150.2A and the NASA Software Assurance Standard NASA-STD-8739.8. Current versions of these documents are NPR 7150.2C and NASA-STD-8739.8a [19]. These documents provide the minimum set of requirements that projects must perform for all phases of software development. The requirements identify considerations for best practices for software development, but do not prescribe how to implement them. Discretion is up to the project on exact implementation of the practice. The rigor of the requirements varies with the software classification, with Class A software (Human-Rated Space Software Systems) requiring the strictest practices. NASA defines Class B Software as "Non-Human Space Rated Software Systems" and Class C Software as "Mission Support Software," which were the two major software classes for LADEE FSW development.

The software requirements in NPR 7150.2A were developed by the NASA Office of the Chief Engineers Software Working Group, and the NPR included the requirement that Class B software be acquired, developed, and maintained by an organization with a non-expired Capability Maturity Model Integration for Development (CMMI-DEV) rating as measured by a Software Engineering Institute-authorized lead appraiser. Since the LADEE flight software was classified as Class B, the project participated in several CMMI appraisals during the course of the mission that resulted in the project's home division at NASA Ames successfully achieving the required CMMI Maturity Level 2 rating in 2010 and again in 2013. The ground support software was identified as Class C software, CMMI-Level 2 processes were developed for that software base as well. The philosophy was to fully comply with the NPRs, and use them to identify the range of activities for ensuring quality software that integrated well with the spacecraft, payloads, and ground support systems.

Software Overview

LADEE utilized a model-based development approach layered on hand-code, Government Off-The-Shelf (GOTS), and Commercial Off-The-Shelf (COTS) software elements. The core of the spacecraft control and simulation software was modeled in MathWorks Simulink [20] software, version 2010B. Figure 4 shows that LADEE models were broken into two major components: Onboard Flight Software (OFSW – Class B) and the Ground Side Simulation Equipment (GSSE – Class C). The OFSW was modeled independently of the simulation of the spacecraft and was auto-coded separately for use on different processors in the hardware test environment. The simulation of the spacecraft was developed to sufficient fidelity to develop and test the flight software (it was not originally intended to be an authoritative source (digital twin) for the behavior of the vehicle), but it was eventually developed to high fidelity and used in Mission Operations to train the operators for nominal and off-nominal conditions, as well as to develop operational scripts to command the spacecraft. It included environmental effects; spacecraft dynamics; models of thermal, electrical, power and propulsion systems; and the ability to insert failures in the spacecraft systems.

Software reuse was a driving philosophy in the LADEE mission. Core Flight Executive (cFE) and Core Flight Services (cFS) [21] are a platform-independent, mission-independent, reusable flight software product line from Goddard Space Flight Center (GSFC) that provides executive services such as power-on, processor, and application resets, and time and scheduler services, and it incorporates many applications such as data and file management. It would be both impractical and

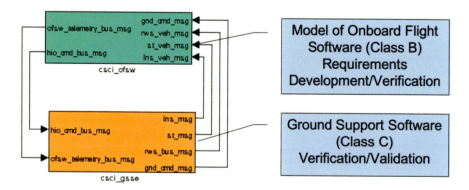

Fig. 4 Onboard flight software modeled separately from spacecraft model

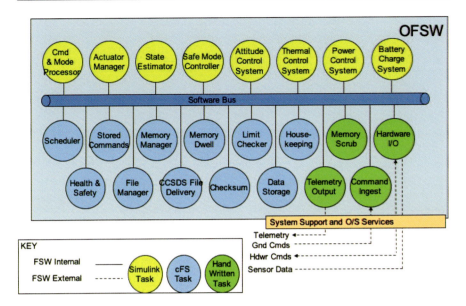

Fig. 5 Onboard flight software architecture

infeasible to redevelop these sorts of capabilities for each mission. As such, LADEE adopted the cFE/cFS stack as the core execution layer.

The overall layered architecture for the onboard flight software is shown in Fig. 5. The applications auto-coded from Simulink modules are shown in yellow, with each control application being output as a separate object file. These components were integrated with the cFE and cFS (blue) using a hand-coded Simulink Interface Layer. Other hand-coded components developed for the project included memory scrub, telemetry, and command and hardware input/output drivers shown in green. The real-time operating system used here is VxWorks. One requirement on the architecture of the flight software was that each component be amenable to V&V techniques such as those described later in this chapter.

Prototyping the Process

The model-based development process for LADEE was prototyped using a Hover Test Vehicle (HTV) [22]. The HTV was a prototype of a modular common bus lunar landing configuration, with a payload module, propulsion module, and landing leg module. It could take off and hover for several seconds before landing. The HTV used a thrust equivalent pressurized cold-gas propulsion system (repurposed scuba tanks), rather than a bipropellant system. The tanks were refilled in under an hour with a high-pressure air feed system that was resident in the test facility. The tests thus allowed fast iterative development of the prototype design, control system software, and MBSE methodology.

This period was used to perform a trade study between two different graphical programming environments: Matlab/Simulink and MATRIX x [23]. With the version of MATRIX x available at the time, it was difficult to modularize the control block diagrams, making it challenging for multiple developers to work on the model simultaneously. The support was limited for full bidirectional traceability and lacked hooks for automated test development with integrated reporting. The trade study identified Matlab and Simulink as the better toolset for achieving the requirements. It was easy to develop modular diagrams for each component of the flight software and ground support software. The MathWorks Automotive Advisory Board modeling guidelines were used as a basis for development of clean models that would easily auto-code. The Matlab Report Generator could be custom programmed to provide an integrated test and report harness for providing traceability between requirements, models, tests, and results.

Matlab/Simulink also enabled the rapid iteration of software changes between hover test runs of the HTV. During tank refilling and resetting of the HTV on the test platform, control-system changes could be developed in the modeling environment, and unit testing of the changed elements would be performed. Once the changes complied with system requirements, the models would be auto-coded, compiled, and downloaded to the avionics unit. This method became a model for the development process employed on the spacecraft.

Development Cycle

A Concept of Operations and Build Specification document was developed for LADEE FSW that laid out the course of development over multiple build cycles. It was envisioned that LADEE FSW would be developed in five build cycles, with the final two builds being delivered to the spacecraft for testing, scenario development, and operation. Build 1 emphasized end-to-end operations with basic Guidance, Navigation, and Control (GN&C) operations with reaction wheels and basic Command and Data Handling (CD&H). During this phase, the testing infrastructure and philosophy as well as the required sets of flight software plans were developed. Build 2 extended the GN&C command set to incorporate thrust maneuvers and extended CD&H to Consultative Committee for Space Data Systems (CCSDS) protocols, onboard data management, and housekeeping. Build 3 emphasized thermal and electrical management, along with extending the number and capability of GN&C modes and incorporating verification with a Hardware-in-the-Loop (HIL) Testbed. Build 4 was focused on Fault Detection, Isolation, and Recovery (FDIR), payload interfaces, and development of a Command and Telemetry (C&T) dictionary, and was delivered to Mission Operations for initial integration and testing with the spacecraft. Build 5 was reserved for bug fixes and verification and certification of all software, payload, and instrument interfaces, FDIR, software users guides, and other documentation. The final acceptance testing for release to the spacecraft was performed on Build 5.2, also known as the "Golden Load" (Fig. 6).

Fig. 6 Hover test vehicle used to prototype MBSE approach

During each one of these build cycles, a standard systems engineering cycle modified for the model-based system was used. Figure 7 shows the development cycle that was executed multiple times during each build cycle. The Concept of Operations (ConOps) would identify requirements that would be targeted and verified during each build cycle. Algorithms and designs would be developed and

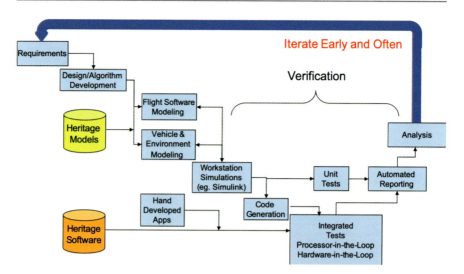

Fig. 7 Development cycle for LADEE FSW

documented and then modeled within Simulink or hand-developed, as necessary. Heritage models and software (such as cFE/cFS) were reused as much as possible to mitigate risk and shorten the development cycle. Initial verification of the requirements was performed using the Workstation Simulation Models (WSIM). Once the WSIM unit tests were satisfactory, the model was auto-coded to "C" and integrated using the Simulink Interface Layer (SIL). The code would be loaded onto Processor-in-the-Loop (PIL) and Hardware-in-the-Loop (HIL) systems where an automated test suite would exercise unit and integrated tests as well as a series of scenario tests. The test results and scenario telemetry were then processed and incorporated into automated report suites and analyzed at weekly test meetings. Test failures were evaluated for test script problems, vehicle model inaccuracies, flight software model errors, or incorrect requirements. Recommendations would be made to the software or spacecraft Configuration Control Boards (CCBs) for variations on the requirements needed to make them feasible.

Requirements

Core to the software development effort are specifications of clear and concise requirements for the behaviors to be verified. Too often projects overspecify requirements, allow turbulence in the formulation of requirements or accept desirements as real requirements. As the LADEE software development process stood up early in the project lifecycle (before the physical system specifications were known), the opportunity was used to prototype candidate requirements along with the system models and software. A key concept was that the requirements identified the critical behaviors that needed to be tested at each level of the flight software. One tendency

32 MBSE Validation and Verification

projects have is that elements of design are often documented in a design document and then also specified in the requirements. This leads to an extensive set of requirements in which the key drivers of system performance are hidden among design elements. Design documents are equivalent in importance with formal requirements and the as-built code or model is inspected relative to the as-designed to rectify any differences.

On the LADEE mission, system requirements were decomposed across the hardware and software. The software Level-4 requirements governed the behavior of the integrated flight software (e.g., the OFSW), while Level 5 specified the behavior of each subsystem (e.g., the Attitude Control System). These requirements could be further decomposed to specify unit-level behavior. The requirements were maintained in Excel spreadsheets and parsed using custom Matlab Report Generator scripts. This was simpler than using a high-overhead formal requirements management system such as DOORS [24]. Modern versions of Matlab have the ability to directly relate requirements to models, and this ability would be investigated in any new efforts. It would hopefully make it easier to provide bidirectional traceability among the artifacts rather than custom development of that capability in the reporting scripts for LADEE.

On LADEE, a conscious decision was made to formulate requirements as performance-related test drivers to scope the testing activities. For example, a representative controls requirement was: "The OFSW shall be capable of controlling the s/c to have a pointing stability to within 0.4 mrad over 20 s when commanded to hold an inertially-fixed attitude." The test suite used three scenarios that demonstrated this fine pointing control for instrument operations. Figure 8 shows several requirements that would normally be considered inspection items, but they were used to drive particular tests on the build. Predictability (FSW-3) drove interface, random, static, and testing of components. Metric units (FSW-5) drove testing of the

Summary Statistics.

```
There are 144 requirements to verify in this build out of a total
of 144 Level 4 requirements.
Number of TBRs: 0
Number of TBDs: 0
Number of tests that pass:  135
Number of tests that fail:  0
Number of Build requirements that partially pass: 9
Number of Build requirements that are tested but insufficient data to verify: 0
Number of Build requirements that have tests with execution errors: 0
Number of Requirements with stubbed tests: 0
Number of Uncategorized requirements: 0
Number of remaining requirements to verify in future builds:  0
```

ID Number	Requirement	Status
FSW-3	The FSW should be predictable in its operation.	PASS
FSW-5	The FSW implementations shall use standard metric units (kilogram [kg], meter [m], second [sec.], degrees centigrade [deg C], etc.) as the standard unit convention. Controlled use of hybrid units will be allowed per LADEE Systems Engineering Management Plan (Doc # C03.LADEE.SEMP).	PASS
FSW-6	The FSW shall define quaternions as vectors where the fourth element is the scalar value with a range >=0 and <=1.	PASS
FSW-10	The OFSW shall be designed for a minimum mission duration of 200 days.	PARTIAL

Fig. 8 Level-4 requirements testing summary matrix

Command and Telemetry Dictionary for naming conventions and units compliance and quaternions (FSW-6) drove a unit test suite. The 200 day minimum mission requirement led to testing on all counters to ensure proper rollover behavior. The testing strategy for each of the different classes of requirements is further described in later sections of this text.

Simulation Hardware

The Integrated Avionics Unit (IAU) for the LADEE mission was a Broad Reach RAD 750 computer. Several different test environments were used to simulate the IAU depending on the level of requirements being verified and focus of the test being applied. Workstation Simulation (WSIM), Processor-in-the-Loop (PIL), and Hardware-in-the-Loop (HIL) systems were used in different phases of development and testing. The WSIM environment includes only the Simulink modules of the model of the FSW and the model of the spacecraft. This environment is used early in the process to perform algorithm development, requirements analysis, unit testing, and low-level integration testing of the model performance. Inexpensive commodity flight-like processors were used for the PIL hardware. The Simulink models were auto-coded and integrated with the Real-Time Operating System (RTOS), Broad Reach drivers, and the cFE/cFS software running on the real-time processors. PILs were useful for testing the software integration, timing, and communications as well as the impact of communication delay on control algorithm performance. The HIL systems incorporated flight avionics Engineering Development Units (EDUs) that provided definitive answers on integrated performance and resource utilization and were the highest fidelity simulations for V&V activities. The "Big-HIL" (BHIL) also incorporated hardware simulators to stimulate thermal, battery, and solar panel functions to a high level of fidelity.

One important risk-reduction activity was the development of the Traveling Road Show (TRS), whereby an EDU would be integrated with the flight software in a mobile chassis. Members of the FSW team would travel to sites where payloads and other subsystems were being developed to test the interfaces long before integration of the hardware on the spacecraft. These tests clarified the Interface Control Documents (ICD) and payload requirements and caught defects at a time when they could be more easily fixed. Engineering units of other spacecraft subsystems were also used for early integration testing. One lesson learned from LADEE is that when these tests were not performed, critical defects were found in spacecraft integration that had to be rapidly corrected and verified. For instance, there was not an engineering unit of the Star Tracker to use for the integration tests, and the ICD for the Star Tracker was ambiguous as to the frequency and timing of the camera heads. The incorrect interpretation of the Star Tracker operation was not discovered until integration of the device on the spacecraft itself. This event required an emergency revision of the Star Tracker logic and retest of the flight software during a period with very tight schedules. The Star Tracker turned out to have "emergent behavior" during the flight, so this revision turned out to be valuable practice for in-flight updates of the state-estimator software.

Test Infrastructure and Traceability

The LADEE development effort emphasized testing of both the models and software to take the greatest advantage of the MBSE approach throughout the software life cycle. It is usually hard to test and debug internal modes and interfaces at a system level and impossible to test systems integration issues at a unit level. It can be expensive and difficult to debug latent defects that arise during integrated testing. Thus, an early emphasis was placed on the unit test suite infrastructure, with an eye toward permanently capturing the unit test effort as a regression test suite.

Testing on the LADEE program was enhanced by the modular and layered system architecture. The NPRs specify that one must maintain bidirectional traceability among code and models, requirements, design, and test artifacts. This was performed in a low overhead manner using a system of naming conventions. For example, for the modeling environment, pertinent naming conventions are:

- < name >_lib.mdl: Simulink Model library
- < name >_hrn.mdl: Simulink test harness
- < name >_test.m: Test script driver associated with model library

The < name > included a unique identifier for the model library that could be cross-referenced with requirements and other external documents. Each developer was responsible for providing all the necessary artifacts and developing the unit test suite associated with their model libraries. By adhering to the naming conventions, a set of regression test suites would interrogate the LADEE model and generate a manifest of all expected test artifacts. Custom Matlab scripts were developed to walk the manifest, identify units that did not comply with the naming conventions, and parse the names to identify all expected testing artifacts. The scripts were also used to exercise the entire test suite under development, evaluate the progress on verifying requirements, gather statistics, and provide a high-level summary matrix of test status. Simulink Report Generator was used as a platform to both drive the test suites and capture the related requirements and modeling diagrams in order to provide bidirectional traceability between low-level requirements, models, test suites, and metrics.

Model-Based Unit Testing

The developers were responsible for the development of Matlab-based unit tests for each of their Computer Software Units (CSUs) relative to the L5+ requirements. The L5+ requirements were targeted at basic functional responses needed by each CSU. Examples of behaviors that were verified included responses to out-of-range input values, mode commands, stability, and response for GN&C modules, modes for electrical and thermal subsystems, and functionality of utility modules such as quaternion conversions.

Each Computer Software Component (CSC) was comprised of multiple CSUs, each had their own unit test suite, as described above. Low-level integration tests were conducted at the CSC level to verify behaviors of the component as a whole. These tests exercised the returns of the CSCs and made sure that appropriate signals and responses were propagated through the system. These low-level model-based integration tests were more akin to the unit testing than the rest of the integration testing, so they were rolled up in the unit test suite reporting.

The test suite also exercised the GSSE side of the LADEE model and produced a separate report, metrics, and summary matrix for the class C software elements. Assumptions were documented, as were ranges of acceptable operation for the model. Depending on the type of model, documentation included refinement studies, sensitivity analyses, comparison with analysis, subsystem performance specifications, or outside recognized tools. For instance, the propulsion subsystem was validated with a variety of means, including comparison of internal states with one-dimensional fluid and heat transfer theory, calibration curves supplied with the nozzles, and with the simple response model developed by the propulsion team.

GOTS Components

To reduce the software development schedule time and budget while improving reliability, LADEE incorporated a significant amount of hand-developed traditional GOTS software from the cFE and cFS packages developed at Goddard Space Flight Center. The LADEE team reviewed an extensive set of documented test results provided with the software. Those test results were considered sufficient verification evidence that the software functioned as required in a stand-alone test environment. It was not considered a value-added exercise to independently re-execute the provided test suite on the lab's computer systems. Instead, the concern was with the operation of each reused unit within the context of the fully configured option set for LADEE FSW. The LADEE FSW L5 requirements driving the integrated test suites covered the functionality of each cFE and cFS component as configured for the flight software. Functional requirements were tested in the scenarios, and housekeeping behaviors were tested in the Command and Telemetry Dictionary (C&T) tests.

Command and Telemetry Dictionary

One of the L4 requirements directed that each command in the C&T be tested. There were 587 commands in the C&T, so this was an extensive task to perform. It was decided to verify the requirement by developing a test suite that checked the behavior of commands in the PIL/HIL environments, including interactions with the ground station software, Integrated Test and Operations System (ITOS). Patterns of testing were identified among common classes of commands and implemented in an automation language for writing Systems Test and Operation Language (STOL) scripts. An example of behaviors that were tested in all units was proper incrementing and clearing of all command and error counters. While most of the commands could be partially tested in this manner, some command executions could only be demonstrated (not formally tested) through scenario-based testing. In other cases, required behaviors were not observable or masked by other conditions and thus were considered unverifiable.

In order to provide traceability, the data from the test suite is associated three ways: relative to requirements, on a per command basis, and on a per test basis. Metrics are provided for each basis to assess the progress of the command coverage and verification of Level-5 requirements. As results from the test suite were interrogated to provide evidence for higher-level requirements, they were output in parseable html files. Figure 9 shows the high-level summary information for the test suite. The summary information is composed of "clickable" html elements that can be expanded to review the test results in fine detail.

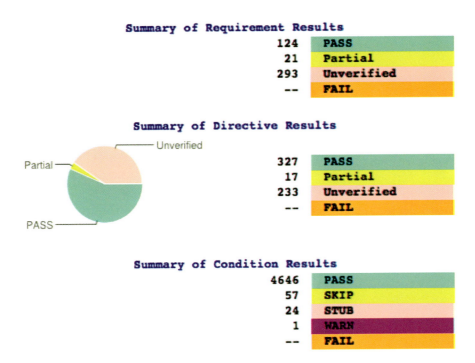

Fig. 9 Command and telemetry dictionary summary page

In the end, approximately 350 of the commands were able to be fully or partially tested using the C&T suite. About 120 commands were demonstrated during the course of scenario testing. In the end, approximately 110 commands were neither tested nor demonstrated, and no commands failed the testing. Overall pass rate for the suite was about 80%, which was considered an acceptable pass rate for the requirement, and which led to a classification as partially verified. The remaining unverified commands were reported to Mission Operations as being risky to use. Fortunately, they were not part of the normal operation of the spacecraft.

ConOps-Driven Testing Methodology

The unit test suites were used as early indicators that pertinent requirements are being met, but they did not test the end-to-end functionality of the software. The driving philosophy was "Test as you Fly – Fly as you Test," so the same methodology that the Missions Operations team would use to command the spacecraft was also used for flight software testing: using the ITOS test station, and verifying requirements through post-processing of the telemetered data. This verification was performed in a PIL/HIL environment through the use of scenario-based testing. The scenarios were developed to test anticipated operations using the spacecraft ConOps document described earlier. These scenarios included C&T, data handling, and spacecraft operations such as separation and activation, orbital maneuvering, science operations with the payloads, and fault management-related scenarios. Representative Absolute Time Sequences (ATS) and Relative Time Sequences (RTS) were developed using the STOL language to command the spacecraft.

A master script controlled the execution of the test suite. It put the spacecraft in the appropriate mode for testing, issued the command scripts for the flight software and simulation sides, telemetered the data off the PIL or HIL at the end of the scenario, and post-processed the binary data files into comma-separated value files as well as textual log files. The files needed for verification of the Level-4 requirements were copied into a database for further post-processing. This was a highly I/O bound process, and several strategies were used to optimize the wall-clock time associated with it. Initially, complete bus messages were downloaded for post-processing, but much of the data was only used internally on the spacecraft, so the data was compressed to just the data needed for the Level-4 verification.

Hand-encoded housekeeping packets proved useful for minimizing the amount of data brought off the spacecraft, but the process of coordinating the telemetry record files (.rec) with display pages (.page), sequential print formatting (.sprt), and the C code that copied specific telemetry locations into the new packets was time-consuming and highly error-prone. A Matlab script was developed that auto-generated the rec, page, sprt, and copy table.c files given a simple text specification of the telemetry point and original record it could be found in. Despite the growth in the number of scenarios over the build cycles, use of the housekeeping telemetry packets enabled us to plateau at only about 70 Gb of data for verification of the Level-4 requirements. While useful for minimizing the data foot print for FSW testing, the routines had the added side benefit that it was easy for the mission operations team to optimize packets that fit within the narrow bandwidth requirements for safe mode telemetry.

Once the scenarios were run, a custom Simulink report generator script was used to post-process the data and capture the output. The script first read external imported data, such as requirements, spreadsheets, and Interface Control Documents. It then processed each of the Comma Separated Variable (CSV) files to extract needed variables for the test suite. This is once again an I/O intensive process, so if a fresh set of CSV files was found, a Matlab binary file was written with the extracted information. The data was further compressed by only storing only change points in the data. Significant compression can be achieved for buses with mode information or slowly changing status bits. On successive runs of the test suite, the extracted binaries were used in the initialization process, but the original CSV files were retained to facilitate the debugging process if errors were found.

Each chapter of the reports incorporated one requirement and ran the associated test. Naming conventions were once again used to associate the various artifacts. Each of the test scripts had a common interface to assist in the automation of the system. This is shown in Fig. 10 test driver code for a Level-4 requirement on dwell limits during safe mode. The file is named according to the convention, and a structure containing the entire processed scenario data is passed into the test routine. The routine returns a text message, test status, and the handle to an HTML file with detailed data. Any images that were generated during the course of the test were captured and incorporated in the output of the chapter. The test driver sets the maximum limit to be tested, provides the units and the variable to be tested, tells the gnc limit test routine that the absolute value of the variable is to be tested and that scenario 11 is to be used to provide the data. The philosophy was to have simple descriptive driver code which called more complex abstracted test code that implemented the common testing patterns.

An example of the detailed information included in the HTML file is shown in Fig. 11. The routine outputs the maximum observed value for the variable in the PIL

```
 1  function [status, msg, html_file] = fsw_20_test(scn_struct)
 2
 3  % Performance requirement during safemode
 4
 5  limit = '>174';
 6  units = 'mrad';
 7  var = 'error.pointing.smc.max_total_angle';
 8  abs_var = 1;
 9  scn = [11];
10
11
12  [status, msg, html_file] = gnc_limit_test(scn_struct, var, units, abs_var, scn, limit);
```

Fig. 10 Sample test driver code for Level-4 GN&C requirements

Fig. 11 Sample test output for Level-4 GN&C requirements

```
Processing scenario number 11 type = pil
Number of periods this mode observed = 1   1
Observed Value equals 15.5523 mrad
within threshhold of >174 mrad: PASS
```

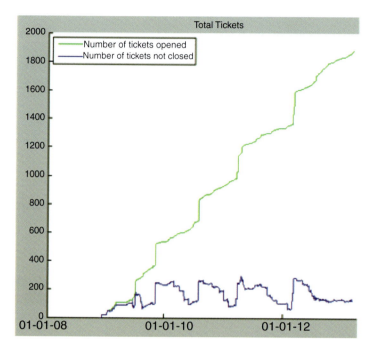

Fig. 12 Burndown chart for LADEE TRAC tickets

run of scenario 11 and compares it with the limit to provide PASS information for this particular PIL run. If multiple scenarios were used, all scenarios had to pass the test. If some expected data was missing and some passed, the test was considered a partial pass. Any failure caused the entire test to fail. If the scenario failed, the information provided was sufficient to pinpoint the pertinent underlying CSV data to determine if the test itself was failing, or if indeed the performance parameter did not meet the requirements. This inspection of the data would be performed in the weekly test meetings so the cause of the failure was well understood.

A summary verification matrix, shown in Fig. 8, was also generated and was extremely useful for quickly identifying failures within the extensive documentation generated. Summary statistics were also used to track progress against goals. For instance, in addition to historical requirements, burndown charts, ticket burndown charts were also maintained (Fig. 12) to show total effort related to closure of work tickets. In this chart, one can see the initiation and subsequent reduction in tickets for each of the five major build cycles. Residual tickets at the end of build cycle 5 include those for maintenance activities, risk assessment, and closeout activities.

Test Patterns for Verification of Level-4 Requirements

The flight software Level-4 requirements are directed at the operation of the system, and the requirements, statements are crafted to be verifiable using a variation of formal methods. Optimally, one would use formal methods on the underlying Simulink model to ensure that the logic would always provide the expected results. However, the Simulink model is only one early artifact of the flight software. It is the auto-code generated from the model in conjunction with propagation of the signals through the interface layer and the core services that are of interest for verification of the overall flight software. At the time of development for the LADEE FSW, the state of the art in formal methods was typically limited to component models, and formal analysis of fully integrated flight software was intractable. The telemetered data represented the authoritative source for verifying that the OFSW performed as required in the context of the mission. Thus, concepts from the formal methods communities were used, but they were applied to the telemetered data stream in a lightweight manner suitable for regular regression testing.

Typically, the requirements were encoded as Matlab strings in a top-level script, as seen in Fig. 10. The encoded assertions would be evaluated in detailed test scripts for each of the testing patterns on the telemetered data using dynamic programming techniques. One subtlety associated with correlating telemetry was the clock stamp behavior on each of the telemetry packets. The telemetry was time-stamped as it was generated within each application and different applications were run at different times. There was dither and drift in the clocks between the flight processor and the simulation. Telemetry could occasionally be lost, and, as stated previously, the telemetry was compressed to change points. The data reduction algorithms were designed with these conditions in mind, but occasionally false positives occurred, usually due to a telemetry point being on the cusp of the decision logic as to whether it was a member of the telemetry set or not. These cases were generally resolved by inspection of the underlying data. In many early cases, though, failures of the assertion indicated mode logic problems within the flight software itself.

Several different classes of testing were abstracted from the form of the requirements; these fall into the major categories below.

Guidance, Navigation, and Control

GN&C requirements typically stated while in a particular mode that the absolute values of an error term not exceed a limit. The scripts gathered the maximum value of the error term for all pertinent scenarios and test platforms and the limit criterion was applied.

Commanding Modes and Transitions

Requirements in this category typically specify behaviors such as "while in a flight mode, with active control disabled, a particular parameter shall also be disabled" or "on transition to a particular mode, a parameter shall be disabled." A simple Matlab-based temporal language was developed to specify the required behaviors and underlying scripts applied the assertions to the telemetry stream.

Fault Management

The spacecraft simulation model was developed with hooks to test various fault management scenarios that were considered to be correctable by flight software. A series of scenarios were developed with commands sent to the simulated spacecraft that ranged from out-of-range temperatures to stuck thrusters. A fault management spreadsheet was developed which identified critical signals, latency, persistence, and the type of response. A recursive algorithm was developed to find the period of time in which all signals were true, and to scan the log files to find the expected response. Time stamps were compared to determine if the response happened within an acceptable window. One complication was that if the software transitioned to safe mode, telemetry rates are restricted and data from the critical signals could be lost. For this restricted-rate logging, the software monitored to see if the desired response was found in the telemetry even if the trigger could not be specifically identified. These cases were identified as passing tests, but with a note that they were demonstration only.

Commands and Telemetry

Several of the Level-4 requirements were directed at ensuring that variable dimensions were in metric units, commands had explicit states, or that certain items of telemetry existed. Generally, these sorts of inspection requirements would be delegated to a review process for verification, given the vast amount of information to be reviewed, and effort was made to automate and regression test these requirements. Regular expressions were used extensively to correlate variables, commands, and telemetry with interface control documents that were used to identify required behavior. This turned out to be a remarkably effective technique to monitor design and requirements changes in the lower-level software.

Testing Cadence During Development

Of the 144 Level-4 requirements on the flight software, three were met through the design of the software, and four had to be manually tested. Testing of the payload interfaces was partially automated, but the full verification was accomplished through the TRS process. The remaining requirements were tested using the automated techniques described above. For readiness testing, the scenario suite would be executed over a weekend, and the automated processing of the telemetry would take an additional day. This approach allowed us to hold regular weekly reviews of the test data and quickly target defects in the flight software.

Formal Inspection

Formal inspection has been a useful technique to drive out defects and promote consistency among the various software artifacts. Rather than just walking through code, reviewers were assigned to various roles and asked specific questions to focus their reviews of the material. Advanced software engineering tools were used to focus the inspection. Results of a static analyzer (Polyspace [25]) would be used by one reviewer to direct attention to unproven aspects of the code. Another reviewer would use code-coverage analysis generated over the entire test suite to identify missing test cases or driving requirements. The models were compared with designs, requirements, test cases, and ICDs to verify consistency across the artifacts. Code and models were compared with style guides and coding standards for compliance. After the designer removed defects found in the inspection, the moderator would point-by-point review the changes with the designer and track compliance. Through the flight software development process, a total of approximately 1,150 h of inspection time with a cumulative defect rate capture of about 0.65 per hour were logged.

Illustrative Examples

The general rule of thumb during flight is that uploads to a spacecraft are risky and should be minimized as much as possible. Smaller and simpler satellites, such as CubeSats, sometimes neglect software maintenance features when developers decide to roll their own executive system. This approach presents a different risk vector when defects are found in flight. The use of cFE on LADEE enabled a range of update features such as a parameter table upload feature upload capability for each cFE application, as well as executive services that manage boot partition, application resets, and processor resets. Different risk profiles are associated with each of these software maintenance activities, with table upload considered least risky, and software image upload with processor reboot most risky. The type of model-base driven V&V activity conducted on the artifacts varied with the system risk and extent of changes. The model-based methodology was also useful during validation activities in Integration and Test (I&T) for correcting defects, Mission operations for the Operational Readiness Training (ORT), performing incident investigations, and for validating uploads to the spacecraft during flight. Each of these areas are discussed in this section.

Training and Mission Operations

Originally, the LADEE model was intended only for use within the flight software team. As the capabilities and fidelity of the model grew, other teams incorporated it into their operations. Mission Operation Systems (MOS) used the LADEE model extensively for development and validation of their flight sequences. MOS set up a lab with multiple WSIM, PIL, and an authoritative HIL to use as a truth article for

ORT. ORTs were conducted prior to the LADEE mission where faults would be injected into the model of the spacecraft or ground environment for operators to debug. Both nominal and fault scenarios were devised for the immersive simulations [26] using a hazard analysis of the operational control loops. The scenarios were remarkably helpful at training the operators, and one fault scenario was eerily predictive of an incident with the reaction wheels discussed later in this section.

One of the freebie faults involved operators inadvertently modifying a fault management table from a previous build of the code and uploading it to the simulated spacecraft. This resulted in Fine Pointing Divergence error based on a misconfigured set of action points and many hours of analysis to determine the root cause. This was a difficult error to find, because the tables had no embedded strings encoded in the object file associating it with a Subversion (an open source version control system) version number for the software build. After this event, the build process was modified to automatically include the revision number, but MOS had no tools to amend the revision number when they did normal maintenance using a table. Comparison of hash codes and complex configuration management techniques were used to identify the proper versions of tables. For future missions, the table management tools will be updated to provide the capability to amend table version numbers embedded in strings in the binary.

Ambiguous ICDs

Heater Tanks

One of the key causes of defect escape for the spacecraft was misunderstood or ambiguous ICDs. One classic example of this occurred during I&T when the spacecraft was being thermally cycled to check heater operation and performance. This is a critical phase of testing with potential for component damage, a rapid cadence of system tests, and a firm schedule. The spacecraft was configured with two oxygen tanks and two fuel tanks. Each tank in a pair was labeled A or B in the spacecraft ICD. In the flight software, these were designated 1 and 2, respectively, in the Electronic Data Interface Control Document (EDICD). During the testing, some of the heaters exhibited runaway behavior and an urgent investigation was initiated. The WSIM model of the spacecraft was compared with the behavior of the flight hardware, and it was determined that the numbering on one pair of the tanks in the ICD was misunderstood, and instead should be switched such that $(A, B) = (2, 1)$.

Fortunately, with the MBSE approach, the thermal model numbering was not hardwired. Instead, it was automatically configured by reading the updated EDICD. Then, all WSIM, PIL, and HIL tests associated with the thermal system were performed, as well as a select set of other scenario tests to ensure the changes had not affected the rest of the system. PIL and HIL tests were performed with the baseline software on the spacecraft, with only a new thermal binary and associated configuration table uploaded and activated. The test suite showed nominal results, and with this evidence, the Configuration Control Board (CCB) authorized upload of the new thermal control system binary and table to the spacecraft. This step corrected the thermal runaways, and spacecraft testing was able to proceed as planned. This turned out to be a good dress rehearsal for investigation, repair, and V&V of defects found in flight.

Reaction Wheels

Just after separation from the Minotaur launch vehicle, LADEE's four reaction wheels were disabled by the fault management system. Telemetry reported that electrical currents in all four reaction wheels exceeded the "Current Limit Flag." The reaction wheels were a critical component in safe mode for thermal control of the vehicle. In safe mode, the spacecraft assumed a slow "barbecue roll" such that it maintained power positive attitude for the solar panels while rolling to limit temperature rise in any particular part of the vehicle. The incident CCB determined that the ICD for the reaction wheels was misread, and the "Current Limit Flag" was a warning flag, not an error flag. A new fault management table was developed and tested with the model in the PIL environment to disable the action points associated with that flag. The table was uploaded to the spacecraft and proper safe mode performance was restored upon enabling.

Star Tracker System

Due to funding limitations, the project was unable to purchase an engineering development unit of the Star Tracker system. This was an accepted risk maintained on the LADEE risk register. There was a software model of the system as well as the documented ICD to develop a functional model of the system. The ICD for the Star Tracker stated that the system operated on an 8 Hz clock. That was interpreted to mean that the heads reported data on alternating 4 Hz cycles and developed the state-estimation model to utilize the data on that cycle. When the flight unit received testing initiated, it was found that the Star Tracker heads reported synchronously at 4 Hz. The state-estimation model and bus interfaces had to be updated to use the synchronous data. This was a case where the initial development had used the wrong system model and therefore incorrectly verified state estimation functions. The Star Tracker continued to present challenges during the flight phase of the spacecraft.

Emergent Behavior

The LADEE Star Tracker system software performed in a way unanticipated by developers of the system, exhibiting emergent behavior when a camera head came close to a Big Bright Object (BBO) like the earth or the moon. BBOs induced the internal Star Tracker software to take significantly longer to produce spacecraft attitudes than normal. When these "stale" attitude measurements were provided to the LADEE state estimator module, they caused control of the spacecraft to be destabilized. Initially, it was found that the as-designed position of the Star Tracker differed slightly from as-built. A table was uploaded with the corrected position, and though this did improve the performance of state estimation, it did not entirely correct the problem. In fact, the closer the spacecraft orbited to the moon, the more delayed the quaternion updates were during a BBO event. The fault management system would transition the spacecraft to safe mode because the estimated rates (but not the real rates) were in excess of fault management settings.

To debug the root cause, the model of the Star Tracker in the WSIM was modified to include the delayed quaternion estimates during BBOs. Once the anomaly was fully understood, the State Estimation module was modified to reject the "stale" attitude measurements of the Star Tracker. Since the state estimation system is core to the behavior of the GN&C functionality and contributed to the fault management behavior being controlled, all V&V, core software maintenance, GN&C, and fault management-related requirements were tested before uploading the patch to the spacecraft.

One key element to performing software patches was to ensure that the software bus structures associated with each application were consistent with those of the software load on the spacecraft. The state-estimation model had been developed in Simulink and was auto-coded out to C code for inclusion into the flight software. It was critical to configuration manage the bus interfaces in an interface control database and externally apply them to the Simulink modules so that their busses remain consistent with the software version on the spacecraft. In addition, when new object code would be uploaded for a Simulink model, one had to be careful to upload the associated parameter table with the object code. The order of parameter table elements could change from build to build. Eventually, two software patches and associated parameter tables had to be installed on the spacecraft to compensate for the Star Tracker behavior. In this way, it could be verified that the proper patch had been applied and accepted during the application reset by issuing a No Operation (NOOP) command to the application. The NOOP return event included the Subversion revision control system number with which the software was created with.

On LADEE, the busses were optimized to contain the minimum number of signals possible so that possible performance problems related to bus traffic could be avoided. In retrospect, there should have been a small amount of spare capacity in the busses to accommodate the need for additional signals. It would have been useful for debugging to add a signal with a real-time assessment of the Star Tracker latency rather than creating post-incident models of the behavior of signals unexposed to the interfaces.

System Reboot

One of the payloads on the LADEE spacecraft was the LLCD, an optical communications device. During the mission, the LLCD was demonstrated several times by downloading Synchronous Dynamic Random-Access Memory (SDRAM) partitions from the spacecraft. It was during one of these operations that the 622 Mbps record download rate from the moon occurred. The LLCD operations turned out to be useful for evaluating Error Detection and Correction (EDAC) events [27] during flight and whether they had permanently affected the SDRAM. The only unanticipated reboot of the mission occurred during LLCD operations at the end of the download of a SDRAM partition.

The reboot turned out to be difficult to debug. A careful investigation of the time stamp of the last data transmitted, the onboard software history frame content, Deep Space Network (DSN), and ground-system time stamps indicated that the reboot event was strongly correlated with the interrupt at the end of an LLCD data transmission. Localizing the investigation to the driver software showed that a task lock had been used in an interrupt service layer.

Figure 13 shows the sequence of events that was discovered to have caused the reboot. At the time that the LLCD issued the reboot, tasks could be operating in KERNEL SPACE equal to either TRUE or FALSE. The LLCD process had previously issued several interrupts at the end of each transmission, but apparently the interrupted tasks had not been operating in kernel space. In this case, the KERNEL SPACE must have been true before the interrupt. Inside the interrupt layer, a task-lock function had been called instead of an interrupt lock, with the key difference here being that task lock always issues a KERNEL SPACE = FALSE upon exit. When control was returned to the interrupted task, it assumed that it was still in

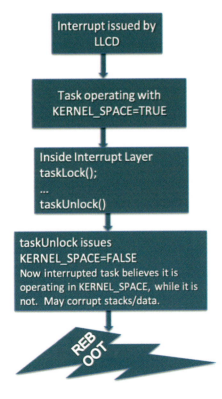

Fig. 13 Sequence leading to unanticipated reboot

kernel space and apparently corrupted a critical stack or data region, which resulted in an immediate reboot.

This is a classic example of defect escape. This error was in hand code outside the model-based paradigm, the condition was difficult to catch using static analysis, and the formal inspection program used on LADEE missed the context of task locks being illegal within interrupt service layers. This defect could not be patched in the normal sense used in the onboard applications, since it was part of the base image for the software.

The Configuration Control Board (CCB) for LADEE was presented with several options to remedy the defect:

- Accept the risk of further operations with the current spacecraft load. Since there were only limited LLCD operations contemplated, this was considered the correct path forward during the regular mission.
- Apply a kernel patch devised through experimentation with the HIL simulators using shell commands to modify memory to skip over the offending task lock. The CCB was highly averse to the risk of this option, and it was declined.
- Upload a whole new image of the flight software.

Final Load

After the mission had completed all of its primary science goals, the spacecraft was authorized to conduct extended science operations but also would also expose it to endure a lunar eclipse that it had not been designed to survive. The eclipse became an interesting engineering experiment to evaluate the robustness of the spacecraft to the challenging cold and battery drain in the eclipse environment. One weakness of the spacecraft found prior to the eclipse was a bug in the interaction between Battery Control System logic and the underlying hardware battery control algorithm when recovering from a load-shed state. This provided a compelling reason to update software fixing this, the task lock bug discussed earlier, and other small defects encountered over the course of the mission. The decision was made to upload a new software load prior to entering eclipse. For the software team, this was an extended engineering test of the maintenance capabilities of the spacecraft.

The final software upload, denoted as the Platinum Load, underwent an almost complete V&V cycle. The V&V process took approximately a month to complete, including running all WSIM, PIL, HIL automated testing, and hand-test elements, neglecting only the testing of the diagnostic load, which was not being modified or uploaded. The evidence was presented to the CCB, and the software load was approved. The new software was written to the second boot slot, and after reboot, the engineering evaluation of the new load was successful. The Platinum Load, in fact, operated flawlessly until the planned impact of the spacecraft with the moon. The final build slot configuration for the spacecraft is shown in Fig. 14. The avionics build partition had substantially more capacity than was used over the course of the mission, in part because the software applications could be easily reconfigured and so few critical defects were found over the course of the mission. However, the

X Not Write-able	9 EST patches	10 Blank	11 Blank	12 Blank	13 Blank	14 Blank	15 Blank
0 Golden Load	1 Golden Load	2 Platinum Load	3 Blank	4 Blank	5 Blank	6 Blank	7 Diag-nostic Load

Fig. 14 Final boot slot configuration for LADEE

testing and V&V of the LADEE software used all available slots in the HIL environment over the course of the five build cycles, as it was convenient to be able to transition to previous software loads when investigating anomalies.

Conclusions

The FSW team learned many valuable lessons for MBSE over the course of development and maintenance of the software. From the development perspective, it was a significant advantage for GN&C engineers to be able to model their software in their "native language," (Simulink) while the software engineers automated a high-quality auto-coding process. The models made it easy to communicate these designs, algorithms, and data flow with other spacecraft subsystems, management, and stakeholders. Early definition of requirements and the data dictionary made it possible to formalize a test infrastructure early in the development cycle, even before the system functionality had been developed. The cFE/cFS software layers used for this mission proved to be effective, capable executive and services layers for providing necessary software maintenance during an extended mission. For the LADEE mission, MBSE practices, including the early incorporation of an automated V&V infrastructure, enabled us to drive the software development with test-based evidence. This approach also helped pull in potentially late software schedules by reducing the time for full acceptance test cycles of the flight software to half of the planned amount. Model-based development is effective when used in a highly disciplined manner. Functionally complete flight software was delivered to the spacecraft for integration and test on schedule. When defects were found in space-craft integration, the highly automated verifications have made full testing and delivery of the new software a fast and consistent process.

The LADEE mission was extraordinary in achieving its science goals over the course of its mission. During the 188 days of lunar orbit, it returned approximately 200% of the planned science data to earth with all science goals being met. The lowest regular science operations were conducted around 2 km above the lunar surface, so dust surveys covered a significant range of altitudes. The science operations led to a discovery that micrometeorite bombardment is a significant factor in the generation of a dust atmosphere on the moon. The LLCD experiment demonstrated a 622 Mbps downlink rate from lunar orbit, breaking all previous communication records. Other significant events included the imaging of LADEE from the Lunar Reconnaissance Orbiter Camera (LROC) despite nearly perpendicular orbits to each other (https://www.nasa.gov/content/goddard/nasas-lro-snaps-a-picture-of-nasas-ladee-spacecraft/). Initially, unplanned images of the moon's surface were taken using the Star Tracker camera heads. The spacecraft also survived an extended solar eclipse that it had not been designed to withstand, with only the failure of one propulsion-system heater. It brought both sadness and great pride for those of us on the MOS team when the telemetry showed that it had successfully deorbited by impacting the ridge of Sundman V Crater on the night of April 17, 2014.

LADEE flight software enabled very complex spacecraft behaviors during the course of the mission. On-orbit table uploads were regularly performed for control of both the vehicle and payload systems. The payloads required a complex set of nods, rolls, and slews for data taking while maintaining thermal and acceleration limits on the spacecraft. Two software patches were uploaded to the spacecraft to account for emergent Star Tracker behavior. One unanticipated reboot occurred over the course of the mission due to an error in interrupt handling. During the extended mission, a successful upload and reboot into a new software load was performed. No defects were found during the approximately 1 month's continuous operation on the new load. In the end, the LADEE FSW team was able to deliver quality software on budget and on schedule for a highly ambitious mission by using MBSE in a highly disciplined manner.

Cross-References

▶ Digital Twin: Key Enabler and Complement to Model-Based Systems Engineering
▶ Exploiting Digital Twins in MBSE to Enhance System Modeling and Life Cycle Coverage
▶ Model-Based Hardware-Software Integration

Acknowledgments The author would like to thank the LADEE FSW team for their friendship, creativity, immense skills, and outstanding work ethic for the LADEE mission.

References

1. Delory, G.T.; Elphic, R.; Morgan, T.; Colaprete, T., Horanyi, M;, Mahaffy, P.; Hine, B.; Boroson, D, The lunar atmosphere and dust environment explorer (LADEE), 40th Lunar and Planetary Science Conference, (Lunar and Planetary Science XL), held March 23–27, 2009 in The Woodlands, Texas, id.2025.

2. Nozette, S., Rustan, P., Pleasance, L.P., Kordas, J.F., Lewis, I.T., Park, H.S., Priest, R.E., Horan, D.M., Regeon, P., Lichtenberg, C.L. and Shoemaker, E.M., 1994. The Clementine mission to the Moon: Scientific overview. Science, 266(5192), 1835–1839.

3. Collier, M.R., Mahaffy, P.R., Benna, M., King, T.T. and Hodges, R., 2011. Neutral mass spectrometer (NMS) for the lunar atmosphere and dust environment explorer (LADEE) Mission.

4. Colaprete, A., Vargo, K., Shirley, M., Landis, D., Wooden, D., Karcz, J., Hermalyn, B. and Cook, A., 2014. An overview of the LADEE ultraviolet-visible spectrometer. Space Science Reviews, 185(1–4), 63–91.

5. Horanyi, M., Sternovsky, Z., Lankton, M., Dumont, C., Gagnard, S., Gathright, D., Grn, E., Hansen, D., James, D., Kempf, S. and Lamprecht, B., 2014. The lunar dust experiment (LDEX) onboard the lunar atmosphere and dust environment explorer (LADEE) mission. Space Science Reviews, 185(1–4), 93–113.

6. Boroson, Don M., et al. "Overview and results of the lunar laser communication demonstration." Free-Space Laser Communication and Atmospheric Propagation XXVI. Vol. 8971. International Society for Optics and Photonics, 2014.

7. Hine, B., Spremo, S., Turner, M. and Caffrey, R., 2010, March. The lunar atmosphere and dust environment explorer mission. In 2010 IEEE Aerospace Conference (pp. 1–9). IEEE.

8. Pettit, R.G., Mezcciani, N. and Fant, J., 2014, June. On the needs and challenges of model-based engineering for spaceflight software systems. In 2014 IEEE 17th international symposium on object/component/service-oriented real-time distributed computing (pp. 25–31). IEEE.

9. Trevino, L.C., Berg, P., Johnson, S. and England, D., 2016. Modeling in the state flow environment to support launch vehicle verification testing for mission and fault management algorithms in the NASA space launch system. In AIAA SPACE 2016 (p. 5223).

10. Berg, P. and Garcia, S. 2019. SLS risk reduction through rapid prototyping and modeling of Mission and fault management algorithms, 2019 Flight Software Workshop.

11. Hamon, G. and Rushby, J., 2004, March. An operational semantics for Stateflow. In international conference on fundamental approaches to software engineering (pp. 229–243). Springer, Berlin, Heidelberg.

12. Jamison, B.R., Hannan, M.R., Kaidy, J.T., Orphee, J.I. and Olson, N.S., 2019. Pilot evaluation of model based design tooling for guidance, navigation, and control flight software development, 2019 flight software workshop.

13. https://www.mathworks.com/content/dam/mathworks/mathworks-dot-com/solutions/aerospace-defense/standards/npr7150-2c-workflow.pdf

14. https://nodis3.gsfc.nasa.gov/displayDir.cfm?t=NPR&c=7150&s=2B

15. Rierson, L., 2017. Developing safety-critical software: A practical guide for aviation software and DO-178C compliance. CRC Press.

16. Ricco, A.J., Santa Maria, S.R., Hanel, R.P. and Bhattacharya, S., 2020. BioSentinel: A 6U Nanosatellite for Deep-Space Biological Science. IEEE Aerospace and Electronic Systems Magazine, 35(3), 6–18.

17. Cannon, H., 2017. Application of CFS to a lunar rover: Resource prospector (RP).

18. Sanchez, H., McIntosh, D., Cannon, H., Pires, C., Sullivan, J., DAmico, S. and OConnor, B., 2018. Starling1: Swarm technology demonstration.

19. https://standards.nasa.gov/standard/osma/nasa-std-87398

20. MATLAB and Simulink 2010b, The MathWorks, Inc., Natick, Massachusetts, United States.

21. Wilmot, J. Implications of responsive space on the flight software architecture, 4th Responsive Space Conference, RS4-2006-6003, April 2006.

22. Bell, J., Brown, J., Christa, S., Daley, E., Cannon, H., Durais, G., Gundy-Burlet, K., Hine, B., Kennon, J., Kulkarni, N. and Mallinson, M., 2008. Hover testing of a prototype small planetary spacecraft. In 59th International Astronautical Congress 2008, IAC 2008 (pp. 4371–4387).

23. Walker, R., Gregory, C. and Shah, S., 1982. MATRIX x: A data analysis, system identification, control design and simulation package. IEEE Control Systems Magazine, 2(4), 30–37.

24. Hull, E., Jackson, K. and Dick, J., 2002. DOORS: A tool to manage requirements. In Requirements engineering (187–204). Springer, London.
25. Abraham, J., MathWorks, Natick, MA, 01760.
26. Owens, B.D. and Crocker, A.R., 2015, March. SimSup's loop: A control theory approach to spacecraft operator training. In 2015 IEEE aerospace conference (pp. 1–17). IEEE.
27. Limes, G., Christa, S., Pires, C. and Gundy-Burlet, K., 2015, March. EDAC events during the LADEE Mission. In 2015 IEEE Aerospace Conference (pp. 1–8). IEEE.

Dr. Karen Gundy-Burlet, PMP, is the Director of Aviation Systems at Crown Consulting, Inc. She is a technical area lead for the NASA Academic Mission Services (NAMS) contract in the Aviation Systems Division at NASA Ames Research Center. She was the software quality and flight software lead for the Lunar Atmosphere Dust Environment Explorer mission. Other areas of past research include Computation Fluid Dynamics, Modeling and Simulation, Intelligent Flight Control, and Machine Learning.

MBSE for System-of-Systems

33

Daniel DeLaurentis, Ali Raz, and Cesare Guariniello

Contents

Introduction: System-of-Systems and Imperative for Model-Based Approach	988
Systems and System-of-Systems	989
Implications for a Model-Based Approach for System-of-Systems	990
State of Art: Three Model-Based Functions Supporting SoSE	990
Sharing of Information via Models	990
Discovery of Information via Models	991
Risk Detection and Response via Models	992
Best Practice Approach: Structured Approach to Modeling System-of-Systems	993
Overview	993
Preliminary Phase for SoS Problem Scoping	994
Definition Phase	996
Abstraction Phase	998
Implementation Phase	1000
Illustrative Examples: Research Vignettes in System-of-Systems	1001
The Past	1001
The Present	1003
The Future	1009
Chapter Summary	1011
Cross-References	1011
References	1011

D. DeLaurentis (✉)
Purdue University, West Lafayette, IN, USA
e-mail: ddelaure@purdue.edu

A. Raz
Systems Engineering and Operations Research, George Mason University, Fairfax, VA, USA
e-mail: araz@gmu.edu

C. Guariniello
Purdue University, West Lafayette, IN, USA
e-mail: cguarini@purdue.edu

© Springer Nature Switzerland AG 2023
A. M. Madni et al. (eds.), *Handbook of Model-Based Systems Engineering*,
https://doi.org/10.1007/978-3-030-93582-5_59

Abstract

A System-of-Systems (SoS) is a special kind of complex system in which new capabilities arise from interacting components that are controlled with varying degrees of independence by multiple owner/operators. Thus, the capabilities generated by an SoS are not merely a sum of the capabilities of individual constituent systems, but instead are an emergent property stemming from their interactions. Thus, SoS behavior over time is highly dependent on the dynamic and evolving character of those interaction. Further, the architecture of an SoS consists not only of the constituent systems, but of their interfaces and the communication and exchanges occurring between the systems. In this context, model-based system engineering methods, processes, and tools are especially important for supporting desired evolutionary development of the SoS as the architecture is almost always in a state of flux. This chapter draws on a combination of theory and experience with practical projects to describe three examples of important model-based functions in SoS Engineering (SoSE), a tailored approach for building effective SoSE models, and vignettes of these approaches in practice. Summary observations on the critical role of MBSE in support of SoSE close the chapter.

Keywords

System-of-systems · Complex systems · Aerospace systems

Introduction: System-of-Systems and Imperative for Model-Based Approach

This chapter motivates the need for, and demonstrates value of, Model-Based Systems Engineering (MBSE) approaches for an important class of systems – a System-of-Systems (SoS). Rather than treating one specific case study, the chapter leverages the experience of the authors across a number of different vignettes associated with a particular subset of SoS applications – distributed architectures for command, control, and communications. Before presenting these vignettes, in this introduction section of the chapter we describe the essence of SoS and means to understand the implications of its distinguishing features. In addition, we present a short introduction to command and control, considered as an SoS architecture, including a preview of the military and space exploration instances that will appear later in the chapter.

Our objective is that this chapter will assist the reader in building an appreciation for the unique modeling and design challenges presented by the SoS class and how MBSE may play an important role in tackling them, now and in the future. Further, this appreciation should motivate deeper investigation of key emerging MBSE tools and techniques addressed in other chapters of this volume.

Systems and System-of-Systems

According to Blanchard and Fabrycky [1], "A system is an assemblage or combination of functionally related elements or parts forming a unitary whole, such as a river system or a transportation system." The interactions among the elements give rise to capabilities of the "unitary whole." Over many centuries, engineered systems have evolved from simple tools to powered machines to today's cornucopia of complex machines, massively connected devices, and everything in between. Further, many of these largest systems are now socio-technical – the machines and the humans interact in different ways for the system to work (or not work!). Altogether, it is quite difficult to design and predict outcomes (at the scale of the whole) in these contexts. One particular challenge is that these interacting components of the whole system are often controlled by multiple independent operators, and in these cases the term "System-of-Systems" has emerged to distinguish this class of systems.

Historically, two approaches to understand this special class of systems called System-of-Systems (SoS) have emerged: by definition or by characterization using distinguishing criteria. As expected, a single universally accepted definition of an SoS remains elusive; however, the discussion and at times argument in the community about these definitions are indeed useful for, at a minimum, agreement on what SoS are not. The ISO/IEC/IEEE 21839 [2] definition is "Set of systems or system elements that interact to provide a unique capability that none of the constituent systems can accomplish on its own." A note follows the definition indicating that the system elements can be necessary to facilitate the interaction of the constituent systems in the System-of-Systems. This definition may appear to some as not powerful enough in distinguishing from the definition of the larger class of systems quoted above from Blanchard and Fabrycky. Other definitions such as those offered by DeLaurentis [3] offer similar characterization and point to the need for ongoing research and collaboration to develop practical tools that deal with the particular opportunities (and challenges) brought by the SoS concept. Boardman and Sauser's perspective that emphasized the trait of "belonging" among the component systems provides a natural description of the SoS essence that has enriched the community's understanding [4].

The most prominent and widely cited approach via distinguishing characteristics was offered my Mark Maier. Maier first postulated five distinguishing traits (operational and managerial independence, evolutionary and emergent behavior, and geographic distribution), but later in Ref. [5] he clarified that it is the operational and managerial independence pair that provides the essential distinction and thus these two traits are the core of a useful taxonomic node. Operational independence means that each of the constituent systems of the SoS is capable of independent operation, even when isolated from the other constituent systems; managerial independence means that each of the constituent systems can be managed independently from other constituent systems, that is, multiple decision-making authorities can be present.

These insights from the distinguishing characteristics of SoS have guided the development of methods, tools, and approaches for SoS Engineering (SoSE). Jamshidi's book is a good snapshot of the state of the art at a particular point in time for SoSE [6].

Implications for a Model-Based Approach for System-of-Systems

Both the ISO definition and the SoS distinguishing features described in the previous section highlight how the capabilities associated with an SoS are not merely a sum or a combination of the capabilities of individual constituent systems, but they stem from their interactions and the corresponding SoS behavior. Therefore, the architecture of an SoS consists not only of the constituent systems, but of their interfaces and the communication and exchanges occurring between the systems. The primary differences between Systems Engineering and SoS Engineering is well-documented in the Systems Engineering Body of Knowledge (SEBoK) (System of Systems (SoS)) [7]. In essence, SoSE tends to be more complicated because of the multiple stakeholders (remember the operational and managerial independence!) involved in everything from requirements, objectives, interfaces, risk profiles, robustness levels, operational modes, etc.

A model-based approach, including conceptual models, graphical representations, requirements, analysis activities, essentially most of the contents of this volume, can be a powerful asset to manage these particular complications of SoS. Further, model-based methods are especially important for supporting evolutionary development of the SoS as the architecture is almost always in a state of flux due to the dynamics of particular systems and, occasionally, the SoS architect [8].

The remainder of the chapter draws on a combination of theory and experience with practical projects to describe three examples of important model-based functions in SoSE (section "Three Model-Based Functions Supporting SoSE"), to overview a tailored approach for building effective SoSE models (section "Structured Approach to Modeling System-of-Systems"), and finally to offer vignettes of these approaches in practice. Summary observations on the critical role of MBSE in support of SoSE close the chapter.

State of Art: Three Model-Based Functions Supporting SoSE

As illustrated in the other chapters of this volume, MBSE is useful for a multitude of functions in SE, from defining and validating transactions, to architecting cybersecurity and cyber-resilience, creating robust network architectures, establishing clear use case definitions, etc. However, in this section we introduce three functions that are particularly important in the context of SoSE.

Sharing of Information via Models

According to INCOSE, MBSE is *the formalized application of modeling to support system requirements, design, analysis, verification, and validation activities beginning in the conceptual design phase and continuing throughout development and later life cycle phases* [9]. This entire volume dedicated to the state of the art and practice in MBSE is rich with examples of how advances in methods, tools, and culture related to MBSE have produced great fruits for system engineering and architecture. In the context of SoS, the sharing of information (reliably and efficiently) is the basis for all those SE functions that are supported in the above definition.

MBSE was developed with the goal of improving communication and providing efficient and reliable knowledge capture about systems and their behavior. However, first and foremost the use of models in systems engineering provides a unique source of truth about systems to all the involved stakeholders. Considering the characterizing traits of SoS, it is obvious how helpful MBSE can be in support of SoS engineering. While preserving the operational and managerial independence of the component systems, appropriate models can be built and used to share the required type and amount of information allowing each stakeholder to analyze the behavior of the SoS (and eventually form a value proposition for remaining "involved"). Managers and operators of the constituent systems can run simulations and perform verification and validation on models of systems that they do not directly operate or manage and use models to share information about their own systems with other SoS stakeholders.

Interesting work on information sharing in SoS modeling context was offered by Fry and DeLaurentis [10]. This work developed a framework to understand what information should be exchanged, as a complement to determining what information could be exchanged and how. The framework was eventually combined with complexity concepts and metrics to propose decision aids in challenging SoSE architecture settings like ballistic missile defense and air transportation [11]. The findings essentially produced a strong case for the need of MBSE for SoSE, just as MBSE as an accepted term and concept was coming into clear view. It also provided impetus for the development of one of many agent-based modeling techniques that were founded on solid MBSE-related software principles [12].

Discovery of Information via Models

Another powerful feature of MBSE is that it proposes different types of models representing different viewpoints and characteristics of a system. Appropriate use of diagrams and views provided by languages for MBSE (e.g., UML, SysML, and other languages with even stronger semantics) and by architecture tools such as the DoD Architecture Framework, DODAF [13], equip SoS stakeholders with information according to their own needs and to the needs of the entire SoS project.

For example, a stakeholder might want to gain knowledge about interfaces, but limiting this knowledge only to the interfaces relating other constituent systems to his/her own systems, rather than looking at all interfaces in the system. Likewise, each stakeholder will be interested to obtain information that assists in performing verification and validation and to run simulations pertaining to his/her own goals. Such an approach may also enable privacy/security where needed such that each stakeholder gains access only to a relevant part of data, sufficient for running simulations and analysis of the constituent systems that affect their portion of the SoS, while protecting the data of other stakeholders.

Perhaps most important in the SoS context, behavior models may allow SoS participants to discover the form and function of the operational and managerial independence that distinguishes SoS. This information is often the most difficult to determine, especially in a non-model-based world, but also most crucial to successful collaboration.

Risk Detection and Response via Models

The compelling benefit of SoS – the development of novel capabilities, and possibly robustness which would not be achieved by component systems operating alone – is always in conflict with the elevated risk inherent in collaboration (or worse, in "perceived collaboration" that in reality is not present). SoSE approaches must be able to detect such risks (often termed "Emergent Behavior") and identify means to address them. Promising approaches to do just this are maturing, see for example Refs. [14, 15].

The first risk we highlight is simply the risk of misclassification – believing one is working with a monolithic system when in reality it is an SoS or vice versa. This risk was a central motivation for the seminal classification and characterization paper by Maier [5]. If one forms a model of the system architecture neglecting the operational and managerial independence of the component systems, then the likelihood (and possibly the severity) of the emergent outcomes from those features is high. Even if one correctly recognizes an SoS, and approaches modeling with this in mind, it is possible to mischaracterize the kind of SoS at hand. Modeling behaviors associated with operational/managerial independence can be tricky business! One central reason for this difficulty is the multiplicity of perspectives component systems may possess and project, depending on the circumstance. Examples and implications of such multiplicity was presented in Ref. [16] at a time when discussion about MBSE was only at the beginning. The very challenges called out in that paper, especially those related to scope creep and context shifting events, are ones that are well-addressed with a model-based approach to SoS architecting problems. Behaviors captured in standard behavioral models typical in MBSE languages go a long way to reducing risk of misclassification or being fooled by the multiplicity of perspectives.

The second risk we highlight is that associated with dynamic integration, or fusion, of information to enable the hoped-for SoS-unique capability. Combining the right information at the right times lies at the heart of command, control, and communication architectures. In this sense, fusion is the last activity in the triad of discovering information, sharing information, and then combining information. Vignettes offered in section "Research Vignettes in System-of-Systems" will show how models can indeed assist in this very important area and help reduce this risk of combining the wrong information, or the right information at the wrong time, and thus improve the effectiveness and efficiency of the SoS.

Best Practice Approach: Structured Approach to Modeling System-of-Systems

Overview

This section describes a structured approach to modeling SoS focused on facilitating collaboration between stakeholders when using MBSE artifacts to model SoS. First of all, the stakeholders need a common lexicon to represent the entities involved in modeling an SoS in different categories and at various hierarchical levels. The combination of the two dimensions of categories of SoS entities and hierarchical level form a table that can be filled by the stakeholders and that will provide the common structure to build SoS models. The hierarchical levels begin at the basic building blocks of the SoS, that are the lowest-level independently operated and managed systems that constitute the SoS. Collections of these elements belong to a higher hierarchical level, and so on. The four categories in this lexicon include Resources, Operations, Policies, and Economics (ROPE). Resources include all physical and nonphysical components or systems that form the SoS. Operations are the procedures that define the dynamics of SoS and the interactions between systems. Policies are the rules that govern the management and operation of the SoS. Economics define the needs of the stakeholders involved in the SoS [17].

Figure 1 is a visual representation of the four categories of resources, operations, policies, and economics, and four levels of hierarchy indicated by sequential Greek letters. The pyramid shape represents the relative number of elements that exist at the various levels. The α-level represents the base level of entities for a given SoS problem scope and the higher levels are successive networked aggregations. Having a such a common lexicon helps the stakeholders build consistent MBSE models.

Based on the ROPE table lexicon, a three-phase procedure has been proposed for SoS modeling [18]. The method, summarized in Fig. 2, is called *Definition-Abstraction- Implementation* (DAI). The purpose of the Definition phase is to understand and define the SoS problem space and scope. In the second phase, the Abstraction, the goal is to identify appropriate strategies to translate the knowledge gained in the definition phase into relevant resources for modeling and implementation. In the third phase, the Implementation, methods, processes, and tools are used to generate SoS solutions.

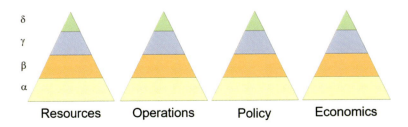

Fig. 1 The ROPE categories and levels of hierarchy

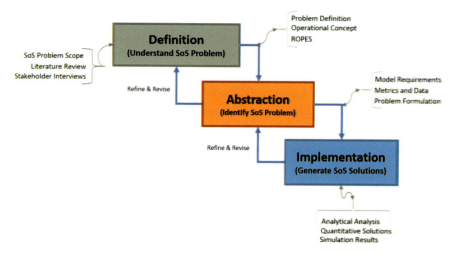

Fig. 2 Overview of the DAI process

Preliminary Phase for SoS Problem Scoping

The DAI process begins with a preliminary phase that scopes precursors to establish the context of how knowledge is acquired, how the SoS is modelled, the execution of analysis in the DAI phases, and the nature of SoS solutions to be constructed. The following preliminary elements are considered.

Purpose of SoS Analysis and Time Horizon
Prior to any given decision point of initiating an SoS evolution, a clear understanding of the objectives to be accomplished during the evolutionary epoch is needed. Relevant technical and programmatic knowledge, objectives, constraints, stakeholders, concept of end solutions, and other such knowledge artifacts need to be established for future SoS development activities to be realized. Prior researchers [8, 19, 20] allude to similar considerations in the incipient stages of the SoSE process; however, we propose a simple categorization of the SoS evolution's primary intent, as follows.

Fully Specified Evolution: In this case, the objective(s) of the SoS evolution is clear, there is a need to be fulfilled, determined by a measurable goal. For example, a planned upgrade of the ballistic missile defense system [21, 22] would entail a fully specified problem across the armed services, as there are key objectives, measures of performance, risks, stakeholder viewpoints, and other dimensions that define the problem space adequately. While complex system considerations across multiple stakeholders need to be performed, there are nevertheless existing methods, processes, and tools throughout the evolution that can be reasonably orchestrated, using existing processes to accomplish the evolution. Furthermore, the full specification may already assume a backbone construct of using existing SE artifacts that will aid in the SoS evolutionary activities, such as the use of MBSE artifacts (e.g., UML,

SysML), assuming that actors within the SoS are universally dependent on these (e.g., DoD use of DoDAF artifacts with UML/SysML). While development timelines may be long due to unforeseen scheduling issues (e.g., program delays in developing platforms), the projected timeline for the SoS evolution decisions is intended to be short to medium span.

Exploratory Evolution: This form of evolution must recognize and address many diverse uncertainty sources that complicate the establishment of clear value objectives; this is due to the need to explore ill-defined spaces that could lack the necessary data, technical and programmatic expertise, and make identification of clear value to be pursued. For example, an SoS that needs to be evolved following a long-term, innovation-driven path will not only need to generate strategies to best generate novel solutions, but also have necessary foresight on future directions and quantify value associated with the yet-to-be realized solution. An example may be a multinational corporation that seeks to predict future value, and take action on, the introduction of radically new technologies to market, across a spectrum of interconnected products. The lack of adequate data to scope market reactions for long-term innovations and the need for realignment of the SoS objectives to be "innovation" centric will require a very different set of processes (Note: as businesses become more agile, so will the need for strategies to scope such uncertainties quickly across the SoS).

Hybrid Evolution: This is a hybrid of strategies from Fully Specified and Exploratory problem scopes. For example, an SoS entity may have fully specified evolutions for a subsection of the overall SoS, but may wish to have exploratory objectives for other elements in the SoS. However, the scope of the exploration may impact the intended specified portions of the SoS, and vice versa.

We relate the envisioned time horizon for the SoS solution with the features (purpose and type) of evolution in Table 1 below.

SoS Stakeholder Control and Organizational Structure

During an SoS evolution epoch, there needs to be an awareness of the organizational structure and nature of the span of control of each stakeholder in the SoS in both contributing to the SoS effort and in participation of the envisioned end solution. While prior frameworks, such as those by Gorod [23], Dahmann [8], and Bartolomei [20], account for these elements, we propose the inclusion of early stage assessment

Table 1 SoS evolutionary change time horizons

SoS time horizon	Features	
	Evolutionary purpose	SoS problem space
Short term	Specific, measurable goals	Structured, complex but quantified uncertainties. Short-term solutions favor lower uncertain solutions
Mid term	Blend of short-term and long-term needs.	Blend of defined and high uncertainties for subsets of SoS
Long term	Strategic innovations	High multidimensional uncertainties in prediction. Both aleatoric and epistemic uncertainties driven by time

of incentive structures and cultural biases that exist throughout the SoS based on its structure and each participant's control.

The connections, alignment of objectives, and span of control that a stakeholder has will dictate how the stakeholder is incentivized to participate in the evolutionary process. While works by Ross et al. [24] and Tamaskar [25] account for stakeholder utilities in the technical design space for complex systems and SoS constructs, none account for how a stakeholder will engage during the SoSE process. For example, during an SoS evolution, the alignment of an SoS constituent's local objective may not directly align with the objective of the overall SoS. This implies that participation and sharing of data by the stakeholder is done on a "constraint basis" where the dissemination is treated a constraint to be satisfied as opposed to be a shared objective to be pursued; this relates to the observations of Honda [26], where such differences in connections between systems impacts the end product, only abstracted to SoS activities. Furthermore, the nature of end solutions synthesized will need to be compatible with the stage of evolution in the SoS, the participant's willingness to participate in the SoS architecture update, and the compatibility of the SoSE solution. Otherwise, it could lead to conditions where proposed architectural solutions are not adopted in the long run by a subset of stakeholders in the SoS such as reflected in work by Austin-Breneman [27].

Definition Phase

In SoS, the operational and managerial independence of constituent systems can bring un certainty in the definition of problem scope and requirements. To enable SoS stakeholder to build useful MBSE models, they must first define the boundaries of the SoS, the behavior of its component systems, and the desired objectives. The stakeholders will maintain independent in designing their own systems, but in this phase, they will define the interdependencies between the systems that will be reflected in the models.

In general, the primary task in this phase is to understand the problem and its context, with the following objectives:

- Identify the operational context including the domain, timescales, and goals for the SoS
- Describe the problem using appropriate common lexicon
- Identify barriers to preferred behavior in the status quo

The elements identified in this phase will be used to define the desired interactions between constituent systems, including interface standards and incentives for participation. Instruments like the ROPE table in Fig. 1 support this process with a structured approach to guide the stakeholders. The rest of the definition process is usually iterative and can include interviews with the stakeholders, exchange of documentation, and study of existing literature. This phase of the SoS modeling process, characterized by uncertainty and flexibility, can be considered analog to the conceptual design phase in systems engineering. The descriptions obtained at the end of this phase are mostly qualitative but provide a clear problem statement and an understanding of the whole SoS problem under consideration.

The definition phase characterizes the key elements of the SoS and identifies barrier to preferred behaviors in the SoS. The phase involves an activity of mapping relevant inputs provided on the SoS architecture to two levels, as initially described in the proto-method by DeLaurentis [28, 29]: The first level spans the breadth of the SoS space and is represented by several primary dimensions that include Resources, Operations, Economics, and Policy (ROPE) – these can be extended to include additional dimensions as appropriate to the SoS under analysis. The phase includes identification of relevant stakeholders at each stage of these dimensions which are examined in a hierarchical sense across each level of an SoS, ranging from the alpha level (lowest level component construct of an SoS) to the epsilon level of the hierarchy (representing the highest level of abstraction – the SoS itself). The flexibility of the ROPE dimensions applies also to the hierarchical levels, which can be adapted to the specific problem and user needs. That is, not all levels might be considered, or more levels could be added as necessary. The framework intuits the important point that the behavior of the SoS is dominated by the structure and organization at higher levels within the SoS [28, 29]. Table 2 below shows an example of the ROPE table instantiation for the transportation system, as presented by DeLaurentis [28].

The ROPE table serves as the primary mechanism to capture salient variables (endogenous, exogenous) related to each level of the SoS. While there are many perspectives to an SoS, the activity of populating the ROPE table provides a structured, categorical, and sequential approach to guiding stakeholder explorations on constructing the SoS mental map. The discovery process, through the mapping activity, naturally gives rise to many issues on ensuring that an effective ROPE table is constructed to add value to subsequent steps in the DAI process. The challenge is to extend the existing guidance of the ROPE table to practically accommodate informational artifacts in a robust and organized fashion for real-world problems. As we inspect Table 2, we can imagine that each level of hierarchy and the corresponding category may contain very large and diverse types of information. Additionally, there exists a rich context of connections between such informational artifacts and relevant actors in the SoS, where individual artifacts may be public or private, depending on the prior established purpose, organizational structure, and stakeholder control. Informational control and compartmentalization is heavily influenced both by organizational policies and by utility seeking behaviors of units within the organization (here, the SoS), as evidenced by prior empirical works [26, 27].

Furthermore, the size and diversity of information collected may prompt the need of specific frameworks to be brought to bear in marshalling the information collected. However, consideration for stakeholder adoption of how the information is captured needs to be considered even in this incipient Definition stage. For example, while MBSE perspectives may be advantageous to capture information, the SoS under consideration may have many stakeholders who are not vested in an MBSE

Table 2 ROPE table from Ref. [28]

Level	Resources	Operations	Economics	Policy
α ($\approx 10^6$)	Vehicles and infrastructure (e.g., aircraft, truck, runway)	Operating a resource (aircraft, truck, etc.)	Economics of building/ operating/ buying/selling/ leasing a single resource	Policies relating to single resource use (e.g., type certification, flight procedures, etc.)
β ($\approx 10^4$)	Collection of resources for a common function (an airport, etc.)	Operating resource networks for common function (e.g., airline)	Economics of operating/ buying/selling/ leasing resource networks	Policies relating to multiple vehicle use (e.g., airport traffic management, noise policies, etc.)
γ ($\approx 10^2$)	Resources in a transport sector (e.g., air transportation)	Operating collection of resource networks (e.g., commercial air ops)	Economics of a business sector (e.g., airline industry)	Policies relating to sectors using multiple vehicles (safety, accessibility, etc.)
δ ($\approx 10^1$)	Multiple interwoven sectors (resources for a national transport system)	Operations of multiple business sectors (i.e., operators of total national transportation system)	Economics of the total transportation system (all transportation companies)	Policies relating national transportation policy
e ($\approx 10^0$)	Global transportation system	Global operations in the world transportation system	Global economics in the world transportation system	Policy relating to the world transportation system

infrastructure yet, and therefore, may not see immediate value in its use for capturing information artifacts.

Abstraction Phase

Once the SoS scope and objectives have been clearly defined, the abstraction phase serves as the *bridging phase* to facilitate the transition from the definition to the implementation phase. In this phase, abstraction techniques are used to understand how to practically build the desired MBSE models of the SoS. The stakeholders identify interrelations between constituent entities, define the required inputs, outputs, and metrics that will be included in the model, clarify the meaning of variables, and formulate hypotheses about the SoS problem. In this phase, the stakeholders move from the textual, descriptive nature of the definition phase to symbolic, logical, and graphical representations. In general, the abstraction phase has the following set of objectives, each enhanced by MBSE:

- Elaborate characterization of the main classes of actors, effectors, disturbances, and interdependency networks
- Focus on interrelations among entities in hierarchy
- Identify the big-picture dynamics

It is important to begin this phase at the highest level of abstraction, avoiding the risk of being biased by preconceived implementations or solutions, and to gradually add details towards the construction of the right model. A good strategy towards this goal is to reason in terms of three generic classes of design variables in the abstraction phase:

- Composition: which systems should be present in the model? How are they connected? What are their functions and resources?
- Configuration: what are the operational interdependencies between the systems? What information is passed between the systems? How does connectivity evolve over time?
- Control: what level of autonomy do the systems have? What are the incentives to influence systems and stakeholders?

The abstraction phase uses informational artifacts collected from the *Definition* phase to construct subdomains and identify appropriate inputs (e.g., resources, constraints) and outputs (e.g., measures of performance, effectiveness, -illities) associated with each subdomain. The objective in this phase is to identify an appropriate domain decomposition of the SoS and to identify a problem definition germane to each subdomain. Furthermore, the typical size, scope, and multi-domain nature of SoS problems means that multiple decompositions may be possible – the challenge is to objectively select a decomposition that is consistent with the nature of informational artifacts from the definition phase, and established *purpose, organizational structure,* and *stakeholder control* from the preliminary phase.

Proper abstraction guides the stakeholders in adding more complexity as needed, until there is appropriate fidelity to model the SoS. The outcome of this phase defines the design variables, uncertainties, constraints, and objectives that will appear in the SoS models. The abstraction phase circumvents a natural tendency to tailor an SoS problem formulation to fit pre-conceived implementation methods and solution approaches to siloed development cycles. The artifacts of the definition phase, particularly the ROPE table, identifies all the various entities which remain applicable to the SoS. The abstraction phase begins to build relationships by describing the various networks between the entities and developing entity descriptors as explicit-implicit and endogenous-exogenous. These entity descriptors are largely tied to the SoS problem scope (i.e., fully specified evolution, exploratory evolution, etc.) and guide the development of the SoS problem formulation and narrowing down of the SoS solution space.

The ongoing challenge for the abstraction phase is identification and utilization of tools which facilitate effective development, management, and communication of abstracted artifacts for the SoS analysis. In the past, we have found SoS abstractions based on network theory and systems dynamics to be of particular importance when building explicit-implicit and endogenous-exogenous relationship between entities in SoS. For example, Refs. [30, 31] demonstrate a network-theoretic modeling of air transportation system, whereas [32] develops an SoS abstraction based on system dynamics model for airline fleet allocation problem. Furthermore, multitiered approaches which include a combination of Design Structure Matrix [33, 34] to represent the relationships between entities followed by leveraging MBSE constructs such as block definition diagrams (BDDs) to build input-output models of variables can also be used. Reference [35] demonstrates the use of BDDs and internal BDDs to develop abstract representations of SoS architectures based on functional and physical decomposition of SoS problem space.

Implementation Phase

The last phase of this formal SoS process completes the purpose of modeling and analysis. Once the abstraction phase establishes a conducive decomposition of the SoS problem space that is consistent with prior steps, the objective of this phase is to implement a solution to enable analysis of the SoS; this is accomplished by allocating an appropriate solution strategy to subdomains in the SoS involved for an overall analytical tractability. When using MBSE, in this phase the actual models are built by the stakeholders and utilized to perform verification and validation, and simulations in order to identify possible emergent behavior. Once an initial set of models has been implemented, verified, and validated, it can be used by stakeholders to run simulations of the behavior of the SoS. The simulations provide possible solutions to the development and use of the SoS to achieve its intended objectives, as well as the objective of the stakeholders. The analysis also provides insight into technical and programmatic risks. The models can be iteratively modified and expanded as needed, both before an SoS is actually built, and throughout its life cycle.

Several critical considerations come to mind when choosing an SoS-level solution strategy:

Equity in Participation: Implementations need to incentivize stakeholder participation and utilization, particularly in the case of SoS organizational structures with independent stakeholders (e.g., virtual or collaborative SoS). For example, a solution that is complex may only be amenable to a limited set of stakeholders, and therefore discourage widespread adoption. A common root cause (among many others) of poor solutions includes human factor considerations such as UX/UI in software design.

Complexity of Solution: Choice of technical solution should not encumber original functions and tasks of SoS stakeholders. Oftentimes, a solution with lesser capability/lower complexity is chosen due to other desirable characteristics such as compatibility with legacy hardware/software platforms, ease of calibration effort, robustness, lower maintenance, and service costs. Highly specialized solutions may require expertise levels that are beyond the stakeholder expectations of ease of use; acceptance of a solution with lesser capability but higher usability is preferred.

Technical and Programmatic Tractability: While a solution may be technically feasible, there needs to be careful consideration on the programmatic elements, including life cycle considerations of the solution. For example, modular material concept solutions for the DoD under the Modular Open Systems Approach gave rise to problems in the defense acquisition life cycle processes [36].

The considerations listed above are consistent with Maier's views of promoting stable intermediate forms and encouraging collaborative behaviors [5].

Illustrative Examples: Research Vignettes in System-of-Systems

In this section, we share our experiences by briefly illustrating how past SoS applications might have benefited from the use of MBSE, how current applications are making use of some MBSE elements, and how we envision that our own work (and that of the larger SoSE community) could expand the benefits derived from the application of MBSE during SoSE efforts in the future.

The Past

Lunar C3I
As illustrated in section "Three Model-Based Functions Supporting SoSE," due to the complexity and variety of the topics involved, projects and problems in System-of-Systems can utilize some of the advantageous features of MBSE. The use of standardized, consistent models between different projects allows for comparison of systems and their interactions even if they belong to very different areas. The production and use of models that constitute a *single source of truth* within a single project is key to objective evaluation and to reduction of personal interpretation, especially when operational and managerial independence is strong. In a study funded by NASA within the constellation program, Sindiy et al. [37, 38] compared different communication and information architectures in space programs and terrestrial missions. These architectures have been used as analogs to guide the analysis and synthesis of potential System-of-Systems architectures for Command, Control, Communication, and Information (C3I) in Lunar campaigns.

While the study provided the objective utility, its reliance mainly on documents, surveys, and expert opinion (versus digital models of information flow) limited correlation among each project or mission. The researchers involved in these studies had to extrapolate the required information through interpretation of the existing models and conversion of the data into a form that would be consistent between the various projects, so as to allow for adequate comparison (Fig. 3).

These studies would have greatly benefited by the adoption of formal MBSE techniques and artifacts and by the use of standard diagrams. If specific MBSE

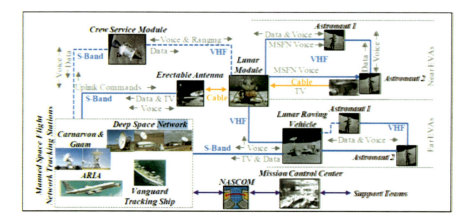

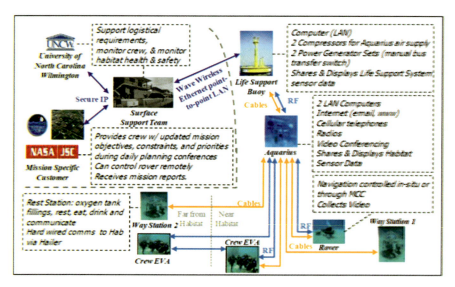

Fig. 3 (Top) Apollo communication network. (Bottom) NEEMO mission communication and data exchange network. (Courtesy NASA. Figure from Ref. [37])

models (e.g., models for architecture and operation, even if built in different languages) had been built by the original designers of each mission and program, this would have guaranteed consistency between different missions without the need for an additional step to make the information from each mission easily comparable. Since Sindiy et al. used analogs not always developed according to MBSE principles, and since there was not enough maturity for the combination of MBSE with SoSE, they used the approach of interpretation and comparison. The execution of similar studies could now greatly benefit from the development of MBSE models for each of the projects under study. This approach would increase objectivity and reusability of the models by reducing the amount of personal interpretation and by providing baselines for the applicability to other analogs, thus implementing the supporting functions described in section "Three Model-Based Functions Supporting SoSE."

The Present

SysML Diagrams to Help Characterize System-of-Systems

In 2014, Bonanne [39] used SysML language to demonstrate a formal use of an MBSE language applied to the Wave Model for defining, analyzing, and evolving an SoS [8]. The Wave Model is a representation of evolution of an SoS where the stakeholders perform analysis of the SoS architecture and, if necessary, plan and implement SoS updates at intervals. Each applicable step in the Wave Model is performed within SysML. Three different SoS types (directed, acknowledged, and collaborative) were studied within the domain of a distributed sensor management problem. The three types are differentiated by the amount of presence of a centralized authority: directed SoS have stronger centralization, where the long-term operations of the SoS are guided by a central authority; acknowledged SoS have recognized objectives for the SoS, but the stakeholders maintain more independence; collaborative SoS have a structure where stakeholders collaborate more or less voluntarily to a central objective. Bonanne's project discussed the applicability of SysML to each step of the Wave Model, as SoS were established, evaluated, and updated. The outcome showed that SysML is capable of defining, analyzing, and evolving an SoS via the processes described in the Wave Model. SysML excels at strictly defining and organizing the elements and features of an SoS, while requiring more development in the analysis portions of the SoSE process. This limitation is mostly due to the lack of analytical SoS tools built on purpose for the use of MBSE. However, the importance of this work resides in the fact that Bonanne demonstrated how SysML diagrams which capture important aspects of SoS, such as hierarchy in elements from a ROPE table (Fig. 4, adapted from Ref. [39]) in a block definition diagram (BDD), were effective building blocks for generating current SoS state and potential future adaptations in the next "wave." Figure 4 shows not only the hierarchy of element of the SoS, in this case resources, but also compositions and aggregations of these elements.

An additional valuable addition of this work was the examination of how the underlying SysML models would take shape in three different cases: Directed SoS, Acknowledged SoS, and Collaborative SoS. Each has their own distinctive aspects in operations. Bonanne found that the allocation of activities to different systems is easily done with the use of allocation lanes on activity diagrams or directly with the allocation relationship between an activity and a block or its ports. This allocation is independent of level of abstraction, allowing any combination of activities and blocks. The independence of the allocation relationship allows for vastly different levels of centralization in different SoS models drawing on the same hierarchy of activities, a feature that is useful in capturing different SoS types. Additional details on this topic are found in Ref. [39].

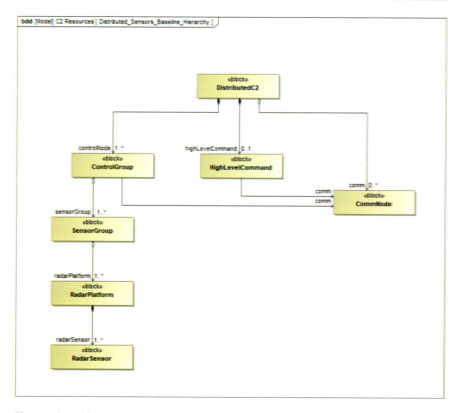

Fig. 4 Hierarchical structure of resources in a distributed C2 architecture, represented in a block definition diagram in SysML

MBSE for Modeling and Analysis of Space Exploration SoS

Several years later, Zusack [40] built on the work of Bonanne and in conjunction with NASA's MBSE Pathfinder project to forge more progress via connection of SysML models with an SoS analysis tool, the Systems Operational Dependency Analysis (SODA) tool, for space SoS analysis. SODA [41] is part of a suite of tools to model and analyze System-of-Systems, called SoS Analytic Workbench (AWB) [42]. Tools in the AWB have been developed specifically for the study of SoS; therefore they address specific features of this category of systems. However, integration of the various tools has been hindered by the different network representations used by tools and by the need for the users to provide inputs in a specific format.

Zusack's application explored means by which MBSE facilitates generation and analysis of In Situ Resource Utilization architectures in Mars missions. The importance and innovation of this work resides in the enhancement of the visual depiction provided by MBSE with a software that actively shows changes in relationships as architectural modifications are made. Therefore, this approach facilitates the discovery of information through the models. Another important contribution was the use of SoSE-specific methodology on a network that was designed using MBSE, as shown in Fig. 5.

MBSE as an Enabler for System-of-Systems Integration

Information Fusion System (IFS) is comprised of distributed and heterogeneous systems that generate, process, and fuse information to accomplish low-level and high-level information fusion (LLIF and HLIF) functionality [43]. LLIF functions pertains to acquiring, collecting, fusing sensed or preprocessed information (e.g., heat signature from infrared sensors or raw images from a social media feed), whereas the HLIF functions accomplish sense-making and action plan by developing situation awareness and assessment. A simple example of an IFS is a health monitoring system for patient where multiple low-level sensors (i.e., heart rate monitor, oxygen sensors) collect the raw and preprocessed information which is collectively analyzed and fused to monitor health condition (a HLIF functionality). For large-scale systems, such as industrial Internet of things (IoT) and/or military command and control systems, these LLIF and HLIF functions, along with the physical systems that provide this functionality, are developed independent from one another but require collaboration to achieve the IFS mission objectives [44]. The distribution, independence, and heterogeneity of systems, in addition to the multiplicity of LLIF and HLIF functions, creates an Information Fusion System-of-Systems (IF-SoS) with an extensively large design space [45]. SoSE architecting processes, when combined with MBSE, provide integrated allocated architectures of IF-SoS that combine the physical and functional architectures. MBSE techniques in this application help constrain the design space of SoS and elicit functional and physical attributes for use in SoS analytics and enable comprehension of the performance implications of various architectures. The following paragraphs summarize an example application of utilizing MBSE to create IF-SoS allocated architectures.

Consider an example of a multi-sensor multi-target tracking system illustrated in Fig. 6 to motivate SoSE architecting of IF-SoS [35]. In this example, physical systems such as radars, manned and unmanned aircraft, satellite, ship, and a command-and-control center are used to detect threats. In the SoS lexicon (introduced earlier in section "Structured Approach to Modeling System-of-Systems," both the different aircraft and the radar can be considered as α-level resource (i.e., the base entities that cannot be further be decomposed), the satellite and the ship as β-level systems as they enable a network of α-level systems, and finally the command-and-control center as the γ-level system. Identification of distribution of these physical resources along with their connectivity constitutes the IF-SoS physical architecture.

However, in order to achieve the objective of the multi-sensor multi-target tracking system, these physical systems are required to accomplish certain data fusion functions that are determined by the functional architecture. These required data fusion functions are well-documented in the data fusion literature and are termed as the Joint Directorate of Laboratories Data Fusion Model (JDL-DFM) levels [43, 46]. A functional description of these levels is provided in Fig. 7a (it is

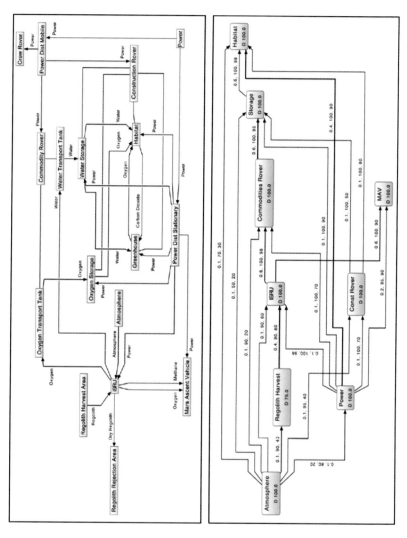

Fig. 5 (Top) MBSE model of Mars In Situ Resource Utilization. (Bottom) SODA network based on the MBSE model. (Courtesy NASA. Figure from Ref. [40])

33 MBSE for System-of-Systems

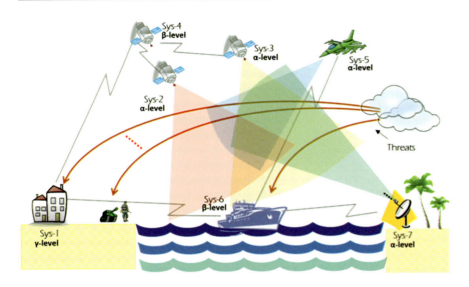

Fig. 6 IF-SoS operational concept and physical systems

important to note that Level 0 and Level 1 are classified as LLIF while all other levels are considered HLIF). An inconspicuous issue with the description of data fusion functionality as levels of JDL-DFM is that no interdependence or feedback between levels is implied which is not correct [47]. When MBSE techniques are applied and a functional architecture of the JDL-DFM levels is constructed, the interdependence between levels and feedback becomes evident. Fig. 7b illustrates a functional architecture of data fusion functionality built in Integrated Definition for Functional Modeling (IDEF-0) diagram that provides a model-based representation [48, 49].

Allocation of the functions to the physical resources to meet the mission objectives instigates the allocated architecture that can entirely be described via MBSE. Figure 8 provides an example of two notional IF-SoS allocated architectures depicting a conceptual illustration as well as a MBSE implementation in CORE (CORE is MBSE software implementation tool developed by Vitech Corporation, https://www.vitechcorp.com/core_software/) to ensure functional and physical architecture consistency. The two IF-SoS allocated architectures differ from each other with contrasting considerations in systems and networks that constitute the allocated architecture. The IF-SoS allocated cartoons depict a conceptual allocation of data fusion functionality to the various systems and instantiate a conceptual allocated architecture. These IF-SoS allocated architectures build from the SoS lexicon described earlier in section "Structured Approach to Modeling System-of-Systems" and combine the operational concept and functional architecture depicted in Figs. 6 and 7b, respectively. The α-level resources in the IF-SoS Allocated Architecture 1 perform only JDL-DFM level 0 functionality while the α-level systems in the IF-SoS Allocated Architecture 2 perform level 0 and level 1 functionality. Since the SoS as a whole has to accomplish all of the JDL-DFM levels, the

(a)

JDL-DFM levels.	
JDL-DFM levels	Scope
Level 0: source pre-processing	Pixel and signal level data characterization. Signals and features that are determined by measurements and observations
Level 1: object assessment	Object location, parametric and kinematic information. Establishment of tracks, IDs and classification. Combination of multi sensor data
Level 2: situation assessment	Contextual interpretation of objects, events, situations and their relationships
Level 3: threat assessment	Future projections of current situations and consequence determination
Level 4: process Refinement	Resource management and adaptive fusion control in support of mission objectives
Level 5: user Refinement	Fusion system and human interaction

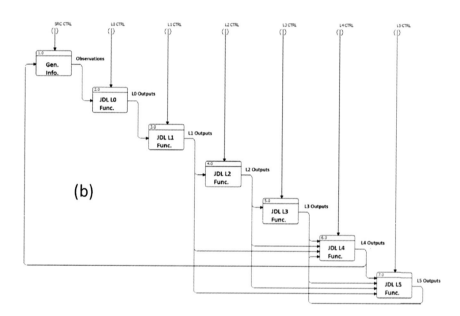

Fig. 7 IF-SoS functional architectures and data fusion functions

decisions made at α-level of the SoS architecture impact other hierarchical levels of the SoS. Here the MBSE representation and tools ensure functional flow, and resource constraints requirements are satisfied in creating these architectures. The MBSE allocated architecture representations, in conjunction with the functional architecture, identify the information and functional flows between the different systems that differ between the different allocated architectures. Subsequently, the MBSE representation serves as the basis for building implementation and performance evaluation models (e.g., agent-based models) of the IF-SoS architectures that

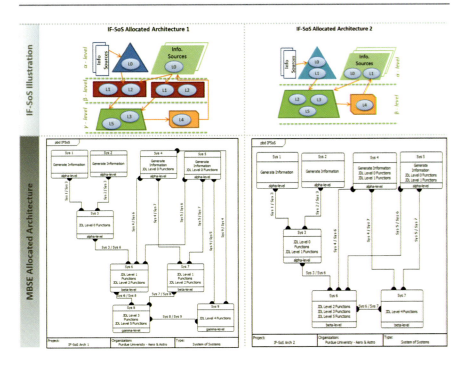

Fig. 8 Building allocated architectures of IF-SoS in MBSE

can then be analyzed with SoS analytic techniques to evaluate trade-offs between allocations. Raz et al. have described the process of building the allocated architectures of IF-SoS [35] and utilizing Design of Experiments and Machine Learning techniques [50] for analytical evaluation of different architectures for mission effectiveness.

The Future

MBSE as Enablers for Better SoSE Analysis and Synthesis Activities

As already seen in other application areas, the adoption of MBSE in support of SoSE will depend squarely on the realized value in terms of enhanced quality and efficiency of key SoSE activities. A particular example is the SoS Analytic Workbench [42] and associated tools. These tools currently utilize a document-based approach to provide input data. This approach does not offer a simple, graphical way to visualize and modify the models, and both the information provided by the user and the output data by the tools are somehow disconnected from the real world of the systems and their connections and interactions. The use of MBSE as data enablers for SoS tools would serve the double purpose of providing a user-friendly interface and of connecting the SoS modeling and analysis to standard system models, thus making it possible to understand the relationship between the system in the SoS world, and the same systems in the many different viewpoints that can be represented with MBSE models. In 2018, Guariniello et al. [51] suggested that DoDAF [13] views and SysML diagrams would be suitable to support various stages of the application of the AWB tools to SoSE.

Artificial Intelligence as Enabler for SoS MBSE Model Definition

The process of collecting data that resides in models for use in SoS analysis brings several additional challenges beyond the basic level. One of the challenges is that MBSE models do not currently have built-in properties that would represent unique SoS traits. Another challenge is that the extraction of appropriate data from MBSE model to feed SoSE tools can be complex and prone to individual shortcomings and even biases.

Artificial Intelligence (AI) may be an enabler for SoS MBSE model definition and for automated data extraction.

A recent study applied to design of satellite constellations [52, 53] demonstrated the combined used of AI, SoSE, and MBSE to increase the level of automation in the analysis and synthesis of SoS in the aerospace domain. For the modeling part, MBSE representation has been used both as source and as archive of SoS-related information, which has then been used to build models for the tools in the AWB. Then, the researchers used AI techniques to identify the relationships between SoS features of the aerospace systems architectures and metrics of interest, with the goal of understanding some of the patterns and features which result in better values of the performance metrics (Fig. 9). The relationship model achieved by training a Bayesian-Regularized Artificial Neural Network and later a fully stochastic Bayesian Neural Network provided capability to generate new architectures exhibiting

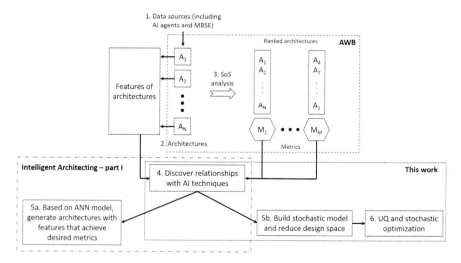

Fig. 9 Use of AI to support automated SoS analysis and design and to enable SoS model definition and construction

desired features. These architectures, automatically generated, can then be represented again with MBSE models.

This effort still used human-in-the-loop for data extraction from MBSE models. However, a parallel effort demonstrated a promising approach where AI can extract the necessary information to feed SoS tools directly from MBSE models as well as from verbal descriptions. A description of these techniques can be found in Ref. [54]. The use of AI for both extracting information from MBSE tools and for producing models of SoS architectures is showing potential for exploiting functions brought by MBSE and create a "full-fledged MBSoSE."

Chapter Summary

This chapter examined the need, status, and opportunities for MBSE in the particular problem domain of System of Systems (SoS), a special class of complex systems. The distinguishing traits of operational and managerial independence of systems in an SoS present many challenges for engineering analysis and synthesis, that is, SoS Engineering (SoSE). MBSE presents an opportunity to improve the discovery of information, sharing of information, and risk detection in SoS applications. In general, information awareness of all kinds is central to coordination and control of SoS and required for effective design and architecture. Several past and present vignettes describe particulars of some attempts to inject MBSE into SoSE and how many opportunities were discovered for value not original considered. However, there are current limitations of graphically oriented MBSE languages that must be overcome. For example, complexities in modeling complex information sharing constructs (e.g., cloud comprising multiple service providers and multiple, asynchronous client requests) that represent several large-scale SoSs will challenge current MBSE languages, processes, and tools. New extended capabilities emerging, such as SysMLv2, may provide progress in this vein. Overall, the imperative to create a "full-fledged MBSoSE" is still a strong one and the increased attention of both research and practitioner communities should lead us far toward this destination.

Cross-References

▶ Introduction to the Handbook

References

1. B. Blanchard and W. Fabrycky, *Systems Engineering and Analysis*, 5th ed. New Jersey: Prentice-Hall, 2011.
2. ISO/IEC/IEEE, "ISO/IEC/IEEE 21839:2019." ISO/IEC JTC 1/SC 7 Software and systems engineering, 2019. Accessed: Sep. 07, 2021. [Online]. Available: https://www.iso.org/cms/render/live/en/sites/isoorg/contents/data/standard/07/19/71955.html

3. D. De Laurentis *et al.*, "A Case for an International Consortium on System-of-Systems Engineering," *IEEE Systems Journal*, vol. 1, no. 1, pp. 68–73, Sep. 2007, https://doi.org/10.1109/JSYST.2007.904242.

4. J. Boardman and B. Sauser, "System of Systems - the meaning of of," in *2006 IEEE/SMC International Conference on System of Systems Engineering*, Los Angeles, California, USA, 2006, pp. 118–123. https://doi.org/10.1109/SYSOSE.2006.1652284.

5. M. Maier, "Architecting principles for system-of-systems," *Systems Engineering*, vol. 1, no. 4, pp. 267–284, 1998.

6. M. Jamshidi, *Systems of systems engineering: principles and applications*. CRC Press, 2010. Accessed: Aug. 30, 2013. [Online]. Available: http://books.google.com/books?hl=en&lr=&id=YvxUon2vAfUC&oi=fnd&pg=PP1&dq=Jamshidi,+M.,+System-of-systems+Engineering%E2%80%94Principles+and+Applications,+Taylor+Francis+CRC+Publishers,+Boca+Raton,+FL,+2008&ots=1JhYbJ1V2f&sig=LbFOrkdcDZlH8PK8koNdCLecVMU

7. "Systems of Systems (SoS) - SEBoK." https://www.sebokwiki.org/wiki/Systems_of_Systems_(SoS) (accessed Jan. 27, 2019).

8. J. Dahmann, G. Rebovich, J. Lane, R. Lowry, and K. Baldwin, "An implementers' view of systems engineering for systems of systems," in *Systems Conference (SysCon), 2011 IEEE International*, 2011, pp. 212–217.

9. INCOSE, *INCOSE Systems Engineering Handbook: A Guide for System Life Cycle Processes and Activities*, 4 edition. Hoboken, New Jersey: Wiley, 2015.

10. D. N. Fry and D. A. DeLaurentis, "Measuring net-centricity," in *2011 6th International Conference on System of Systems Engineering*, Jun. 2011, pp. 264–269. https://doi.org/10.1109/SYSOSE.2011.5966608.

11. D. N. Fry, "Cost, Performance, and Networked Information Sharing in a Ballistic Missile Defense System," PhD Thesis, Purdue University, West Lafayette, IN, 2015.

12. D. N. Fry, R. Campbell, and D. A. DeLaurentis, "Modeling Systems-of-Systems from Multiple Design Perspectives: Agents, Interfaces, and Architectures," Kissimmee, FL, 2015. Accessed: Aug. 03, 2015. [Online]. Available: http://arc.aiaa.org/doi/abs/10.2514/6.2015-0908

13. DODAF, "DOD Architecture Framework Version 2.02 - DOD Deputy Chief Information Officer," 2010. https://dodcio.defense.gov/Library/DoD-Architecture-Framework/ (accessed Aug. 13, 2020).

14. L. Rainey and M. O. Jamshidi, *Engineering Emergence: A Modeling and Simulation Approach*. CRC Press, 2020. Accessed: Sep. 07, 2021. [Online]. Available: https://www.routledge.com/Engineering-Emergence-A-Modeling-and-Simulation-Approach/Rainey-Jamshidi/p/book/9780367656119

15. A. K. Raz, J. Llinas, R. Mittu, and W. F. Lawless, "Engineering for emergence in information fusion systems: A review of some challenges☆," in *Human-Machine Shared Contexts*, W. F. Lawless, R. Mittu, and D. A. Sofge, Eds. Academic Press, 2020, pp. 241–255. https://doi.org/10.1016/B978-0-12-820543-3.00012-2.

16. H. Polzer, D. A. DeLaurentis, and D. N. Fry, "Multiplicity of Perspectives, Context Scope, and Context Shifting Events," in *2007 IEEE International Conference on System of Systems Engineering*, Apr. 2007, pp. 1–6. https://doi.org/10.1109/SYSOSE.2007.4304262.

17. N. Davendralingam, A. Raz, K. Moolchandani, C. Guariniello, S. Tamaskar, and D. DeLaurentis, "A DAI Process for System-of-Systems Engineering - antecedents, status quo and path forward," presented at the Council of Engineering Systems Universities (CESUN), Tokyo, Japan, Jun. 2018.

18. D. DeLaurentis and R. K. Callaway, "A system-of-systems perspective for public policy decisions," *Review of Policy research*, vol. 21, no. 6, pp. 829–837, 2004.

19. M. C. Jackson and P. Keys, "Towards a System of Systems Methodologies," *The Journal of the Operational Research Society*, vol. 35, no. 6, pp. 473–486, 1984, https://doi.org/10.2307/2581795.

20. J. E. Bartolomei, D. E. Hastings, R. de Neufville, and D. H. Rhodes, "Engineering Systems Multiple-Domain Matrix: An organizing framework for modeling large-scale complex systems," *Systems Engineering*, vol. 15, no. 1, pp. 41–61, Mar. 2012, https://doi.org/10.1002/sys.20193.

21. R. K. Garrett, S. Anderson, N. T. Baron, and J. D. Moreland, "Managing the interstitials, a System of Systems framework suited for the Ballistic Missile Defense System," *Syst. Engin.*, vol. 14, no. 1, pp. 87–109, Mar. 2011, https://doi.org/10.1002/sys.20173.

22. "MDA - The Ballistic Missile Defense System." https://www.mda.mil/system/system.html (accessed Jan. 26, 2018).

23. A. Gorod, B. Sauser, and J. Boardman, "System-of-Systems Engineering Management: A Review of Modern History and a Path Forward," *IEEE Systems Journal*, vol. 2, no. 4, pp. 484–499, Dec. 2008, https://doi.org/10.1109/JSYST.2008.2007163.

24. A. M. Ross, D. E. Hastings, J. M. Warmkessel, and N. P. Diller, "Multi-Attribute Tradespace Exploration as Front End for Effective Space System Design," *Journal of Spacecraft and Rockets*, vol. 41, no. 1, pp. 20–28, Jan. 2004, https://doi.org/10.2514/1.9204.

25. S. Tamaskar, "Managing complexity of aerospace systems," *Ph.D. Thesis*, 2014, Accessed: Feb. 06, 2018. [Online]. Available: http://adsabs.harvard.edu/abs/2014PhDT.......248T

26. T. Honda, F. Ciucci, K. Lewis, and M. Yang, "Comparison of information passing strategies in system-level modeling," *AIAA Journal*, vol. 53, no. 5, pp. 1121–1133, 2015.

27. J. Austin-Breneman, T. Honda, and M. C. Yang, "A Study of Student Design Team Behaviors in Complex System Design," *J. Mech. Des*, vol. 134, no. 12, pp. 124504-124504–7, Nov. 2012, https://doi.org/10.1115/1.4007840.

28. D. DeLaurentis, "Understanding Transportation as a System-of-Systems Design Problem," Jan. 2005. https://doi.org/10.2514/6.2005-123.

29. D. A. DeLaurentis and R. K. Callaway, "A System-of-Systems Perspective for Public Policy Decisions," *Review of Policy Research*, vol. 21, no. 6, pp. 829–837, 2004.

30. J.-H. Lewe and D. A. DeLaurentis, "Entity-centric abstraction and modeling of future transportation architectures," 2004.

31. T. Kotegawa, D. A. DeLaurentis, and A. Sengstacken, "Development of network restructuring models for improved air traffic forecasts," *Transportation Research Part C: Emerging Technologies*, vol. 18, no. 6, pp. 937–949, Dec. 2010, https://doi.org/10.1016/j.trc.2010.03.004.

32. K. Moolchandani, P. Govindaraju, S. Roy, W. A. Crossley, and D. A. DeLaurentis, "Assessing Effects of Aircraft and Fuel Technology Advancement on Select Aviation Environmental Impacts," *Journal of Aircraft*, vol. 54, no. 3, pp. 857–869, May 2017, https://doi.org/10.2514/1.C033861.

33. D. Steward, "The Design Structure System: A Method for Managing the Design of Complex Systems," *IEEE Transactions on Engineering Management*, vol. 28, no. 3, Aug. 1981.

34. A. Yassine and D. Braha, "Complex Concurrent Engineering and the Design Structure Matrix Method," *Concurrent Engineering: Research and Applications*, vol. 11, no. 3, pp. 165–176, Sep. 2003.

35. A. K. Raz, C. R. Kenley, and D. A. DeLaurentis, "A System-of-Systems perspective for information fusion system design and evaluation," *Information Fusion*, vol. 35, pp. 148–165, May 2017, https://doi.org/10.1016/j.inffus.2016.10.002.

36. U. S. G. A. Office, "Defense Acquisitions: DOD Efforts to Adopt Open Systems for Its Unmanned Aircraft Systems Have Progressed Slowly," no. GAO-13-651, Jul. 2013, Accessed: Feb. 11, 2018. [Online]. Available: https://www.gao.gov/products/GAO-13-651

37. O. V. Sindiy, K. L. Ezra, D. A. DeLaurentis, B. S. Caldwell, T. I. McVittie, and K. A. Simpson, "Analogs Supporting Design of Lunar Command, Control, Communication, and Information Architectures," *Journal of Aerospace Computing, Information, and Communication*, vol. 7, no. 5, pp. 151–176, May 2010, https://doi.org/10.2514/1.47542.

38. Oleg V. Sindiy, *Model-based system-of-systems engineering for space-based command, control, communication, and information architecture design.* Thesis PhD–Purdue University, 2010.

39. K. Bonanne, "A Model-Based Approach To System-Of-Systems Engineering Via The Systems Modeling Language," *Open Access Theses*, Jul. 2014, [Online]. Available: https://docs.lib.purdue.edu/open_access_theses/407

40. S. Zusack, C. Guariniello, and D. DeLaurentis, "Operational dependency analysis of a human mars architecture based on the SODA methodology," in *2018 IEEE Aerospace Conference*, Mar. 2018, pp. 1–12. https://doi.org/10.1109/AERO.2018.8396813.

41. C. Guariniello and D. DeLaurentis, "Supporting design via the System Operational Dependency Analysis methodology," *Res Eng Design*, vol. 28, no. 1, pp. 53–69, Jan. 2017, https://doi.org/10.1007/s00163-016-0229-0.

42. D. DeLaurentis, N. Davendralingam, K. Marais, C. Guariniello, Z. Fang, and P. Uday, "An Sos Analytical Workbench Approach to Architectural Analysis and Evolution," *INSIGHT*, vol. 20, no. 3, pp. 69–73, 2017, https://doi.org/10.1002/inst.12166.

43. J. Llinas, C. Bowman, G. Rogova, A. Steinberg, E. Waltz, and F. White, "Revisiting the JDL data fusion model II," DTIC Document, 2004. Accessed: Apr. 12, 2013. [Online]. Available: http://oai.dtic.mil/oai/oai?verb=getRecord&metadataPrefix=html&identifier=ADA525721

44. E. Blasch, É. Bosse, and D. Lambert, *High-Level Information Fusion Management and System Design*, 1st ed. Artech House, 2012.

45. A. K. Raz and D. A. DeLaurentis, "A System-of-Systems Perspective on Information Fusion Systems: Architecture Representation and Evaluation," Jan. 2015. https://doi.org/10.2514/6.2015-0644.

46. C. L. Bowman, "The dual node network (DNN) data fusion & resource management (DF & RM) architecture," presented at the AIAA Intelligent Systems Conference, Sep. 2004.

47. A. N. Steinberg and L. Snidaro, "Levels?," in *2015 18th International Conference on Information Fusion (Fusion)*, Jul. 2015, pp. 1985–1992.

48. D. M. Buede, "Appendix B: IDEF0 Model of the Engineering of a System," in *The Engineering Design of Systems*, John Wiley & Sons, Inc., 2009, pp. 455–474. [Online]. Available: https://doi.org/10.1002/9780470413791.app2

49. Federal Information Processing Standards (FIPS), Pub. No. 183., "Integration Definition for Function Modeling (IDEFO),." U.S. Dept. of Commerce, Washington, DC., 1993.

50. A. K. Raz, P. Wood, L. Mockus, D. A. DeLaurentis, and J. Llinas, "Identifying Interactions for Information Fusion System Design Using Machine Learning Techniques," in *2018 21st International Conference on Information Fusion (FUSION)*, Jul. 2018, pp. 226–233. https://doi.org/10.23919/ICIF.2018.8455429.

51. C. Guariniello, Z. Fang, N. Davendralingam, K. Marais, and D. DeLaurentis, "Tool suite to support model based systems engineering-enabled system-of-systems analysis," in *2018 IEEE Aerospace Conference*, Mar. 2018, pp. 1–16. https://doi.org/10.1109/AERO.2018.8396642.

52. C. Guariniello, L. Mockus, A. K. Raz, and D. A. DeLaurentis, "Towards Intelligent Architecting of Aerospace System-of-Systems," in *2019 IEEE Aerospace Conference*, Mar. 2019, pp. 1–11. https://doi.org/10.1109/AERO.2019.8742173.

53. C. Guariniello, L. Mockus, A. K. Raz, and D. A. DeLaurentis, "Towards Intelligent Architecting of Aerospace System-of-Systems: Part II," in *2020 IEEE Aerospace Conference*, Mar. 2020, pp. 1–9. https://doi.org/10.1109/AERO47225.2020.9172585.

54. C. Guariniello, T. B. Marsh, R. Porter, C. Crumbly, and D. A. DeLaurentis, "Artificial Intelligence Agents to Support Data Mining for SoS Modeling of Space Systems Design," in *2020 IEEE Aerospace Conference*, Mar. 2020, pp. 1–11.https://doi.org/10.1109/AERO47225.2020.9172802.

Dr. Daniel DeLaurentis is professor in Purdue University's School of Aeronautics and Astronautics, where he also directs the Center for Integrated Systems in Aerospace (CISA) and its main component, the System of Systems Lab. His primary research and teaching interests include problem formulation, modeling, design and control methods for aerospace systems and systems-of-systems, all from a model-based perspective. Dr. DeLaurentis annually teaches a graduate course "System of Systems Modeling and Analysis" at Purdue, now regularly with enrollment over 100 students. DeLaurentis has supported many students and professional staff who have made impacts across diverse domains including air transportation, defense/security, civil infrastructure, and space exploration. Dr. DeLaurentis also serves as the Chief Scientist of the U.S. DoD's Systems Engineering Research Center (SERC). He is Fellow of the International Council on Systems Engineering

(INCOSE), Associate Fellow of the American Institute of Aeronautics and Astronautics (AIAA), and senior member of the Institute of Electrical and Electronics Engineers (IEEE).

Dr. Ali Raz is assistant professor in the Systems Engineering and Operations Research Department at George Mason University. His research and teaching addresses the collaborative nature of autonomy and developing systems engineering methodologies for integrating autonomous systems. Raz's research brings a systems engineering perspective, particularly inspired by complex adaptive systems, to information fusion and artificial intelligence/machine learning technologies that form the foundations of collaborative and integrated autonomous systems. Prior to joining George Mason University, Raz was a visiting assistant professor at Purdue University School of Aeronautics and Astronautics. Raz is a co-chair of the International Council of Systems Engineering (INCOSE) Complex Systems Working Group and a Certified Systems Engineering Professional (CSEP). He is also a senior member of the American Institute for Aeronautics and Astronautics (AIAA) and the Institute of Electrical and Electronics Engineers (IEEE).

Dr. Cesare Guariniello is a research scientist in Purdue University's School of Aeronautics and Astronautics. He holds two Master's degrees, in Automation and Robotics Engineering and in Astronautical Engineering, from the University of Rome "La Sapienza," and a PhD in Aeronautics and Astronautics from Purdue University. His research interests include System-of-Systems design and architecting, space applications, cybersecurity, dynamics and control, and planetary geology. Dr. Guariniello has been involved in research projects with NASA, the US DoD, the US Navy, MITRE corporation, and the NSF. Recently, he expanded his role to mentoring and managing students and lecturing graduate-level classes. Dr. Guariniello is a senior member of the American Institute of Aeronautics and Astronautics (AIAA) and the Institute of Electrical and Electronical Engineers (IEEE), and a member of the International Council on Systems Engineering (INCOSE), and the American Astronautical Society (AAS).

NSOSA: A Case Study in Early Phase Architecting

34

M. W. Maier

Contents

Introduction 1018
    Concept Exploration 1019
    Concept Development and Development Strategy 1019
    System Development and Production 1019
Concept Exploration Phase Goals 1020
NSOSA Background and Goals 1022
Models Relevant to the Problem 1026
A 42010 and National Academies Guide to Model Selection 1028
    Functional Description 1029
    Physical Descriptions: Models Versus Design Vectors 1029
    Performance and Value Modeling 1030
Choosing the NSOSA Value Model 1030
NSOSA Design Vector Models 1034
NSOSA Modeling Processes 1037
NSOSA Results 1039
From Early Phase to Acquisition 1041
Chapter Summary 1042
References 1042

Abstract

Systems engineering happens throughout a system's lifecycle, including at its earliest conceptual phases. Systems engineering can be based on models in the earliest phases. Choice of model and method must adapt to the development phase, and the goals are different and the nature of the information to be modeled will differ, sometimes dramatically. From 2014 to 2017, the National Oceans and Atmosphere Administration (NOAA) conducted a study of the future of its environmental monitoring satellite architecture. This study made use of model-

---

M. W. Maier (✉)
The Aerospace Corporation, Hill AFB, Ogden, UT, USA

© Springer Nature Switzerland AG 2023
A. M. Madni et al. (eds.), *Handbook of Model-Based Systems Engineering*,
https://doi.org/10.1007/978-3-030-93582-5_79

based approaches throughout its duration. This case-study illustrates how those models were chosen and integrated with each other, to best achieve the goals of a concept exploration phase study on constellations of satellites.

Keywords

Architecture · Architecting · Concept exploration · Satellites

Introduction

MBSE, and its implementation in digital engineering (DE), can be applied throughout a system's lifecycle. Much of the discussion of MBSE application has been to the development period, when an overall design concept is being transformed into physical designs and implementations. There are clear incentives for application during physical development and transition to production. Cost increases greatly in this phase versus concept development. In some systems, design and production may be the bulk of lifecycle cost, as, for example, when the developer and manufacturer sells the system to consumers who incur the operational costs. The organizations who conduct system design and manufacturing are also the primary implementers of systems engineering. Time to develop (as opposed to time spent on concept search) is very visible. Attention is paid to trying to cut development time and it is considered a major competitive advantage. It is unsurprising that design and manufacturing activities get the bulk of the MBSE attention.

MBSE can also be applied to the early phases of system development, to concept development or architecting activities. While the potential for direct cost reduction here is low, because relative expenses in this phase are relatively small, the possibilities for leverage on later effects are large. The nature of MBSE needed both overlaps and is different from MBSE needs later in the lifecycle. Some aspects start in the early phases and carry over; others are relatively unique to the early phase.

This chapter explores a case study in the application of MBSE methods to very early phase systems engineering work. By very early phase we mean not only before a classic Milestone A, but before there is an acquisition program, or even a decision to have an acquisition program(s). The case study is the NOAA Satellite Observing Systems Architecture (NSOSA) study, a study conducted from 2014 to 2018 to explore overall approaches to the next generation of NOAA environmental monitoring satellites (weather satellites, in more familiar terms). See references [1, 2] for further discussion of the background of the study and its methods. This case study illuminates practical decisions on what types of model are most useful in the earliest phase and which can be bypassed as being primarily applicable to later phases. It also highlights which model types are most useful for setting up the foundation for follow-on activities.

For the purposes of this chapter, we will divide system development activities into four phases. Terminology is not standard across mission areas or industries, so there is some inevitable confusion. But these four phases are recognizable in almost all situations.

Concept Exploration

In this phase we are simultaneously exploring the problem space and the solution space. No decision has been made to build any system at all, though there is a presumption that a physical system of some kind is desired (or else why spend resources exploring?). The outcome is a decision to pursue development of a system (or systems), or not to, and a general definition of what that system would be. Similar terms are pre-Phase-A and pre-Formulation.

Concept Development and Development Strategy

In this phase, we develop the specific system concept of interest and determine who to develop the system. This might result in traditional definition products (like requirements sets) and a decision to run a competitive acquisition. It might result in a decision to pursue a very incremental and agile development approach or some other approach. This roughly corresponds to Phase A or Formulation activities in acquisition models.

System Development and Production

In this phase we are actively developing the system of interest, whether that involves physical construction, software development, or other activities. It also encompasses production of the system of interest, assuming we want multiple copies.

This case study is of a concept exploration activity. We first provide context on the goals of concept exploration model types to support those goals and NOAA's specific goals for the NSOSA study and how they structured what was done. Then we review models developed in the study that are specific to the goals and domain of the study, satellite constellations for operational environmental remote sensing. The case study illustrates making specific, context-dependent choices of models within a generic framework. For example, generically a context model of the outside environment and existing systems-of-interest are of high priority in a concept exploration study. In the NSOSA case a fly-out availability model, an analytical model that draws from physical models of systems-of-interest is specific and essential instance. Likewise, a value model is an essential element in concept exploration and concept development (though often replaced in system development). The case study provides a description the specific environmental remote sensing observational value model used. We review how these were chosen in the NSOSA study (the possible choices were far from unique and required judgment) and provide examples of their employment in the study and what conclusions could be drawn.

Concept Exploration Phase Goals

What are the goals of the earliest phases of a system development effort? What kinds of models are most useful to achieving those goals? Early phase efforts are decision centric. A government team will have to decide whether to advocate for developing a system at all and, if so, what its essential characteristics should be. A commercial firm exploring new product possibilities faces the same basic decision problem, should we build something new, and if so what are its basic characteristics? In more detail:

1. Should we build a new system at all? Alternatively, revise an existing system, rely on someone else, or exit the market or operational domain?
2. If we do build a new system, what sort of capabilities and performance should it have? What is the performance envelope in which we would search for a specific solution system? Where should it fit within the cost performance envelope?
3. If we do build a new system, what architecture should it have? Here "architecture" means overall but not detailed design. Architecture could be thought of here as class of systems within which live a great many possible systems that all share some significant implementation traits but are far from identical.

The goals of other system development phases obviously differ. Concept development emphasizes narrowing the trade space of physical systems (the overall decision to build one having already been made) and building the information necessary to conduct the type of system development eventually decided on. System development emphasizes actual system construction, whether through classical system-subsystem decomposition and integration or sequential evolution of partial systems. Just as the goals differ, the role of models differs. Table 1 illustrates how the goals and roles of models vary between phases. This approach builds on the more abstract representations of different points of emphasis in different phases; see Table P.1 in [3].

The role of modeling support to systems engineering in concept exploration is decision support, specifically supporting decisions as framed above. We are interested in building models to the extent that they support making better decisions relative to concept exploration goals. In other phases, the roles of models shift as the goals shift. Consider the construction of typical models written in the SysML language. There is obviously a great deal of variety in what could be written, but typical SysML models focus on the physical decomposition of the system-of-interest into blocks and associated functional descriptions. In system development activities, these are detailed and central, as outlined in Table 1. They specify the physical realization that is being built and functions the realization is required to support. The features of the language support smooth refinement from system level of disciplinary level design activities, very valuable in minimizing rework costs and organizing verification. But in the concept exploration phase, there may be many, even hundreds, of distinct physical configuration alternatives. Precise specification of each is

34 NSOSA: A Case Study in Early Phase Architecting

Table 1 Comparison of aspects, including model types, in development phases

Aspect	Concept exploration	Concept development	System development
Goals	Scope the problem space, choose acquisition approach, identify effective architectures	Specify the problem, choose the architecture, prepare for acquisition	Develop and deliver the system of interest
MBSE focus	Problem space	Problem-solution integration	Solution space
Nature of analysis	Breadth over depth	Depth in critical areas	Depth over breadth
Optimization?	Identify best classes of solutions	Best alternative within the class	Optimize subsystems against the whole
Interface modeling	Feasibility of external interfaces (context) and essential internal	Detailed context specification	Elaboration of all interfaces (internal)
Value modeling	Expansive of problem space	Focused problem space trade analysis	Subsystem solution trades and requirements
Physical modeling	Context, external interface focused	Reference physical architecture	Flow down to disciplinary engineering
Functional modeling	High-value thread identification	Reference functional architecture	Elaboration to subsystems, agile development integration
Information modeling	Implementation invariant	Invariant/variant boundary identification	Physical data and object modeling

of little value since almost all will eventually be dismissed for more detailed development of a small subset in concept development (and eventually system development). Models will be of greatest value in concept exploration when they can capture concept independent information or are simple enough to support automated generation of many alternatives for assessment. For example, physical models of the system-of-interest's invariant context or external interfaces. Using models that smoothly pass between design phase is not likely to be highly valuable in this phase since the leverage on economizing on model processing is almost certainly small compared to the leverage on making better decisions.

The nature of the of concept exploration decisions is broad trades. The trades will have wide scope, crossing between cost, capability, value (presumably derived from capability), risk, and schedule. The vehicle of those trades is whatever the large-scale choices are in the structure of the system of interest. During concept exploration, we are identifying alternative architectures in the sense of an architecture being the representation of the common aspects of a large group of alternative system concepts.

NSOSA Background and Goals

For the case study, we examine these issues as they appeared to NOAA in 2014 and 2015 when the study was structured and then see how they were realized during the course of the study to 2018. For this it is important to understand some background in weather forecasting and environmental monitoring from space in support of weather forecasting. NOAA, which contains the National Weather Service (NWS) that is broadly responsible for weather forecasting in the USA, collaborates with other space faring nations and international organizations to coordinate their efforts to fly and operate weather satellites. Weather satellites perform important functions across the range of weather forecasting missions [4].

1. Data from US and European polar orbiting satellites, primarily in the form of microwave and infrared vertical soundings, is the main source of data that initializes the global numerical weather prediction (NWP) codes. The NWP codes are the primary source of forecasts 3–5 days out and thus play a primary role in mitigating the impact of hurricanes and winter storms.
2. Imagery from geostationary satellites is a major contributor to 24-hour regional forecasting. This plays a large role in establishing watches for severe weather and precipitation events as well as day-to-day forecasting.
3. Data from geostationary equatorial orbit (GEO) satellites and satellites at the Earth-Sun L1 Lagrange point is critical to warnings of space weather events, such as geomagnetic storms, that can affect Earth infrastructure. The Lagrange points are points where the forces of bodies orbiting each other balance, and it is possible to place a spacecraft that can stably maintain the position (at the cost of some continuous station keeping). The L1 point between the Earth and Sun is a point along the Earth-Sun line where the Earth and Sun gravitational attraction balance. The L1 point is used by satellites for continuous solar observation and to allow in situ (magnetic fields and solar wind flux) measurement of events moving from the Sun to the Earth. The L4 and L5 points maintain a constant geometric relationship between the Earth and Sun, one moving ahead of the Earth's orbit, while the other trails a constant distance behind. In situ sensors at L1 are the only reliable source of advance warning for geomagnetic storms that can damage the power grid.

NOAA has built and flown GEO and Low Earth Orbiting (LEO) polar satellites for many decades. The GEO satellite series is known as Geostationary Operational Environmental Satellite (GOES) and supports regional real-time weather imaging as well as other missions (e.g., lightning mapping) over the western hemisphere. The European Meteorological Satellite (EUMETSAT) Agency flies a similar system over the European hemisphere and the Japanese Meteorological Agency (JMA) flies its own over the western Pacific. The LEO series was originally known as Polar Orbiting Environmental Satellites (POES), and the current series is the Joint Polar Satellite System (JPSS). The JPSS satellites have multiple instruments, including microwave and infrared sounders for measuring temperature and humidity profiles globally and imagers for global weather imagery. EUMETSAT likewise flies its own LEO satellite with similar instruments in an offset orbit, timed to maximize coordinated joint global coverage.

The overall international constellation, from the NSOSA perspective, is depicted in Fig. 1 as it is currently planned for 2025. The Program of Record in 2025 (POR2025) is an important reference point for all studies since it represents the nominal package of capabilities provided by current funding commitments. There are, of course, many other satellite systems flying or planned for the 2025 period than are shown in the figure, but they are not considered operational contributors to NOAA needs, either because they are one-off scientific systems (no sustained data supply) or they are flown by nations without an operational partnership agreement with the US. Readers should note, "not an operational contributor" does not mean that NOAA does not or would not make use of the data, as there is a broad strategy for making use of data from sources that are not considered operational, but it means that the data source is not counted against operational requirements for planning purposes.

The current GOES and JPSS programs have limited production runs (four and five satellites, respectively). Based on current performance and design life, the

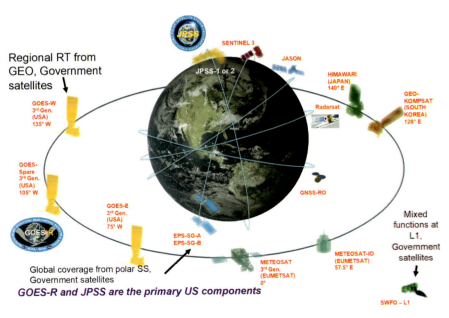

Fig. 1 Graphical representation of the current weather satellite program of record, from the NSOSA perspective, as of 2025. It includes US and partner components. Many other components exist from other countries, but they are not considered part of the program of record because they do not contribute to satisfying NOAA operational needs. Adapted from [6]

capabilities of these satellites will need to be replaced as early as 2028 and certainly by the mid-2030s. The current program for space weather measurements at the L1 point requires replacement on the same time frame. Both EUMETSAT and JMA also have issues in how their constellations will evolve, but both are at points in their program lifecycles where they have reasonable commitments to current or in-development replacement capabilities to approximately 2040. The situation is illustrated in Fig. 2 [1, 5]. Because satellite development and manufacturing timelines are long, typically a decade, NOAA needs to decide on its approach for the generation of satellites to replace those currently in production and launch.

NOAA's key decision on the next-generation satellite constellation study was to scope the decisions the study would support. The study could have been scoped narrowly by assuming that the existing constellation architecture (US government-owned satellites in specific GEO and LEO orbits and/or specific performance requirements) was fixed. Then the decisions would revolve only around the best choices for those satellites. However, NOAA chose to expand the study trade space and thus the decision space, and this had specific consequences for the models needed. The hierarchy of decisions was:

1. Do we retain the overall CONOPS of today where satellites sweep up regional and global data largely without specific tasking and feed that data into models to be assimilated and to human forecasters who choose what to use from what is presented?

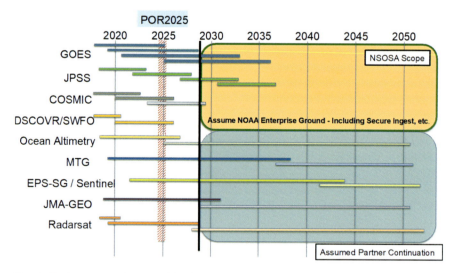

Fig. 2 Trade space and assumed background of the NSOSA study in time and contributed operational systems. Adapted from [6]

2. If yes to #1, then do we make radical replacement decisions on the current satellite architecture, for example, by abandoning one of the orbital regimes or replacing both with a uniform or tailored constellation in other orbits? Current excitement around proliferated LEO constellations is typically this type of replacement, where a uniform LEO constellation replaces GEO communications satellites.
3. If we do not make a radical change, and instead decide to retain the basic GEO/LEO/L1 construct of today, then we still have some additional structural decisions to make within that:

   (a) Space weather observations are currently handled by piggybacking instruments on GEO weather satellites and having a standalone, relatively simple, capability at the L1 point. Many other partitions are possible, including standalone GEO satellites, comprehensive satellites at L1 (instead of GEO), and augmenting satellites at other Earth-Sun Lagrange points, such as L5.
   (b) There has been interest in highly inclined, high altitude orbits (similar to GEO altitude), such as Tundra, as components of an environmental monitoring constellation. If they were added, they could fulfill an augmentation role (providing just new data) or could be used in ways that overlap existing capabilities and thus lead to re-allocation of capability from GEO or LEO to the highly inclined, high orbit.
   (c) The current constellation contains a set of user-facing communication services that date back several decades. These cover specialized high speed and disadvantaged user data distribution and servicing field collectors. The explosion of terrestrial and satellite communications capabilities in the last 20 years calls maintaining these specialized capabilities into question. What should be done about them in future satellite generations?
   (d) There are business alternatives for both GEO and LEO satellites, including a variety of commercial instrument hosting options. What role should alternative business arrangements play, a dominant role or only an occasional supplement?
   (e) Finally, if legacy continuation GEO and LEO satellites are core parts of the future constellation, where should they be situated in the performance space? Should performance of current observations be maintained, improved, or allowed to fall back? Should the emphasis be in increased performance for existing observations or adding in new observations that have not yet reached operational status?

NOAA set out a formal terms of reference (TOR) for the NSOSA study that set out the issues to be studied and those to be assumed. The TOR specified that the study should look for best value alternatives, within a budget constraint, without constraining the orbital architecture. Any orbits, any satellite types, and any business arrangements could and should be considered, but must be compared on a common basis in value and cost to NOAA. This direction was deeply influential on what models were constructed. Had the study been more constrained, the models needed would have been substantially different. This is the first and probably most important observation on the application of MBSE to early phase concept development.

The nature of the concept trade space to be explored determines the systems engineering models needed. The choice of models must flow from the goals of the exploration.

The sections to follow define the modeling approach chosen for NSOSA, which focused on three model types: A value model, constellation level cost models, and design models focused on alternative constellations (rather than on alternative satellites). These models overlap those that would be most appropriate for concept development or system development. For example, no matter which phase there would have to be a design model (or models) representing the physical systems to be built or purchased. In the NSOSA concept exploration the focus was on alternative constellations packaging different sets of observational capabilities. In a system development that wide alternative space would be drastically narrowed to a small number of satellite configurations that would be intensively explored.

Models Relevant to the Problem

In discussing the models needed, we will distinguish between relatively conventional models, the kind of models that appear on standard lists and standards of practice and those that are very specific to this problem area. Based on the goals in the study TOR, an analysis of the approach showed the team would have to develop:

1. A value model that compared the cross-mission impact of different baskets of observational performance. This model did not assume any specific constellation and was agnostic to how a constellation provided the observational performance.
2. A cost model that included development, production, and launch and extended over potentially multiple program lifetimes. The cost model had to capture the long-term cost of maintaining a given level of mission performance not just the cost of the next set of programs.
3. A design model that generated disparate alternative constellations, allowed computation of the performance of those constellations against the value model, and allowed estimation of the cost using the cost model. The stress on the design model was the need to cover many (perhaps 100's) of alternative constellations rather than depth in any one constellation.

An important issue is that the costs were required to be justifiable through a well-documented cost model. The team could not use an arbitrary approximate cost model; it had to be traceable to validated sources. The team had such cost models available, rooted in various NASA cost models [7, 8]. However, using these cost models created rippling effects. The cost models required a significant level of design maturity. Thus, the designs generated in #3 above had to include a minimum level of fidelity matched to the cost model, as well as describing all of the performance attributes required in the value model. This triplet of requirements, being a self-consistent design model, adequately filling out the cost model, and also matching the requirements of the value model, placed strong constraints on level of detail that had to be included in the design.

Value, cost, and design models are relatively standard. Most early phase studies need such models, though they may be different in the domain-relevant details. The NSOSA study also needed very specialized, contextually specific models, within the design model area. For example, it needed a fly-out model that showed how a production policy (when we build satellites to a state ready to launch), coupled to a launch policy (when we launch the satellite and how we operate it once launched), determines the probability distribution for various levels of on-orbit observational capability over time.

The NSOSA team referred to these as fly-out plots. The idea is that combining a production and launch policy with a statistical characterization of the likelihood of successfully arriving on orbit and the statistical lifetime on orbit yields a probability distribution over time for the number of functional satellites on orbit (thus the level of capability provided). See Fig. 3 for a typical example of such a plot. The NSOSA study took existing discrete event simulation capability and combined it with constellation characteristics and empirical modeling of satellite lifetimes to create realistic models of how production policies correlated to delivery of capabilities over time [9]. The example results of Fig. 3 were repeated for various combinations of

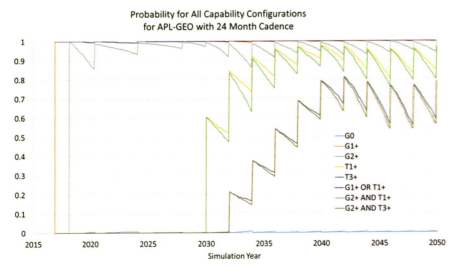

Fig. 3 Typical fly-out plot, this for a relatively complex policy where orbital destination is chosen 12 months before launch and the impact of various empirical satellite lifetime distributions can be studied. G0/1/2+ refers to levels of GEO capability; T1/2/3 refers to levels of tundra orbit capability. Adapted from [9]

policy and optimism versus pessimism in development assumptions (e.g., launch cadence, lifetime probability distributions, orbital placement policy timelines). This allowed, as in Fig. 3, exploration of issues such as what assumptions would allow dynamic placement policies to generate on-orbit capabilities above expectation for a given launch (and thus production) cadence.

A 42010 and National Academies Guide to Model Selection

The overall objective in the NSOSA case was to determine the most cost-effective constellation architecture for NOAA environmental monitoring from space. The architecture is a set of decisions about the structure and content of the constellation. Those decisions are expressed through models, more precisely, through a model-based architecture description. Two sources of guidance for architecture descriptions were especially useful, the ISO 42010 standard for architecture description [10, 11] and the National Research Council study on early pre-phase A systems engineering [12]. Models thus play dual roles. Some models provide the analytical basis to support architecture decisions, and models (ideally the same systems models) document the decisions once made.

The 42,010 standard was used both as a template for building relevant architecture descriptions and more generally as a guide to content. The NSOSA team wrote an architecture description document for the Program of Record 2025 (POR2025, the officially funded satellite environmental monitoring program through 2025, which served as a key performance benchmark) using ISO 42010 as a document template [13]. More generally, the information developed in the study was organized around required and recommended 42010 elements. The ISO 42010 standard requires that an architecture description contains:

1. Explicit identification of the system-of-interest's stakeholders considered in forming the architecture and the concerns of those stakeholders.
2. Description of the architecture as a series of models. The models are arranged into discrete Views, each view dealing with the whole system from the perspective of a set of related concerns.
3. Each View must have an explicit Viewpoint, a definition of the modeling approach, rules, and associated stakeholders/concerns addressed in the View.
4. Some required analytical and explanatory content, including identification of known model inconsistencies and explicit rationale for selection of architecture alternatives.

The MBSE approach used in the NSOSA study fell within the 42010 template. The approach drew on the NRC recommendations for early phase systems engineering in where to place model focus and the depth of exploration. The study recognized the importance of deferring detailed model development and explication in non-critical areas to follow-on acquisition activities.

Functional Description

The 42010 framework implicitly assumes that the architecture description will contain some form of functional view, but does not require any specific model or notation. Hypothetically, a functional model could be left out, though that would be nearly impossible in most circumstances given stakeholder concern coverage requirements. But unlike description framework standards like the DoD Architecture Framework [14], 42010 does not demand any particular functional model be provided. This was convenient in the NSOSA case as the functional description at the level of architectural decisions was best folded into an implicit model within the value model. As described later, the value model subsumed the functional description in that the essential functions for each observation were "vacuum cleaner collection." Each observation, with very few and limited exceptions, is a fixed task continuous activity. The sensor scans the area it can observe and transfers the data to the ground, where the data are then out of the scope of study.

Implementing a constellation will, of course, be far more complex functionally and require extensive modeling. But from a value-delivering standpoint, the value is in the performance attributes on the data not on variations in the data gathering functional behavior. The data gathering functional behavior is essentially fixed.

It is important to realize that this was itself an architectural decision and one that was made deliberately and the rationale documented. One could have heavily tasked environmental observations. In such a scenario the sensors would be tasked to collect in certain areas with certain performance levels. Observations could be intensified in areas of high interest (e.g., where models show high frequency change, either temporally or spatially) and deprecated in areas of low interest. However, the stakeholder assessment was explicitly that such tasked approaches were not of interest within the primary period of concern for the study. If they ever became of interest, it would not be until a considerably later period. The choice to make the functional model implicit would have been incorrect in other scenarios where the functional content of the system is complex and trades on functionality are an important part of the earliest phase trade process.

Physical Descriptions: Models Versus Design Vectors

The physical system to eventually be built (satellites, sensors, orbits, operations concepts) was clearly of high stakeholder concern. This was both because performance (the primary source of value) and cost flows directly from the physical models. The NSOSA architecture description clearly must contain physical models of systems of interest. The NSOSA team modeling approach was oriented around "design vectors" rather than graphical system representations. A design vector is a parameterized representation of a particular solution given in discrete form in a way to facilitate generation of multiple alternatives by varying the parameter values. The design vector can be used by a design process to generate more detailed design representations, including cost. In a subsequent section, we define the design vector used and describe how it was chosen as a linking intermediate between the value, cost, design, and risk legs of the study.

- The design vectors were chosen to match the design synthesis capabilities of The Aerospace Corporation's Concept Design Center [15, 16], which could take in a design vector and produce a corresponding physical design including cost. If desired that physical design could be rendered into representational models, such as SysML or something more domain specific.
- By leaving some parameters of a design vector and allowing others to vary the team could produce families of solutions that shared common architectural choices. This allowed comparison not just of point solutions but of architecture families, which was in keeping with the overall goal of selecting cost-effective architectures and not just individual solutions.
- The design vector was also linked to the value model. The parameters of the design vector were sufficient to derive performance values in the Environment Data Record (EDR) Value Model (EVM) used in the study to compare alternative architectures.
- The elements of the design vector mapped directly to assessments of technical and business risk.

Performance and Value Modeling

Neither ISO 42010 nor the NRC guidelines explicitly recommend having a quantitative value model, in the sense of value or utility model as used in decision analysis [17, 18]. However, the goal of cost-effectivity trade-offs in the NSOSA made such a model essential, and the NRC guidelines at least implicitly recommend having a value model to enable transparency in the trades. Without an explicit value model, cost and effectiveness cannot be traded, nor overall cost-effectiveness judged. If trade decisions are made then a value model must be in place, even if only implicitly. A value model of some form is inherent in an MBSE approach. The value model might be based entirely on requirements with binary satisfaction or be as complex as an arbitrary multi-attribute utility model. The NSOSA value model was built on the principles of multi-attribute utility theory (MAUT) [18], but some complex choices had to be made in forming the model, discussed in the next section.

Choosing the NSOSA Value Model

The first step in forming a value model is to examine how stakeholders realize value from the system-of-interest, whether or not that occurs on a consistent level, and how differing views of value might be reconciled. The NSOSA study engaged in a wide-ranging assessment of stakeholder needs and considered formation of value models at several different levels [19, 20]. The process and alternatives considered are exemplary of what is necessary in choosing a value model (a key component of the MBSE set) for an early phase study. The operational purpose of collecting environmental data from space is primarily to support weather and space weather forecasting. At the top stakeholder level (the US Congress and rest of government and the US public), weather forecasts are valuable to the extent that modifying behavior in response to a forecast delivers value. There are numerous examples, from highly consequential to routine.

- Based on hurricane forecasts, officials evacuate vulnerable areas, preemptively harden civil and industrial infrastructure, and pre-position relief supplies. The accuracy of the forecasts determines areas evacuated and the timeliness of actions.
- Advance warnings of geomagnetic storms from solar coronal mass ejections can mitigate or eliminate physical damage to power networks by allowing adjustment of load factors, activation of protective modes, and preemptive shutdown of the most vulnerable infrastructure.
- Tornado warnings save lives by getting people moved into sheltered spaces in advance of the tornado arrival.
- Winter storm forecasts can mitigate the duration of disruption by allowing for road shutdowns prior to impassable conditions (easing snow clearing by avoiding the presence of disabled vehicles) and for pre-positioning of snow removal assets.
- Routing adjustments and preemptive re-scheduling in response to aviation weather forecasts improves network performance and reduces accidents.
- Temperature and wind forecasts allow for improved electricity network management by forecasting loads and renewable generation effectiveness.
- Individuals adjust their behavior based on weather forecasts in everything from commuting times to choice of activity.

The National Weather Service (NWS) tracks and reports the most consequential forecast metrics as part of the Government Performance Reporting Act (GPRA) [21], while satellite data is integral to realizing value as in this list the pathway is indirect. Satellite data is only consumed directly in a small number of key end-user forecast metrics. Satellite data is several layers of assimilation and modeling removed from the performance metrics visible to the highest-level stakeholders.

- Three-day and longer forecasts are produced by global numerical weather prediction (NWP) programs. Global NWP systems are the largest consumers of satellite data, and their performance is heavily dependent on the quality and quantity of satellite data [22–24]. However, the link between data quality and forecast performance is highly nonlinear and only partially understood.
- Much satellite data, for example, imagery, is consumed by human forecasters who use it along with many other sources in producing specialized forecasts. Some of these human products may be directly delivered (e.g., common in aviation weather forecasting); in other cases, such as severe storms and tornadoes, the production chain is a complex mixture of remote sensing, numerical modeling, and human exploitation of local data sources such as weather radars.

- Climate and other science researchers make use of many specialized data products derived from satellite data. For example, sea surface temperature data is of fundamental importance in seasonal and beyond forecasting and climate science. Global sea surface temperature data comes largely from satellites but is derived from multiple sources with complicated calibration chains.
- Current satellite sensors are the source for hundreds of derived observational products that are continuously produced and delivered from satellite ground systems. Each of these has a user audience, albeit sometimes very narrow.

In any real application where a value model is required, the first and often most difficult problem is to choose the appropriate level at which to establish the measures of value [25]. The NSOSA case was no different.

- Metrics relevant to high level stakeholders, like hurricane track forecast accuracy, were high appealing but were unworkable.
  - The linkage between such high-level metrics and what the system-of-interest produces is poorly understood. We know with certainty that the performance of the system-of-interest drives performance on the metrics [22], but the cause-and-effect relationships are nonlinear and only roughly understood.
  - The standard for estimating forecast performance of a possible constellation is an observing system simulation experiment (OSSE) [26, 27]. However, the level of effort (human and computational) required to conduct an OSSE means 6 months to a year are required to assess an alternative. This is unusable for exploring a trade space.
  - The other end-users (human forecasters and scientists) are also important stakeholders, and they consume satellite products directly.
- Metrics on the satellite products themselves are also a poor choice for doing a wide-ranging trade analysis.
  - There are hundreds of such products, an unwieldly number for conducting a wide-ranging trade analysis.
  - Most of the products are derived from overlapping sensor sources, so their performances are highly correlated. This makes standard MAUT techniques hard to use properly without introducing "double-counting" effects [25].

As a result, the NSOSA study developed a set of genericized observational objectives that covered the stakeholder concerns at both the high-level (were the observations that fed NWP accuracy) and were the sources for deriving individual user products. We found they could be chosen to be largely independent of each other, both in that performance on one said little or nothing on performance on another and in that preference for one had little effect on preference for another. This type of independence is very desirable in a decision analysis framework. From a functional perspective, the picture was also simplified. Each observation was the output of a generic function that was applied (usually) with little or no tasking or feedback. Finally, each was well characterized by the performance of well-defined sensor types, and the performance of those sensors was known to be strongly correlated with physical design constraints and costs.

The resulting value model, known as the EVM, had 19 terrestrial weather-related observational objectives, 19 space weather-related observational objectives, and 6 strategic objectives. The strategic objectives were non-functional objectives of the sponsoring organization. The details of this model have been published elsewhere [28], and the choices of specific characteristics and performance levels are domain specific and not germane here in a discussion of modeling.

From a modeling perspective, each observational or strategic objective was decomposed into from 1 to about 10 performance attributes. Typical performance attributes were horizontal resolution, coverage update time (for an observation-dependent coverage area), spectral coverage, and accuracy. The choice of performance attributes was observation dependent. For example, horizontal resolution is highly relevant to an imagery observation but irrelevant to a radio occultation observation.

Each performance attribute had three values determined by the Space Platform Requirements Working Group (SPRWG):

- The **study threshold** value, a lower value that would result in an alternative exclusion if it failed to meet the value. Some optional observations had a study threshold value of "none."
- The **Expected** value, which represented the SPRWG consensus on what performance, was expected in 2030 by the user community and was compatible with projections for developments in NWP and other processing elements.
- The **Maximum Effective**, which was pegged to the highest level of performance where these was scientific evidence that it would have operational value. The operational value constraint was to ensure that performance ranges were grounded in operational need and not "wish lists."

Finally, following MAUT practice, the NSOSA team elicited swing weights and a splined curve to integrate all of the objectives into a scalar value model. This actually allowed multiple value models to be formed, matched to different audience needs. For example, a model that included trades of observational against strategic objective, a model that considered only terrestrial weather forecasting objectives, etc. The elicitation process was conducted in a way to allow the model to maintain uncertainty measures [29] so that the team could judge significant and insignificant differences.

From a modeling perspective, the EVM was a tree that took in a performance vector (performance against the observation performance attributes) computed up to first objective level performance then value against terrestrial weather, space weather, and organizational strategy. The performance attributes could be all computed from the design vectors (see next section). Finally, all could be combined to a scalar measure, when and if needed. From a tool perspective, the data was maintained in spreadsheet tables that could be exchanged between some custom written tools. A more targeted tool might have been useful, but in practice spreadsheets supported the analysis needed and made it easier to implement data exchange between tools and groups.

NSOSA Design Vector Models

The NSOSA study, like any analysis of alternatives, needed alternatives to compare. To meet the needs of the study, the alternative set had multiple goals to achieve:

1. The alternative set had to be diverse in orbits, satellite size, and business models. An approach that exhaustively explored modest variations on the legacy approach would not be sufficient. The diversity needed to encompass high interest current space technology trends, such as proliferated LEO constellations.
2. Alternatives needed to be comparable in performance on the value model (discussed previously) and in cost. The cost estimates needed to be traceable to credible and well-documented sources.
3. The alternative set needed to be organized into architectures or sets of related and consistent structural decisions. The presumption (to be examined) was that there were distinct architectures, and within an architecture, many alternatives of varying cost and performance could be generated.
4. It needed to be possible to generate performance assess and cost a large enough set of alternatives to broadly explore the trade space. The team could not know how many might be required, but it was clear the number was likely to be on order of 100. Much larger sets have been reported in the literature.

The NSOSA approach was to define as set of design vectors. A design vector was a standardized set of parameters with which to express a constellation of interest. The NSOSA design vectors had two aspects: the satellite configurations and the constellation configuration.

Satellite configurations were coupled to both performance and cost and linked to the capabilities of The Aerospace Corporation Concept Design Center (CDC) [16]. The CDC is capable of rapid satellite design and costing taking as inputs a set of instrument specifications and orbital placement. From these the CDC has rapid design tools to derive and cost both a satellite bus concept and a select a launch vehicle. The input to the CDC was a list of instruments with their accommodation requirements (available in NSOSA from the instrument catalog task). The CDC processes and tools use the inputs to derive a satellite bus design, identify a compatible launch concept, and cost both from standard cost databases. The design vector for each satellite configuration contained the list of instruments from the NSOSA instrument catalog, the orbit (or family of orbits where it would be placed), and production block size. The CDC returned the overall satellite mass, power, and communication needed and the non-recurring and recurring cost for development and production of the defined block size.

A constellation design vector assigned satellite configurations into orbits with sufficient information to allow constellation level performance calculations and cost over the time interval of interest (several production and launch cycles). Each constellation in the NSOSA study was defined by a set of constellation components. Each constellation component consisted of:

- A satellite configuration, one of the configurations described previously.
- An orbit or orbits and number of orbital planes. This could be geostationary, a specific LEO (like 832 km sun synchronous with 1330 local time of ascending node (LTAN)), a small family like 832 sun synchronous with varying LTAN, or many other cases. In some cases, multiple orbits, like geostationary and tundra, might be specified, if that was appropriate to the occupancy and launch policy (see next item). The number of planes drove selection of launch concept and on-orbit management.
- Orbital occupancy and launch policy. For each of these, there is a baseline launch cadence in months that drives production rates and capability availability. The launch policy would lead to a nominal number of satellites available at any given time, an important parameter to cross-checking with external system models. Many combinations were used, but the most important were:
  - Fixed cadence launch with 2 active and 1 spare or residual (commonly used in GEO as it is the legacy policy).
  - Fixed cadence with 1 active and residual (common for fixed LEO orbital occupancy).
  - Fixed cadence and multiple orbits with prioritized rotating replacement. For example, this can be used with GEO and tundra orbits where positions are occupied in priority order depending on the experienced on-orbit lifetime of previously launched satellites. The method also applies to proliferated MEO and some partially disaggregated LEO approaches.
  - Fixed cadence with multiple satellite block replacement, used in some proliferated LEO constellations.
  - Event-based replacement instead of fixed cadence, typically used in legacy-like GEO or LEO configurations.
- Target date for first launch.
- Production block size (how many satellites are produced and launched before recycling the development process).
- Communications latency, to determine what changes (if any) would be required in the space-ground communications systems, which were otherwise outside the scope of the study.

See Table 2 for a typical constellation design vector. The short codes, G4c, L4f, etc. are indexes to satellite configurations that were built from their own design vectors at the satellite level.

Taken together, and as discussed in detail in the next section, the satellite and constellation and design vectors allow the performance and cost models to fully specify the characteristics needed to close the modeling loop:

Table 2 Example typical constellation design vector

Satellite config.	G4c	GH4	L4f	L3j	S3b
Orbit	GEO	Hosted GEO	LEO, polar SS	LEO polar SS	L1
Plane(s)	1 (GEO)	1 (GEO hosted)	2 (0530, 1330)	1 (SS1330)	1 (L1)
Launch policy	32 mon st 4/1/2029	7 years, st 1/1/2031	46mon (x2) st 7/1/2028	54mon st 7/1/2032	79 mon st 2/1/2028
tot # sats on-orbit	2 + 1 spare	1	2 (+ residual)	1 (+1 residual)	1 (+1 residual)
Block size	4	1	6	4	4
Latency	5 min	5 min	30 min	80 min	1 min

- The constellation design vectors, processed through launch and fly-out models, define the statistical distribution of the defined satellites in orbit. This provides geographic coverage, update rates, and availability.
- The constellation design vectors also define the rates of satellite block development, production, and launch. The CDC models provide the cost of each of those elements (development, production, and launch) and so allow computation of cost over time.
- The satellite configurations define the instrument sets, and combined with the constellation results and the instrument catalog values, this defines the full set of observation performance values.

Constraints on the constellation design vector define a constellation architecture. For example, the US legacy program of record:

1. Has a GEO launch cadence to maintain two active and one spare or residual satellite in GEO. The GEO satellite is in a single configuration with a comprehensive set of terrestrial and solar observation capabilities.
2. Has a launch cadence to the agreed sun synchronous LEO to maintain a single active and (with high confidence) a residual satellite. The LEO satellite is in a single configuration with imaging and sounding capabilities.
3. Has committed to a sustained launch cadence to maintain a satellite at the L1 Lagrange point with in situ space weather observation capabilities, though this is still in development.

Taking these three components as fixed, but allowing for additional augmentation, we have an extensible Legacy Continuation constellation architecture. If we start the constellation configurations with just elements 1–3 above and choose satellite configurations built on instruments with performance equivalent to current systems, then we can generate additional alternatives with varying cost performance by manipulating the design vectors:

- Leave the three elements 1–3 above fixed but change which satellite configuration is assigned to each one and change the base launch cadence. Change the satellite configurations to alternatives with higher or lower performance instruments from the instrument catalog or to configurations with additional instruments (e.g., replace the GEO imager with a time-shared imager-sounder and add a microwave imaging sensor to the LEO element). Adjust the launch cadence to a faster or slower value. Higher performance instruments will increase costs while improving observations and thus EVM performance. Changing the launch cadence up or down will improve or degrade availability (a strategic objective in the EVM) and thus also change the overall scored value.
- Add elements to the 1–3. For example, add additional, smaller LEO satellites restricted to sounding measurements to additional orbits to improve sounding update rate and availability. Add satellites to tundra orbits for more global real-time coverage. These will result in higher performance on the EVM (and so higher value scores) and higher costs.

Following similar logic, other design vector changes will yield alternative architectures, some with modest differences, and others radical. For example, we could break the GEO entry into multiple GEO entries with different satellite configurations. One configuration could be for terrestrial observation, the other for solar observation instruments. Likewise, the LEO component could be broken into smaller satellites. This would be a Disaggregated Legacy Continuation constellation architecture. A radical alternative would drop both of the first two legacy components and replace them with a large LEO or MEO constellation. We take up how NSOSA dealt with the process of constellation synthesis and integrated into the modeling tools in the next section.

Contrast for a moment the design vector approach with other, typical, approaches to design representation in MBSE methods. The design vectors are tabular form; there is no graphical representation. Of course, any given alternative could be represented graphically, and it was occasionally useful to do so in presentations. But the primary purpose of the alternative generation was to feed an analysis process. The process was driven from the contents of the design vector. The human intervention was limited and to have people examine a set of graphical artifacts was not part of the design analysis process. Much as a graphical model in a language like SysML can be alternatively represented in non-graphical form; the design vectors in NSOSA could be transformed into alternative forms. The form was chosen for maximal convenience in the analysis rather than visual appeal or human presentation. In a different phase where much greater design detail was necessary and different processes were needed, the choice would be different.

NSOSA Modeling Processes

MBSE in the NSOSA study was based on integrating multiple analysis threads. Each thread contributed a product built on a structured template. The tools in each thread ingested the products of the other threads to produce their own products, requiring varying degrees of human intervention. Given the scale of the study (44 objectives in the value model with many quality attributes per objective and around 200 discrete alternatives), the emphasis was on partial automation. Some elements were automated, while others used tools that required manipulation by subject matter experts. The flow of the products and tools is shown in Fig. 4.

Three threads are the primary feeders of the iterative process: instrument catalog development, value model development, and constellation synthesis. Products from all three were fed into integration activities. The scoring activity assesses the EVM quality factors against constellation and observation performance determined from the instrument catalog and the constellation synthesis activity. The satellite synthesis activity builds design models, and cost estimates, for the satellite configurations needed by the constellation design vectors. Those costs for the satellites are then spread over time based on the development rules in the constellation design vectors.

Value model development was discussed in a preceding section. The EVM was a spreadsheet model of the observations and their associated quality factors. It was accompanied by a scoring rubric that was used to define the relationships with the constellation synthesis models.

Instrument catalog development was a standalone activity that produced a catalog of notional environmental observation instruments for the 2030 era. Each entry included a standardized set of performance specifications and estimates of the instruments size, mass, power, communication requirements, and cost. The models used to elaborate instrument catalog entries were variable. In many cases the team could use pre-existing payload studies or industry inputs to define the instruments. In some cases, instrument-specific design models available to the team member's organizations were invoked to do a targeted instrument design study.

The constellation synthesis activity built and processed the design vectors discussed previously. Key elements included computing coverage and update rate

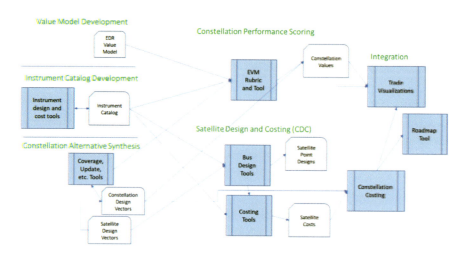

Fig. 4 Flow of design and analysis data among the various systems engineering models

for all observation types and availability (a measure of reliability) for the constellations. The tools involved were pre-existing orbital simulation, and satellite fly-out model tools, based either on Monte Carlo simulation or analytical models, were available. This process was, in practice, a targeted combination of synthesis and analysis. The team had in mind in each cycle target performance points and constellation structures of interest. They used subject matter expertise to start constellation designs and then refine them, sometimes heuristically and sometimes using optimization techniques [30, 31]. Promising pathways were explored until there was a set that covered the performance range of interest for the design cycle. There was a general objective of stepwise improvement from cycle to cycle, each cycle seeking to best the cost performance benchmarks from the previous cycle.

Satellite designed used the pre-existing tools of the Aerospace CDC. The most important tools were a set of satellite bus relationships that allowed synthesis of a bus for any particular payload design point [32] and the associated cost databases that allowed projection of development and production costs from the payload component costs and the overall size, mass, power, and orbital target of the satellite.

These threads converged in producing analysis products for decision support. The most generally useful were efficient frontier charts showing overall patterns of trades and the relationships between constellation architectures. We review a sampling in the next section.

The team also built a specialized tool known as the "roadmap tool" that took as input the design vectors and component satellite costs and produced a multiple decade schedule of flights and associated annual costs by year (often referred to as a "sandchart"). This tool could operate either automated to produce a quick look picture or could be more hand tuned by a subject matter expert in production programs to adjust dates and seek overall goals such as budget leveling.

NSOSA Results

The NSOSA study's purpose was to support decision-making on the future NOAA constellation architecture. The use of MBSE was a tool to support decision-making. As a result, the depth of modeling, and the integration in digital engineering environments, was limited to the decision-making scope.

Several other papers provide a more detailed look at how the models supported NSOSA decision-making. The paper [2] is a broad summary of the methods and results. The paper [33] discusses how the study assessed radical alternatives, alternatives characterized by design vectors that dropped the legacy continuation components entirely (rather than just re-aggregating them) and replaced them with some other, dissimilar constellation like a proliferated LEO or MEO. Figure 5 shows a summary efficient frontier plot with radical alternatives and legacy continuation and augmented legacy continuation alternatives. The radical cases are only able to approach the cost-benefit frontier of the augmented cases when combined with a radical change to the business model (the "Hosted Almost All-LEO" alternative). Unfortunately, this sole competitive case was created as an extremal case, essentially

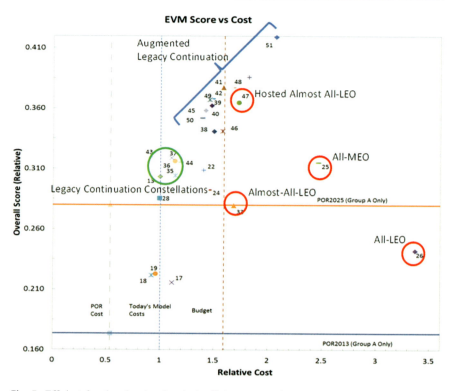

Fig. 5 Efficient frontier plot showing the inefficient cost performance of radical alternatives [33]

stringing together all of the most favorable assumptions for radical change, something we would have no confidence in for real programs.

Figure 6 shows the ability to trace out sets of alternatives and the use of variance analysis in conjunction with the methods to assess decision robustness.

As discussed in the section on using design vectors within the modeling framework, many alternatives were generated from a base case with appropriate rules. The NSOSA study referred to these sets or classes of alternatives as its constellation architectures. The figure traces the efficient frontier, the convex hull of maximum value for a given cost, of four such constellation architectures. Several key insights come from these curves. The two curves labeled Business Alternative and Hybrid are systematically superior to the curve labeled Legacy, except at the high cost end. The two closely spaced curves have a systematic cost advantage over the Legacy case, except at high annual costs where the gap greatly diminishes. In contrast the two curves Business Alternative and Hybrid show a small gap between each other.

To really understand if a gap is "large" or "small" requires a calibration. The vertical and horizontal "error bars" are that calibration. They are not strictly speaking error bars since there are no empirical errors to compile. There are uncertainties in both value estimation and cost estimation, and those can be visualized in a way similar to error bars. See [29] for more details. Where the gaps are small compared to

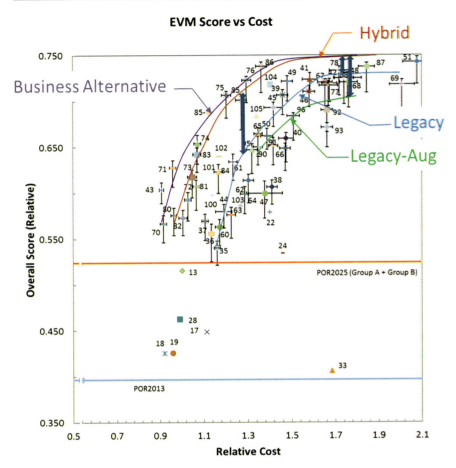

Fig. 6 Efficient frontier plot with uncertainty bars and showing the trace of various architecture alternatives as the design vector values vary over ranges of interest to generate cost-capability regions [2]

the uncertainties, the ranking of one architecture or another is not reliable, and it is appropriate to look to other decision sources. Where the gap is large compared to the uncertainty bars, then we have confidence that the model-identified superiority of one architecture over the other is robust.

From Early Phase to Acquisition

Very early phase studies, like NSOSA, are intended to have a closure point but to lead to follow-on activities. The nature of the follow-on activities will govern how the systems engineering models generated in the study transition and may influence what models are generated during the study.

If the study results in the selection of a specific constellation alternative, that alternative, at the architecture level, may be captured as a Government Reference Architecture (GRA). Here a "Government" Reference Architecture since the government is the sponsor and ultimate acquisition customer for whatever would be built. The GRA can be an MBSE reference point on which subsequent competitive bidders can construct their own proposal-phase MBSE models of their particularized solution. If this is the intent, then the early phase work may emphasize components of the model that fit within required acquisition frameworks.

In the NSOSA study, the intent, and the reality, is that a single satellite and programmatic alternative not emerge, but instead the study identified several most promising candidates. These included traditional government acquisition programs, as well as commercial service acquisitions and re-structured partnerships. In this case a single GRA is less appropriate, and a more open-ended definition of what constraints are set and what areas are left open works better.

In either case it is helpful to capture what was done in a standardized architecture description framework. The NSOSA study used the ISO 42010 framework for published architecture description documents. A released example is the architecture description document written for the pre-existing program of record [13].

Chapter Summary

Model-based systems engineering can apply to the earliest phases of system development, such as concept exploration. Since the goals for early phases are different than for system development, the models will likewise vary. At their core there are essential parallels. Whatever phase the development is in systems engineering requires a value model, models of the physical system concept(s), and cost models. Those are central in the early phase as they are central in full-scale development. While there are similarities in the basics there are deep differences in the specific choices of model technique and focus. System developments are focused on bringing a single concept into existence while controlling performance, cost, and schedule appropriately to the type of acquisition that is being conducted. In the early phases there is no single concept to build; the emphasis is on breadth of exploration and identification of drivers.

These differences were explored in this chapter through the NSOSA program as a case study. It illustrated the centrality of value models, physical models, and cost models with the overall MBSE framework, but showed the differences in the specific ways models in those categories may be chosen.

References

1. Volz, S., M. Maier, and D. Di Pietro. *The NOAA Satellite Observing System Architecture study.* in *2016 IEEE International Geoscience and Remote Sensing Symposium (IGARSS).* 2016. Beijing, China 10–15 July 2016: IEEE.

34 NSOSA: A Case Study in Early Phase Architecting

2. Maier, M., et al., *Architecting the Future of Weather Satellites*. Bull. Amer. Meteor. Soc.,102.3 (2021) E589-E610.

3. Maier, M.W., *The art of systems architecting, Third Edition*. Third ed. 2009, Boca Raton, FL: CRC press, Taylor & Francis Group.

4. Reining, R., et al. *NOAA Observing System Integrated Analysis (NOSIA): development and support to the NOAA Satellite Observing System Architecture*. in *AGU Fall Meeting*. 2016. San Francisco, CA, 12–16 December 2016.

5. St Germain, K., *Overview of the NOAA Satellite Observing Systems Architecture (NSOSA)*, Commerce, Editor. 2018, Department of Commerce: Washington, D.C.

6. NOAA-NESDIS. *NOAA Satellite Observing System Architecture Study Draft Report*. 2018; Available from: https://www.regulations.gov/docket?D=NOAA-NESDIS-2018-0053.

7. Habib-agahi, H., *NASA Instrument Cost Model (NICM) Version IV*. Pasadena. CA: NASA JPL, 2010.

8. Mahr, E. and G. Richardson, *Development of the small satellite cost model (sscm) edition 2002*, in *IEEE Aerospace Conference*. 2003, IEEE: Big Sky, MT. p. 336–6791.

9. Maier, M.W., et al. *Launch and production schedule modeling for sustained earth observation constellations*. in *2018 IEEE Aerospace Conference*. 2018. IEEE.

10. Emery, D. and R. Hilliard. *Every architecture description needs a framework: Expressing architecture frameworks using ISO/IEC 42010*. in *Software Architecture, 2009 & European Conference on Software Architecture. WICSA/ECSA 2009. Joint Working IEEE/IFIP Conference on*. 2009. IEEE.

11. ISO, *ISO/IEC/IEEE 42010.WD4 Software Systems and Enterprise – Architecture Description*. 2019. p. 69.

12. NRC, *Pre-milestone A and Early-phase Systems Engineering: A Retrospective Review and Benefits for Future Air Force Systems Acquisition*. 2008, Washington, DC: National Academies Press.

13. Maier, M. and E.B. Wendoloski, *Weather Satellite Constellation As-Is and To-Be Architecture Description: An ISO/IEC/IEEE 42010 Example*, in *Aerospace Technical Report*. 2020, The Aerospace Corporation: El Segundo, CA.

14. DoD, D., *Department of Defense Architecture Framework (DoDAF) Version 2.0*. DoD Deputy Chief Information Officer. Available via: http://cionii.defense.gov/sites/dodaf20/ [accessed August 16, 2010], 2010.

15. Aguilar, J.A. and A. Dawdy, *Scope vs. detail: the teams of the Concept Design Center*, in *Aerospace Conference Proceedings, 2000 IEEE*. 2000, IEEE. p. 465–481.

16. Aguilar, J.A., A.B. Dawdy, and G.W. Law, *The Aerospace Corporation's Concept Design Center*, in *INCOSE International Symposium*. 1998, Wiley Online Library. p. 776–782.

17. Keeney, R.L., *Decision analysis: an overview*. Operations research, 1982. **30**(5): p. 803–838.

18. Keeney, R.L. and H. Raiffa, *Decisions with multiple objectives: preferences and value trade-offs*. 1993: Cambridge university press.

19. Anthes, R.A., et al., *Developing Priority Observational Requirements from Space Using Multi-Attribute Utility Theory*. Bull. Amer. Meteor. Soc., 2019. **100**(September): p. 1753–1774.

20. Maier, M. and R. Anthes, *The EDR Value Model for the NSOSA Decision Process*, in *American Meteorological Society 98th Annual Meeting, 14th Annual Symposium on New Generation Operational Environmental Satellite Systems*. 2018: Austin, TX, 7–10 January 2018.

21. NWS. *Performance Management Homepage*. 2018 [cited 2018; Available from: https://verification.nws.noaa.gov/services/public/index.aspx.

22. Bauer, P., A. Thorpe, and G. Brunet, *The quiet revolution of numerical weather prediction*. Nature, 2015. **525**(7567): p. 47.

23. Palmer, T., *The primacy of doubt: Evolution of numerical weather prediction from determinism to probability*. Journal of Advances in Modeling Earth Systems, 2017. **9**(2): p. 730–734.

24. Radnóti, G., et al., *ECMWF Study to Quantify the Interaction Between Terrestrial and Space-based Observing Systems on Numerical Weather Prediction Skill*. 2012: European Centre for Medium-Range Weather Forecasts.

25. Keeney, R.L., *Value-focused thinking: A path to creative decision making*. 1996, Cambridge, MA: Harvard University Press.
26. Atlas, R., et al., *Observing system simulation experiments to assess the potential impact of new observing systems on hurricane forecasting.* Marine Technology Society Journal, 2015. **49**(6): p. 140–148.
27. Prive, N., *Observing System Simulation Experiments.* 2015.
28. Anthes, R.A., et al., *Developing Priority Observational Requirements from Space Using Multi-Attribute Utility Theory.* Bulletin of the American Meteorological Society, 2019. **100**(9): p. 1753–1774.
29. Maier, M.W. and E.B. Wendoloski, *Value Uncertainty in Architecture and Trade Studies.* IEEE Systems Journal, 2020. **To Appear**.
30. Ferringer, M.P. and D.B. Spencer, *Satellite constellation design tradeoffs using multiple-objective evolutionary computation.* Journal of spacecraft and rockets, 2006. **43**(6): p. 1404–1411.
31. Whittecar, W.R. and M.P. Ferringer. *Global coverage constellation design exploration using evolutionary algorithms*. in *AIAA/AAS Astrodynamics Specialist Conference*. 2014.
32. Mosher, T., et al. *Integration of small satellite cost and design models for improved conceptual design-to-cost*. in *1998 IEEE Aerospace Conference Proceedings (Cat. No. 98TH8339)*. 1998. IEEE.
33. Maier, M.W. *Is there a case for radical change to weather satellite constellations?* in *2018 IEEE Aerospace Conference*. 2018. Big Sky, MT, 3–10 March 2018: IEEE.

Dr. Mark W. Maier is a Technical Fellow at The Aerospace Corporation and an author and practitioner of systems architecting (the art and science of creating complex systems). He is co-author, with Dr. Eberhardt Rechtin, of The Art of Systems Architecting, Third Edition, CRC Press, the mostly widely used textbook on systems architecting, as well more than 50 papers on systems engineering, architecting, and sensor analysis. Since 1998 he has been employed by The Aerospace Corporation, a non-profit corporation that operates a federally funded research and development center with oversight responsibility for the US National Security Space Program, where he holds the position of Technical Fellow, the highest technical rank in the company. At Aerospace he founded the systems architecting training program (an internal and external training program) and applies architecting methods to government and commercial clients, particularly in portfolios-of-systems and research and development problems. He received the BS and MS degrees from the California Institute of Technology and the Engineer and PhD degrees in Electrical Engineering from the University of Southern California, while at USC, he held a Hughes Aircraft Company Doctoral Fellowship, where he was also employed as a section head. Prior to coming to The Aerospace Corporation, he was an Associate Professor of Electrical and Computer Engineering at the University of Alabama at Huntsville.

Cybersecurity Systems Modeling: An Automotive System Case Study

35

Mark L. McKelvin

Contents

Introduction and Terminology 1046
    Origins of Cybersecurity 1047
State of the Art 1050
    Threat Modeling with Attack Trees 1051
    State-Based Security Modeling Approaches 1052
    Simulation-Based Security Modeling 1053
    Security Viewpoints in Unified Architecture Framework 1055
    Chassis 1056
    SysML-Sec 1057
    MB3SE 1057
    MBSEsec 1059
Best Practice Approach 1061
Illustrative Example 1063
    HSUV Security Issues 1063
    Instantiating the MBSEsec Security Domain and Profile 1064
    Applying the Best Practice Approach 1065
Chapter Summary 1072
Cross-References 1073
References 1073

> **Abstract**
>
> Modern automotive systems are increasingly complex and interconnected, providing enhanced services and functionalities, such as improving driver experience and safety. Characteristics of modern automotive systems also increase vulnerabilities to attacks. Consequently, modern automotive systems are prone to security threats. Automotive systems are characterized as cyberphysical systems (CPS). CPS couple computational and physical processes through networks of sensors, actuators, and automated controls. These characteristics drive the need

---

M. L. McKelvin (✉)
University of Southern California, Los Angeles, CA, USA
e-mail: mckelvin@usc.edu

© Springer Nature Switzerland AG 2023
A. M. Madni et al. (eds.), *Handbook of Model-Based Systems Engineering*,
https://doi.org/10.1007/978-3-030-93582-5_61

for systematically engineering security into the system development from the beginning of the system life cycle. Model-based security-engineering approaches of cyberphysical systems are a means to address system security design challenges. This chapter provides an overview of models for security engineering and applies a model-based methodology to an a Hybrid Sports Utility Vehicle as an example case study to show how security concerns are incorporated into the system development life cycle.

Keywords

Model-based · Cybersecurity · Safety · Systems engineering · Automotive · Cyberphysical

Introduction and Terminology

Technology advancements have led to a significant increase in cyberphysical systems (CPS) comprising a tight coupling of computational and physical processes connecting to sensors, actuators, and automated controls [1]. CPS provides networking and computational capabilities to the physical world and is highly susceptible to cyberattacks. Often, CPS supports guaranteed real-time performance that enables use in systems that protect human lives and in safety-critical applications such as the electrical power grid, intelligent buildings, industrial control systems, and healthcare systems [2]. An automotive system is another example of a system that shares characteristics of a CPS, including a reliance on safety-critical data, sensors, and actuators deployed over open networks. The use of networks to provide critical functionality and enhanced features to exchange and operate on critical information increases the susceptibility to malicious and deliberate attempts to disrupt automotive services [3, 4]. The many interfaces on the boundary of a CPS provide ripe opportunities for compromise by an attacker [5]. Proactive and coordinated systematic efforts are needed to strengthen security that justifies increased CPS dependability [6] as illustrated by the following examples.

The September 2020 ransomware attack on critical information in a university hospital system demonstrates the need for increased cybersecurity. A ransomware attack occurs when malicious software inserted into a computer system denies an organization or individual access to information. The 2020 attack caused network outages leading to the loss of emergency health services and the transfer of patients to other hospitals. One patient died as a result of delayed medical attention [7].

In another example, Honda automobiles had a vulnerability that allowed malicious actors to capture transmissions from an owner's key fob. The attacker unlocked the vehicle and started the engine by playing back the captured signals [8]. This example highlights the vulnerability of system interfaces to an attack resulting from a growing number of services and network connectivity within an automobile [9].

In 2021, a hacker remotely accessed a computer system at a water treatment plant and attempted to increase the amount of sodium hydroxide in the water supply to potentially dangerous levels [10]. Though the intrusion was detected, the incident shows the potential for harm when the cybersecurity of computational, networking, and physical processes are treated individually rather than as a holistic system.

Origins of Cybersecurity

"Creeper," the first computer worm, was developed as an experiment in 1971 by researcher Bob Thomas at Bolt Baranek and Newman. Creeper was groundbreaking at the time and is generally accepted as the first computer worm. Creeper traversed ARPANET nodes leaving a breadcrumb trail wherever it went so that its activity could be traced [11]. A later version of Creeper written by Ray Thomlinson enhanced Creeper making it self-replicating rather than just moving from node to node. Creeper did not cause harm to the computers it visited but instead printed a message on the computer teletype that read, "I'M THE CREEPER: CATCH ME IF YOU CAN." In 1972, "Reaper" another self-replicating program spread across the ARPANET. Reaper looked for and deleted Creeper and then deleted itself, essentially becoming the first antivirus software. The first denial of service attacks, "Rabbit" and "Wabbit," also appeared in 1972, causing overloaded computer systems to crash. Malware has gone from a tool used by individuals looking to inflict petty revenge to a means by which professional criminals and state-supported hackers extort money, steal information, or otherwise inflict damage. The cat-and-mouse battle between hackers and antimalware providers fuels the need for increasingly more sophisticated cybersecurity.

The National Institute of Standards and Technology defines cybersecurity as the ability to protect or defend the use of cyberspace from cyberattacks [12]. The term "cyberspace" emerged from science fiction and was popularized by William Gibson in his novel "Neuromancer." The novel describes a world connected via the Internet and digital technology as a world apart from reality. Cyberspace now describes interconnected digital technologies, including telecommunication networks, information technology infrastructures, and computer systems.

A cyberattack is an attempt to expose, alter, disable, destroy, steal, or gain unauthorized access to a computer system [12]. Cyberattacks emerged from a fascination with cyberspace and software that accesses computer systems remotely. Films in popular culture introduced the idea of computer hacking to the mainstream with movies such as "Hackers" and "War Games," in which software causes damage after gaining unauthorized access to computing systems [13].

Cybersecurity Goals. Fundamental goals of cybersecurity include confidentiality, integrity, availability, and loss prevention.

Confidentiality protects valuable information from being accessed by unauthorized people or systems [14]. For example, failing to maintain the confidentiality of bank records could result in someone without privileges gaining access to a customer's personal information.

Access control mechanisms are one means for preventing unauthorized disclosure of information. Access control defines rules and policies that grant individual privileges and accessibility to a system's resources [15, 16]. One access control mechanism is data encryption which scrambles data making it unintelligible [17]. Deciphering or decrypting the scrambled data requires using a cryptographic key that is known only to legitimate users.

Integrity is defined as protecting information from unauthorized modification or destruction and ensuring the authenticity of the information and its source [14]. One example of an integrity failure is a successful attack that alters essential sensor data needed by a CPS control function. Another example of an integrity failure is a malicious website redirect in which an attacker inserts code into a legitimate website that forwards transactions to a malicious website.

Prevention mechanisms seek to maintain integrity by blocking any unauthorized attempts to change the data or any attempts to change the data in unauthorized ways. Detection mechanisms detect unauthorized modification of information after the information is compromised [18]. For example, checksums provide a degree of safety but not security. Other coding methods include Hamming codes, cyclic redundancy codes, hash-based digital signatures, and RSA codes [86]. Maintaining multiple data copies may be used in voting schemes that mask corrupted information [19].

Availability ensures timely access to information and system functions by authorized users [14]. Relevant to cybersecurity, an attack might deny access to information or prevent a system from performing its intended function. A common cyberattack against availability is a denial-of-service attack [20].

Loss prevention guarantees reliable access to valuable information by authorized people. Physical protection and information redundancy ensure reliability and consistency. Housing critical information and systems in secure locations safeguards against physical damage. Redundant computation, storage, data transmission, information redundancy, and contingency plans and procedures protect against information loss.

Threat Terminology. *Cyber threat* is a generic term associated with anything threatening a cyber system [22]. Cybersecurity threats (*attacks*) are malicious acts that have the potential to violate one or more security goals [21]. A *vulnerability* is an exploitable weakness or flaw in a system's design, implementation, operation, and management that could violate the system's security policy [23].

A cyberattack on a system is possible if the system and a vulnerability are exposed to an attacker. Unreachable systems cannot be attacked, and a vulnerability does not mean an attack is inevitable, especially if an attacker does not have the resources to exploit an existing vulnerability. An attacker's access to a system is also considered a factor in a threat's ability to manifest into a successful attack.

The notion of *risk* captures the uncertainty associated with the possibility that a cyberattack is successful. Risk is a function of the impact of an attack and the likelihood that the attack takes place. Attack likelihood depends on an attacker's ability to take advantage of a vulnerability and whether the attack achieves the attacker's goals. In practice, it is difficult to predict impact or likelihood accurately.

Hence, an approach to address cybersecurity threats is to reduce their potential impact by assessing the likelihood that a risk will occur based on subjective qualities and the system impact using predefined ranking scales.

Security Policy. A security policy is a set of assumptions, behaviors, practices, and procedures designed to protect system resources from cyber threats. Policies define and address secure and nonsecure system states [24] and the operational constraints, functions, data flows, external systems, potential adversaries, and necessary security required in those states [12]. Mathematically precise axioms may define policies, but usually, policies consist of natural language requirements and statements.

Lack of adherence to security policies increases the risk of successful cyber-attacks [22]. A security policy reduces risk by requiring employee training for recognizing and avoiding fraudulent email that contains malware (phishing) and prepares organizations to quickly detect and respond to any attack while protecting critical information. An example information security template is available from the Office of the National Coordinator for Health Information Technology (ONC) [25].

Security Mechanisms. Prevention, detection, and recovery mechanisms are technical and procedural methods, tools, and procedures for enforcing security policies [26].

Prevention. Prevention mechanisms close cybersecurity vulnerabilities making attacks impossible or less likely to succeed. Cryptography and access control are examples of prevention mechanisms.

Detection. Detection mechanisms determine whether an attack is occurring or has occurred and report it. These mechanisms provide data on the nature, severity, and results of cyberattacks by monitoring system messages and events during and postattack. Continuous monitoring of critical resources reduces the risk of unauthorized access and damage.

Recovery and Repair. When possible, recovery and repair mechanisms react to and repair the damage caused by an attack. Typical responses include stopping the attack in progress or continuing operation during the attack with extensive logging that allows tracing the attacker. Repair includes identifying the vulnerability and needed prevention mechanisms, such as patching the software in the system, determining the damage caused by the attack, and then repairing the damage caused by the attack. One approach to recovery places the system into a known secure state or state where the data are last known to be good. Using fault tolerance methods [22], safety-critical and high-availability systems must restore damaged data and resume normal or, in some situations, useful degraded service as quickly as possible.

Assurance and Trust. Security assurance, or assurance, is evidence-based confidence that a system meets its intended security requirements. Assurance may be achieved through informal, semiformal, or formal means. Informal techniques are the least rigorous and use natural language specifications and verification evidence. Semiformal approaches use natural languages for specification and justification but impose some process rigor. Formal methods use mathematics or rigorously defined semantic languages compatible with correctness proofs.

The ability to access needed information and preserve the security of that information is called information assurance. Information assurance differs from security assurance in focusing on threats and information protection.

Trust is the belief or assumption that a system or some of the system components are capable, honest, and reliable. Trust beliefs and assumptions can significantly impact the security of a system [28]. System or component trust can make or break a security policy [27].

All system components participate in providing assurance and trust, which drives the need for system-level security analysis [29, 30]. Although assurance and trust cannot be measured precisely, security assurance techniques such as penetration analysis, covert channel analysis, formal specification, and verification can improve the confidence that the system will perform as desired [31].

State of the Art

The application of models in security for engineered systems, particularly CPS, is a promising approach to incorporate cybersecurity considerations across the system life cycle [87–89]. Models facilitate requirements engineering, architecture and design, implementation, verification, operation, and maintenance. In CPS security engineering, models capture stakeholder protection needs and security concerns and help identify risks throughout the system life cycle.

Models at all system, software, and hardware development levels help bridge the gap between problem identification and implementation. Exemplar security models include threat models, attack models, models of information assets, countermeasure models, security controls, and system architecture and behavior models. Modeling introduces several significant benefits for security engineering [32]:

- Models allow assessing functional and nonfunctional system properties, such as confidentiality, integrity, and availability, which are considered simultaneously with business logic, performance, and other quality attributes, early in the development process.
- Models enable system reasoning in support of verification and validation methods that increase the confidence that the system achieves its specified requirements and conditions. For example, formal model checking can verify security properties.
- Verified models lead to higher confidence by reducing errors in manual or automatically generated code and security configuration [33].
- A model-based approach improves understanding of how the system operates and its potential misuse by a hacker.
- Lastly, models used within a structured methodology enhance performing systematic analysis across the CPS components and interactions.

This section illustrates threat modeling using attack trees for threat modeling, state-based models for analyzing security properties, and simulation models that

35 Cybersecurity Systems Modeling: An Automotive System Case Study

enable an analysis of software and hardware implementations. This section then concludes with examples of frameworks and methodologies that support model-based security engineering.

Threat Modeling with Attack Trees

Attack trees are among the most widely applied means for evaluating cyber threats. An attack tree is a combinatorial model that systematically describes a security breach as a failure to protect an asset [34–36]. Attack trees provide a systematic and reusable way to describe security vulnerabilities, thus making it possible to assess risks and make security decisions. Attack trees are model and programming language-independent, enabling the use of many implementation options. Attack trees support hierarchical construction in which higher-level attack trees embed lower-level attack trees. Although useful, attack trees do not model dynamic behavior, limiting their utility.

Traditionally, attack trees are stand-alone threat models but may find use in conjunction with other threat-modeling techniques. Like fault tree analysis, attack trees are tree structures having a discrete attack goal as the root node and a logical combination of different paths from the leaf nodes that achieve that goal. Each leaf node represents an attack goal (or subgoal), and leaf nodes are atomic attacks. There are two kinds of nonleaf nodes: *AND* nodes and *OR* nodes. An *AND* node represents an attack goal for each incident subgoal to be achieved for the attack to succeed. The *AND* node's children represent the attack subgoals. An *OR* node represents an attack goal that can be achieved in several ways, as represented by the *OR* node's children.

An attack tree is not a methodology, but a model used within methodologies to help evaluate security from the perspective of modeling an attacker's actions and system mitigations to specific attack actions. For example, an attack tree is used to perform security analysis in the context of a probabilistic risk assessment [37] and within the frameworks such as STRIDE and the Process for Attack Simulation and Threat Analysis (PASTA) [38]. Hence, attack trees are used in the requirements determination phase of security engineering.

Figure 1 shows an example of an attack tree, where the goal is to gain access to a car. An attacker accomplishes this goal by picking the car door lock or learning the signal from the remote key fob that emits a radio frequency signal to unlock the car door. The signal is obtained by eavesdropping – remotely obtaining the signal by recording or copying the original signal – or physically copying the key fob. Eavesdropping is successful if the signal is recorded and obtained by the attacker.

For multiple security goals, an attack tree is created for each goal to model an attack. Multiple trees could be combined if there is a logical dependency between attack goals, meaning that one attack goal depends on another. Each leaf node can be assigned a quantitative value or function that maps the atomic attack goal to a cost for achieving the goal. Values could be the cost of an attacker succeeding in a goal represented by a node, the cost of succeeding as a quantitative value, the likelihood of an attacker accomplishing the goal, or a combination of quantities. A quantity for the root node is computed as a logical combination of leaf nodes, as illustrated in

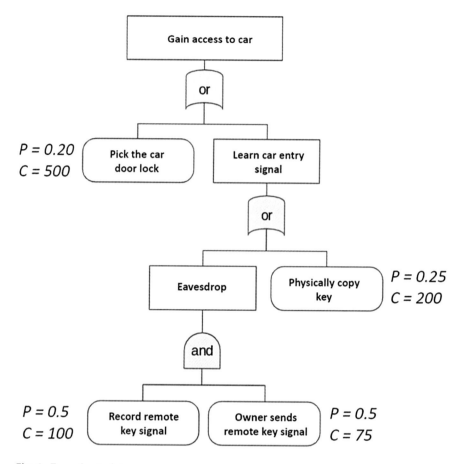

Fig. 1 Example attack tree

Fig. 1, by propagating the value of leaf nodes through the logical combination of nonleaf nodes. For example, in Fig. 1, the probability (P) of an attacker to successfully "listen to conversation" is 0.5, and it is achieved at a cost (C) of 100 for the attacker. The probability of "target says combo" is 0.5 at a cost of 75. Assuming that the goals of the leaf nodes are statistically independent, then the probability of "eavesdrop" is 0.25 at a total cost of 175. Continuing this logic by traversing the tree and computing the probability and cost will result in a probability and cost of the attacker's goal (the root of the tree).

State-Based Security Modeling Approaches

Model-checking is a state-based approach that uses formal methods based on mathematical techniques to specify and verify a system's behavior [39]. Using logical reasoning, formal methods check whether the system model satisfies functional and nonfunctional properties, such as security. A model checker, in this context, is a tool that checks the design for the specified security properties defined in the system security specification – the requirements for the security-relevant portions of the system. The security properties are encoded in a mathematically precise language, such as a Linear Temporal Logic [40], so the system can be evaluated using formal mathematical logic.

Model-checking tools include analyzing computer programs for security flaws [41] and modeling cyberattacks. In [42], the authors successfully analyze large-sized well-known software packages for a set of security flaws using model-checking techniques. In programs they experimented with, failure to drop privilege caused a security vulnerability. In [43], the authors use model-checking to identify vulnerabilities in an OpenRISC processor hardware architecture.

In [44, 45], the authors model cyberattacks on networks, in which the network state includes a description of hosts, network connectivity, and vulnerabilities. An attacker's state includes capabilities and access gained so far during an attack. A state transition occurs when there is a match between a capability in the attacker's state and a vulnerability in the network state, resulting in potential access elsewhere in the network. States that represent an attacker's access to network assets reflect successful exploits. For every asset, one can ask whether it can be compromised and determine the number of paths to states in which the asset is first compromised. The definition of "compromise" depends on the precision of the model. Although model checking effectively finds security specification and design flaws, large state spaces present a challenge.

Pokhrel and Tsokos [46] propose a stochastic model to quantify the risk associated with the overall network using a Markovian process and a vulnerability-scoring technique. Madan et al. [47] use a semi-Markov model to evaluate an intrusion-tolerant system's security properties. As described in [47], states may be associated with availability, integrity, and confidentiality failure. The mean-time-to-security-failure then quantifies security. An example of a state-based model is shown in Fig. 2 to illustrate a set of system security states and transitions between states.

State-based models are typically applied during the requirements determination and security architecture and design phases of security engineering. State-based models for automated model checking are implemented in different languages and require a computation engine to perform the execution. For example, the Promela language [48] captures formal verification models in the SPIN model checker [49], and the PRISM language [50] is used for probabilistic state-based models. The authors in [51] provide a summary of state-based security models used to check various properties, such as security protocols to identify protocol vulnerabilities. Identifying these vulnerabilities informs security requirements on protocol design.

Simulation-Based Security Modeling

A simulation is an artificial construct that imitates the behavior of a real-world process or system. Unlike state-based methods that exhaustively explore the state space, simulations are stimulated by a set of defined input vectors that drive the

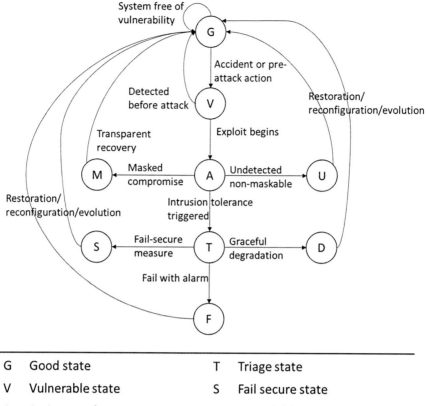

G	Good state	T	Triage state
V	Vulnerable state	S	Fail secure state
A	Active attack state	D	Graceful degradation state
M	Masked compromised state	F	Failed state
U	Undetected compromise state		

Fig. 2 Example of a state-based model for an intrusion-tolerant system

evaluation of system security properties. Simulation is not as mathematically tractable as state-based evaluation methods, but simulation evaluations have the ability to capture the real-time response of the system that is modeled. Models for simulations depend on the level of fidelity that is needed for the purpose of the model. Thus, the validity of the model is measured against the purpose using one of several methods, including making a subjective decision on model validity by the model developers, users, or third-party teams that are independent of the model users and developers. Simulations use programming languages instead of modeling languages to capture properties, rules, and the execution of a simulation.

The most common use of simulation in cybersecurity is evaluating network attack traffic. Network simulations use computer systems and networks that perform a variety of attacks to explore countermeasures and the impact of attacks on computing systems. Simulation models help quantify cybersecurity metrics during attacks, countermeasures, and training scenarios [52–55].

Network attack and defense strategies are specific to the network environment in which they are executed, and in general, many scenarios must be considered. The cost of testing these multiple scenarios in real-world systems is often prohibitive [56]. Consequently, evaluating and selecting effective network policies and defense mechanisms is challenging. The use of simulation expands the scope and scale of cybersecurity evaluations due to the ability to assess multiple scenarios repeatedly.

Simulation approaches using discrete-event simulations allow evaluating multiple scenarios under mathematical and behavioral constraints that imitate the modeled environment. One implementation of discrete-event simulation uses agent-based simulation modeling. In [57, 58], the authors apply agent-based modeling, a computational formalism that models a system as a collection of autonomous decision-making entities called agents [59], to capture emergent system dynamics resulting from multiple network agents' interactions. Attackers, defenders, the network environment, and system behaviors are captured as agents. The agent-based system produces metrics and visualizations that provide insights into network security and guide the search for efficient policies and controls [60].

Other simulation techniques include combining real, emulated, and simulated components to analyze a networked information system's security features and components [61]. The security component may be the hardware and software with all the surrounding components represented in simulation or by surrogate devices. The combination of simulation with surrogate devices and the actual software and hardware in an interconnected, isolated environment enables realistic evaluations in support of cybersecurity analysis [61].

In [59], a discrete network simulator predicts the impact of denial of service, malware propagation, and man-in-the-middle attacks on supervisory control and data acquisition systems (SCADA) used in industrial control systems. Liu and Srivastava [62] utilize a testbed that simulates hardware and software in a hardware-in-the-loop capability to evaluate the effects of attacks on the power grid. A survey of industrial control system testbeds for cybersecurity evaluations is summarized in [63]. The simulation approach to security engineering emphasizes the implementation of the CPS and does not explicitly address the requirements definition and architecture modeling and design.

Security Viewpoints in Unified Architecture Framework

The Unified Architecture Framework (UAF) is an enterprise architecture framework that provides system views [64] in the form of diagrams with objects and relationships. UAF provides views that capture security features, including *security constraints, security property, security assets, security controls, risk,* and *security impact property.* The properties are aligned with the standards of the National Institute of Standards and Technology (NIST) standards and the United States Department of Defense (DoD) [65]. The standards provide a basis for the unified information security framework for the United States federal government. UAF enables the integration of model elements across the architecture and security domains to address cross-cutting concerns.

UAF does not natively provide security models such as attack trees or state-based formalisms. However, such formalisms can be modeled in UAF by extending the use of the model views and annotating with additional information. In addition, UAF is not a methodology but a set of model views that may be captured at any point in a given process. Thus, UAF is applicable for documenting security requirements, the system and security architecture, and alignment to the software and hardware implementation.

Chassis

The Combined Assessment of Software Safety and Security Requirements (CHASSIS) is a method for determining system safety and security requirements [66]. CHASSIS introduces a method for integrating security and safety when developing systems that account for modern systems whose safe and secure operation is critical. CHASSIS comprises a requirements analysis process for eliciting functional, safety, and security requirements. The requirements are captured in Unified Modeling Language (UML) diagrams [68] representing attacks, attackers, and threats. Risks and weaknesses are identified by using a Hazard and Operability study (HAZOP) analysis [67].

From a modeling perspective, the CHASSIS method introduces two diagram types derived from UML: the *misuse case diagram* and *misuse sequence diagram*. Misuse sequence diagrams are extensions of UML sequence diagrams that describe chains of events and interactions that lead to system misuse. The misuse case diagram extends the UML use case diagram with additional properties that capture how the system could be misused and by whom (misusers). Misusers include human users and external systems that can threaten or cause a system failure. A misuse case is associated with a use case, and there may be more than one misuse case per use case.

The CHASSIS method starts by identifying functional requirements as a basis for eliciting safety and security requirements. The CHASSIS method describes users, functions, and services in UML diagrams. UML sequence diagrams model the detailed interactions between the system and users.

Next, the elicitation of safety and security requirements is carried out. An approach to elicit requirements is through a brainstorming session with safety and security experts to identify potential misuses of the system. A HAZOP analysis is also performed to obtain potential misuses of the system. The HAZOP analysis informs the misuse cases. Misuse case diagrams are drawn using a graphical notation, and additional information about the misuse case are captured using text. From the misuse cases, scenarios that describe how the system is misused are captured in a set of scenarios for each misuse case.

SysML-Sec

The SysML-Sec method is a methodology that promotes collaboration between system designers and security experts across design and development phases. SysML-Sec includes requirements capture, attack scenarios, functional model, hardware and software partitioning, designs of software components, and validation of safety and security properties [69]. As shown in Fig. 3, SysML-Sec relies on a modeling approach that emphasizes partitioning of the functionality of the underlying hardware and software architecture that implements the functionality [70]. SysML-Sec enables exploring design options created from different allocations of security features between the hardware and software architecture.

The SysML-Sec methodology consists of three phases: analysis, design, and validation. The analysis phase defines security requirements, attack scenarios, the main functions, and conceptual system architecture. In the system design phase, security requirements are refined with security properties, and security functions are defined. The validation stage assesses whether the security properties captured in the model are valid and verified. Simulation models are used to perform the validation and verification.

In the SysML-Sec methodology, security requirements use a requirement model object from SysML. The requirement model object is extended by a SysML stereotype, giving the required security attributes such as confidentiality, access control, and integrity. Stereotyping allows users to distinguish security requirements from other requirements. Attack trees are specified using a SysML parametric diagram. An attacker is modeled using SysML block and state machine diagrams.

MB3SE

The Model-Based System, Safety, and Security (MB3SE) integrates security engineering with safety engineering processes [71]. The method is primarily used for

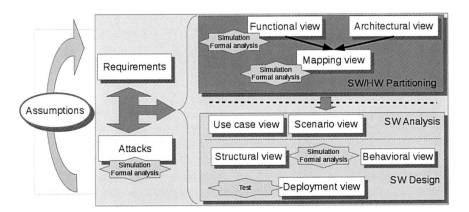

Fig. 3 Overview of the SysML-sec method

developing medical devices. The method comprises two main components: model-based system architecture development and safety and security risk analyses convergence. Model-based system architecture development is a two-step process based on model-based systems engineering practices.

Step 1 (Functional Analysis). The first step in the architecture design is functional analysis, which is foundational for expressing system needs leading to functional architectures. Needs come from use case analyses, requirements, and interviews. High-level functional architectures can be decomposed into lower levels suitable for eventual allocation to system components.

Step 2 (Component Analysis). The second step in the architecture design is component analysis. This step transforms the functional architecture from step 1 into a physical architecture consistent with nonfunctional constraints such as safety and security risk. Physical architecture components are abstractions of system hardware and software items. This step produces a model of requirements, functions, and components for the medical device under design. These artifacts are inputs to the security and safety analyses that follow in the next set of steps where safety and security risk analyses convergence supports aligning safety and security risk factors.

Step 3 (Identification of Feared Events). This step identifies *feared events (FEs),* which are potential sources of harm, damage to property, or damage to the environment. Safety-related FEs are identified by performing a safety hazard analysis. Security-related FEs are threat sources identified by searching existing knowledge databases and relying on expert knowledge. Prioritization of safety and security FEs is based on their impact to patient safety.

Step 4 (Identification of Failure Modes, Vulnerabilities, and Preventions). This step comprises a Failure Modes, Effects, and Criticality Analysis (FMECA) [72] which determines the failure modes (FMs) that can lead to safety-related FE identified in step 3. FMECAs are failure analyses that establish cause and effect chains between potential failures (failure modes), the impact on the mission (effects and criticality), and the causes of the failures (failure modes). Security FEs are evaluated by examining exploitable component vulnerabilities and potential threats. Security evaluation may be supported by Failure Modes, Vulnerabilities, and Effects Analysis (FMVEA) [73] and databases.

Step 5 (Identification of Critical Scenarios). This step develops a set of safety and security scenarios from fault tree analysis, extended by including malicious events and threat scenario analyses [74]. These analyses describe how failures and malicious events propagate inside the system architecture. Fault tree and attack tree analyses are used in this step to identify the critical failure and threat scenarios that are used in subsequent steps for risk assessment.

Step 6 (Risk Assessment). Risk assessment is the process of identifying, analyzing, assessing, and mitigating potential system issues, e.g., as defined in ISO 14971 [75]. The security-related risk is identified by exploiting the cybersecurity vulnerability and the severity of the security-related FE to patient harm. Safety risks are collected into a risk matrix in which risk likelihoods are row headings, and risk severity are column headings. A matrix cell is assigned one of three acceptance values: *acceptable, tolerable,* or *unacceptable,* and represents all risks with the same likelihood and severity. Similarly, security risk is a function of the likelihood that a vulnerability is exploited and the severity of the exploitation to a patient. This step results in a safety risk analysis and threat scenario analysis to identify, analyze, assess, and mitigate system issues.

Step 7 (Security Risk with Potential Safety Impact). This step evaluates whether the exploitation of cybersecurity vulnerabilities results in controlled or uncontrolled safety risks using a qualitative assessment of exploitation likelihood and severity.

Step 8 (Implemented Controls, Residual Risks, and Actions Plan). After security risks and the impact of security risks are identified, this step implements controls for risks identified in the previous step, and courses of actions are devised to address residual risks. The controls that are identified must meet the risk acceptance criteria. The implementation of controls are reviewed to ensure that safety controls do not affect security controls and vice versa. If any effects on safety due to security control implementation or effects on security given safety control implementation are discovered, then the effects are identified as residual risks that must be addressed. Once controls are implemented, a team of experts review the safety and security control implementations for residual risks that may arise as a result of implementing controls. If any residual risks are found, then plans to mitigate those risks are identified and documented for execution in another iteration of steps 6, 7, and 8 until security and safety concerns are addressed.

The set of software applications for modeling, analyzing, and documenting artifacts in the MB3SE method include multiple open-source applications that are used collectively to capture the system architecture and analyze safety and security aspects. The functional analysis and component analysis are achieved using Capella [76] and Papyrus with the CHESS plugin [77]. Following the component analysis in step 2, the system functions, components, and requirements levied on the medical device are provided as inputs into safety and security software application tools. The Safety Architect [78] application is used to identify failure modes, malicious events, and perform safety analyses as identified in steps 3 through 7 of the MB3SE method. The functions, components, and requirements that are security related in the component and functional analyses are provided as inputs into the Cyber Architect software application [78] to perform security analyses, as identified in steps 3 through 7 of the MB3SE method. Unlike other methodologies in this chapter, these tools implement MB3SE by embedded, custom, domain-specific language rather than standard UML or SysML.

MBSEsec

The MBSEsec methodology, as shown in Fig. 4, describes another model-based methodology for developing secure cyberphysical systems [79]. The MBSEsec method comprises model artifacts for incorporating security into requirements development, system architecture, and design using SysML and UML modeling languages. The methodology extends SysML and UML enabling the modeling of security concepts using a security profile. A security profile captures the security

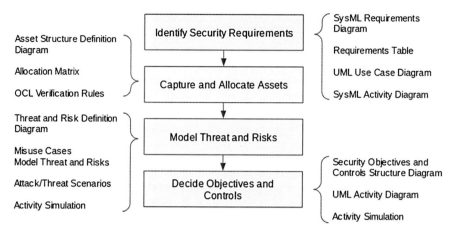

Fig. 4 Phases of the MBSEsec method

concepts in SysML so that the security concepts can be applied to SysML model elements and diagrams. A profile is a set of model elements that extends UML and the SysML language with custom domain concepts using model elements called stereotypes. A stereotype is a mechanism for extending SysML and UML-based models with precise natural language constructs. SysML model elements are decorated with stereotypes to identify security-related artifacts that result from phase activities of the methodology. As shown in Fig. 4, the methodology comprises four phases: *identifying security requirements*, *capturing and allocating assets*, *modeling threats and risks*, and *deciding objectives and controls*.

Phase 1 (Identify Security Requirements). The first phase of the MBSEsec method identifies security requirements in conjunction with the system's functional and nonfunctional requirements. MBSEsec does not prescribe requirement elicitation, but common practices include interviews with stakeholders and users, identifying business needs, and considering security policies. The methodology captures requirements in SysML Requirements Diagrams and tables. SysML Use Case and Activity diagrams can clarify and refine requirements.

Phase 2 (Capture and Allocate Assets). In the second phase, physical component and software component models of assets are created. The model includes assets to secure, such as data and software assets. These assets are then allocated to logical system components captured in the model a priori. Logical system components and assets are represented as SysML Blocks. MBSEsec employs a diagram type called the Asset Structure Diagram for capturing the structure of assets. The Asset Structure Diagram is an extension of the SysML Block Definition Diagram that captures assets and allocations to logical system components. Making the allocation explicit using the allocation relationship in the model enables verifying that the model is constructed correctly according to syntax rules imposed by MBSEsec utilizing SysML constructs to check syntactical rules. Thus, MBSEsec utilizes SysML tools to perform syntax checks.

Phase 3 (Model Threat and Risks). This phase captures system threats and risks. MBSEsec prescribes the use of Misuse Cases and the UML Activity diagram for modeling attack scenarios. The Threat and Risk Definition diagram in the MBSEsec method creates and relates model elements that represent risks, threats, and vulnerabilities. The Threat and Risk Definition diagram is an extension of the SysML Block Definition Diagram by annotating basic SysML elements with stereotypes, and the diagram captures the relationships between risk-related information. Risk-related information can be summarized in a HAZOP table.

Phase 4 (Decide Objectives and Controls). The final phase defines security controls and objectives in a Security Objectives and Controls Structure diagram. This diagram is an extension of the UML Class diagram that is employed in the MBSEsec method to capture security objectives and controls. The model elements that appear on the UML Class diagram are extended with stereotypes that identify the elements as part of the security controls and objectives.

The outcomes of each phase are continuously reviewed and updated to ensure consistency and adapt to changes. For example, after identifying the security controls in Phase 4, the Threat and Risk Definition diagram may be updated to reflect changes in risk. Similarly, if security requirements change, models and artifacts created in subsequent phases should be checked for consistency with the changes.

Best Practice Approach

Current literature lacks a commonly accepted security approach that spans an entire system life cycle. However, a promising approach is described by Uzunov et al. [80]. The approach conforms to the International Organization for Standardization (ISO) and the International Electrotechnical Commission (IEC) standard ISO/IEC 2700, which specifies requirements and practices for establishing, implementing, operating, monitoring, reviewing, maintaining, and improving information security management systems [81].

The approach in Uzunov et al. outlines a broad set of steps grouped into phases where security is included systematically across the system life cycle. The approach is a process that incorporates security needs, concerns, architecture, and implementations. The process provides a structured means for addressing security concerns while being flexible enough to include a choice of security models, modeling languages, and tools. Figure 5 illustrates the phases.

Security requirements determination. As previously noted, the trustworthiness of a system is the belief or assumption that a system or some of the system components are capable, honest, and reliable. Implicit in this definition is a system's ability to

Fig. 5 Security engineering phases

fulfill its requirements. Security requirements are elicited from system stakeholders, such as customers, security experts, users, and developers. Requirements come from other potential sources, including organizational policies, government regulations, and the intended environment in which the system operates. The process for gathering security requirements is a discipline that consists of steps such as identifying system assets to protect, identifying system vulnerabilities, threat modeling, and ensuring the requirements are specific, necessary, understandable, accurate, feasible, and testable.

A model-based approach in this phase must have the ability to capture security requirements and support the requirements engineering process to include the ability to help elicit security requirements, document the requirements, analyze the requirements for consistency, and manage the requirements throughout the system life cycle. Additionally, a model-based approach must enable capturing, documenting, and managing threats and their associated risks. Moreover, the model must support change management when threats, risks, or requirements change. Threat modeling methods are used in this phase to develop a profile of potential attackers, their goals, means, and to catalog potential threats that may arise.

Security architecture and design modeling. This phase involves defining the system architecture and design in preparation for hardware and software implementation. The architecture is implemented by adopting a modeling language, incorporating security attributes, analyzing the security design, and visualizing security attributes and analysis results within the system models. Explicitly capturing the security properties and associated requirements into the system architecture and design model increases the likelihood that essential security considerations that the stakeholders agreed to are not lost, and it helps enable systematic analysis of the security design. In addition, documenting the security encapsulates the reasoning and judgment of the engineers, thus, helping to preserve and transfer knowledge between experts over time.

A model-based approach captures key features of the system and security architecture such as behaviors; structure; functional attributes; security attributes, controls, and mechanisms; and other nonfunctional attributes of the system that contribute to the security architecture. A model must also capture and manage the relationships between security-related attributes such as security requirements, risks, and system threats. Additionally, a model must enable reasoning over the architecture to help inform the system design to enhance system security.

Implementation. This is the stage in which security requirements, security controls, and architecture are implemented in software and hardware. Representing security information in the design provides a tighter coupling to the implementation such that changes in the implementation are reflected more accurately to the design model. Missing a security-related design decision might result in incomplete or incorrect implementation, thereby, introducing vulnerabilities in the system. Therefore, maintaining consistency between the implementation, architecture, and requirements is essential.

Configuration and monitoring. The final stage is responsible for maintaining the consistency of the final product with the initial requirements and design. By maintaining consistency, the design model can be used to predict implementation behavior sufficiently to troubleshoot issues with the implemented system or to provide a starting point for reuse on subsequent designs. This phase also offers support during the system run-time by providing the capability to monitor software and hardware services. Enabling the monitoring of threats and vulnerabilities during run-time execution benefits system security. Configuration focuses on documenting the system, along with its security properties. Configuration also includes updating the documentation when changes to the system or security features occur. Updates can be applied when threats change or vulnerabilities are exposed by keeping track of software versions and hardware parameters.

Illustrative Example

As an illustration of the best practice approach, the MBSEsec methodology is applied to a model of a Hybrid Sport Utility Vehicle (HSUV) [82] within the context of the security engineering process described by Uzunov et al. [80]. The MBSEsec methodology incorporates the use of SysML as a language to capture artifacts of the process. SysML is commonly used as the de facto modeling language for the systems engineering community, and the security engineering artifacts are constructed by extending the language to capture security engineering concepts.

HSUV Security Issues

Modern vehicles are subject to cyberattacks through their internal vehicle network interfaces. For instance, a modern vehicle's Controller Area Network (CAN) does not ensure confidentiality and authentication in its data transmission protocol [83]. As a result, it is subject to attack through its On-Board Diagnostics II (ODB-II) interface, an interface for diagnostic reporting of vehicle state and health. Diagnostic tools broadcast CAN data frames without encryption and authentication to check the functions of the Electronic Control Units (ECUs) during a diagnostic process. Suppose an adversary has access to a diagnostic tool. In that case, they can easily get access to the CAN data frames and use them to inject malicious data that disrupts the operations of the vehicle ECUs. This case study shows how the ODB-II and Power ECU security issues can be identified, analyzed, and mitigated using the MBSEsec methodology. Figure 6 shows the attack surfaces for the HSUV, including the Power Control ECU and the ODB-II as highlighted. The attack surfaces are interfaces where attackers can access the HSUV system.

Fig. 6 Attack surfaces for the HSUV case study

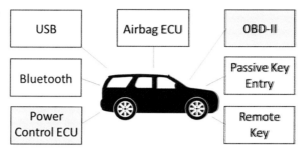

Instantiating the MBSEsec Security Domain and Profile

The MBSEsec method defines a Security Domain Model that captures security concepts. The concepts in the domain model are implemented in SysML by creating a profile that enables the model user to create model elements, relationships, and diagrams using the domain concepts. The security domain model in Fig. 7 illustrates the concepts that are defined and applied using the MBSEsec method.

The concepts of the Security Domain Model in Fig. 7 are implemented in the security profile shown in Fig. 8. The *Configuration* group contains stereotypes for the "Risk Assessment Configuration." This element has a property named "Criteria for Accepting Risks." The property is used to set the criteria for accepting risks. The property is represented as an integer, where the integer maps to a risk acceptance level. The *Security Requirements* group contains the definition of an element to capture security requirements. A "Security Requirement" is a stereotype that is extended from the SysML Requirement element through the use of a UML generalization relationship. The *Assets* group consists of stereotypes for defining and allocating system assets. According to the security domain model, the "Software," "System," and "Data Assets" are defined as specializations of the "Asset" stereotype. This allows for distinguishing between the different types of assets that are defined in the security domain model. The *Threat and Risks* group defines stereotypes for "Risk," "Risk Impact," "Probability," "Threat," and "Vulnerability." Moreover, the dependency-based stereotypes are created for all possible relations, including the "Misuse," "Cause," "Characterize," "Use," and "Applicable To." The property of "Level of Risk" for "Risk" is added to capture the concept of risk levels. The *Security Objectives and Controls* group refers to the objectives and options for the risk treatment.

Before starting the security requirements definition phase for the MBSEsec method, assume a risk assessment document exists. The risk assessment document describes how risks are identified, evaluated, and documented throughout the system life cycle. A reference to the risk assessment document is contained in the "Risk Assessment Configuration" model element as a link to the document. An example of a Risk Assessment Configuration for the HSUV is illustrated in Fig. 9.

35 Cybersecurity Systems Modeling: An Automotive System Case Study

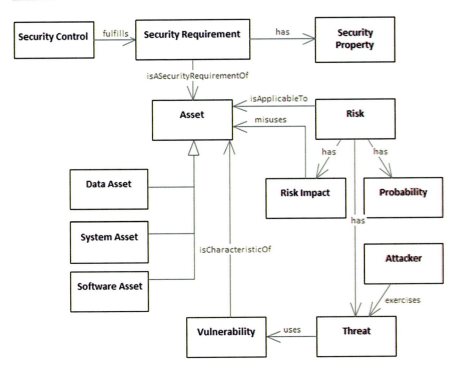

Fig. 7 Concepts captured in the security domain model of the MBSEsec method

Applying the Best Practice Approach

Phase 1: Security Requirement Determination. It is assumed that requirements elicitation has resulted in a top-level security requirement that addresses the overarching need to protect the HSUV Power Control ECU from physical access. The top-level security requirement for this case study is "Limit external access to Power Control ECU." The requirement derives two additional requirements, "Limited signals" and "Obfuscated access." The requirements are captured in a SysML Requirements diagram, as shown in Fig. 10. If there is a need to refine the security requirement, a diagram such as the SysML Activity Diagram can be used to elaborate the intent of the requirement further.

Phase 2: Security Architecture and Design Modeling. The MBSEsec method supports the architecture and design modeling phase of the security-engineering process by executing the following three phases: *Capture and Allocate Assets*, *Model Threats and Risks*, and *Decide Objectives and Controls*. These phases lead to the implementation of a security architecture. In the *Capture and Allocate Assets* phase of the MBSEsec method, the following assets must be secured: Power Control ECU hardware, embedded software, and the interface between the hardware and embedded software. An Asset Definition Diagram is used to model elements that represent assets of the HSUV to protect. The assets are then assigned to system

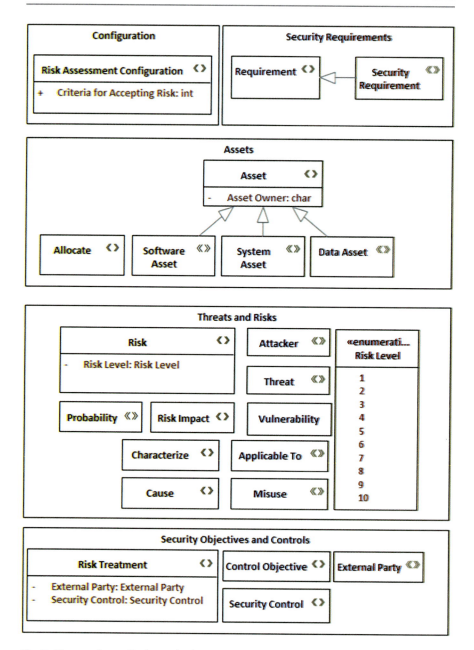

Fig. 8 The security profile for use in the MBSEsec method

Blocks that represent the logical components of the HSUV using SysML Allocate relationships. The relationships are created on the Asset Definition Diagram, a SysML Block Definition Diagram that shows the allocation of assets to logical system components, as shown in Fig. 11.

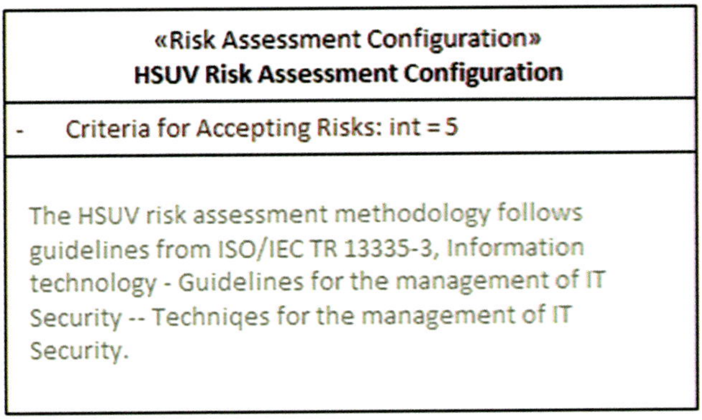

35 Cybersecurity Systems Modeling: An Automotive System Case Study

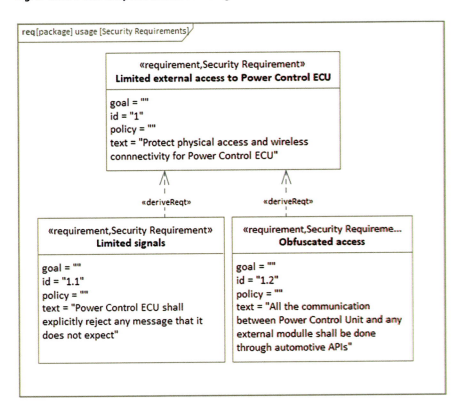

Fig. 9 HSUV case study risk assessment configuration

Fig. 10 Security requirements for the Power Control ECU of the HSUV

The next phase of the MBSEsec method is to *Model the Risks and Threats*. This case study uses a malicious mobile application to attack the HSUV by inserting malicious code into a vehicle through the ODB-II interface, as further described in [83]. The attack scenario describes a possible fatal malfunction of rapidly

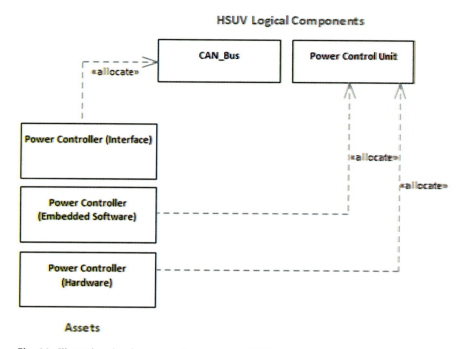

Fig. 11 Illustration showing assets allocated to the HSUV components

accelerating up to 200 kilometers per hour due to the abnormal control data that is transmitted from the malicious application. Using the MBSEsec method, the first step identifies the risks and threats. In this step, threat models are applied. Although not shown, a model such as an attack tree could be used to identify threats and how attacker actions can lead to a violation of the security requirements. For this case study, a Misuse Case Diagram, as illustrated in Fig. 12, is used to help identify how an attacker can attack the HSUV. The Misuse Case Diagram helps inform system security risks. The Misuse Case Diagram identifies the actors as the Driver and the Attacker. The figure shows the high-level interactions between the Driver, Attacker, and lists the misuse cases under consideration.

As illustrated in Fig. 13, a SysML Activity Diagram captures the attack scenario for one of the misuse cases [83]. In Fig. 13, swimlanes represent the actions of the malicious application and the actions of the HSUV components. Although not shown, the model for the attack scenario could be extended to enable simulation of the attack behavior and validation that the attack scenario model correctly represents expected behavior.

For the final step in the *Model Risks and Threats* phase of the method, a Threat and Risk Definition Diagram is used to model the risk, risk impact, probability, threat, and vulnerability. For the HSUV, the risk is captured as "an attacker takes

35 Cybersecurity Systems Modeling: An Automotive System Case Study

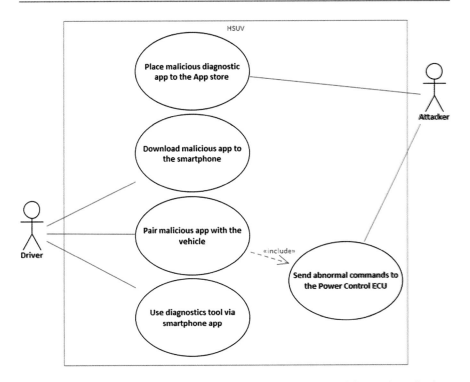

Fig. 12 Illustration of Misuse Case Diagram to reflect malicious usage of diagnostic application

over a Power Control ECU via the ODB-II port, reprogram it, and execute functions of Power Subsystem." The impact of that risk is "lost control of HSUV acceleration," and the impact of the risk is captured as a SysML relationship between the risk and the impact. The likelihood of this risk occurring is set to "low." Based on the probability and risk impact estimation, the "Level of Risk" property in the model is set to 5. The possible threat is "fault injection on automotive diagnostic protocols," and it uses the vulnerability of "Control Area Network (CAN) protocol." The Threat and Risk Definition Diagram in Fig. 14 shows the relevant security elements and their relationships.

The *Decide Objectives and Controls* phase of the MBSEsec method helps identify objectives for security control and risk mitigation control. In this case study, the control objective for the HSUV is "System shall prevent unauthorized access to the Power Control ECU," as illustrated in Fig. 15. This objective is associated with a corresponding security control, as shown in Fig. 16 as a SysML Activity Diagram. The security control in the figure captures a sequence of actions to achieve the security control objective using a multilayered protection approach.

With the model elements for the security architecture modeled, automated analyses could be performed to provide insight to stakeholders about security

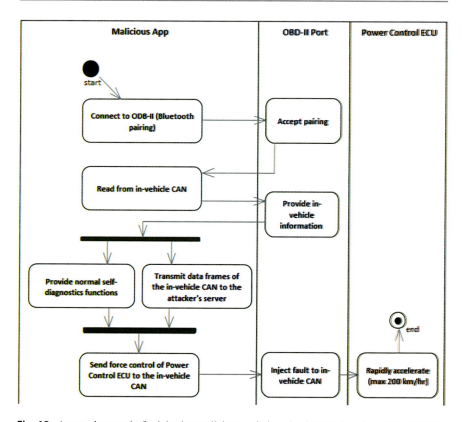

Fig. 13 An attack scenario for injecting malicious code into the OBD-II interface of the HSUV

architecture metrics. For example, in the HSUV case study, an impact analysis is performed to show stakeholders which system and security elements need to be reviewed if the initial system requirements are changed. Since traceability from requirements to the system and software assets are explicitly captured in the model, the analysis enables stakeholders to trace the security architecture back to the original system and security requirements.

Phase 3: Implementation. In the security-engineering phase, implementation is the development of hardware and software that conforms to the requirements and security architecture from the previous phases. The MBSEsec method does not specify how to implement the security controls from the architecture and modeling phase. However, techniques can be developed to conform to the security architecture specifications. For example, to satisfy the security control, and "limit unauthorized commands and resource access," a software and hardware implementation of a cryptographic key management technique provides encryption and authentication solution [84, 85].

Phase 4: Configuration and Monitoring. The MBSEsec method uses the model as developed in SysML to document the requirements, security architecture, and

35 Cybersecurity Systems Modeling: An Automotive System Case Study

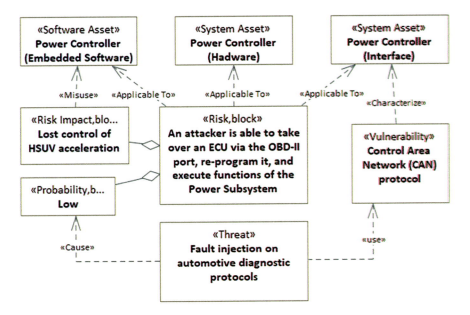

Fig. 14 The Threat and Risk Definition Diagram for the Power Control ECU

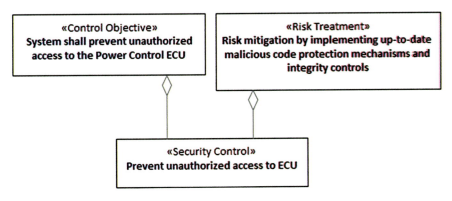

Fig. 15 Security objectives and controls structure for the Power Control Unit

controls in the model and diagrams that are described in the case study. The method does not prescribe how the models are used in monitoring the HSUV during the run-time of the security solution. However, suppose the configuration of the software and hardware implementation is traced to the security controls and system architecture that has been modeled. In that case, it is possible to provide monitoring capabilities using state-based security models that could potentially be used to monitor indicators that might imply malicious behaviors. This is one way to use the architecture model to monitor the execution of the security protocol solution.

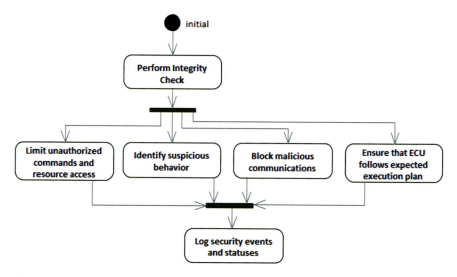

Fig. 16 The security control for preventing unauthorized access to the ECU

Chapter Summary

This chapter presents a case study that applies a model-based methodology using SysML and the MBSEsec method to show the use of models in security engineering for a HSUV. The methodology is guided by a best practice approach that incorporates security engineering across the system life cycle. The best practice approach introduces four phases: Security Requirement Determination, Security Architecture and Design Modeling, Implementation, and Configuration and Monitoring. The MBSEsec method is then employed on the HSUV case study using these four phases of the best practice security-engineering approach. The application of the MBSEsec method combines systems engineering and security-engineering disciplines at an early stage of system development. The method consists of a security profile and domain model that is based on SysML. The method covers phases starting from security requirements identification; capturing assets and modeling threats and risks; and finally, deciding on security control objectives and appropriate controls. The method and approach are presented in the context of a case study that extends a model of a Hybrid Sport Utility Vehicle (HSUV) that is presented in the OMG SysML specification. The HSUV case study showed relevant security artifacts that are created in the sysML model including security requirements, misuse cases, asset structure definition, attack scenarios, threat and risk definition, security controls, and supporting diagrams for documentation and communication with stakeholders.

Cross-References

► MBSE Methodologies
► Overarching Process for Systems Engineering and Design
► SysML State of the Art

References

1. E. A. Lee, "Cyber physical systems: Design challenges." In 2008 11th IEEE International Symposium on Object and Component-Oriented Real-Time Distributed Computing (ISORC), pp. 363–369. IEEE, 2008.
2. V. Gunes, S. Peter, T. Givargis and F. Vahid, "A Survey on Concepts, Applications, and Challenges in Cyber-Physical Systems." KSII Trans. Internet Inf. Syst. 8 (2014): 4242–4268.
3. Y. B. Reddy, "Security and design challenges in cyber-physical systems." In 2015 12th International Conference on Information Technology-New Generations, pp. 200–205. IEEE, 2015.
4. R. Rajkumar, I. Lee, L. Sha, and J. Stankovic, "Cyber-physical systems: the next computing revolution," In Design Automation Conference, pp. 731–736. IEEE, 2010.
5. S. Y. Dorbala and R. S. Bhadoria, "Analysis for security attacks in cyber-physical systems," Cyber-Physical Systems: A Computational Perspective. Chapman & Hall, 2015:pp. 395–414.
6. Y. Ashibani and Q. H. Mahmoud, "Cyber physical systems security: Analysis, challenges and solutions." Computers & Security 68 (2017): 81–97.
7. R. Hackett, "Ransomware attack on a hospital may be first ever to cause a death." https://fortune.com/2020/09/18/ransomware-police-investigating-hospital-cyber-attack-death/ (accessed Jul. 22, 2022).
8. N. Wasson, "Honda and Acura Replay Attack Lets Hackers Remotely Unlock and Start These Cars." hothardware.com. https://hothardware.com/news/honda-acura-replay-attack-lets-hackers-remotely-unlock-start-cars (accessed Jul. 22, 2022).
9. F. Youd, "Cybersecurity, a growing threat for the automotive industry." just-auto.com. https://www.just-auto.com/interview/cybersecurity-a-growing-threat-for-the-automotive-industry/ (accessed Jul. 22, 2022).
10. N. Statt, "Hackers tampered with a water treatment facility in Florida by changing chemical levels." theverge.com. https://www.theverge.com/2021/2/8/22273170/hackers-water-treatment-facility-florida-hacked-chemical-levels-changed (accessed Oct 2, 2022).
11. T. M. Chen and J. Robert, "The evolution of viruses and worms." Statistical methods in computer security 1 (2004).
12. C. Paulsen and R. Byers, "Glossary of key information security terms," NISTIR 7298 Rev. 3. National Institute of Standards and Technology, June 2019.
13. E. A. Lee, "The past, present and future of cyber-physical systems: A focus on models." Sensors 15, no. 3 (2015): 4837–4869.
14. Federal Information Processing Standards Publication, "Standards for security categorization of federal information and information systems." NIST FIPS PUB 199, 2004.
15. R. Tourani, S. Misra, T. Mick, and G. Panwar, "Security, privacy, and access control in information-centric networking: A survey", IEEE communications surveys & tutorials, 20(1), 566–600, 2017.
16. J. Lopez and J. E. Rubio, "Access control for cyber-physical systems interconnected to the cloud." Computer Networks 134 (2018): 46–54.

17. A. J. Menezes, P. C. Van Oorschot, and S. A. Vanstone, Handbook of applied cryptography. CRC press, 2018.
18. V. S. Pless, W. Huffman, and R. A. Brualdi, Handbook of coding theory. Vol. 1. Amsterdam: Elsevier, 1998.
19. N. B. Gunti, A. Khatri, and K. Lingasubramanian. "Realizing a security aware triple modular redundancy scheme for robust integrated circuits." In 2014 22nd International Conference on Very Large Scale Integration (VLSI-SoC), pp. 1–6. IEEE, 2014.
20. Y. Yuan, Q. Zhu, F. Sun, Q. Wang and T. Basar, "Resilient control of cyber-physical systems against Denial-of-Service attacks," 2013 6th International Symposium on Resilient Control Systems (ISRCS), San Francisco, CA, 2013, pp. 54–59.
21. C. P. Pfleeger, Security in computing. Pearson Education India, 2009.
22. A. Avizienis, J. Laprie, B. Randell, and C. Landwehr, "Basic concepts and taxonomy of dependable and secure computing." IEEE Trans. Dependable and Secure Computing 1, no. 1, 2004.
23. R. Shirey, Internet security glossary, version 2. RFC 4949, August, 2007.
24. M. Bishop, "Introduction to computer security." (2005).
25. Office of the National Coordinator for Health Information Technology, "Information Security Policy Template," healthit.gov. https://www.healthit.gov/resource/information-security-policy-template (accessed July 17, 2022).
26. M. Bishop, "What is computer security?." IEEE Security & Privacy 1, no. 1 (2003): 67–69.
27. R. Anderson, Security engineering: a guide to building dependable distributed systems. John Wiley & Sons, 2020.
28. C. B. Haley, R. C. Laney, J. D. Moffett, and B. Nuseibeh, "Using trust assumptions with security requirements." Requirements Engineering 11, no. 2 (2006): 138–151.
29. D. G. Firesmith, Common concepts underlying safety security and survivability engineering. No. CMU/SEI-2003-TN-033. Carnegie-Mellon University Pittsburgh, PA Software Engineering Inst, 2003.
30. J. D. Moffett, C. B. Haley, and B. Nuseibeh, "Core security requirements artefacts." Department of Computing, The Open University, Milton Keynes, UK, Technical Report 23 (2004).
31. R. A. Kemmerer, "Cybersecurity." In 25th International Conference on Software Engineering, 2003. Proceedings., pp. 705–715. IEEE, 2003.
32. P. H. Nguyen, S. Ali, and T. Yue, "Model-based security engineering for cyber-physical systems: A systematic mapping study." Information and Software Technology 83, 2017:pp. 116–135.
33. T. Weigert and F. Weil, "Practical experiences in using model-driven engineering to develop trustworthy computing systems," IEEE International Conference on Sensor Networks, Ubiquitous, and Trustworthy Computing (SUTC'06), 2006.
34. B. Schneier, Secrets & Lies: Digital Security in a Networked World (1st. ed.). John Wiley & Sons, Inc., USA, 2000.
35. C. Ericson, "Fault tree analysis – a history," In Procs. Of the 17th International Systems Safety Conference, 2010.
36. P. Brooke and R. F. Paige, "Fault trees for security system design and analysis," Computers and Security 22.3, pp. 256–264, 2003.
37. F. Xie, T. Lu, X. Guo,L. Jingli, Y. Peng, and Y. Gao, "Security Analysis on Cyber-Physical System Using Attack Tree," In Proc. Ninth International Conference on Intelligent Information Hiding and Multimedia, pp. 429–432, 2013.
38. N. Shevchenko, T. A. Chick, P. O'Riordan, T. P. Scanlon, and C. Woody, "Threat modeling: a summary of available methods," Carnegie Mellon University Software Engineering Institute Pittsburgh United States, 2018.
39. E. M. Clarke, E. A. Emerson, and A. P. Sistla, "Automatic verification of finite-state concurrent systems using temporal logic specifications." ACM Transactions on Programming Languages and Systems (TOPLAS) 8, no. 2 (1986): 244–263.

40. M. Y. Vardi, "An automata-theoretic approach to linear temporal logic," Logics for concurrency, 1996, pp.238–266.
41. F. Besson, J. Jensen, D. L. Métayer, and T. Thorn, "Model checking security properties of control flow graphs," J. Computer Security, vol. 9, no. 3, pp. 217–250, 2001.
42. H. Chen, D. Dean, and D. Wagner, "Model checking one million lines of C code," in Proceedings of the 11th Annual Network and Distributed System Security Symposium, San Diego, CA, 2004.
43. B. Kumar, A. Jaiswal, V. S. Vineesh, and R. Shinde, "Analyzing Hardware Security Properties of Processors through Model Checking," 2020:pp.107–112.
44. R. W. Ritchey and P. Ammann, "Using model checking to analyze network vulnerabilities," in Proceedings of the IEEE Symposium on Security and Privacy, May 2000, pp. 156–165.
45. O. Sheyner, J. Haines, S. Jha, R. Lippmann, and J. Wing, "Automated generation and analysis of attack graphs," in Proceedings of the 2002 IEEE Symposium on Security and Privacy, May 2002, pp. 273–284.
46. N. R. Pokhrel and C. P. Tsokos, "Cybersecurity: a stochastic predictive model to determine overall network security risk using Markovian Process," In Journal of Information Security, vol. 8, no. 2, 2017.
47. B. Madan, K. Goševa-Popstojanova, K. Vaidyanathan, and K. Trivedi, "Modeling and quantification of security attributes of software systems," in Proc. Int. Conf. Dependable Systems and Networks, 2002, pp. 505–514.
48. G. J. Holzmann, "The model checker spin," IEEE Trans. Software Eng., vol. 23(5), 1997: pp.279–295.
49. N. B. Henda, "Generic and efficient attacker models in spin." In Proceedings of the 2014 international SPIN symposium on model checking of software, 2014:pp. 77–86.
50. M. Kwiatkowska, G. Norman, and D. Parker, "PRISM 4.0: Verification of Probabilistic Real-time Systems," In Proc. 23rd International Conference on Computer Aided Verification (CAV'11), volume 6806 of LNCS, pages 585–591, Springer, 2011.
51. D. Basin, C. Cremers, C. Meadows, "Model Checking Security Protocols". In: Clarke, E., Henzinger, T., Veith, H., Bloem, R. (eds) Handbook of Model Checking. Springer, Cham, 2018.
52. V. Venkataraghavan, S. Nair, and P. M. Seidel, "Simulation-based validation of security protocols." Proceedings of OPNETWORKS 2002 (2002).
53. D. Apostal, T. Foote-Lennox, T. Markham, A. Down, R. Lu, and D. O'Brien, "Checkmate network security modeling." In Proceedings DARPA Information Survivability Conference and Exposition II. DISCEX'01, vol. 1, pp. 214–226. IEEE, 2001.
54. V. Gorodetski, I. Kotenko, and O. Karsaev, "Multi-agent technologies for computer network security: Attack simulation, intrusion detection and intrusion detection learning." Comput. Syst. Sci. Eng. 18, no. 4 (2003): 191–200.
55. N. B. Shourabi, "A model for cyber attack risks in telemetry networks." International Foundation for Telemetering, 2015.
56. S. Jain and C. R. McLean, "Components of an incident management simulation and gaming frame-work," Simulation, 84(3), 2008.
57. J. Kim and H. Kim, "DEVS-based Modeling Methodology for Cybersecurity Simulations from a Security Perspective," KSII Transactions on Internet and Information Systems, vol. 14, no. 5, pp. 2186–2203, 2020.
58. J. M. Couretas, An Introduction to Cyber Modeling and Simulation, Wiley Series in Modeling and Simulation. Hoboken, NJ: Wiley, 2019.
59. E. Bonabeau, "Agent-based modeling: Methods and techniques for simulating human systems." In Proc. of the National Academy of Sciences 99, no. suppl_3, 2002:pp. 7280–7287.
60. N. Wagner, R. Lippmann, M. Winterrose, J. Riordan, T. Yu, and W. W. Streilein, "Agent-based simulation for assessing network security risk due to unauthorized hardware." In Proceedings of the Symposium on Agent-Directed Simulation, pp. 18–26. 2015.

61. B. P. Van Leeuwen, V. Urias, J. Eldridge, C. Villamarin and R. Olsberg, "Cyber security analysis testbed: Combining real, emulation, and simulation." 44th Annual 2010 IEEE International Carnahan Conference on Security Technology (2010): 121–126.
62. R. Liu and A. Srivastava, "Integrated simulation to analyze the impact of cyber-attacks on the power grid," 2015 Workshop on Modeling and Simulation of Cyber-Physical Energy Systems (MSCPES), Seattle, WA, 2015, pp. 1–6.
63. Y. Geng, Y. Wang, W. Liu, Q. Wei, K. Liu, and H. Wu, "A survey of industrial control system testbeds." In IOP Conference Series: Materials Science and Engineering, vol. 569, no. 4, p. 042030. IOP Publishing, 2019.
64. Object Management Group (OMG), 2020, Unified Architecture Framework version 1.1. [Online]. Available: https://www.omg.org/spec/UAF.
65. J. E. Richards, "Using the Department of Defense Architecture Framework to Develop Security Requirements." (2014).
66. C. Raspotnig, P. Karpati, and V. Katta, "A combined process for elicitation and analysis of safety and security requirements." In Enterprise, business-process and information systems modeling, pp. 347–361. Springer, Berlin, Heidelberg, 2012.
67. International Electrotechnical Commission, "IEC 61882: Hazard and operability studies (HAZOP studies) - Application guide," 2001.
68. Object Management Group (OMG), "The Unified Modeling Language," 2001.
69. Y. Roudier and L. Apvrille, "SysML-Sec: A model driven approach for designing safe and secure systems." In 2015 3rd International Conference on Model-Driven Engineering and Software Development (MODELSWARD), pp. 655–664. IEEE, 2015.
70. D. D. Gajski and R. H. Kuhn, "New VLSI tools." Computer 12 (1983): 11–14.
71. M. Sango, J. Godot, A. Gonzalez, and R. R. Nolasco, "Model-Based System, Safety and Security Co-Engineering Method and Toolchain for Medical Devices Design." In Frontiers in Biomedical Devices, vol. 41037, p. V001T07A001. American Society of Mechanical Engineers, 2019.
72. B. S. Dhillon, "Failure modes and effects analysis—bibliography," Microelectronics Reliability 32, no. 5 (1992): pp. 719–731.
73. C. Schmittner, T. Gruber, P. Puschner, and E. Schoitsch, Security application of failure mode and effect analysis (FMEA), SafeComp, 2014.
74. R. M. Blank, "Guide for conducting risk assessments," National Institute of Standards and Technology, 2011.
75. ANSI/AAMI/ISO 14971:2007, Medical devices – Application of risk management to medical devices. Association for the Advancement of Medical Instrumentation, 2007.
76. Capella, http://www.polarsys.org/capella/ [Accessed 22 Jul 2022].
77. Papyrus/CHESS, https://www.polarsys.org/chess/ [Accessed 22 Jul 2022].
78. Cyber Architect, http://www.all4tec.com/cyber-architect [Accessed 22 Jul 2022].
79. D. Mažeika and R. Butleris, "MBSEsec: Model-Based Systems Engineering Method for Creating Secure Systems." Applied Sciences 10, no. 7 (2020): 2574.
80. A. Uzunov, E. Fernandez, and K. Falkner, "Engineering security into distributed systems: A survey of methodologies," 2012.
81. ISO/IEC 27001. Information Technology - Security Techniques - Information Security Management Systems - Requirements, Technical Report, ISO: Geneva, Switzerland, 2013.
82. OMG Systems Modeling Language. Version 1.5. Available online: https://www.omg.org/spec/SysML/1.5/PDF (accessed on 22 July 2022).
83. S. Woo, H. J. Jo, and D. H. Lee, "A practical wireless attack on the connected car and security protocol for in-vehicle CAN," In IEEE Trans. Intell. Trans. Syst. vol. 16, 2015:pp. 993–1006.
84. M. D. Pesé, J. W. Schauer, J. Li and K. G. Shin, "S2-CAN: Sufficiently Secure Controller Area Network," In Annual Computer Security Applications Conference, pp. 425–438, December 2021.
85. J. N. Luo, C. M. Wu, and M. H. Yang, "A CAN-Bus Lightweight Authentication Scheme." *Sensors* no. 21, p. 7069, 2021.

86. R. Rivest, A. Shamir, L. Adleman, "A Method for Obtaining Digital Signatures and Public-Key Cryptosystems," Communications of the ACM, 21(2), pp. 120–126, Feb. 1978
87. A. Tantawy, s. Abdelwahed, A. Erradi, and K. Shaban, "Model-Based Risk Assessment for Cyber Physical Systems Security," Computers and Security Vol. 96, 2020
88. S. Adepu, and A. Mathur, A. "Generalized attacker and attack models for cyber physical systems," Proceedings - International Computer Software and Applications Conference, vol. 1, pp. 283–292, 2016
89. J. Zografopoulos, X. Ospina, and C. Konstantinou, "Cyber-Physical Energy Systems Security: Threat Modeling, Risk Assessment, Resources, Metrics, and Case Studies," *IEEE Access*, vol. 9, pp. 29775–29818, 2021

Dr. Mark L. McKelvin, Jr., Lecturer, University of Southern California. Dr. McKelvin is a Lecturer in the System Architecting and Engineering graduate program at the University of Southern California, Viterbi School of Engineering where he teaches courses in Model Based Systems Engineering and Systems Engineering Theory and Practice. He is also a Senior Project Lead in Digital Engineering at The Aerospace Corporation. In this role, he serves as the technical authority and the Aerospace team lead for the digital transformation and implementation of Enterprise Systems Engineering across the National Space Systems customer base. Prior to joining the Aerospace Corporation, he led the development of model-based engineering technology and techniques for space system development at the National Aeronautics Space Administration Jet Propulsion Laboratory as a software systems engineer and fault protection engineer on major flight systems. He is a Senior Member of the American Institute of Aeronautics and Astronautics, and he serves on the Board of Directors for the International Council on Systems Engineering, Los Angeles Chapter. He earned a PhD in Electrical Engineering and Computer Sciences from the University of California, Berkeley, and a Bachelor of Science in Electrical Engineering from Clark Atlanta University.

Assistive Technologies for Disabled and Older Adults

Models of Use Cases, Market Economics, and Business Cases

36

William B. Rouse and Dennis K. McBride

Contents

Introduction	1080
Population of Interest	1081
Assistive Technologies	1082
Mobility	1082
Sensory Disabilities	1083
Older Adults	1083
Social Inclusion	1083
Summary	1084
Wearable Coach™	1084
Accessible Driverless Cars	1085
Model-Based Approach	1086
Model Integration	1088
Models of Use Cases	1089
Economic Modeling	1090
Market Economics Model	1091
Economic Projections	1094
Investment Strategies	1095
Models of Business Cases	1097
Discussion	1099
Conclusions	1099
Cross-References	1100
References	1100

W. B. Rouse (✉)
McCourt School of Public Policy, Georgetown University, Washington, DC, USA
e-mail: wr268@georgetown.edu; dkmcbride@vt.edu

D. K. McBride
Intelligent Systems Division, Hume Center for National Security and Technology, Virginia Tech National Security Institute, Blacksburg, VA, USA

© Springer Nature Switzerland AG 2023
A. M. Madni et al. (eds.), *Handbook of Model-Based Systems Engineering*,
https://doi.org/10.1007/978-3-030-93582-5_65

Abstract

Having a sense of purpose, maintaining social connections, and staying mobile are often challenges for disabled and older adults who want to work and/or "age in place" but struggle to perform acceptably, both for employment and for activities of daily life. This chapter focuses on AI-based assistive technologies for cognitive disabilities. Two case studies are reported. First, the concept of a human-centered wearable coach is presented that includes both a job coach and counseling coach. Second, accessibility of driverless cars by disabled and older adults is considered. The human-centered needs of these populations are summarized, including implications for assistive technologies, in terms of use case models. These case studies set the stage for consideration of the economic value of such interventions. Market economics models are employed to estimate potential savings in disability payments and costs of assisted living and nursing home accommodations. Potential revenues are projected for sales of assistive technology products and services. Contributions to GDP by those benefitting from these offerings are also estimated. In the US, with 100 million disabled and older adults, the annual economic value of these offerings is estimated to be between $150 billion and over $1 trillion, depending on assumptions regarding market penetration. Finally, business case models are used to select among alternative investments by automotive OEMs.

Keywords

Use case models · Economic models · Business case models · Disabled adults · Older adults · Driverless cars

Introduction

Health has been said to include having a sense of purpose, maintaining social connections, and staying mobile [11]. This is often a challenge for disabled and older adults who want to work and/or "age in place" but struggle to perform acceptably, both for employment and for activities of daily life. This chapter focuses on AI-based assistive technologies for cognitive disabilities.

We describe two classes of human-centered care. Wearable Coach™ is an intelligent application that assists people to perform tasks, either at work or in the home. It has two modes – job coach and counseling coach. The second class of human-centered applications is intelligent support for accessing and using driverless cars by disabled and older adults. For this example, we present an analysis of their needs and technologies to meet these needs.

The latter half of this chapter addresses three models of these human-centered offerings – use case models, market economics models, and business case models. The use case models articulate the vision for assistive technologies. The market economics models estimate the value these technologies provide, in terms of both cost savings and revenue and profits to providers of products and services. As will be seen, the economic case is compelling. This suggests that the marketplace will be highly motivated to make the investments needed to deliver the great benefits assistive technologies can provide to disabled and older adults. However, the business case models show that prudent investments are warranted.

This chapter proceeds as follows. We next describe the population of interest – 100 million people in the USA, probably more than a billion globally. We then review the nature of assistive technologies, including numerous examples. Our two case studies – Wearable Coach™ and Driverless Cars – are then summarized. Use case models are used to portray the benefits of these concepts. This sets the stage for our market economics model, economic projections, and alternative investment strategies. Business case models address alternative investments by automotive OEMs. We conclude with a discussion of policy issues, including the "who pays?" question for assistive technologies.

Population of Interest

It is essential to begin by describing the population of interest. We are addressing disabled and older adults. Population projections for these two groups are shown in Fig. 1. Note that there is significant overlap between the two groups in that roughly half of people over 65 have one or more disabilities.

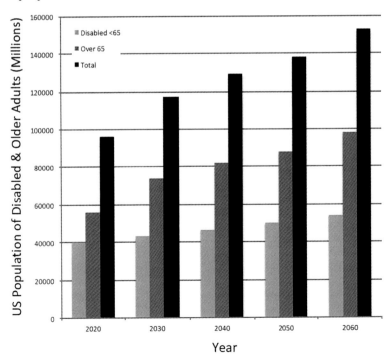

Fig. 1 US population of disabled and older adults

The US Census Bureau [5] and US Centers for Disease Control and Prevention [28] project 60 million people with disabilities in 2020, while Cornell [9] and Pew [30] report 40 million. All four sources agree that 20 million of the disabled in 2020 are older than 65. It appears that this difference is due to how disabilities are reported, e.g., subjective reports versus medical diagnoses. We have employed the larger number as we expect that people who perceive themselves to be disabled, despite not being medically diagnosed as disabled, will nevertheless be attracted to assistive technologies.

Assistive Technologies

Assistive technologies (AT) encompass products or services that are used to enable, maintain, or improve the functional capabilities of people with disabilities and older adults. There is an enormous literature associated with studies of assistive technologies. The NIH [26] reports on types of assistive technologies and how they are used. They provide a useful list of technologies with links to references on each assistive technology. Rouse and Spohrer [39] review intelligent system technologies. Anderson et al. [1] address economic and policy issues associated with autonomous vehicles.

Giampapa et al. [12] review promising technologies and technology gaps for wayfinding and navigation, intelligent transportation systems and assistive technologies, automation and robotics, data integration, and enhanced human service transportation. Millan et al. [22] consider combining brain-computer interfaces and assistive technologies for communication and control, motor substitution, entertainment, and motor recovery. The state of the art and possible future developments are reviewed. Representative examples are briefly reviewed in the remainder of this section.

Mobility

Assistance with mobility is a well-studied area. Garcia et al. [10] review the needs of people with physical, sensory, and cognitive impairments in terms of intelligent services required to access public transit, including stations, stops, and in vehicles. Layton [18] considered barriers and facilitators to community mobility for assistive technology users. Among many specific findings, the most common was the need for universal design, e.g., doors and doorways. Zwald et al. [50] report that "People who used public transportation at least once in the previous week were more likely to meet moderate physical activities recommendations by walking for transportation."

Gray [13] considers whether autonomous vehicles will be accessible to people with disabilities. Needs addressed include designing vehicles and hardware with adaptive uses in mind, creating accessible user interfaces, and providing disabled-friendly car service interfaces. Saripalli [43] argues for "machine learning and AI to enable vehicles to understand spoken instructions, observe nearby surroundings and communicate with people." Poon [31] discusses a smart system, based on IoT technologies, to assure "commuters are exactly where they are supposed to be." Jameson and Monhan [15] address the potential for self-driving wheelchairs for people who do not have use of their hands.

Sensory Disabilities

Technologies to support people with sensory disabilities are also often reported. Nguyen et al. [24] studied patient-reported quality of life outcomes in severely visually impaired individuals using assistive technology. The user wears glasses with a video camera mounted that, when activated, livestreams to a human agent who assists the user in the specified task. Initial results have been very positive. Lersilp et al. [19] assessed technology use by students in special education schools. Students had physical, sensory, or intellectual disabilities. Students with visual disabilities most used AT; those with physical disabilities least used AT. Difficult to use AT were rejected by all students.

Older Adults

With our aging population, assistive technologies for older adults have received increased attention. Shrestha et al. [45] report on the European Union GOAL project (growing older and staying mobile). Issues studied were affordability, availability, accessibility, and acceptability. They provide detailed requirements for five older adult lifestyle scenarios. Schulke et al. [44] address intelligent control of lighting for elderly patients. They address assurance, compensation (aiding, not payment), and assessment and consider principles of non-harm, autonomy, welfare, and equality.

Czarnuch et al. [7] have developed models to predict the levels of independence of people with dementia during 20 activities of daily life. Knowledge of task independence can inform the development of assistive technologies for people with dementia to improve applicability and acceptance. Burleson et al. [6] discuss an assistive technology system that provides personalized dressing support for people living with dementia. The system recognizes clothing and assesses the extent that it is worn and secured correctly, providing feedback when incorrect, e.g., worn backward or inside out.

Social Inclusion

Beyond enabling task performance, including the activities of daily life, there is great interest in the extent to which these technologies enable greater social inclusion. Owuor et al. [29] consider whether assistive technology contributes to social inclusion for people with intellectual disabilities. They are in the process of consolidating evidence on the interaction between intellectual disability, assistive technology, community living, and social inclusion.

Routhier et al. [41] studied the use of mobility assistive technologies to understand how these technologies are used, barriers and facilitators of their use, impact of environmental accessibility, and consequences for work and social engagement. Manzoor and Vimarlund [20] reviewed hundreds of studies of digital technologies for social inclusion of individuals with disabilities. Lack of standardization of terminology made it difficult for them to reach generalizable conclusions.

Summary

This small sampling of AT developments portrays a wealth of inventions, some of which will likely become market innovations. The sample also portrays the enormous diversity of AT offerings and the difficulties of evidence-based assessments. The economic argument presented later in this chapter suggests that there will be ample motivation for this market to mature.

Wearable Coach™

SourceAmerica has developed an approach to assistive technologies to support its 50,000 disabled employees; deployment for older adults is a work in progress [38]. Figure 2 illustrates the overall functioning of a Wearable Coach™ [27]. A beta

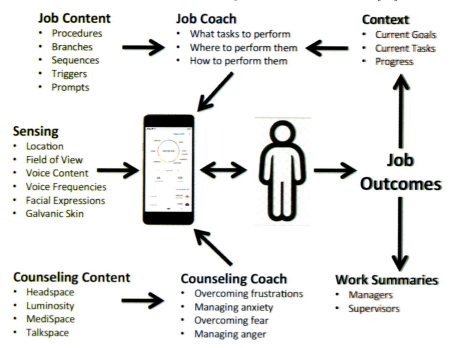

Fig. 2 Functions of the Wearable Coach

version of the Job Coach functionality (for guidance on what tasks to perform, where and how to perform them) has been completed and evaluated; the Counseling Coach functionality (for managing anxiety, anger; overcoming frustration, fear) is in development.

Wearable Coach™ is designed to be worn and, for the most part, used without looking at the screen, for unobtrusive coaching with appropriate prompts and reminders throughout the day. Users typically provide input to their device via gestures or voice. The device "branches" through varying sequences based on information in each sequence combined with user input, time elapsed, time of day, location, downloaded information about what others may have accomplished, and anything else the phone may be able to sense [25]. Information is stored in the cloud, with a copy of relevant information on the user's device. Real-time updates about completed work and productivity, worker whereabouts, or schedule changes are uploaded or downloaded, making them accessible to both the worker and management most of the time.

Typical work sequences and prompts can be maintained in standard libraries, and companies or caretakers can use them to generate custom procedures and prompts. Simplified creation of work or activity sequences makes the system easy and inexpensive to maintain.

The sequence for a workday is buildable in a simplified interface that allows one to compile and nest sequences for a worker's day along with the conditional branching needed to keep the person on track. Schedules can be rearranged in real time during the day to respond to changing needs, with such changes pushed to user devices.

The Commissary Helper is a version of the Wearable Coach™ that has been deployed in beta test mode at some 30 military commissaries worldwide [47] to help employees with disabilities manage grocery store shelves, for example, in terms of scanning expirations, restocking inventory, and retrieving stock from warehouse locations. Initial evaluation of Commissary Helper has been quite positive in terms of worker performance and, of particular importance, worker retention.

It is easy to imagine a version of this coach, perhaps called Grocery Shopping Assistant, being deployed to help older adults navigate in grocery stores, find their desired purchases, and take advantage of specials and discounts. Over time, use of this assistant would result in it also knowing the inventory in the user's cupboards. This could be a key enabler for a cooking assistant.

Accessible Driverless Cars

The Auto Alliance hosted a series of three workshops on "AVs & Increased Accessibility" [2]. We helped plan and conduct these workshops. The first workshop addressed needs of disabled and older adults. The second workshop focused on technologies to meet needs. The third addressed the policy and economic implications of the findings of the first two workshops.

Participants in Workshop 1 suggested a large number of needs. We clustered these needs into 20 categories. Eight categories covered 70% of the suggestions. Definitions of these categories are as follows:

- **Displays and controls** concern information that users can see, hear, touch, etc. and actions they can take.
- **Locating and identifying vehicle** concerns users knowing where their ride is waiting and recognizing the particular vehicle.
- **Passenger profiles** include secure access to information about passengers, in particular their specific needs.
- **Emergencies** concern events inside and outside the vehicle that may require off-normal operations and user support.
- **Adaptation to passengers** involves adjusting the human-machine interface to best support particular users with specific needs.
- **Easy and safe entry and egress** concerns getting into and out of the vehicle as well as safety relative to the vehicle's external environment.
- **Trip monitoring and progress** relates to providing information as the trip proceeds, particularly with regard to route and schedule disruptions.
- **Onboard safety** concerns what happens in the vehicle as the trip proceeds, assuring minimal passenger stress and injury avoidance.

An example mapping from needs to technologies is shown in Table 1. Technologies required include hardware, software, sensing, networks, and especially enhanced human-machine interfaces. Human-machine interfaces need to enable requesting vehicle services, locating and accessing vehicles, monitoring trip progress, and egressing at destinations to desired locations.

The wealth of AT and supporting technologies in Table 1 suggest a substantial need for seamless technology integration to avoid overwhelming disabled and older adults or indeed anybody. We expect that AI-based cognitive assistants may be central to such integration. The question of who might provide which pieces of an overall integrated solution is addressed later.

Model-Based Approach

There are three models involved in our integrated approach to developing market strategies for assistive technologies for disabled and older adults. Figure 3 shows how these three models fit together. We are interested in several questions:

- What does this market want?
- What is the likely economic value of providing it?
- What competitive strategies are most attractive?
- What investments are required to execute chosen strategies?

Table 1 Needs versus technologies

Needs	Technologies				
	Hardware	Software	Sensors	Networks	HMI
Displays and controls	Hardware for displays and controls	Tutoring system for HMI use	Use and misuse of displays and controls	Access to device failure information	Auditory, Braille, haptic, tactile and visual displays
Locating and identifying vehicle	Vehicle-mounted sensors	Recognition software	Integration of sensed information	Sensors of external networks	Portrayal of vehicle and location
Passenger profiles, privacy	Phone or smart phones, tablets	App to securely provide profile information	Recognition of passenger	Access to baseline info. on disabilities	Portrayal to assure recognition
Emergencies	Controls to stop vehicle and move to safe space	Recognition and prediction of situation	Surrounding vehicles, people, and built environ.	External services – police, fire, health	Portrayal of vehicle situation
Adaptation to passengers	Adjusting entry, egress, seating	Learning passenger preferences	Sensing reactions to adaptations	Access to baseline info. on adaptations	Portrayal to enable change confirmations
Easy and safe entry and egress	Sufficient space to maneuver	Capturing data on space conflicts	Surrounding vehicles, people, and built environ.	Networked access to, e.g., bldg. directions	Portrayal of surrounding objects
Trip monitoring and progress	Speedometer, GPS, maps	Predictions of progress, points of interest	Surrounding vehicles, people and built environ.	Access to traffic information, e.g., accidents	Portrayal of trip and progress
Onboard safety	Securement of wheelchairs and occupants	Capturing data on securement conflicts	Sensing and recording safety risks	Access to best practices on safety risks	Portrayal of securement status

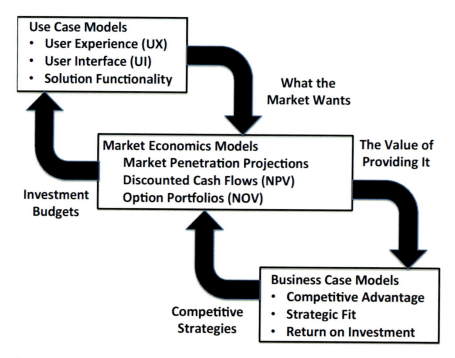

Fig. 3 Integrated model-based approach

In answering these questions, several trade-offs were addressed. What market needs should you choose to target in terms of risks vs. returns? As is later discussed, the greatest economic opportunities were found to be too risky, requiring non-core competencies.

Who will you partner with for total solutions in terms of what revenues and profits you will likely sacrifice to the partners? We found that the most appealing partnership might not happen. Hence a second hedge is formulated to address the risks that preferred partners choose other alliances.

Who will you likely compete with in terms of your competencies vs. their competencies? Competitors' brands may be seen as more innovative, but have no vehicle manufacturing experience. At least one strategy in the portfolio needs to leverage manufacturing competencies.

Model Integration

The three classes of models are integrated as shown in Fig. 3. The use case models yield the functionality that the market – or the mission – requires and desires to accomplish chosen objectives. The list of functions, as well as technologies for providing these functions, emerges from detailed analyses of the use case models once they have been validated by market/mission stakeholders. Later analyses address the uncertainties associated with these needs and technologies.

The market economics models address the likely demands for solutions embodying the functions derived from the use case models. This includes detailed consideration of likely profiles of market penetration by these solutions. These profiles are strongly dependent on assumed offerings by competitors or adversaries. Typical metrics applied to these economic projections include net present value and net option value.

The market economics models project the value of providing what the market/mission wants. Given competitors or adversaries, one cannot assume that one's offerings will capture all of the market. Investments in R&D, production capacities, supply chain capabilities, and marketing and sales will likely be needed to compete. Decisions about these investments require business case models.

Such models can be framed as multi-stakeholder, multi-attribute utility models, where corporate, finance, engineering, production, and marketing typically have different attributes of greatest concern. These stakeholders often usually differ in terms of their appetite for risks, which are reflected in the probabilities or probability density functions associated with projected attribute levels of alternative strategies. Overarching all of this are the utility functions of customers or sponsors in different market segments.

Use of the business case models leads to an initial choice of market strategy or perhaps a portfolio of strategies as our later example illustrates. The initial choice of competitive strategies leads back to the market economics models to revisit assumptions underlying economic projections. Once these projections are refined, the use case models are revisited to determine investment budgets for design and development of the functions the enterprise has chosen to address.

Models of Use Cases

Both of our examples involve task coaching and counseling coaching. The task coaching can be reasonably proceduralized. Counseling when, for example, anxiety or fear has emerged is not so straightforward. We modeled this functionality using stories of humans interacting with AI-based cognitive assistants.

We have employed this technique in several domains including team-based cancer care [36], automotive engineering [35], and this application addressing disabled and older adults [37]. The latter resulted in the story of Fred and Alice.

Fred is a young man with Asperger's who works as a software programmer and uses a driverless car service to get to and from work. Alice is his AI-based cognitive assistant, who is with him regardless of the particular vehicle he is in. She also evolves to help him with organizational issues at work.

This four-page story served to illustrate to stakeholders our vision for how AI would help someone like Fred. Stakeholders included advocates for disabled and older adults, health professionals, and automotive executives. They critiqued the story, made suggestions, and wanted to know when such capabilities would be available.

We have had similar responses from stakeholders in the other domains studied. Story-based models are much more compelling to a wider range of stakeholders than block diagrams and lists of functions. Once a story is finished, we analyze it in detail to determine what the cognitive assistant needs to know and needs to be able to do. Many of the insights in Table 1 resulted from this analysis

Economic Modeling

A key to useful economic modeling is careful assessment of the time series of costs and benefits, especially when considering ambitious and crosscutting initiatives [42]. Broader views are better than narrower views, particularly when investing in people [34]. In this section, we define the central phenomena to be modeled and alternative assessment metrics. In the subsequent section, we outline a model applied to AT investments.

There are several cash flows of interest:

- Savings due to investments
- Income due to investments and taxes paid on this income
- Revenues due to sales of solutions
- Revenues due to sales of services
- Profits resulting from sales and taxes paid on these profits

Estimates of these cash flows are used to define three time series:

- Costs over time
- Revenues over time
- Net of revenues minus costs

Money received in the future is not as valuable as money received today, for example, because one has to borrow money to sustain waiting for future monies. For this reason, future cash flows are discounted by the interest or discount rate to calculate a net present value (NPV).

NPV assumes that the projected costs and revenues occur regardless of intervening circumstances. In many situations, however, the results for the first year or two may cause reconsidering the investment, perhaps even exiting the investment. In such situations, one should consider the first year or two of investment as having purchased an "option" on the subsequent years. In this situation, one should calculate a net option value (NOV).

An investment portfolio may include some investments characterized by NPV, some characterized by NOV, and some with both NPV and NOV. Boer [4] suggests how to value a portfolio that includes some investments characterized by both metrics. He argues for strategic value (SV), which is given by

$$SV = NPV + NOV$$

The NPV component represents the value associated with commitments already made, while the NOV component represents contingent opportunities for further investments, should the options be "in the money" at a later time. We elaborate the distinctions between the two types of investments in the next sections.

A very significant issue in these types of analyses is the distinction between who bears the costs of investments and who realizes the returns on these investments. If it is the same entity, interpretation of the results is fairly straightforward. In contrast, if one entity invests and a different entity realizes the returns, the investing entity will tend to see expenditures as costs and try to minimize them [34].

Finally, health economic analyses often include quality-adjusted life years (QALYs) as an additional metric. In some cases, QALYs are monetized. The technologies discussed early in this chapter will certainly improve quality of life for disabled and older adults. However, we lack data for projecting increases in lifespan due to assistive technologies.

Market Economics Model

There is a range of economic issues associated with the impacts of AT on disabled and older adults. First, consider savings due to more people with disabilities working and increased numbers of older adults aging in place.

Roughly 25% of people with disabilities receive benefits from the Social Security Disability Insurance (SSDI) Program; one in nine also receives benefits from the Social Security Supplemental Security Income (SSI) Program [48, 49]. The average total benefit is $1335 per month. The total for all beneficiaries is $162 billion annually.

Among adults over 65 years old, 1.4% live in assisted living facilities at an average monthly cost of $3600 and 4.2% live in residential nursing homes at an average monthly cost of $8100 [23]. Total annual costs are $281 billion. Medicaid pays roughly half of this amount. Medicare pays for the first 100 days.

The two examples of AT illustrated earlier can enable people with disabilities to work and, therefore, no longer qualify for SSDI and SSI, creating savings for SSA (Social Security Administration). These forms of AT can also enable older adults to age in place and avoid, or at least delay, assisted living or nursing homes, resulting in savings to CMS (Centers for Medicare and Medicaid Services). Figure 4 projects these saving versus percent utilization of these AT capabilities.

The annual total savings range from $44 billion to $222 billion for utilization ranging from 10% to 50%. We do not expect that AT will enable every person with disabilities to work, nor every older adult to avoid assisted living or nursing homes. However, something in the 10–30% range may be achievable.

Next, consider the revenue generated by the AT. People with disabilities, as well as their caregivers now working, will generate income and pay at least Federal Insurance Contributions Act taxes. In 2016, the median earnings of people with

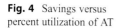

Fig. 4 Savings versus percent utilization of AT

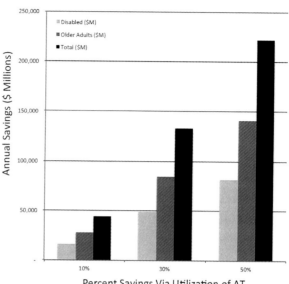

disabilities ages 16 and over in the USA was $22,047, about two-thirds of the median earnings of people without disabilities, $32,479 [17]. At the lower income, 2018 income taxes were 10% plus 7.65% for FICA and Medicare. For the caregiver earning the higher amount, 2018 income taxes were 12% plus 7.65% for FICA and Medicare.

Thus, the disabled person, now employed, paid the Internal Revenue Service (IRS) $3891 (17.65% × $22,047) in taxes, while the caregiver returning to employment paid $6382 (19.65% × $32,479) in taxes. Together they pay $10,273 in taxes. Thus, they were receiving $16,020 (12 × $1335) in benefits and are now paying $10,273, an annual swing of $26,293 due to employment. This would further enhance the projections in Fig. 3, although the savings are for SSA and the gains are for the IRS.

Substantial revenues will come from sales of wearable coaches, vehicle services, vehicle maintenance, and vehicles. Perhaps wearable coach services will be purchased like Amazon Prime at, say, $500 per year, where we assume people purchase their own digital devices. If market penetration grows from 10% the first year to 50% the fifth year, revenues will grow from $1 billion to $5 billion. The profit margins could be substantial for this software as a service (SAAS) type of business. However, these numbers pale in comparison to sales of autonomous vehicles and especially transportation or mobility services.

Table 2 shows how we estimate these revenues. These high-tech vehicles will be expensive, but services dominate revenues because a fare must be paid with each use. Hundreds of millions of trips per week easily add up to billions of annual revenues as shown in Fig. 5.

Table 2 Calculations for vehicle and service revenues ($M)

Variable	Source	Example
Market size (M)	Input	100
Market penetration (%)	Input	50%
Trips per person per week	Input	10
Trips per week (M)	Calculated	500
Trips per AV per week	Input	200
No. of AVs (M)	Calculated	2.5
Life of AV (Y)	Input	2
No. of new AVs/year (M)	Calculated	1.25
Price per AV ($)	Input	$60,000
AV vehicle revenue ($M)	Calculated	75,000
Price per AV trip ($)	Input	$15
AV service revenue ($M)	Calculated	390,000

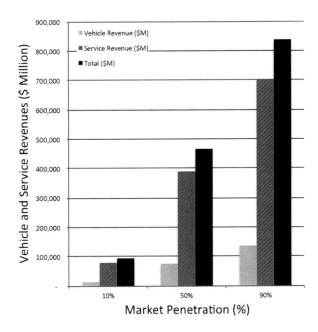

Fig. 5 Vehicle and service revenues versus market penetration

This is clearly an enormous opportunity, with revenue potentially approaching $1 trillion annually if one or more players can successfully address this large underserved market. The successful competitors will earn on the order of $100 billion profits annually, yielding over $20 billion in corporate tax revenues. Further, the $900 billion in costs to a great extent represent salaries and wages that will generate perhaps $200 billion in personal tax revenues. Of course, it is important to keep in mind that it will likely take several years to achieve the levels of market penetration needed to yield these results.

Economic Projections

The foregoing provides a look at the sensitivity of savings and revenues to adoption of coaching and mobility services. In this section, we provide what we feel are prudent projections, including the rationale for these projections. Figure 6 provides projections of market penetration. Coaching services are much easier to adopt than mobility services and much less expensive.

Revenue projections are shown in Fig. 7. We assumed 3% annual inflation and a 5% discount rate. Mobility service revenues follow the formulation in Table 2, adjusted for inflation. Coaching services are assumed to start at $500 per year and grow at 20% per year as a steady stream of upgrades provides enhanced functionality. It is much more difficult to enhance transportation from point A to point B. Indeed, coaching services may be a primary enhancement to mobility services.

In terms of discounted cash flows, the net present values (NPVs) for coaching and mobility services are $332 billion and $1136 billion, respectively. These are impressive numbers, but the key question, at this point, is what investments would be required to secure these results.

Before pursuing this question, it is useful to consider the federal tax revenues that would likely results from these revenues. Assuming current corporate and personal tax rates, the NPV tax revenues would be $66 billion and $227 billion for coaching and mobility services, respectively. These revenues, when combined with the projected savings discussed earlier, make a strong case for societal investment in AT for disabled and older adults.

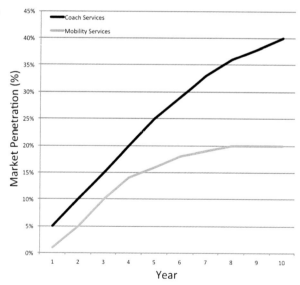

Fig. 6 Market penetration for coaching and mobility services

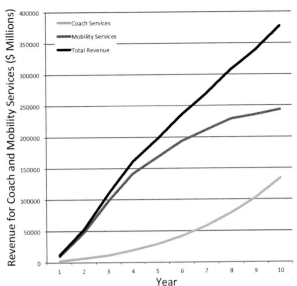

Fig. 7 Projected revenues for coaching and mobility services

Investment Strategies

This opportunity is laced with uncertainties. We are least uncertain about whether people would benefit from coaching and mobility services. We are somewhat uncertain about how long it will take for adoption to become substantial, as reflected in Fig. 6. We are most uncertain about which players will dominate, particularly for mobility services.

Will it be Ford or GM, Uber or Lyft, or Amazon, Apple, Google, or Microsoft – or somebody else? Figure 7 portrays the enormity of the market, but who will dominate. This substantial uncertainty suggests that none of these players should "bet the company" on their assumed success. In this section, we explore a more prudent strategy.

Buying "options" on possible futures makes sense when there is great uncertainty and many competing players. Options can be "purchased" by investing in R&D, alliances, acquisitions, etc. These options can be "exercised," or not, at a later date when uncertainties are better understood. Thus, purchasing an option gives one the right but not the requirement to later exercise the option.

We used option-pricing theory to evaluate possible futures [3, 21]. More specifically, we used real options models to project the value of alternative investments [4, 34]. Options are "real" to the extent they involved tangible assets, e.g., technologies, rather than financial securities.

Our investment portfolio will include positions in three market segments – coaching services, mobility services, and vehicle sales. Figure 8 shows our

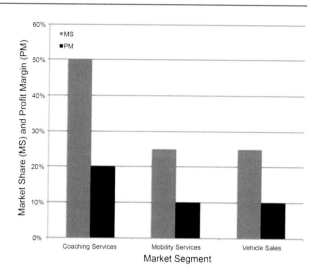

Fig. 8 Market share (MS) and profit margin (PM) objectives

aspirations in each segment. We plan to be most aggressive in coaching services because that is a core competency, not for cognitive assistants in general but for cognitive assistants for disabled and older adults.

We formulated options for each of these investments as follows:

- Coaching Services: Option purchase of $500 million per year for Years 1 and 2; option exercise of $1000 million per year for Years 3 and 4; 50% of the profits from Fig. 7 for Years 5–14.
- Mobility Services: Option purchase of $1000 million per year for Years 1 and 2; option exercise of $4000 million per year for Years 3 and 4; 25% of the profits from Fig. 7 for Years 5–14.
- Vehicle Sales: Option purchase of $500 million per year for Years 1 and 2; option exercise of $2500 million per year for Years 3 and 4; 25% of the profits from Fig. 7 for Years 5–14.

Inputting these parameters into our Black-Scholes call option calculator yielded the results in Fig. 9. Coaching services and mobility services are clearly superior investments to vehicle sales. Of course, the vehicles are needed to enable the mobility services.

Notice that the NPV for vehicle sales is negative while NOV is positive. This is due, in part, to the capitally intensive nature of vehicle development. The NOV is positive, about $1 billion, because the option approach explicitly represents the possibility of exiting the investment (not exercising the option) after Year 2 if technical and/or market uncertainties are unacceptable going forward. As noted earlier, NPV assumes you proceed regardless of these assessments.

If one were an automotive OEM (original equipment manufacturer), the investments in core competencies of vehicle design, development, and manufacturing

Fig. 9 Investments and net present/option values for three segments

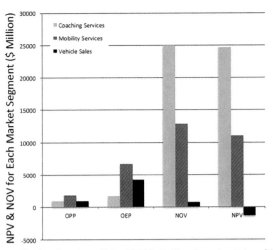

Option Purchase Price (OPP), Option Exercise Prices (OPP), Net Option Value (NOV) & Net Present Value (NPV)

might be key to participating in the other two market segments. On the other hand, a technology company might lead with mobility services, perhaps with one or more vehicle partners.

In summary, at least two of the three market segments are quite attractive. The vehicle sales segment is less attractive, but a necessary piece of the puzzle. Nevertheless, the overall value proposition suggests that disabled and older adults are very likely to have their needs met.

Models of Business Cases

At this point, we know that the market opportunity is very attractive, but also highly uncertain in terms of numerous competitors and when key technologies will be sufficiently mature to achieve significant market penetration.

We developed an integrated framework for addressing uncertain markets with uncertain technologies [40]. This framework embodies the principles of human-centered design [32, 33], built around set-based design [46], Quality Function Deployment [14], design structure matrices [8], and multi-stakeholder, multi-attribute utility theory [16]. The contents of Table 1 can be easily represented in this framework, as well as stakeholders' varying preferences and competitors varying offerings.

The business case of primary concern is how an automotive OEM (original equipment manufacturer) should position itself in this market and the associated investments needed. OEMs are interested in three primary attributes:

- Competitive Advantage (CA): To what extent will the investment of interest enable value-added pricing, reduce production costs, reduce operating costs, and leverage existing capacities?
- Strategic Fit (SF): To what extent will the investment of interest leverage technology competencies, exploit current delivery architectures, complement existing value propositions, exploit current partnerships and infrastructure, and provide other opportunities for exploitation?
- Return on Investment (ROI): What capital expenditures, technology acquisition costs, and labor expenses will be needed? What revenue and profits will likely result?

Building on the philosophy of set-based design, the hypothetical OEM wants to consider five alternative solutions, or scenarios, because each includes a market strategy as well as a solution:

1. Provide total vehicle package.
2. Provide vehicle platform to host intelligent software.
3. Provide vehicle platform to host user-centered HMI.
4. Provide vehicle platform without alliance.
5. Provide integrated mobility services.

These scenarios represent five alternative models of business cases. Based on discussion with automotive executives, utility functions were assessed for CA, SF, and ROI, and probabilities were established based on the following line of reasoning:

- Competitive Advantage: The utility of CA is high if providing total solution, moderate if only providing vehicle; the probability of CA is low without strong partners, not just branding partners.
- Strategic Fit: The utility of SF is high if only providing vehicle, moderate if also providing intelligent software; the probability of SF is high if only providing vehicle, moderate if integrating partners' intelligent software.
- Return on Investment: The utility of ROI is high if providing the total solution, moderate if partnering, low if only providing vehicle; the probability of ROI is low if providing total solution, moderate if partnering or only providing vehicle.

Weightings on the three utility functions were varied in sensitivity analyses. Three scenarios survived this analysis. Scenario 4 was retained as a status quo investment. Scenarios 2 and 3 were retained as hedges. These scenarios are risky, as the cultivation of partners may not succeed. However, the upside is substantial if these options end up "in the money."

With the set of three scenarios chosen, the next step involves more detailed analysis of the functionality in Table 1, including the addition of other stakeholders, e.g., service customers, service providers, etc. Thus, we started with the vision embodied in the use case models, formulated a model of the market economics, and, given that the results were attractive, developed models of business cases to arrive at the set of solutions that next can be addressed in detail.

Discussion

There are a variety of policy implications that need to be addressed. Several issues surround the legal and ethical implications of disabled and older adults availing themselves of coaching and mobility services. How is the validity of the coaching evaluated? What if users follow bad advice? What if users are injured while using mobility services? What training is provided to help users avoid such consequences?

Perhaps the biggest policy issue is payment for these services. The targeted populations have limited abilities to pay the full costs. Such services, albeit much lower tech, are often subsidized by various levels of government. For example, public transit for older adults in Washington, D.C. is one dollar per trip. However, few municipalities could afford the costs underlying the revenue projections shown earlier.

The needed monies exist, but they are not in the right buckets. Savings of $100–200 billion for SSA and CMS are easily imaginable. Tax revenues to the IRS are clearly of the same order of magnitude. But, translating these monies to subsidies for disabled and older adults will be far from straightforward. We expect that legislation will be needed to enable this.

Fortunately, the projections are so attractive that all stakeholders will push to make this opportunity a reality. This is a situation where everybody benefits by proceeding.

Conclusions

The traditional overarching question for investments such as outlined in this chapter is, "Who benefits and who pays?" Clearly, disabled and older adults will be enormous beneficiaries in terms of enhancing their sense of purpose, maintaining social connections, and staying mobile. Assistive technologies will be key enablers of these benefits.

The economy and society will also benefit from addressing this underserved market. Millions of jobs will be created. For example, the AI-based cognitive assistants will need human backups, likely remote, with skills like those of the Geek Squad or Genius Bar as well as abilities to interact with disabled and older adults – an interesting new interdisciplinary job.

Another example stems from driverless cars likely needing daily maintenance. Experts have suggested they will need daily maintenance to calibrate sensors as well as clean these sensors. At the end of each day of use, probably late at night or very early in the morning, each vehicle will find its way back to a maintenance depot for inspection and recalibration. This may be partially automated, but eventually maintaining millions of vehicles per day will require people highly skilled with using this automation and interpreting measurements.

Vehicle and service providers will greatly benefit from this opportunity, but maybe not all of them. As noted earlier, it could be Ford or GM, Uber or Lyft, or Amazon, Apple, Google, or Microsoft – or somebody else. There is enormous uncertainly in this nascent market. Some companies will be big winners, perhaps the ones that best know how to manage the inherent uncertainty.

Who pays? We need a systemic perspective on costs, savings, and revenues. The resulting payers will probably include a mix of employers, government agencies, and individuals. However, a significant portion of people may have limited abilities to contribute. It seems reasonable to argue that a portion of savings and revenues should contribute to payment, perhaps only indirectly, e.g., via taxes.

Somewhat simplistically, the bottom line is 100 million people see enhanced quality of life; perhaps 1 million new jobs are created, and up to $1 trillion is added to the annual GDP. Yet, a prudent automotive OEM should not bet the company on this opportunity. The models discussed in this chapter enable evidence-based formulation of a portfolio of traditional investments and options that hedge the contingencies of this market.

Cross-References

▶ Model-Based Human Systems Integration
▶ Role of Decision Analysis in MBSE

References

1. J.M. Anderson, N. Kalra, K.D. Stanley, P. Sorenson, C. Samaras, and O.A. Oluwatala, *Autonomous Vehicle Technology: A Guide for Policymakers*. Santa Monica, CA: RAND Corporation, 2016.
2. Auto Alliance. *Assessing Transportation Needs of People With Disabilities and Older Adults: Report of Workshop 1*, Washington, DC: Auto Alliance, 2019.
3. F. Black and M. Scholes, "The pricing of options and corporate liabilities," *Journal of Political Economy*, 81, 637–659, 1973.
4. F.P. Boer, *The Valuation of Technology: Business and Financial Issues in R&D*. New York: Wiley, 2008.
5. M.W. Brault, *Americans With Disabilities: 2010*. Washington, DC: US Census Bureau, 2012.
6. W. Burleson, C.Lozano, V. Ravishankar, J. Lee, and D. Mahoney, "An assistive technology system that provides personalized dressing support for people living with dementia," *JMIR Medical Informatics*, 6 (20), e21, 2018.
7. S. Czarnuch, R. Ricciardelli, and A. Mihaildis, "Predicting the role of assistive technologies in the lives of people with dementia using objective care recipient factors," *BMC Geriatrics*, 16 (143), 2016.
8. S.D. Eppinger and T.R. Browning, *Design Structure Matrix Methods and Applications*. Cambridge, MA: MIT Press, 2012.
9. W. Erickson, C. Lee, and S. von Schrader, *2017 Disability Status Report: United States*. Ithaca, NY: Cornell University Yang-Tan Institute on Employment and Disability, 2019.
10. C.R. Garcia, A. Quesada-Arencibia, T. Cristobal, G. Padron, R. Perez, and F. Alayon, "An intelligent system proposal for improving the safety and accessibility of public transit by highway," *Sensors*, 15, 20279–20304, 2015.
11. A. Gawande, *Being Mortal: Medicine and What Matters in the End*. New York: Metropolitan Books, 2014.

36 Assistive Technologies for Disabled and Older Adults

12. J.A. Giampapa, A. Steinfeld, E. Teves, M.B. Dias, and Z. Rubinstein, *Accessible Transportation Technologies Research Initiative (ATTRI)*. Pittsburgh, PA: The Robotics Institute, Carnegie-Mellon University, 2017.

13. L. Gray, "Will autonomous vehicles be accessible to people with disabilities?" *Shared-Use Mobility Center*, October 31, 2017.

14. J.R. Hauser and D. Clausing, "The house of quality," *Harvard Business Review*, 63–73, May–June, 1988.

15. E. Jameson and C. Monhan, "Self-driving care tech can help another transport: Wheelchairs," *Transportation*, November 18, 2017.

16. R.L. Keeney and H. Raiffa, *Decisions with Multiple Objectives: Preference and Value Tradeoffs*. Cambridge, UK: Cambridge University Press, 1993.

17. L. Kraus, E. Lauer, R. Coleman, and A. Houtenville, *2017 Disability Statistics Annual Report*. Durham, NH: Institute on Disability. University of New Hampshire, 2018.

18. N. Layton, "Barriers and facilitators to community mobility for assistive technology users," *Rehabilitation Research and Practice*. https://doi.org/10.1155/2012/454195, 2012.

19. S. Lersilp, S. Putthinoi, and N. Chatpitak, "Model of providing assistive technologies in special education schools," *Global Journal of Health Science*, 8 (1), 2016.

20. M. Manzoor and V. Vimarlund, "Digital technologies for social inclusion of individuals with disabilities," *Health and Technology*, 8, 377–390, 2018.

21. R.C. Merton, "Theory of rational option pricing," *Bell Journal of Economics and Management Science*, 4 (1), 141–183, 1973.

22. J.D.R. Millan, R. Rupp, G.R. Muller-Putz, R. Murray-Smith, C. Giugliemma, M. Tangermann, C. Vidaurre, F. Cincotti, A. Kubler, R. Leeb, C. Neuper, K-P. Muller, and D. Mattia, "Combining brain-computer interfaces and assistive technologies: State-of-the-art and challenges," *Frontiers in Neuroscience*, https://doi.org/10.3389/fnins.2010.00161, 2010.

23. NCHS, *Long-Term Care Providers and Service Users in the United States, 2015–16*. Atlanta, GA: Centers for Disease Control and Prevention. Office of Vital and Health Statistics, 2019.

24. B.J. Nguyen, Y. Kim, K. Park, A.J. Chen, S. Chen, D. Van Fossan, and D.L. Chao, "Improvement in patient-reported quality of life outcomes in severely visually impaired individuals using the Aira assistive technology system," *Translational Vision Science & Technology*, https://doi.org/10.1167/tvst.7.5.30. 2018.

25. D. Nield, "All the sensors in your smart phone, and how they work," *Gizmodo*, July 23, 2017.

26. NIH, Types of assistive technologies and how they are used. *Health Topics – Rehabilitation Technology*. Bethesda, MD: National Institute of Child Health and Human Development, 2019.

27. P. Nishman, *The Wearable Coach*. Vienna, VA: SourceAmerica, 2018.

28. C.A. Okoro, N.D. Hollis, A.C. Cyrus and S. Griffin-Blake, Prevalence of Disabilities and Health Care Access by Disability Status and Type Among Adults – United States, 2016. *MMWR Morbidity Mortality Weekly Report,* 2018.

29. J. Owuor, F. Larkan, B. Kayabu, G. Fitzgerald, G. Sheaf, J. Dinsmore, R. McConkey, M. Clarke, and M. MacLachlan, "Does assistive technology contribute to social inclusion for people with intellectual disability? A systematic review protocol," *BMJ Open*, 8, e017533, 2018.

30. Pew Research Center, *7 Facts About Americans with Disabilities*. https://www.pewresearch.org/fact-tank/2017/07/27/7-facts-about-americans-with-disabilities/. Accessed 06/24/19, 2017.

31. L. Poon, "Designing a smart city transit systems for commuters with disabilities," *CityLab*, September, 2016.

32. W.B. Rouse, *Design for Success: A Human-Centered Approach to Designing Successful Products And Systems*. New York: Wiley, 1991.

33. W.B. Rouse, *People and Organizations: Explorations of Human-Centered Design*. New York: Wiley, 2007.

34. W.B. Rouse, Ed., *The Economics of Human Systems Integration: Valuation of Investments in People's Training and Education, Safety and Health, and Work Productivity*. New York: John Wiley, 2010.

35. W.B. Rouse, "AI as system engineering: Augmented intelligence for systems engineers," *Insight*, 23 (1), 54–56, 2020.
36. W.B. Rouse and M.M.E Johns, *Clinicians With Cognitive Assistants*. Atlanta, GA: School of Medicine, Emory University, 2018.
37. W.B. Rouse and D.K. McBride, *Fred and Alice*. Vienna, VA: SourceAmerica, 2018.
38. W.B. Rouse and D.K. McBride, "A systems approach to assistive technologies for disabled and older adults," *The Bridge*, 49 (1), 32–38, 2019.
39. W.B. Rouse and J.C. Spohrer, "Automating versus augmenting intelligence," *Journal of Enterprise Transformation*, https://doi.org/10.1080/19488289.2018.1424059, 2018.
40. W.B. Rouse, D. Verma, D.S. Lucero, and E.S. Hanawalt, *Strategies for Addressing Uncertain Markets and Uncertain Technologies*. Hoboken, NJ: Systems Engineering Research Center, Stevens Institute of Technology, 2020.
41. F. Routhier, W.B. Mortenson, L. Demers, A. Mahmood, H. Chaudhury, K.A. Martin Ginis, and W.C. Miller, "Mobility and participation of people with disabilities using mobility assistive technologies: Protocol for a mixed methods study," *JMIR Research Protocols*, 8 (40), e12089, 2019.
42. A.P. Sage and W.B. Rouse, *Economic System Analysis and Assessment*. New York: Wiley, 2011.
43. S. Saripalli, "Are self-driving cars the future of mobility for disabled people?" *The Conversation*. October 5, 2017.
44. A.M. Schulke, H. Plischke, and K.B. Kohls, "Ambient assistive technologies (AAT): Socio-technology as a powerful tool for facing the inevitable sociodemographic challenges," *Philosophy, Ethics & Humanities in Medicine*, 5 (8), 2010.
45. B.P. Shrestha, A. Millonig, N.B. Hounsell, and M. McDonald, "Review of public transport needs of older people in European context," *Population Aging*, 10, 343–361, 2017.
46. D.K. Sobek, A.C. Ward, and J.K. Lifer, "Toyota's principles of set-based concurrent engineering," *Sloan Management Review*, 40 (2), 67–83, 1999.
47. SourceAmerica, *Commissary Helper: Installation and User Guide*. Vienna, VA: SourceAmerica, 2018.
48. SSA, *Annual Statistical Report on the Social Security Disability Insurance Program, 2017*. Washington, DC: Social Security Administration, Office of Retirement and Disability Policy, 2018.
49. SSA, *Annual Statistical Report on the Social Security Supplemental Security Income Program, 2018*. Washington, DC: Social Security Administration, Supplemental Security Income Program, 2019.
50. M.L. Zwald, J.A. Hipp, M.W. Corseuil, and E.A. Dodson, "Correlates of walking for transportation and use of public transportation among adults in St. Louis, Missouri in 2012," *Preventing Chronic Disease*, 11, July 3, 2014.

William B. Rouse is a Research Professor in the McCourt School of Public Policy at Georgetown University, as well as a Senior Fellow in the office of the Senior Vice President for Research and Professor Emeritus and former Chair of the School of Industrial and Systems Engineering at the Georgia Institute of Technology. His research focuses on mathematical and computational modeling for policy design and analysis in complex public-private systems, with particular emphasis on healthcare, higher education, transportation, and national security. Recent books include *Transforming Public-Private Ecosystems* (Oxford, 2022), *Failure Management* (Oxford, 2021), and *Computing Possible Futures* (Oxford, 2019). He is a member of the National Academy of Engineering and fellow of IEEE, INCOSE, INFORMS, and HFES. Rouse received his B.S. from the University of Rhode Island, and his S.M. and Ph.D. from MIT.

Dennis K. McBride is a Research Professor at Hume Center for National Security and Technology, Intelligent Systems Lab, at Virginia Tech. He is President Emeritus at the Potomac Institute for Policy Studies and served for 5 years as VP Strategy and Innovation at SourceAmerica, creating jobs for people with disabilities. Dennis retired as Navy Captain after completing a 20-year career as an engineering psychologist, specializing in the development of human-machine systems at five advanced labs, with two tours as program manager at the Defense Advanced Research Projects Agency (DARPA). His awards include the Legion of Merit and Superior Service Medals. Dennis has served as a Full Professor in colleges of engineering, arts & sciences, policy, and medicine. He earned the B.S., M.S., and Ph.D. from the University of Georgia, an MPA/MSPA from Troy State University, and an M.S. from the Viterbi School of Engineering, University of Southern California.

Multi-model-Based Decision Support in Pandemic Management

37

A. M. Madni, Norman Augustine, C. C. Madni, and Michael Sievers

Contents

Introduction	1106
State of the Art	1110
Ontologies	1110
SIR Model and Pandemics	1111
Best Practice Approach	1114
Illustrative Examples	1124
Chapter Summary	1129
References	1129

Abstract

The ongoing COVID-19 pandemic has drawn the world's attention to several shortcomings in the planning and decision-making infrastructure, testing facilities, and other pandemic response capabilities and measures in our society. These deficiencies have surfaced in the United States as the country continues struggling to bring the pandemic under control. The pandemic has been analyzed thus far from multiple isolated viewpoints with each analysis being limited in scope. Even so, these analyses have increased our understanding of the pandemic. However, to inform and guide decision-making at multiple levels require a more holistic approach that accounts for interdependencies among various aspects of a pandemic response. From a holistic perspective, a pandemic can be viewed as a

---

A. M. Madni (✉) · M. Sievers
University of Southern California, Los Angeles, CA, USA
e-mail: azad.madni@usc.edu; michael.sievers@usc.edu

N. Augustine
Advisory Services, New York, NY, USA
e-mail: norm.augustine@lmco.com

C. C. Madni
Intelligent Systems Technology, Inc., Los Angeles, CA, USA
e-mail: cmadni@intelsystech.com

© Springer Nature Switzerland AG 2023
A. M. Madni et al. (eds.), *Handbook of Model-Based Systems Engineering*,
https://doi.org/10.1007/978-3-030-93582-5_67

complex system-of-systems (SoS) comprising a large number of interdependencies and inter- and intra-system interactions. For such a SoS, the lack of integrated models adversely impacts decision-making leading to suboptimal policy choices that are influenced by economic, political, and social pressures rather than evidence and logic. As a result, management decisions tend to be short term and late, producing over-reactions that can trigger change cascades often resulting in unintended consequences. Complicating matters further are conflicting reports on the origin, diffusion rate, and lethality of COVID-19. The absence of planning, training, testing of vaccines (preventive measure), and proven medication (reactive measure) further exacerbate the effects of the pandemic, especially in the near term. Against this backdrop, this chapter presents an integrated model-based systems approach for decision support of policy-makers and other decision-makers in pandemic management at the federal, state, and community levels. This chapter emphasizes the need for multiple complementary modeling techniques that address different aspects of a pandemic, explicate assumptions, illuminate interdependencies and uncertainties, and show an illustrative dynamic pandemic systems model. Examples of promising complementary modeling and simulation approaches are presented in the context of decision support for pandemic management. The chapter concludes with a few key insights and recommendations and cautions: holistic thinking cannot be abandoned when taking steps for alleviating local and immediate pandemic concerns; given the complexity of a pandemic, multiple, complementary, and integrated modeling and simulation techniques are needed; and decision support based on modeling and simulation support explore alternate futures and compare outcomes with different assumptions (e.g., technologies, regulations, funding, other constraints); the models will be "living models" that continually grow and change in light of new evidence; and the best use of modeling and simulation is for discerning trends and making comparative evaluations of candidate courses of action.

Keywords

Model-based methods · COVID-19 · Pandemic decision support · System dynamics · Domain ontology

Introduction

The ongoing COVID-19 pandemic is extracting a significant health and economic toll throughout the world. It has caused a global public health crisis that has taken lives and also triggered social upheaval, disrupted business operations and supply chains in most industries, tanked economies, and dramatically altered work patterns across the globe. Employees today must work from home or when at work must abide by social distancing rules, shift work, and other required at-work behavior. In-person meetings, conferences and conventions, sporting events, live performances, restaurant dining, trips to the dentist, watching movies in a theater, visiting friends, classroom teaching, and most of what was once part of our daily lives are now off-limits or highly restricted. People who are suspected of infection or who have had positive tests are required to self-isolate for 14 days [8]. Healthcare workers, grocery store clerks, factory workers, and others who are not provided adequate protection inside the workplace are increasingly contracting and dying from the virus. These factors when combined with confusing news stories, inadequate testing and contact tracing, and disregard of best practices by some of the public are leading to increasing number of business closures, growing unemployment, and increasing numbers of infections and deaths. Shortage of food and essentials and hoarding behavior are further complicating an already difficult situation. The population is becoming increasingly fearful and frustrated.

Complex sociotechnical systems problems such as pandemics defy initial understanding but are better understood with the passage of time. In such highly dynamic and complex systems, it is difficult to understand whether actions taken produce desired results or cause a chain reaction leading to undesirable outcomes, something akin to the Swiss cheese concept when bad outcomes go undetected and eventually result in failure [26, 34]. A related notion is that an action might lead to a local (or short-term) undesirable outcome but ultimately produces a global positive result. i.e., some part of society might need to suffer for the good of the whole. Sometimes we need to take the best of only bad options. A good model and simulation can help find the best of only bad options.

The COVID-19 pandemic has exposed the lack of systems thinking in responses that appear short term and narrowly focused on perception of local health crises and less on other factors such as the economy, food supply, psychological health, need for social interaction, and so forth. A proper response is one that balances short-term actions against long-term impacts while recognizing that short-term and long-term goals differ. Equally important are understanding and collecting the metrics needed for assessing progress at all societal levels [29, 31, 35].

Determining effective responses is difficult because of the need to continually balance costs, risks, and benefits while recognizing that there is no single, perfect strategy. For example, each strategy is likely to impact certain segments of society more than others in a positive or negative way [18]. For example, consider a strategy that calls for overly severe restrictions extending over long periods of time. This approach might save lives initially but, in the process, is likely to bankrupt businesses and farms resulting in unemployment, extreme poverty, and possibly an increase in death rate. Conversely, a strategy that does not impose early restrictions can mean significant numbers of initial deaths but might allow survival of businesses and farms that might then hire the survivors of the pandemic.

Achieving the right balance requires holistic understanding of interdependencies, feedback loops, and multi-level change cascades. It also requires technical understanding of variable forcing functions, probability distributions, causal chains, partial observability (of the environment and state space), and emergent behaviors. While it is impracticable or impossible to simultaneously deal with all of these complexities at a micro level, reasonable and reliable macro understanding of SoS dynamics is achievable using an appropriately constructed Model-Based Systems Engineering (MBSE) framework comprising a SoS network of tightly and loosely interconnected pandemic entities. The latter entities represent parameters, functions, and behaviors such as delays in decision-making and implementing actions, detection thresholds, fraction of infected population, duration of asymptomatic infections, fraction of infectious population, propensity for re-infection, duration of confinement, imposition and relaxation of social distancing constraints, availability of protective equipment, availability of intensive care facilities and staff, trust and acceptance of medical professionals, and so forth. Along with other researchers [Bradley et al. [7]], we concluded that addressing a problem of this magnitude and complexity requires a systems approach. We further concluded that we need a resilient response system for pandemic avoidance, containment, and suppression. A resilient response in this context implies the ability to adapt to the disruption based on incoming data by reallocating resources, rescheduling activities, and reconfiguring and reorganizing teams.

Currently, the pandemic is managed country-by-country and state-by-state with little or no coordination. There is little, if any, acknowledgment or understanding of the impact that one nation's approach or one locality's approach has on its neighbors. States and cities largely must work out their own way through the pandemic.

This situation has brought to light several shortcomings in pandemic planning and decision-making infrastructures at all levels in society [22]. In the United States, lack of a coordinated, national, blueprint for effective response has often resulted in state and local authorities making ill-considered decisions that not only do not produce desired outcomes but, worse, trigger serious side effects [19]. These are the characteristics of "wicked problems" – i.e., problems that morph in response to interventions only to create new and potentially even more difficult problems.

This chapter reviews the state of the art in pandemic management and then presents a complex system-of-systems (SoS) framework and modeling approach that provides decision support for policy-makers and other responsible personnel in pandemic management. The overarching pandemic modeling problem is made more tractable through problem decomposition and subsequent allocation of subproblems to interoperating SoS systems and subsystems. A pandemic management ontology is used that informs SoS dynamics and discrete event simulations paired with machine learning and data analytics. An illustrative prototype model demonstrates the viability of model-based systems approaches to pandemic response tradespace analyses. Even in cases where explicit optional decision choices cannot be derived, such analyses provide an important means of determining sensitivities useful for informed human judgment.

Key Concepts. It often takes a catastrophic event such as the outbreak of a pandemic to jar society out of complacency. This has been the case with the COVID-19 outbreak [19, 43, 44]. COVID-19 is currently posing a multi-faceted threat and a challenge that cuts across human and societal behavior, technology, social and economic factors, human decision-making in the face of uncertainty and risk, political factors that influence decision-making and population behavior, and cultural norms. With misinformation and disinformation rampant, for many, telling fact from fiction has become difficult. This is exacerbated, particularly in the science arena, by the newness of the virus and resulting limited medical options and in the political arena by the desire for returning to normalcy. The recognition of these complexities provides the impetus for developing a decision support framework that enables performing "what-if" tradespace explorations in which options can be explored and the results and sensitivities evaluated. This chapter discusses a framework that allows evaluating relative consequences of decisions and assumptions. Whether optimal solutions exist is highly dependent on what stakeholders want optimized. Moreover, the notion of optimality is likely to be a temporal and highly fluid.

Tackling planning and decision-making in pandemic management requires a multi-model approach. A multi-model approach involves multiple models with each model addressing a specific perspective and employing distinct reasoning and simulation strategies. Oren [33] states that a multi-model comprises multiple sub-models, only one of which is invoked at any point in time. However, for a pandemic, all sub-models need to be active. This is a key difference.

Table 1 summarizes the benefits of multi-modeling approach for decision support in pandemic management. The real benefit of this approach is that a dynamic simulation that cuts across multiple models can reflect the effect of changes in one sub-model to other sub-models in the multi-model approach.

In light of the foregoing, the key capabilities of a pandemic decision support framework include:

- *Planning:* Critical to detecting and arresting the spread of the pandemic. Includes stockpiling critical resources, response training, and public information processes and resources. *Decision-making:* Processes and means for coping with rapidly changing conditions in an uncertain environment.
- *Dynamic Systems Modeling and Prediction:* Core capability for evaluating social, economic, infection trends, morbidity, mortality, resources, fear, impacts of virus spread, and candidate mitigations [2, 15, 25, 27, 28, 42].

Table 1 Benefits of pandemic modeling using multi-model approach

Explain behaviors, surprises
Guide people, processes, data collection
Illuminate dependencies, uncertainties, conflicting/contradictory assumptions
Circumscribe/bound outcomes, plausible ranges
Challenge conventional wisdom, theories, beliefs
Facilitate exploration and comparison of alternate futures
Inform and educate decision-makers and the public
Expose deficiencies in prevailing wisdom in light of new evidence
Help pre-plan responses using "what-if" simulations
Help develop socially aware and temporally sensitive interventions
Help discern and understand patterns and trends

- *Tradespace Exploration:* Enables comparing outcomes (i.e., alternate futures) under different assumptions and interventions.
- *Infection model:* Statistical representation of virus transmissibility and severity [5, 46]. Transmissibility pertains to the ease with which the virus from one person can infect another. This characteristic determines how rapidly a pandemic might spread, when it is likely to peak, its overall magnitude when it peaks and its duration. Severity determines the number of infected people who are likely to become seriously ill or die.
- *Response Timeliness:* Related to planning, evaluates the impact of response delays. Since a response plan and decision-making are highly dependent on real-time situational awareness, gathering the right information as soon as possible in a pandemic for analysis and response generation is key. Timely identification of the pandemic scenario that best characterizes a pandemic outbreak can dramatically accelerate decision-making. Viewing historical pandemic outbreaks as cases, best match pandemic scenarios are determined based on pandemic patterns (e.g., triggering condition, contextual variables, decisions made).

The paper is organized as follows: key concepts and definitions; role of ontologies in reasoning and answering questions; use of complementary models in developing a decision support system for pandemic management; illustrative system dynamics simulation for analyzing infection spread and population impacts; and summary of key points made in the paper along with future directions.

State of the Art

Over the last decade and a half, several advances have been made in model-based methods that provide decision support of pandemic management. In particular, ontologies, simulations, and susceptibility, exposure, infection, and recovery (SEIR) models (Bjørnstad [6]) featured prominently in these approaches.

Ontologies

Ontologies define a domain-specific framework for specifying a controlled, shared set of terms and relationships [9, 12, 20, 21]. Rigorously expressed ontologies are the basis of high-quality models and database schemas that comprise interconnected and coherent content. Modelers and researchers depend on ontologies for multi-disciplinary communication, building models, developing simulations, saving and retrieving results and content, calibration, and validation [14]. Many excellent ontologies have been developed for exploring infection, recovery, and treatment aspects of the pandemic domain. Figure 1 shows a snippet of a typical pandemic ontology focused on disease spread and intervention.

A more complete example is the curated, human disease ontology (https://disease-ontology.org, https://disease-ontology.org/media/images/tutorial/DO_Tutorial_Trees.pdf) developed for the biomedical community. The ontology is organized

Fig. 1 Partial OWL disease ontology

in two views: an "is-a" graph of disease terms based on etiology using Open Biological and Biomedical Ontologies (OBO) and an Ontology Web Language (OWL) view that supports disease and content queries and reasoning within the disease ontology.

Other examples include the Coronavirus Infectious Disease Ontology (CIDO) which was developed for big data capture and analysis of coronavirus diseases. CIDO includes phenotypical, host, infection susceptibility, disease process, diagnosis, vaccine, and drug effectivity [20]. An in-pandemic application ontology for COVID-19 was developed for monitoring disease spread, treatment effectiveness, and capturing drug trial data [9]. A Protégé ontology including representations for community, disease, population, and simulation has also been described (https://pdfs.semanticscholar.org/b8d1/aa76d1996b1bf45f5287bdc80b7d6f175852.pdf).

SIR Model and Pandemics

Consider an infectious disease that spreads in a population comprising a certain number of individuals [40]. At any point in time, an individual can be in one of three states: infected and likely to infect others; healthy but susceptible to infection; and recovered and temporarily or permanently immune to the disease. Infection results from contact between an infected and susceptible individual. These assumptions lead to the classical SIR model. It is important to note that the SIR model is necessarily an approximation of the real-world system because the model makes simplifying assumptions. The art of mathematical modeling is reflected in the ability to make simplifying assumptions to achieve just enough fidelity in the models. The level of fidelity will typically vary among the different models. The goal is to have models which are simple enough to allow rapid exploration of outcomes either through software simulations or mathematical methods and yet retain the requisite fidelity to make realistic assessments and comparative predictions about the real-world system within defined bounds. For example, a policy-maker might be interested in understanding how the population of infected people change over time. The system modeler needs to determine which aspects of the real-world system the model should capture to answer questions and enable comparative evaluations. Modelers likely will need a list of competency questions prior to starting a model. Typically, these are domain questions that will inform what is needed in a model and the fidelity of model elements. There are a number of more recent variants of SIR such as SEIR (E = exposed), models that include delays, vaccination models, multiple infection stages, etc.

SIR models employ detail that varies from one disease to the next. In actual modeling, these details are inferred from the available data, and the model is constructed by deriving suitable assumptions from the data. Examples of such assumptions are:

- An infected person will be infectious for a finite time period with a specific mean value and a small standard deviation
- The time between infection and becoming infectious is negligibly small
- After ceasing to be infectious, a person will remain immune to the disease for a finite period
- Contact between an infected and susceptible person during the time interval of infectiousness will result in a new infection with a specific probability

Pandemic models are central to understanding pandemic spread and impact [4, 17, 30]. These models comprise epidemic and social network components, epidemiological data, geographic scope and location parameters, and ongoing interventions. The *epidemic model* comprises susceptibility, incubation times, morbidity (symptomatic and asymptomatic cases), transition probabilities and rates, infection rates, timeframes, immunity rates, morbidity and mortality by age, death rates, population compliance with medical recommendations, and recovery statistics. The *social network* model is geo-referenced. It focuses on factors affecting the pandemic. These include population density, poverty rates, racial composition, and pre-existing condition statistics. It also includes factors such as households (e.g., number, populations in each household), workplaces (workplace mapping, hospitals, schools, hotels), schools (e.g., school assignments), communities (e.g., hotels), group quarters (e.g., dormitories, barracks, prisons, nursing homes), intervention model (e.g., key events, key actions taken), and study design.

The study design includes initial conditions (initial infection location), infection, calibration parameters (outputs to calibrate, historical data, calibration targets), and output variables (economic effects). For SIR-type models, these are factors of infection and recovery.

Simulations

Simulation is an indispensable tool for addressing the complexity of pandemic management [12, 37, 41]. Simulations can evaluate intervention consequences that result from geographic, logistical, social, and cultural factors. An effective simulation allows rapid exploration of changes in assumptions and decisions on pandemic outcomes and answer questions at the right level of detail. Simulation models need to be transparent, readily modifiable, extensible, and flexible with facilities for interchanging disease models, locations, and community models (data-driven, randomized). The simulation needs to be based on a flexible, extensible, and scalable computational framework with multi-perspective, multilevel visualization facilities. Examples of simulation models are community model, disease model, intervention model, and population model. Simulations also employ a variety of algorithms (e.g., a Matlab or Python script to perform numerical analysis). Simulations also employ an intermediate specification language such as XML and different types of simulator engines (e.g., discrete event, agent-based). The models can be captured in software tools such as Protégé with extensions as needed. The simulation engine can be discrete event, agent-based, or system dynamics model-driven. In particular, the simulation architecture of a pandemic should maintain a clear separation between model and implementation. In this regard, ontologies provide a suitable means to circumscribe the scope of the model and an effective representation scheme for epidemiological models. During an actual pandemic, ontology-enabled models are ideally supported by a scenario library comprising a collection of pandemic scenarios and scenario instances. The modeling and simulation environment implementation can be cloud-based.

Preparedness plans and simulation experiments require hypothetical pandemic outbreak scenarios. The development of a scenario requires a strong evidence base for pandemic response planning. In other words, any planning tool should be capable of dealing with past experience as a test case. However, it is likely that the "next" pandemic will not match a previous pandemic because whatever the previous pandemic was has been addressed by vaccines or treatment. In this regard, both mathematical and behavioral models are used to guide response generation to disease outbreaks and support policy decision-makers. Policy-makers and other decision-makers with a stake in pandemic preparedness planning require flexible models for estimating and comparing the intervention effects. Such mathematical tools must easily accommodate different outbreak scenarios and address yearly differences in virus transmission and virulence, lack of understanding of factors affecting the spread of the virus, and lack of feedback on effectiveness of interventions. There are a number of publicly available simulation tools that can be used by policy-makers to model pandemic scenarios in which a virus is transmitted in a population and the problem is to estimate the damage caused by the disease (e.g., number of hospitalizations, number of deaths) in the presence of different disease parameters. Simulation software companies offer public models (e.g., AnyLogic offers Agent Based Epidemic Model, Flattening the Curve at Family Level, Epidemic and Clinic with Accumulation Concerns in 2018; GLEAMviz offers an opensource disease modeling tool). These models address various aspects of a pandemic such as disease spread, and flattening the curve, and charting the next pandemic.

Simulations can potentially play a key role in estimating required operational resource capacity and existing gaps and in assessing the impact on public health during pandemics [32]. However, most models today address only a few types of healthcare resources. For policy-makers and healthcare professionals, having a good estimate of the regional distribution of healthcare resources, potential gaps, and the impact of such gaps on public health outcomes during a pandemic is essential. When confronted with limited resources, decision-makers must trade off and prioritize how and when available resources are deployed. Such tradeoff analysis requires models that provide estimates of resource surplus and shortfalls in various pandemic scenarios. Understanding the dynamics of resources during a pandemic, including the consequences of resource shortages on public health and critical care outcomes, has additional value in pandemic preparedness, planning, and tabletop simulation exercises and mission rehearsal.

Addressing the foregoing problem scenarios requires a simulation capability that among other items includes a flexible resource modeling tool with an easy-to-use, user-friendly interface suitable for use by decision-makers. The user interface should support result displays in a variety of visualization formats such as heat maps, kiviat (radar) charts, 3D views, and customized stakeholder-specified formats. This simulation capability is essentially evidence-based healthcare resource planning. In this regard, we distinguish between a system modeler and a model user. The modeler deals with building and validating the model, analyses and simulation tools, and output formatting and storage. The model user queries the model and performs analyses based on setting up parameters that define scenarios. Model users ask "what-if" questions, i.e., run an experiment and analyze results.

Best Practice Approach

The starting point for the best practice modeling approach presented in this chapter is defining its scope, that is, determining the purpose and use of the model. In this chapter, our initial interest is a model that can be used for evaluating potential effects of pandemic interventions on morbidity, mortality, businesses, and community trust.

To this end, Table 2 presents a partial list of questions to be answered by the model. These questions pertain to preparation; leadership; technology; economic and social factors; key parameters of interest; preventive measures; collaboration at the city, state, and national (and international) levels; and impacts of delayed action.

Table 2 Initial question set

1. How well prepared is a population for a catastrophic pandemic?
2. How much time did a population have to get ready?
3. How effective is the leadership of the population in dealing with preparation?
4. How advanced is medical technology?
5. How effective is the leadership of the population in managing an outbreak when it hits?
6. What are the parameters of the preparation?
7. What are the parameters of the response to an outbreak?
8. What economic and social factors must be considered during preparation and attack response?
9. What is the level of support and collaboration at the local/state/national/international levels?
10. What are the metrics needed for determining efficacy of the response?
11. When can a population return to normal?
12. What is the impact on infections and death of not following preventive measures?
13. What is the impact of personal protective equipment (PPE) availability, intensive care beds, and medical professionals?

Along with the questions in Table 2, we determine the stakeholders and their concerns and key terms we want addressed in the model (Table 3).

From Tables 2 and 3, we next decide which concepts are important for our scope and stakeholder viewpoints. We then create ontology that names, organizes, and relates key terms exemplified in Fig. 2. The core set of concepts in Fig. 2 includes the overall pandemic domain; population demographics; economic factors; disease model; medical interventions; leading indicators; political factors; public information; social tolerance; preparedness factors; leadership factors; and collaborations. Associated with each of these concepts are attributes and their values/thresholds. Over time and as more is known, this ontology is expanded with respect to demography, pandemic life cycle phases, and types of interventions. Models are instantiations of the ontology for specific pandemics. Models are validated by subject matter experts such as public policy experts, virologists, healthcare professionals, FEMA, CDC, WHO, NIH representatives, intelligence reports (bioweapons), etc.

Ontologies employ controlled, domain-specific languages that contain terms and relationships needed for answering domain-related questions. Our approach calls for starting with a relatively sparse ontology (Fig. 2) that answers the questions shown in Table 3.

Multi-model Approach

Complex phenomena such as pandemics require a combination of models to understand different perspectives and analyze how a change can propagate through the entire system. This is called a multi-model approach. A multi-model approach is defined as one in which more than one model, each derived from a different perspective and for a defined purpose and utilizing different reasoning and simulation methods, are employed [13].

Multi-model approaches are used for prediction and decision support. Based on recent history, predictions based on models have turned out to be incorrect, causing the models to be revised [23]. It is not surprising that the general population has begun to lose trust in model-based predictions made by policy-makers and public health officials. Making precise predictions requires extensive validation and testing which requires time, and time is limited when confronting a pandemic. Therefore, the use of models as a prediction tool is not practical especially for novel pandemics [42].

In sharp contrast, models used for decision support have an important role to play. This is not surprising in that the use of models for decision support is quite different than the use of models to make predictions [31]. Specifically, models used for decision support can have selective fidelity and still be useful. With the right expectations and the right perspectives, a multi-model approach can be useful in understanding sensitivities and tradeoffs and are well-suited for decision support of policy makers and public health officials [16]. Public health officials who make pronouncements about the pandemic today generally rely on imperfect mental models. The limitations of imperfect mental models can be overcome with model-based decision support.

Table 3 Candidate factors stakeholders and key pandemic model terms

Goals	
Societal/management goal	Pursue cost-effective, resilience policies that bring about protective structural and behavioral changes in the system and prevent further outbreaks; independent of the current actual perceived number of cases; prevent the pandemic from spreading by instituting proper policies and protective countermeasures and minimizing the impact on those already exposed [45]
Model scope	Develop a dynamic system simulation that is useful in comparing the future outcomes of disease progression and intervention
Stakeholders	
Stakeholder concerns	These influence what goes in the model (views and viewpoints)
Medical and scientific researchers/ practitioners	Help policymakers understand and influence the spread of infection including providing scientific advancements
Politicians	Impact leadership, public information, collaboration, and preparedness
Business owners	Impact jobs, economy, policy enforcement
Collaborations/patients	Impact demand for medications, vaccines, testing clinics and hospitals
Community stakeholders, policy-makers	Decision-makers who need to understand and mitigate the spread of virus across communities through appropriate policies/incentives and inhibitors; they also need to understand implications of decisions and (in)actions at multiple societal levels
City stakeholders, policy-makers	Concerned with local factors that influence the spread of the pandemic; can have some overlapping constraints with state level
State stakeholders, policy-makers	Concerned with factors at the state level that influence spread of the pandemic; can have some overlapping constraints with national and community levels
National stakeholders, policy-makers	Concerned with factors at the national level that influence spread of the pandemic; can have some overlapping constraints with international and state levels
International stakeholders, policy-makers	Concerned with factors at the international level that influence spread of the pandemic; can have some overlapping constraints with national levels
Federal task force	Monitor, make policies to contain, and mitigate spread of virus; keep public informed about significant public health actions such as travel restrictions through advisories
Disease model (progression and mitigation)	
Pathogen	The agent that produces an epidemic or pandemic; has type, attributes (e.g., spread rate, mutation rate, lethality), and phases
Leading indicators	Defines how much advance warning prior to an outbreak in a given region or community

(continued)

Table 3 (continued)

Susceptibility	Portion of population susceptible to infection, includes factors such as workplaces, homes, schools, childcare, and social gatherings; pre-existing conditions that affect infection
Exposure	Duration of time and viral load that lead to infection
Infected	Percentage of susceptible population that becomes infected
Asymptomatic	Percentage of infected population that is asymptomatic and likely infection spreaders
Recovery	Percentage of infected population that recovers
Immunity	Percentage of population that is immune
Death	Percentage of infected population that dies
Community	
Population demographics	Population density by regions, social distancing, protective measures
Community type	Rural, city, densely populated city
Diminishing novelty [36]	Perception of outbreak when it first occurs; declines over time causing reduction in outrage, risk perception, and individual protective behaviors
Economic and business factors	Impact duration and severity of lockdowns for schools, businesses, and government
Social tolerance	Acceptance of action, idea, or person that one disagrees with
Preparedness factors	Availability of required resources and personnel; number of testing centers, hospital beds, ventilators, etc.
Public awareness/information	Level of understanding of the pandemic and countermeasures at a societal level to contain the pandemic
Homelessness	Impact on physical and mental health of homeless due to pandemic; their health problems including HIV infection, alcohol and drug abuse, mental illness, suicide rates
Pre-existing conditions	Medical conditions within population
Racial composition	Percentage of White, African American, Hispanic, Native American, and Asian population
Leadership factors	Open communication, ability to make decisions in face of uncertainty, decisiveness, personal responsibility
Employment	Unemployment rate by regions
ICU availability	Number of ICU locations; number of beds at each location; number of ICU personnel
Medical worker availability	Number of qualified medical workers at each testing center, hospital, clinic
Medical equipment and PPE availability	Amount of medical equipment and PPE at each hospital, clinic, nursing home
Risk perception	Percent population that view virus as high risk; percentage who view virus as low risk

(continued)

Table 3 (continued)

Public biases	Confirmation bias, selection bias, outliers, overfitting, underfitting, etc.
Interventions and immunity	
Protective measures	Decisions/activities taken at individual and societal levels for containing virus spread; if potentially effective protective measures are independent of number of infected people, they can be used to proactively and pre-emptively reduce transmission risk; treatment (reactive) and vaccines (preventive); testing (frequency, testing center density and locations).
Preparedness	Readiness with requisite resources and personnel
Outbreak response lag	Affects recognition of true number of cases; later the response, greater the effort; delayed effects of control measures on detected cases, cause uncertainty about effectiveness of response; gives rise to over-reaction
Government/societal policies	Focus on reducing number of interpersonal contacts and environmental contamination (e.g., access to hand washing and sanitation facilities, physical barriers, increased cleaning of shared spaces, not requiring infected people to work in tightly shared offices, natural ventilation, improved sanitation and food hygiene); these actions provide protection without having to rely on individuals; finding ways to incorporate and sustain such changes can reduce risk of future outbreaks, the incidence of seasonal influenza, and carbon footprints from travel and air conditioning
Political factors	Population bias, media bias, election trends
Detections	Encompasses case finding, contact tracing, and capacity of detection facilities
Multi-models	
Community model	Encompasses geographical, logistical, social, and cultural perspectives from small neighborhoods to nation/ international scope; supported by GIS
Population and demographics model	Model generates population when community data is unavailable; encompasses population demographics, population density, age distribution, pre-existing conditions population, political affiliation
Scenario model	Provides event stream associated with pandemic; characterizes location, initial discover, available resources and events that influence the rate of spread by geographic areas; "what-if" perturbations can be injected
Disease model	Keeps track of outbreaks, their locations, percentage of population who tested positive/recovered/died, rate of spread, mutations
Intervention model	Interventions at all community levels; used in simulation to introduce changes to policies, strategies, tactics to determine impact on outcomes

(continued)

Table 3 (continued)

Community behavior model	A statistical representation of the community that generates decisions to follow (or not follow) guidance at the community level
Healthcare availability, effectiveness, and innovation model	Represents availability and effectiveness of key healthcare services; includes factors (such as rate of healthcare workers to infected population, availability of ICUs, recovery and death rates, availability of medical resources, innovation, resilience
Economic model	Simplified version of reality; allows analysts to observe, understand, and make predictions about economic behavior for a given set of assumptions comprising several sub-models
Food supply model	An especially important model; can be included in a holistic model; e.g., shop closures and their impact on other businesses and people
Behavioral model	Captures reluctance of some people to seek medical attention out of fear of getting infected; so they don't seek medical attention with potential dire consequences; includes hoarding behaviors and fear

Mental models have unarticulated assumptions, cannot be tested for internal consistency, cannot be explored in terms of consequences with different assumptions, and cannot be reconciled with data. Furthermore, humans have cognitive limitations and are ill-equipped to explore side effects and change cascade arising from their actions. These limitations provide the basis for having an explicit, computer-manipulable model for exploring alternate futures, making relative comparisons of future outcomes, and facilitating understanding and communication. Importantly, decision-makers should be able to reason with the model, and the model should be able to "explain" its results. In explicit, ontology-based models, assumptions are spelled out allowing others to replicate the results. Furthermore, domain knowledge calibrated with historical data can be incorporated and updated with current data. The resulting model can then become the basis for decision support and collaborative decision-making among distributed team members.

Another important advantage of explicit models is their ability to support sensitivity analysis and tradeoff analysis – two imperatives for understanding complex phenomena. And, of course, models can be brought to life in simulations that enable exploration of different scenarios over multiple parameter ranges for uncovering robust and brittle regions, important thresholds, and salient uncertainties [11]. In particular, facilities for creating multiple views of system dynamics that incorporate uncertainty and unobservability are especially useful [16]. Discerning high sensitivity to uncertainty in system parameters can become a forcing function for acquiring additional data. Finally, explicit models should indicate which metrics are important and inform data collection.

Considering the foregoing, our proposed model-based decision support architecture employs a "cooperating specialist" construct encompassing federated models

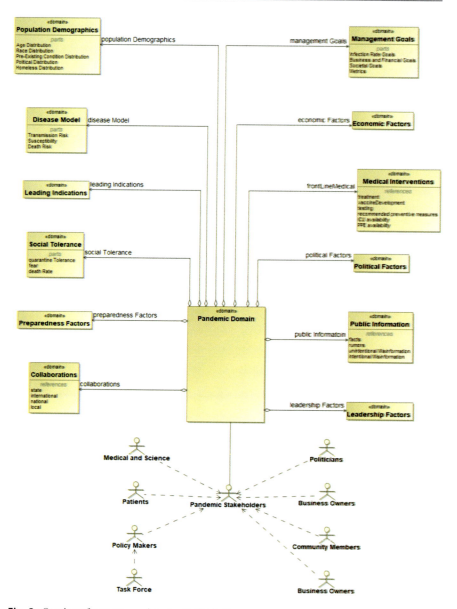

Fig. 2 Ontology fragment consistent with Table 3

within a cloud-based architecture. For a federated model, the ontology has to be modified to reflect the federation.

The two models that need elaboration are the food supply model and the behavioral model. The food supply model can be deceptively complicated in the sense that it cuts across commercial and consumer supply chains and human behaviors. The COVID-19 pandemic has exposed glaring weaknesses in US supply chains especially when it comes to medical and related supplies (e.g., N95 masks, testing swabs, ventilators, and toilet paper). Some of these weaknesses have dissipated as hoarding has been reduced for some items. As the pandemic triggered an economic shutdown, breakdowns appeared in other parts of the supply chain as well. These aspects can be added to the online model (or proposed model). America's food systems are beginning to suffer from both scarcity and excess. We are beginning to see crop dumping because US producers are beginning to destroy excess, perishable crops that they cannot bring to market. Examples of dumping are dairy farmers dumping thousands of gallons of unbought milk into manure ponds, vegetable growers plowing under millions of pounds of vegetables, and breaking and throwing out of hundreds of thousands of chicken-hatching eggs rather than raising the chicken for their meat. Data shows that an estimated 7% of all milk produced in 1 particular week was dumped which equates to 3.7 million gallons a day (a number which is expected to rise). These reports provide a stark contrast to sparsely populated grocery store shelves, severely strained food banks, and endless bread lines. So, what accounts for this logistical failure that is creating artificial shortages today? The answer lies in understanding the behavior of a complex socio-technical, socio-political system at work. Pre-pandemic, much of the food produced on farms was sent to schools and restaurants. But with these destinations shut down nationwide, there is no infrastructure in place to simply redirect the food to grocery stores and food banks, especially for perishable, short shelf-life items like vegetables, meat, and dairy. Some restaurants have begun selling excess supply as groceries, but this is not a scalable solution nationwide. The disruption to the food supply chain is not a short-term proposition. With demand falling precipitously, prices on commodities such as corn and soy have fallen, placing farmers under even greater duress economically. Low prices, low demand, and limited access to markets mean that farmers have to curtail production dramatically. The problem cascades further. Will lenders finance farmers with projected low prices and lower demand? This is the complexity resulting from change propagation. Therefore, when creating a food supply model, we need to account for those factors that largely account for the economics-sensitive behaviors of public hoarding and changes to supply and logistics.

The behavioral model captures the reluctance of some people to seek medical attention out of fear that they might get infected. So, instead of seeking medical help when they have chest pains, they succumb to heart attacks. Also, hoarding behavior and fear are part of this model. Hoarding behavior encompasses food and other commodities. For example, people have begun hoarding disinfectants, and inadequate production means that not everyone can get what they need to protect themselves. Similar behavior occurred with toilet paper. Given that there can be many such models that can be introduced in the future by researchers interested in studying different aspects of the pandemic response problem, we have made provisions for hooks in our holistic online model for introducing additional factors and calculations.

Figure 3 shows a conceptual architecture for a Pandemic Decision Support System based on generic cloud services. A portion of the SaaS functions are implemented and available on the Internet. On the left side of the figure, public health officials, policy-makers, and community organizers who access Reporting, Analysis, and Mining (RAM) applications through web-based user interfaces. A user requests a RAM linked to one or more Data Marts that accesses cloud services and data in the SaaS cloud layer. The SaaS layer delivers integrated, high-level simulation applications that pull content from the Platform as a Service (PaaS) layer. SaaS retrieves the models described above as needed for a user inquiry, invokes the needed simulations and analyses, and then returns the results to a RAM. The RAM creates reports and provides users with the supporting evidence that produced the result.

For data mining applications, a Data Mart comprises a data structure populated by querying the databases and data warehouse shown at the bottom of Fig. 3. A SaaS application sets up database or data warehouse access parameters which are sent to a PaaS application that performs the detailed access. For example, the Population Database includes statistical data related to pre-existing conditions, homelessness, ethnicity, obesity, lung disease, and so forth that impact infection, recovery, and death rates. The SaaS application then populates the Data Mart directly or sends the raw data to a Data Mart application that populates the data structure. Applications requiring more processing may use Data Marts as processing elements to reduce RAM workload.

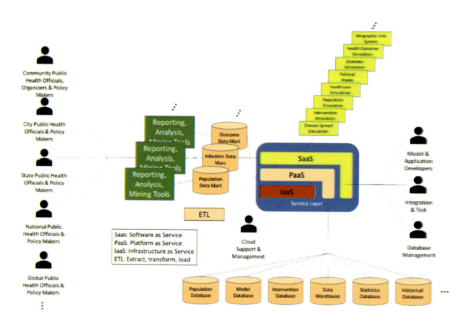

Fig. 3 Conceptual model-based pandemic management system

Transactional databases at the bottom of Fig. 3 are accessible by developers, integration and test, and data entry applications directly from the PaaS. Model and simulation users can only access these repositories through SaaS applications. By implication, only users on the right side of Fig. 3 can create and modify models or update raw data. This restriction is essential for configuration management and protecting important content. This does not mean that SaaS users cannot modify the repositories; however, modifications are performed through SaaS services rather than direct PaaS access. It is also the case that SaaS users will have many "knobs" available for creating scenarios and collecting data; however, SaaS users can only access the knobs already built into models and simulations.

The standard practice of evaluating transactional data and adding new content to the data warehouse is performed by the extract, transform, load (ETL) application within the Platform as a Service (PaaS). Data transformation (cleaning, quality control, redundancy elimination, etc.) is a complicated process, especially when the range of data values and data quality are uncertain. It is anticipated that some form of profiling of source data will be needed that dynamically modifies transformation rules over time [10]. The data warehouse contains information such as a history of lessons learned, time-ordered event occurrences and responses, response effectiveness, medical and community behaviors, infection spread statistics, and drug and treatment efficacy.

In sum, a pandemic response management is a complex adaptive SoS that must continually monitor the progress of a pandemic as well as community behavior and continues to adapt as needed [38]. Such adaptation should account for resource and time constraints, economic implications of adaptation, medical innovations, and cultural predisposition of the community. The level of adaptation is a function of the severity of the disruptions (Fig. 4).

Figure 4 presents the process flow for usage of the pandemic response management system. This flow applies at all community levels. Goal setting, planning, and plan execution are interactive processes open to interventions at all levels by appropriate persons. Shearer et al. [39] also support these concepts.

Ideally, preparations should begin before the pandemic strikes. That is, goals should be established, a plan developed, and exercises performed well in advance of a pandemic. The exercises may result in changes to the plan as shown by the double arrows connecting preparation activities in Fig. 4. Exercises must be conducted no less than annually at all levels (local, state, national, and even international). A key part of the plan should be continual monitoring for outbreaks that trigger execution of the rehearsed plan. Of course, monitoring can be problematic if countries conceal outbreaks; however, major outbreaks cannot be hidden, only delayed. As needed, the plan is tweaked based on observations of its effectiveness. None of this is easy, and yet we know that COVID-19 or its cousin is likely to hit us again. Ultimately, there is a cost for maintaining vigilance, but since we are already spending a huge amount on such needs as military readiness and healthcare, this additional expense would be only a small fraction of that.

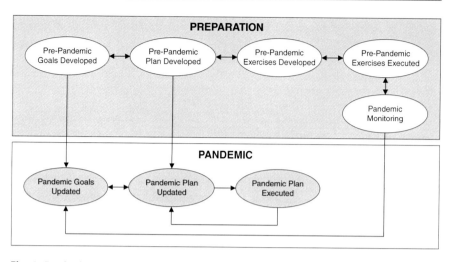

Fig. 4 Pandemic response process management system

Illustrative Examples

System dynamics models provide a useful means for understanding the pandemic SoS [1, 3, 24]. We began with a public version of an SIR model and revised most of the equations to better reflect real-world phenomena and interactions. Additionally, we extended pre-existing conditions and changed the way preventive measures are calculated. This calculation is now based on availability of PPE, critical medications (stockpiles), and how much trust a population places on science information and leadership. The model has "knobs" to turn for most parameters making it straightforward to pose "what-if" questions. There are several additional factors that can be included in the future – for example, as the working population decreases, we already know that food supply is impacted. In addition, there will be fewer healthcare workers and reduced PPE which impacts the number of deaths. The impact of students not being able to continue with their education means further reductions in workforce and possibly fewer skilled individuals contributing to the economy. We added an available ICU equation that is a function of available healthcare workers and beds. We also assume that a fraction of the working population is healthcare workers, which of course declines as they get sick. In theory, there is no end to such dependencies. In practice, we can ignore some of the weaker dependencies to gauge trends, assuming that we correctly understand what the weaker dependencies *are*.

Figure 5 shows an initial version of a working dynamic pandemic model. Ovals in the model represent stateless functions; rectangles are "stocks," that is, elements that retain data and state. Solid lines are flows into and out of stocks, and dashed lines represent instantaneous values used by stateless and stateful functions. Colors have no meaning other than distinguishing stateless functions from stocks. The core of Fig. 5 is a simple susceptibility, infection, and recovery (SIR) construct. The SIR is

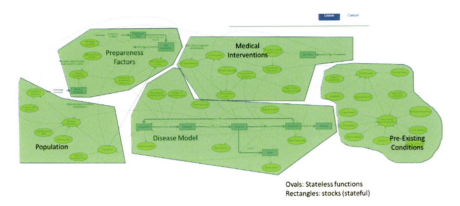

Fig. 5 System dynamics model for infection spread and population impacts

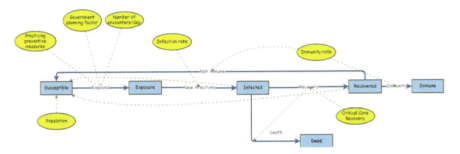

Fig. 6 Disease model

modulated by several factors including population size, infection rate, and fatality rate. Infection rate is a function of percentage of the population practicing prevention, daily number of encounters, and government planning. Recovery is a function of available intensive care units and available healthcare workers. Healthcare workers are a percentage of the population (at a higher or lower rate than the population as a whole) and decline as a percentage of the population dies.

The overarching model in Fig. 5 is provided to show the interactions among five cooperating models: SEIR model; protection model; vulnerability model; medical model; and population and workforce model. Blowups of each model are presented in Fig. 6 through Fig. 10. Each model is described next.

SEIR model: this model represents the basic susceptibility, exposure, infection, and recovery stages. In this model, the entire healthy population is susceptible to an exposure. Within SEIR, the exposure model assumes that some percentage of the population could be exposed, but the actual exposure depends on availability and use of protective measures as well as government-mandated stay-at-home policies. A fraction of those exposed are infected. There may be asymptomatic infections, non-critical infections, and critical infections. Recovery comprises all asymptomatic, all non-critical, and a fraction of the critically infected. This model is presented in Fig. 6.

Protection model: this model represents the availability as well as the percentage of the population that cooperates with protection and isolation recommendations. This model is presented in Fig. 7.

Vulnerability model: this model captures the fact that the elderly and those with pre-existing conditions have a higher probability of becoming critically ill or dying than younger, healthier people. This model, presented in Fig. 8, is live in our online version and seeded with US statistical data where available.

Medical model: this model captures the availability of intensive care unit (ICU) beds and healthcare workers. Sufficiency of facilities and healthcare workers is a factor in the critically ill recovery rate. This model is presented in Fig. 9.

Population and workforce model: how the decline in workforce occurs as the population becomes ill or dies is depicted in this model. Since healthcare workers are a percentage of the workforce, then as more people become ill or die, fewer healthcare workers are available which influences the critically ill recovery rate. The population increases when infected patients recover.

With respect to the overall model, a *dashboard* provides parameter "sliders" that enable the changing of model parameters, thereby effectively creating different scenarios. In other words, what-if scenarios are easily created by changing model parameters. For example, assuming that 1% of the workforce are healthcare workers, and there are roughly nine million ICU beds, the outcomes can be calculated (Fig. 10). In Fig. 11, the cumulative infected and recovery scale is on the left and the death scale on the right. Although the figure does not necessarily represent a realistic scenario, it highlights the point as the number of infection levels off the death rate slows, but also since fewer people are infected, there are fewer recoveries.

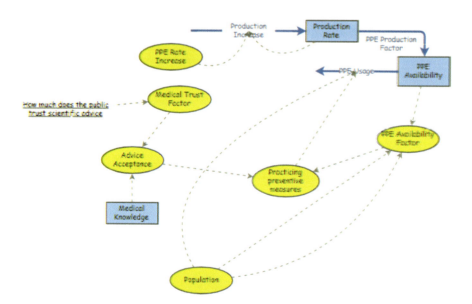

Fig. 7 Preparedness factors

37 Multi-model-Based Decision Support in Pandemic Management

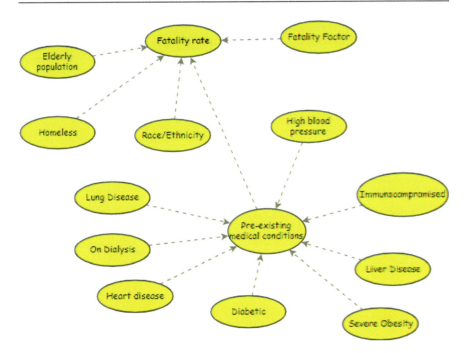

Fig. 8 Pre-existing conditions model

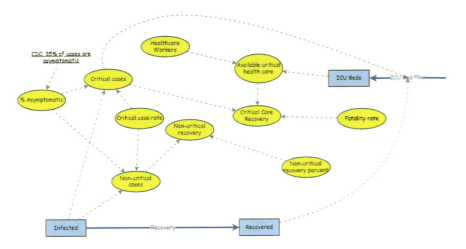

Fig. 9 Medical intervention model

Another interesting result is that unlike polio or scarlet fever, there is high likelihood that there will remain a low-level number of infections which contribute to a low-level death rate. In other words, given the resistance to vaccination and the fact that vaccines do not work on every virus like the H1N1 swine flu, viruses such as

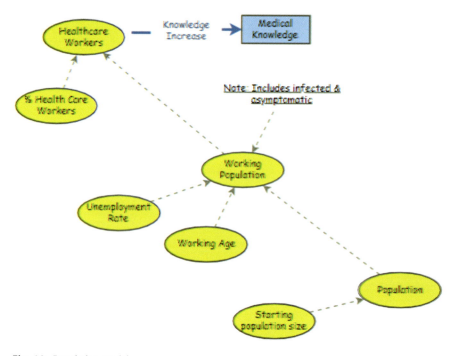

Fig. 10 Population model

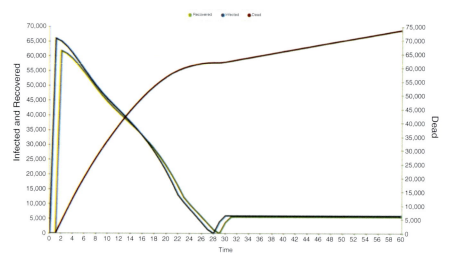

Fig. 11 Illustrative population curves for various population segment

COVID-19 will most likely not go away. Also, it is possible that they might recur in the wild or come back in some mutated form. This is due to resistance to vaccination, less than perfect vaccine effectivity, and likely recurrence in the wild similar to the recurrence of the H1N1 swine flu.

More realistic values can be used based on online statistics. Users can play out "what-ifs" by changing the sliders in the model located here: https://insightmaker.com/insight/197273/COVID-Model. This model is fully functional and does include these features.

Chapter Summary

This chapter has presented a systems approach to pandemic decision support that makes extensive use of modeling and simulation. The approach, which addresses the major concerns of decision-makers at the federal, state, and community levels, relies on a number of complementary modeling constructs to represent pandemics, their ecosystems, and their multiple co-dependencies and interdependencies. A system dynamics simulation of infection spread and population impacts is presented. The role of system modeling to explicate assumptions, illuminate interdependencies and uncertainties, explain behaviors and disease dynamics is discussed. The paper makes a few key recommendations: holistic thinking on the part of system scientists and engineers cannot be abandoned while we take measured steps to alleviate local and immediate concerns; when making decisions in the face of uncertainty, risk, and ambiguity, one must always be aware of the big picture and the undesirable change cascades that can ensue from expedient short-term thinking; given the complexity of a pandemic, it will take multiple complementary models to address the different aspects of a pandemic; and modeling and simulation should be exploited as a decision support tool to explore alternate futures and make comparisons with different assumptions, technologies, regulations, and other constraints. Future work involves collecting available real data for the model and adding new factors as they are reported in the literature. The pandemic management system is expected to be a living model in that it will continue to grow and change as new information becomes available. Although it is reasonable to expect that the model will always lag current events, our goal is to make it relevant and useful for assessing the relative merits of potential decisions under prevailing circumstances.

References

1. Araz, O.M., Damien, P., Paltiel, D.A., Burke, S., Van De Geijn, B., Galvani, A., et al. Simulating school closure policies for cost effective pandemic decision making. BMC Public Health 2012; 12(1):1.
2. Badr, Hamada, S Du, Hongru, Marshall, Maximilian, Dong, Ensheng, Squire, Marietta, Gardner, Lauren, Association between mobility patterns and COVID-19 transmission in the USA: a mathematical modelling study, July, 2020, https://doi.org/10.1016/S1473-3099(20)30553-3
3. Bauch, C., Lloyd-Smith, J., Coffee, M., Galvani, A. Dynamically modeling SARS and other newly emerging respiratory illnesses: past, present, and future. *Epidemiology.* 2005; 16(6): 791–801. https://doi.org/10.1097/01.ede.0000181633.80269.4c. PMID: 16222170
4. Bertozzi, A.L., Franco, E., Mohler, G., Short, M.B., Sledge, D. The challenge of modeling and forecasting the spread of COVID-19, PNAS, 117(29): 16732–16738, July 21, 2020.

5. Black, A.J., Geard, N., McCaw, J.M., McVernon, J., Ross, J.V. Characterising pandemic severity and transmissibility from data collected during first few hundred studies. *Epidemics*. 2017; 19:61–73. https://doi.org/10.1016/j.epidem.2017.01.004. PMID: 28189386

6. Bjørnstad, O.N., Shea, K., Krzywinski, M. and Altman, N., "The SEIRS model for infectious disease dynamics." Nature Methods 17 (2020): 557–558.

7. Bradley, D.T., Mansouri, M.A., Kee, F., and Garcia, L.M.T. "A systems approach to preventing and responding to COVID-19," EClinicalMedicine, published by The *Lancet*, Volume 21, 100325, April 1, 2020. Published March 27, 2020, https://doi.org/10.1016/j.eclinm.2020.100325,

8. Day, T., Park, A., Madras, N., Gumel, A., Wu, J. When is quarantine a useful control strategy for emerging infectious diseases? *Am J Epidemiol*. 2006; 163(5):479–485. https://doi.org/10.1093/aje/kwj056 PMID: 16421244

9. de Lusignan S, Liyanage H, McGagh D, Jani BD, Bauwens J, Byford R, Evans D, Fahey T, Greenhalgh T, Jones N, Mair FS, Okusi C, Parimalanathan V, Pell JP, Sherlock J, Tamburis O, Tripathy M, Ferreira F, Williams J, Hobbs FR. "In-pandemic development of an application ontology for COVID-19 surveillance in a primary care sentinel network," *JMIR* Preprints. 01/07/2020:21434

10. Denny, M.J., et. al., "Validating the Extract, Transform, Load Process Used to Populate a Large Clinical Database," *Int. J. Med Inform*. 2016 October; 94: 271–274

11. Elveback, L.R., Fox, J.P., Ackerman, E., Langworthy, A., Boyd, M., Gatewood, L. An influenza simulation model for immunization studies. American Journal of Epidemiology 1976; 103(2): 52–65.

12. Eriksson H, Morin M, Jenvald J, Gursky E, Holm E, Timpka T. Ontology based modeling of pandemic simulation scenarios. Studies in Health Technology and Informatics 2007;129 (Pt 1):755–759.

13. Fishwick, P.A., Sticklen, J., and Bonarini, A. A Multi-Model Approach to Reasoning and Simulation, IEEE Transactions on Systems, Man and Cybernetics, Vol. 24, No. 10, October 1994.

14. Frank, G., Wheaton, B., Bakalov, V., Cooley, P., and Wagener, D. Ontology for Designing Models of Epidemics, RTI International, July 24, 2009.

15. Fraser, C., Donnelly, C.A., Cauchemez, S., Hanage, W.P., Van Kerkhove, M.D., Hollingsworth, T.D., et al. Pandemic potential of a strain of influenza A (H1N1): Early findings. *Science*. 2009; 324(5934):1557–1561. https://doi.org/10.1126/science.1176062 PMID: 19433588

16. Ge, L., Mourits, M.C.M., Kristensen, A.R., Huirne, R.B.M. A modelling approach to support dynamic decision- making in the control of FMD epidemics. Preventive Veterinary Medicine 2010; 95(3):167–174.

17. Giordano, G., Blanchini, F., Bruno, R., Colaneri, P., Di Filippo, Di Matteo, A., Colaneri, M. Modelling the COVID-19 epidemic and implementation of population-wide interventions in Italy, nature medicine, Letters, April 22, 2020.

18. Gumel, A.B., Ruan, S., Day, T., Watmough, J., Brauer, F., Van Den Driessche, P., et al. Modelling strategies for controlling SARS outbreaks. *Proc Royal Soc B*. 2004; 271:2223–2232.

19. Habibi, R., Burci, G.L., de Campos, T.C., et al. "Do not violate the International Health Regulations during the COVID-19 outbreak," *Lancet*. 2020; 395: 664–666.

20. He, Y., Yu, H., Ong, E. *et al.* CIDO, a community-based ontology for coronavirus disease knowledge and data integration, sharing, and analysis. *Sci Data* **7**, 181 (2020). https://doi.org/10.1038/s41597-020-0523-6

21. Hogan, W.R., Wagner, M.M., Brochhausen, M., Levander, J., Brown, S.T., Milett, N., DePasse, J., and Hanna, J. "The Apollo Structured Vocabulary: an OWL2 ontology of phenomena in infectious disease epidemiology and population biology for use in epidemic simulation," *Journal of Biomedical Semantics* (2016) 7:50.

22. Holmes, E.C., Rambaut, A., Andersen KG. Pandemics: Spend on surveillance, not prediction. *Nature*. 2018; 558(7709):180–182. https://doi.org/10.1038/d41586-018-05373-w PMID: 29880819

23. Jewell NP, Lewnard JA, Jewell BL. Caution warranted: using the Institute for Health Metrics and Evaluation model for predicting the course of the COVID-19 pandemic. Annals of Internal Medicine 2020 April 14
24. Kirkwood, C.W. "System dynamics methods: a quick introduction," 1998. http://www.public.asu.edu/~kirkwood/sysdyn/SDIntro/SDIntro.htm.
25. Lipsitch, M., Santillana, M. Enhancing Situational awareness to Prevent Infectious Disease Outbreaks from Becoming Catastrophic. In: Inglesby T, Adalja A, editors. Global Catastrophic Biological Risks. *Current Topics in Microbiology and Immunology.* Berlin, Heidelberg: Springer; 2019.
26. Larouzee, J., and Le Cose, J-C, "Good and Bad Reasons: The Swiss Cheese Model and its Critics," Science Safety, 126 (2020)
27. McCaw, J.M., Moss, R., McVernon, J. A decision support tool for evaluating the impact of a diagnostic capacity and antiviral-delivery constrained intervention strategy on an influenza pandemic. Influenza Other Resp. 2011; 5(Suppl. 1):202–229.
28. McCaw, J.M., Glass, K., Mercer, G.N., McVernon, J. Pandemic controllability: a concept to guide a proportionate and flexible operational response to future influenza pandemics. Journal of Public Health 2014; 36 (1):5–12.
29. McVernon, J., McCaw, C.T., Mathews, J.D. Model answers or trivial pursuits? The role of mathematical models in influenza pandemic preparedness planning. Influenza Other Resp. 2007; 1(2):43–54.
30. Metcalf, C.J.E., Morris, D.H., and Park, S.W. Mathematical models to guide pandemic response, SCIENCE, Vol. 369, Issue 6502, pp. 368–369, July 24, 2020.
31. Morgan, O. How decision makers can use quantitative approaches to guide outbreak responses. *Phil Trans R Soc B.* 2019; 374:20180365. https://doi.org/10.1098/rstb.2018.0365 PMID: 31104605
32. Ndeffo Mbah, M.L., and Gilligan, C.A. Resource Allocation for Epidemic Control in Meta-populations PLoS One 2011,6(9): e24577
33. Oren, T.I., Dynamic templates and semantic rules for simulation advisors and certifiers, in Knowledge Based Simulation: Methodology and Application, pp. 53–76, Springer Verlag, 1991.
34. Reason, "The Contribution of Latent Human Failures to the Breakdown of Complex Systems," *Philosophical Transactions of the Royal Society of London. Series B, Biological Sciences,* 327(1421), Human Factors in Hazardous Situations, pp. 475–484, April 12, 1990
35. Rivers, C., Chretien, J.P., Riley, S., Pavlin, J.A., Woodward, A., Brett-Major, D., et al. Using "outbreak science" to strengthen the use of models during epidemics. *Nat Commun.* 2019; 10(1):3102. https://doi.org/10.1038/s41467-019-11067-2. PMID: 31308372
36. Sandman, P.M. "Responding to community outrage: strategies for effective risk communication," 2012.
37. Sevilla, N. Open Source Disease Modeling: A Tool to Combat the Next Pandemic, Spatiotemporal Epidemiological Modeler (STEM), Global Biodefense, January 18, 2016,
38. Shea, K., Tildesley, M.J., Runge, M.C., Fonnesbeck, C.J., Ferrari, M.J. Adaptive management and the value of information: learning via intervention in epidemiology. PLoS Biology 2014; 12(10):1–11.
39. Shearer, F.M., Moss, R., McVernon, J., Ross, J.V., McCaw, J.M. Infectious diseases pandemic planning and response: Incorporating decision analysis. *PLoS Med* 17(1): e1003018, January, 2020, https://doi.org/10.1371/journal.pmed.1003018
40. Smith, D. and Moore, L. "The SIR model for spread of disease – The differential equation model," *Convergence,* December 2004
41. Struben, J. The coronavirus disease (COVID-19) pandemic: Simulation-based assessment of outbreak responses and post peak strategies, Version 5.0, medRxiv, June 17, 2020
42. Viboud, C., Sun, K., Gaffey, R., Ajelli, M., Fumanelli, L., Merler, S., et al. The RAPIDD ebola forecasting challenge: synthesis and lessons learnt. *Epidemics* 2018; 22:13–21. https://doi.org/10.1016/j.epidem.2017.08.002 PMID: 28958414

43. World Health Organization. WHO Director-General's opening remarks at the media briefing on COVID-19 – 11 March 2020. 2020. https://www.who.int/dg/speeches/detail/who-director-general-s-opening-remarks-at-the-media-briefing-on-covid-19—11-march-2020. Accessed March 11, 2020.
44. World Health Organization. Pandemic influenza risk management: A WHO guide to inform and harmonize national and international pandemic preparedness and response; Geneva, World Health Organization 2017. https://www.apps.who.int/iris/handle/10665/259893
45. Yaesoubi, R., Cohen, T. Identifying cost-effective dynamic policies to control epidemics. *Statistics in Medicine* 2016; 35(28):5189–5209. https://doi.org/10.1002/sim.7047. PMID: 27449759
46. Yang, Y., Sugimoto, J.D., Halloran, M.E., Basta, N.E., Chao, D.L., Matrajt, L., et al. The transmissibility and control of pandemic influenza A (H1N1) virus. *Science* 2009; 326(5953): 729–733. https://doi.org/10.1126/science.1177373. PMID: 19745114

Azad M. Madni is a University Professor of Astronautical Engineering, holder of the Northrop Grumman Fred O'Green Chair in Engineering, and the Executive Director of University of Southern California's Systems Architecting and Engineering Program. He is also the Founding Director of the Distributed Autonomy and Intelligent Systems Laboratory. He has joint appointments in the Department of Aerospace and Mechanical Engineering and Sonny Astani Department of Civil and Environmental Engineering. He has courtesy appointments in the Rossier School of Education and Keck School of Medicine where he is a faculty affiliate of the Ginsberg Institute for Medical Therapeutics. He is a member of the National Academy of Engineering and Life Fellow/Fellow of AAAS, IEEE, AIAA, INCOSE, IISE, IETE, AAIA, SDPS, and the Washington Academy of Sciences. He is the founder and CEO of Intelligent Systems Technology, Inc., a high-tech company which specializes in model-based and AI approaches to addressing scientific and societal problems of national and global significance. He co-founded and currently chairs the IEEE SMC award-winning technical committee on Model Based Systems Engineering.

He has served as Principal Investigator on more than 95 R&D projects totaling in excess of $100 million in funding. His research sponsors in the government include DARPA, OSD, NSF, DOD-Systems Engineering Research Center, DOD-Acquisition Innovation Research Center, NASA-Ames, MDA, AFOSR, AFRL, DHS, DTRA, AFOSR, AFRL, ONR, NIST, and DOE. His research sponsors in industry include General Motors, Boeing, NGC, Raytheon, SAIC, and ORINCON. He pioneered the field of transdisciplinary systems engineering to address problems that appear intractable when viewed solely through an engineering lens. He is the creator of the TRASEE™ educational paradigm which combines storytelling with the Science of Learning principles to make learning enjoyable while enhancing retention and recall. He is the author of the highly acclaimed book *Transdisciplinary Systems Engineering: Exploiting Convergence in a Hyper-Connected World* (Springer, 2018) and the co-author of *Tradeoff Decisions in System Design* (Springer, 2016). He is the Co-Editor-in-Chief of three proceedings from the Conference on Systems Engineering (CSER).

He has received more than 75 honors and awards from professional engineering societies, industry, government, and academia. His IEEE awards include *2021 IEEE Aerospace and Electronic Systems Judith A. Resnik Space Award*, *2019 AESS Pioneer Award*, and *2020 IEEE Systems, Man and Cybernetics Norbert Wiener Outstanding Research Award*. His INCOSE awards include the *2011 Pioneer Award, 2019 Founders Award*, and *2021 Benefactor Award*. He has also received prestigious educational awards including *2019 AIAA/ASEE John Leland Atwood Award* for excellence in engineering education and research and the *2021 Joint ASEE SED/INCOSE Outstanding Systems Engineering Educator* Inaugural Award. In 2020, he received the *NDIA Lt. Gen. Ferguson Award* for Excellence in Systems Engineering. In 2019, he received the *ASME CIE Leadership Award* and the *Society of Modeling and Simulation International Presidential Award*. In 2016, Boeing honored him with a *Lifetime Achievement Award* and a *Visionary Systems Engineering Leadership Award* for his "impact on Boeing, the aerospace industry, and the nation." He received his Ph.D., M.S., and B.S. degrees in Engineering from the University of California, Los Angeles. He is a graduate of AEA/Stanford Institute Program for Technology Executives.

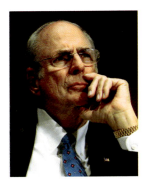

Norman Augustine served as the Chairman of the Board of Lockheed Martin Corp. from August 1997 to March 1998 and Chief Executive Officer from January 1996 to July 1997 and from March 1995 to June 1996. Mr. Augustine previously served as the President and Chairman of the Association of the United States Army. He also served as the President of the American Institute of Aeronautics and Astronautics and the Chairman of American Red Cross. Mr. Augustine served as the Chairman of the National Academy of Engineering, Chairman of the Aerospace Industries Association, and Chairman of the Defense Science Board. He serves as Trustee Emeritus of Johns Hopkins University. In addition, Mr. Augustine is an Advisory Board member to the Department of Homeland Security. He serves as an Honorary Member of Space Foundation. He serves as a Member of the Board of Regents at University System of Maryland. Then Mr. Augustine served as a Director of The Procter & Gamble Company since 1989 and its Presiding Director from 2003 to 2007. He served as a Trustee of Massachusetts Institute of Technology. Augustine also chaired the Obama Administration's Review of US Human Space Flight Committee, a 2009 blue-ribbon panel charged with conducting an independent assessment of the country's planned human space flight activities. He was a Member of the Board of Trustees of MIT and Princeton, his alma mater. Mr. Augustine served as a Lecturer with the rank of Professor Department of Mechanical and Aerospace Engineering of School of Engineering and Applied Science of Princeton University from 1997 to 1999. He served as Under Secretary of the Army. Mr. Augustine served for 16 years on the President's Council of Advisors on Science and Technology. He co-authored *The Defense Revolution* and *Shakespeare in Charge* and is the author of *Augustine's Laws* and *Augustine's Travels*. He was selected by Who's Who in America and the Library of Congress as one of "Fifty Great Americans." Mr. Augustine has been presented the National

Medal of Technology and received the Joint Chiefs of Staff Distinguished Public Service Award and is a five-time recipient of the Department of Defense's Civilian Distinguished Service Medal. He holds 23 honorary degrees. Mr. Augustine received a MS in 1959 and a BS in 1957, both in Aeronautical Engineering from Princeton University.

C. Madni is the Executive Vice President and a Principal Scientist of Intelligent Systems Technology, Inc. She conducts research in modeling and simulation, human-computer interface design, embedded training, and decision support systems. Ms. Madni has led major DOD, DARPA, Army, Navy, and Air Force programs and their successful transition and commercialization. Ms. Madni has over 30 years of R&D and technology experience. She has approximately 100 papers in international and national conferences. She is a Member of IEEE, the Human Factors Society, and the National Association of Female Executives. She received her BS in Engineering from Tulane University and her MS in Engineering Systems with specialization in distributed artificial intelligence from UCLA.

Michael Sievers is a Senior Systems Engineer at Caltech's Jet Propulsion Laboratory and a Lecturer in the System Architecting and Engineering Program at the University of Southern California. He conducts research in model-based systems engineering as well as contributing to several JPL's flight missions. He is a principal investigator on multiple harsh environment, fault-tolerant, high-performance computing research projects. He holds a PhD in Computer Science from the University of California, Los Angeles. He is an INCOSE Fellow, Life Associate Fellow of AIAA, and a Life Senior Member of IEEE.

Semantic Modeling for Power Management Using CAESAR

38

D. A. Wagner, M. Chodas, M. Elaasar, J. S. Jenkins, and N. Rouquette

Contents

Introduction	1136
Problem Statement	1136
JPL, IMCE, and CAESAR	1137
Case Study	1138
Europa Clipper Introduction	1138
Modeling Scenarios	1140
Modeling Waste Heat	1140
CAESAR PEL Application	1141
CAESAR PEL Application	1141
Advantages of the CAESAR PEL Vocabulary	1143
Implementing the CAESAR PEL	1144
CAESAR Authoring Workbench	1145
Power Analysis Workflow	1147
Chapter Summary and Future Updates	1148
Cross-References	1151
References	1151

Abstract

Model-based systems engineering (MBSE) aims to bring rigor to the products and methods used in systems engineering similar to the benefits that computer-aided design (CAD) technologies have delivered to other engineering disciplines. MBSE is most often associated with the Systems Modeling Language (SysML) and its associated tools. While SysML is being developed with the intent of making the practice of systems engineering more rigorous, it isn't the only way to do it. This chapter describes an alternative approach to model-based systems engineering that leverages semantic web technologies to provide a more flexible

D. A. Wagner (✉) · M. Chodas · M. Elaasar · J. S. Jenkins · N. Rouquette
Jet Propulsion Laboratory, California Institute of Technology, Pasadena, CA, USA
e-mail: david.a.wagner@jpl.nasa.gov; mark.chodas@jpl.nasa.gov; elaasar@jpl.nasa.gov;
steve.jenkins@jpl.nasa.gov; nicolas.rouquette@jpl.nasa.gov

© Springer Nature Switzerland AG 2023
A. M. Madni et al. (eds.), *Handbook of Model-Based Systems Engineering*,
https://doi.org/10.1007/978-3-030-93582-5_81

platform for developing and managing descriptive system models in a rigorous manner. We describe a case study, in which this platform has been put into operations on a large project at the Jet Propulsion Laboratory to capture, validate, configuration manage, and report on power system sizing, which is a key systems engineering concern in the design of spacecraft.

Keywords

Spacecraft power management · OML · OWL · Semantic modeling

Introduction

Spacecraft need to be able to operate as autonomous and self-sustaining remote observatories: autonomous in that they must be able to perform their missions of collecting scientific data in situations where operators on the ground will not be able to see the results of their inputs until hours after everything has already happened and self-sustaining in that they must be able to either carry or produce all of the resources they will need to do so. In particular, spacecraft, as electro-mechanical robotic systems, need electrical power to drive the instruments, control electronics to do the work, and heaters to keep everything from freezing in the cold of space.

A key challenge in designing spacecraft is to provide enough electrical power to meet the demands of the mission using a minimum of mass and volume. Although the design of the solution is often assigned to a particular subsystem, it is usually a systems engineering responsibility to assess demands across the system in order to specify the needed performance of the power producing subsystem. At JPL this assessment is referred to by the name of the report that documents this assessment: the Power Equipment List or PEL.

In this chapter we will review a model-based approach to capturing and assessing power demands of a spacecraft using semantic web technologies and tools.

Problem Statement

On the surface, simply adding up the power estimates from a collection of hardware does not sound very hard. This view has generally led projects in the past to use a simple spreadsheet to try to capture the power estimates. Spreadsheets make it relatively easy to add up columns of numbers, the tools are ubiquitous, and there is a very low learning curve. However, several other factors conspire to make the problem more difficult than it may seem. First, estimating the power demands of hardware that has not yet been built is not easy. Early on in the process, the hardware design may not yet exist. But the project cannot wait until all the power-using subsystems have been designed before starting to design the supply. Second, even if the hardware is well-defined, the power demands are going to vary widely with how the system is being operated. Instruments can be turned off when not in use.

Thus, it becomes necessary to refine the demand estimates to also include scenarios describing how the hardware will be operated over time. In extreme cases such as the one reported here, full mission simulations are needed to assess demand over time. But most projects can simplify the problem to define a small number of system "modes" where each mode can be described by a single fixed operating state and load estimate for each demand assembly.

What is it that makes this a good candidate application for a model-based approach? First, let us not dismiss the spreadsheet as a model-based approach. There is still a model there even if it's only a table of data, but the relationships between the numbers and the system they describe are only implied. A more general modeling approach is suggested by several factors. First, the shape of the data is not strictly tabular. In fact, there can be some hierarchical composition: electronic cards are contained in a chassis, and assemblies are aggregated by subsystem. The entries are highly regular, and the relationships are simple, but there can be a lot of entries. Thus, defining a schema or metamodel for this data is not difficult. But validating the consistency of the model is critical, and, since the data are updated frequently, this needs to be done frequently too, suggesting a need to automate the analysis.

JPL, IMCE, and CAESAR

The Jet Propulsion Laboratory (JPL) is operated by the California Institute of Technology to conduct solar system research for the National Aeronautics and Space Administration (NASA). JPL's Engineering and Science Directorate (ESD) manages the implementation of its space missions from formulation through implementation and operations including the systems engineering involved at every level of the process. ESD has funded research and development in model-based systems engineering (MBSE) through the Integrated Model-Centric Engineering project for over a decade. In 2017 ESD transitioned from sponsoring small, independent research tasks that were producing interesting proofs-of-concept to a more coordinated single approach named CAESAR. By combining some common architectural principles and frameworks to ensure that modeling capabilities can be federated along with a focus on developing the new methods and training needed to make improved tools usable in the workplace, CAESAR intends to develop integrated capabilities within the organization to support current and future projects.

CAESAR's first application focused on the problem of spacecraft harness specification and design. As described in [1], this was a ripe target to apply a model to capture the specification and design data in order to make the overall process more efficient because of the large quantity of data having a very regular shape. That application delivered the following features:

- A vocabulary or metamodel that defines the shapes of the data to be captured
- An authoring tool that provides user interface forms for effectively capturing this data
- A model change management system based on Git

- A continuous integration workflow in which the model data can be automatically analyzed to produce review reports and assess formal completeness and consistency criteria

All of these features were developed on top of open-source tools and frameworks (including the Eclipse Modeling Framework) and open-standard languages and interfaces.

As described in the [Metamodeling chapter], large parts of the vocabularies, authoring platform, and continuous integration tooling developed by CAESAR have been open-sourced in the openCAESAR project.

The rest of this chapter will describe the second application CAESAR developed to support spacecraft power demands assessment on the Europa Clipper project at JPL. This application extends vocabulary, model, and tooling from an earlier application described in [1] that is currently supporting spacecraft harness specification on both the Europa Clipper and Psyche projects. The case study will focus primarily on the development of model vocabulary and methods rather than on the details of implementation because the hard part is not in building the tools, but ensuring that the tools are well suited to the problem.

Case Study

Europa Clipper Introduction

The Europa Clipper project is planning to send a spacecraft to study Europa, an icy moon of Jupiter. Europa is an interesting subject of inquiry because earlier missions have observed signs that there may be liquid water beneath its icy surface. Liquid water in the presence of chemical processes driven by an energy source such as Jupiter's gravitationally driven tidal flexing could provide the right conditions to support life. The Clipper mission intends to make more detailed observations of the surface and subsurface conditions through the use of a spacecraft with remote sensing instruments. Europa happens to orbit Jupiter inside a plasma torus produced by an interaction between Jupiter's strong magnetic field and solar wind. To minimize the spacecraft's exposure to this radiation, it will orbit Jupiter such that it spends most of its time outside the plasma torus and only briefly dips in to make close flybys of Europa.

Clipper will be a rather large solar-powered spacecraft – only the second solar-powered mission beyond Mars that NASA has attempted. A key challenge with all spacecraft, but particularly ones using solar power, is to size the power-generating capabilities of the spacecraft to match the demand. Batteries can be used to some extent to buffer the discontinuities in time between when power can be generated and when it is needed, and this strategy is used on Clipper. However, at that distance from the Sun, solar panels have to be quite large to produce enough power to just keep the spacecraft from freezing, not to mention power the instruments and transmit data back to Earth.

Managing the balance of power between the demands of instruments, avionics, and thermal functions on the one hand with the generating capacity of the solar panels is a major concern during the design and development of such a spacecraft. Clipper's challenge is particularly acute due to Jupiter being so far from the Sun which not only reduces its ability to generate solar power, but it increases the amount of power needed to transmit large amounts of data back to Earth. Clipper is a large spacecraft in part because of the number and complexity of its science instruments but also because its solar panels need to be large to collect enough energy to power them.

And then, the orbit that will preserve the life of the spacecraft's electronics by keeping it out of the plasma torus only provides one opportunity to observe Europa each orbit and at the periapsis of the orbit where it will be travelling at very high speeds relative to Europa's surface. This means that all of the close-up observations will occur in a short period of time as the spacecraft speeds by. Taking images at this point will require turning the spacecraft to precisely track the surface – another function that will consume power. And doing this will prevent the spacecraft from keeping solar panels aligned with the Sun at the same time.

All of these factors contribute to making Clipper one of the most power-hungry deep-space spacecraft ever designed. It would seem that supplying that power from solar energy is simply a matter of making solar panels as big as needed to provide the energy. But, the size needed to produce the constant power needed to operate the spacecraft with all its instruments operating would be too big to even get into space. Instead, the mission leverages a set of batteries and the long-period orbit to store energy collected during the outer part of the orbit to use during the busy flyby period.

Since Clipper's power system cannot be sized to simply supply the power needed to operate everything at once, it becomes an energy management problem where the supply needs to be sufficient to keep batteries charged to a safe level in the face of loads that come and go over time. This means that sizing the system can only be done in the context of scenarios that also estimate how the system will be operated over time through the use of simulation.

Sizing a power system is further complicated by the fact that many of Clipper's functional elements are being designed uniquely for this mission. While some common parts such as certain sensors and actuators, computers, radios, and reaction wheels may be acquired off-the-shelf and have known power consumption properties, many other parts including science instruments will be custom made. Power demand drives how much power will have to be produced, so practice dictates that estimating the demand is something that has to begin in the earliest concept studies and continue to be refined throughout the mission.

Before hardware has been manufactured and tested, these estimates must account for the uncertainties associated with design and development. Early designs must account for the uncertainty of precisely how something will be operated, its environment, what parts will be available, and how they will perform. Common practice is to record such estimates using something close to a statistical model by capturing a *current best estimate* (CBE) representing the most likely value based on current knowledge and a *maximum expected value* (MEV) representing the outlying tail of a distribution curve. Since engineers are not good at estimating probabilities such as these, design rules often dictate simple contingency factors as a function of where the project is in its lifecycle to compute the MEV value from CBE. Ideally, the power system would then be designed to the MEV value, and any difference between that and where CBE eventually ends up is design margin.

Modeling Scenarios

The project took a two-pronged approach to modeling and validating operational scenarios. Both approaches relied on using a common set of component power load estimates captured in a system model. The first approach relied on a set of simulation models. More than one simulation model was used because there was no single tool ideally suited for all of the analyses under consideration. One model focused on modeling aspects of plan coordination, and another on aspects of science observations and surface coverage. Some models were able to better leverage existing model fragments or libraries which provided validation for other models.

The second prong of the approach was to use a simpler traditional method to abstract these system behaviors into a small number of system modes where each mode can be described as a set of quasi-steady-state behaviors for the entire system. Some elements can be described as either ON or OFF in a given mode. Other elements with more dynamic operational properties can be described using simple duty cycle descriptions as toggling between ON and OFF or between a set of other steady-state modes. Instead of focusing on behavioral accuracy, these estimates were intended to bound the worst-case power behavior in each system mode as a bounding validation to the simulation models which were much more detailed, took much longer to update and execute, and were much harder to validate on their own.

Modeling Waste Heat

Keeping propellant from freezing and keeping electronics above their minimum operational temperature limit is another major use of energy on a spacecraft in the outer solar system. Direct heating from the Sun does not help much out at Jupiter. Spacecraft use electrical heaters to deliver energy to parts such as propellant tanks that have no other source of energy, and they also rely heavily on the "waste" heat produced by the functioning of electrical components. The heat generated by those components is transmitted through direct conduction or, in the case of Clipper, a thermal fluid loop, from areas where electronics render heat to areas where it is needed. Because sizing the thermal system is as important as sizing the power system and the fact that these two systems are closely coupled, it is helpful to track the amount of electrical power that every component converts to heat in the same model where the power estimates are tracked. Most components dissipate virtually all of their electrical energy input as heat. Radio transmitters, on the other hand, radiate significant amounts of energy away from the spacecraft; this energy is no longer available as local heat.

CAESAR PEL Application

As the Clipper project was approaching the end of its design phase, it began to realize that all of the modeling and simulation capabilities it had developed were costing a lot to maintain. The CAESAR team was brought in to provide institutional resources to identify those parts of what Clipper had developed that could most effectively be made reusable for other projects and to make them so.

Through a collaborative effort with the Clipper project, the CAESAR team identified the modeling and management of power load estimates as a highly reusable function in need of standard tooling and method. Clipper was capturing this information in a SysML model that, due to its size and complexity, could only be updated by expert modelers and, even then, required significant effort on their part to input and validate the information. They had developed methods to project that model data into tabular views using OpenMBEE [2] that the rest of the project could then use to review the data. Editing in SysML was complicated by the fact that the graph of dependencies between thousands of model elements was very difficult to edit through SysML diagrams graphically as the diagrams become too large and unwieldy or, if information is split across multiple diagrams, too difficult to keep consistent between them. This cycle of updating and reviewing the model consumed an excessive amount of resources compared to using a traditional spreadsheet to track power load estimates. However, it did have the advantage of ensuring well-formedness of the model once the updating and reviewing cycle was complete. The CAESAR team set out to provide a solution for tracking power load estimates with the rigor of the SysML model-based approach and the ease of maintenance of the spreadsheet-based approach.

CAESAR PEL Application

The CAESAR PEL vocabulary is designed to address the stakeholder concerns associated with a Power Equipment List (PEL), including providing information about power usage to the power subsystem in order to ensure that sufficient power could be produced and providing information about heat generation to the thermal subsystem in order to ensure that temperature requirements would be met. The vocabulary is only sufficient to describe the quasi-static, mode-based modeling approach that is used to bound the simulation model output.

CAESAR PEL assigns power usage to components that present electrical interfaces which correspond to entities of type *fse:Assembly* in the vocabulary established in [1]. Power usage changes over time, and in keeping with traditional practice, CAESAR PEL characterizes the power usage of an *fse:Assembly* using a finite set of power modes. A *pel:powerMode* is defined as a set of constant-valued constraints on the state variables that describe the power usage of an *fse:Assembly*. CAESAR PEL assumes that the power usage of an *fse:Assembly* is quasi-static in each mode and that an *fse:Assembly* can only be in one mode at any point in time.

Power usage is not precisely known, especially early on in the life cycle. CAESAR PEL reflects that uncertainty using the three traditional resource usage parameters: Current Best Estimate (CBE), Contingency, and Maximum Expected Value (MEV). CAESAR PEL defines CBE as the responsible authority's best estimate of the power usage of an *fse:Assembly* in a certain power mode in the expected operating environment. Contingency is defined as the responsible authority's best estimate of the fractional increase in the power usage of an *fse: Assembly* in a certain power mode incorporating reasonable worst-case assumptions and other uncertainties. MEV is defined as the responsible authority's estimate of the power usage of an *fse:Assembly* in a certain power mode given reasonable worst case assumptions and incorporating other uncertainties. These properties are implemented as constant-valued constraints on state variables corresponding to CBE, Contingency, and MEV power usage estimate state variables that characterize each *fse:Assembly*.

To address thermal concerns, CAESAR PEL can describe the heat that is produced by any *fse:Assembly*. The Waste Heat Fraction is defined as the fraction of the estimated power usage that is dissipated as heat. The Waste Heat Fraction can be applied to either the CBE or MEV power usage value to determine CBE and MEV Waste Heat estimates. For missions with a fluid loop like Europa Clipper, the amount of heat being dissipated on the fluid loop needs to be tracked separately in order to track how heat moves throughout the system. Therefore, CAESAR PEL defines a property called the Fraction of Waste Heat on Thermal Loop. The amount of heat being dissipated on the fluid loop can then be calculated by multiplying either the CBE or MEV power usage estimate by the Waste Heat Fraction and the Fraction of Waste Heat on Thermal Loop. These thermal properties are implemented as constant-valued constraints on state variables corresponding to the CBE and MEV Waste Heat and Heat on Thermal Loop state variables that characterize each *fse: Assembly*.

The CAESAR PEL vocabulary describes several different types of *pel: powerModes*. Modes for which CBE, Contingency, and MEV are directly specified by the user are defined as *pel:SpecifiablePowerModes*. However, the CBE, Contingency, and MEV power usage estimates can also be defined using a mapping to other power modes. For example, a *pel:PowerMode* that characterizes an entire flight system is traditionally defined by specifying the power mode or modes that all components within that system are in. Therefore, CAESAR PEL also defines a type of power mode called *pel:AggregatePowerMode* whose CBE, Contingency, and MEV are defined via a mapping to other power modes. CBE, Contingency, and MEV power usage for *pel:AggregatePowerModes* is calculated using a sum of the power usage properties of the mapped power modes, weighted by a duty cycle.

CAESAR PEL can also capture other information related to the power usage estimate including the voltage assumed when computing power usage and the rationales for why a certain value for CBE or Contingency was chosen by the user.

These other properties are implemented as characterizations of each *pel: PowerMode.*

Some PEL information consumers analyze power usage information in the context of system decomposition trees. For example, thermal subsystems engineers are concerned with where heat is being produced within the system and therefore would like to view the PEL within the context of the mechanical composition tree. Systems engineers are concerned with verifying that power usage estimates are within allocated values and so would like to view the PEL in the context of the Work Breakdown Structure (WBS). To support these and other concerns, CAESAR PEL has the capability to relate power usage to various system composition trees.

The electrical systems engineering vocabulary introduced in [1] is able to describe the mechanical composition of the system by declaring that an *fse:Assembly base:contains* one or more *fse:Assemblies*. Therefore, once power usage information is known for all leaf *fse:Assemblies* in the system (that is, *fse:Assemblies* that do not contain any *fse:Assemblies*), power usage for any other *fse:Assembly* can be calculated by summing along the edges of the mechanical composition graph.

To express the work breakdown structure for a project, several new concepts were introduced as part of the *project* ontology [3]. A *project:WorkPackage* is a discrete unit of project cost, schedule, or activity. *Project:WorkPackages* can *project:authorize* or delegate authority to other *project:WorkPackages*, establishing the Work Breakdown Structure. The responsibility for a Work Package to deliver hardware for a project is described by saying that a *project:WorkPackages project:supplies* an *fse:Assembly*. Therefore, once power usage has been estimated for all *fse:Assemblies* in the system, the power usage of *fse:Assemblies* that any Work Package is responsible for can be determined by summing over the joint graph of *project:authorizes* and *project:supplies* relationships.

Advantages of the CAESAR PEL Vocabulary

The vocabulary described in the previous section is arguably more precise than the SysML equivalent because all of the relations are explicitly meaningful where the SysML implementation merely implied meaning in SysML dependencies. It does this with simpler, more concise syntax, yet the higher precision makes it easier to write portable rules and analysis. And those analyses can be more rigorous at lower cost since they can be reused where analyses written against SysML models with implied meaning cannot. The vocabulary and patterns defined above imply several properties that can be checked automatically using an automated reasoner. The properties can be divided into two categories: range restrictions on values of properties and topological invariants. The properties are all checked by an automated reasoner during analysis of the model as described in a subsequent section.

The definitions of the various power usage estimates and other properties imply limits on the range of the values that they can take. The CBE, Contingency, and MEV properties all need to be greater than or equal to zero. The Waste Heat Fraction and Fraction of Heat to Thermal Loop properties need to be between zero and one inclusive. Similarly, the duty cycle value for any power mode that is mapped to another power mode needs to be between zero and one inclusive. Additionally, for each mapping from a set of power modes to an *pel:AggregatePowerMode*, the sum of the duty values for all modes of each *fse:Assembly* that participates in the mapping needs to be equal to one.

The set of power mappings forms a tree with each system power mode at the root. Trees have topological invariants that can be leveraged to guarantee that the power usage of each leaf component is counted exactly once when summing power usage for a system power mode. A graph is a tree if it is directed, acyclic, connected, and rooted. Directionality is needed to uniquely define the summations along edges. Acyclicity guarantees that each power mode in the graph is counted at most once. Connectedness guarantees that each power mode in the graph is counted at least once. Rootedness guarantees that there exists a level of the tree for which the power summations describe the entire system. If these four properties hold for the graph, then summing the power usage along edges of the graph from leaf components to the root is guaranteed to accurately compute the power usage of the system power mode. While the graph theoretic constraints (e.g., acyclicity) are beyond what can be expressed in description logic (and therefore in OWL 2 DL and OML), the formal representation in terms of nodes and edges makes it convenient to query the graph and subject it to analysis by external tools, e.g., one of many software graph libraries.

The corresponding modeling patterns in SysML can be complex because SysML cannot express all of these relations unambiguously without being extended and, because SysML is authored through diagrams, relationships stretch across multiple diagrams, making them difficult to inspect, verify, and maintain. For example, consider how a mapping between assembly power modes for assemblies that make up an instrument and that instruments' power modes was expressed in SysML. Assembly power modes were defined in SysML using states in a State Machine diagram. Instrument power modes were defined using states in a different State Machine diagram. The assembly power mode to instrument power mode mapping was made with the aid of diagrams (one per instrument mode) populated with swimlanes. These swimlanes were very laborious to update due to the number of relationships, stereotypes, and elements involved. In addition, relationships between these diagrams and instrument modes, work packages, and analysis characterizations had to be maintained.

Implementing the CAESAR PEL

CAESAR PEL extends the model and continuous integration architecture established for harness specification modeling in CAESAR by adding the ability to characterize fse:*Assembly* elements with power modes and relate those power modes in a graph. CAESAR architecture separates authoring from reporting and analysis which is done in a continuous integration process downstream of authoring and outside of the authoring tool.

CAESAR Authoring Workbench

Authoring power modes and associated power estimates has traditionally been performed using whatever ad hoc tools and methods were closest at hand, most often a spreadsheet. The CAESAR team opted to adapt its model authoring tool to this set of viewpoints. The CAESAR workbench is a desktop application based on the popular Eclipse Modeling Framework (EMF) [4] which was itself developed to support the implementation of model-based applications. EMF provides tools needed to encode the underlying vocabulary into a metamodel or schema for the application and then project model information into table-based or form-based views with which users can interact.

The CAESAR workbench had previously been adapted for authoring abstract electrical interconnections used to specify harness design for Europa, including authoring the electrical assemblies to be connected. The same view and underlying model are extended to add the ability to characterize the power modes and associated loads for all electrical assemblies (the model also supports identification of non-electrical assemblies). Figure 1 shows how the composition view presents a list of assemblies grouped by subsystem in a spreadsheet-like presentation showing identifying attributes in the columns.

Additional authoring views are provided for capturing assemblies' power modes and estimates (Fig. 2) and the modes and mode mappings. This view is populated with every assembly defined in the composition view and allows the user to add power modes and associated power estimates to each assembly (not all assemblies consume power so not all will have power modes).

Another view (Fig. 3) enables the creation of subsystem modes. Subsystem modes provide a way to associate behavioral modes of assemblies contained in a subsystem with a common functional behavior in order to somewhat simplify the

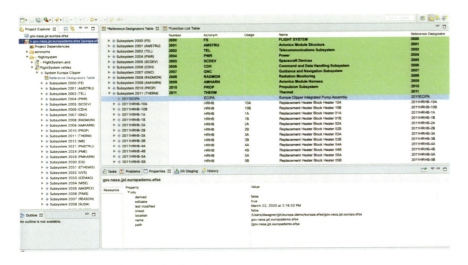

Fig. 1 Composition view (simulated data)

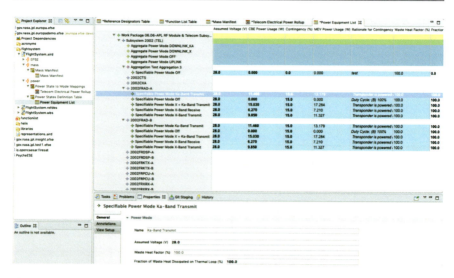

Fig. 2 Assembly mode/estimate input form

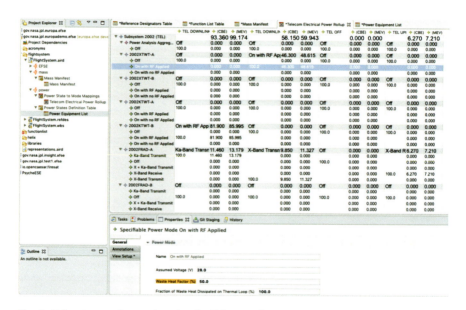

Fig. 3 Subsystem mode mapping view for Telecom subsystem

task of mapping system modes to the behaviors of individual assemblies. For completeness, each subsystem mode needs to map at least one assembly mode for each contained assembly. At the system level, each system mode must then map to at least one subsystem mode for each subsystem or to at least one assembly behavior mode for every assembly in a subsystem where a subsystem mode is not used. In essence, every system mode must map to at least one assembly mode for every

38 Semantic Modeling for Power Management Using CAESAR

assembly that has power behavior. CAESAR PEL allows the modeler to use subsystem power modes as an intermediate step in that mapping.

Note that in all of these views, the user is permitted to enter data via the tabular perspective (spreadsheet-like) or through a form perspective in the Properties tab, below, for the selected row.

Power Analysis Workflow

Continuous integration is supported by the CAESAR architecture through the concept of a workflow that can export one or more source models, transform the data to a common OWL representation, validate consistency of the input models and generate entailments, store the resulting model data in a SPARQL endpoint, and then perform a set of queries, data reductions, and analysis needed to render user-friendly reports and documents. The intent of this is to provide feedback to both the authoring users and their wider information consumers regarding the current content of those models but also their consistency and completeness.

SPARQL query results can be requested in multiple well-known syntactical forms, including CSV and JSON. For this application, JSON results were post-processed into a form suitable for import into the popular data analysis language R. All reductions and reports were produced using various R libraries.

Certain fatal inconsistencies such as mismatched references and cyclic dependencies are trapped in the authoring workbench to prevent these inconsistencies from ever being published or even saved. An integration workflow can begin when a user saves and commits a new formal version of the model through the configuration management view (CAESAR workbench uses git as the underlying CM system). The user can then manually trigger the workflow to begin on the user's working branch of the model. This usually results in a set of reports being generated from that commit of the model (the model validation that takes place in the workflow will occasionally identify an inconsistency that the workbench or other tool failed to notice). These findings generally lead to fixes to the model vocabularies or associated tooling that will reveal errors to users during authoring or simply prevent them from being created.

The workflows are implemented using the openCAESAR tools described earlier, adaptations of those tools that perform custom model transformations, adaptations of the reporting frameworks to create reports, and some scripting needed to containerize the steps and run it all in Kubernetes. As noted earlier, CAESAR makes a distinction between violations of different kinds of constraints. Constraints can be asserted in vocabularies at the linguistic level or in the descriptive model itself. The latter includes constraints expressed by the project that total power consumption for a spacecraft can never exceed a given capacity. The former includes a rule asserting that power load estimates need to express positive quantities in a unit of power or that power modes can only be attributed to assemblies. This layering gives projects control over constraints they need to control while allowing generic constraints to be asserted in reusable vocabularies or libraries. Also, a violation of an analytical constraint within the model is something to be reported to users as work to go. A violation of a linguistic constraint implies that the model is ill-formed and its analysis results are suspect. A few examples are described below (Table 1).

Report views are intended to project model data into presentations that are meaningful to the various consumers of the information. Some reports present information in a form amenable to management review, while others focus on identifying missing, incomplete, or inconsistent with analytical constraints. Some of these reflect forms similar to the authoring viewpoints presented in the workbench but optimized for consumption rather than authoring (Figs. 4 and 5).

In addition to generating all of the reports as HTML documents published to a web server alongside their harness specification peers, all of the PEL reports were also published as OpenMBEE views to the OpenMBEE document Europa users were used to using to find this information. Since the reports are essentially the same HTML data structures and only differ by their containment in a larger document, this was mostly a matter of creating a single new adapter to write the content to OpenMBEE.

Chapter Summary and Future Updates

Formally characterizing different behaviors in a consistent way is only the first step enabling the analysis of those behaviors needed to identify a satisfactory design solution for a spacecraft. Most spacecraft include the ability to store power when it

Table 1 Rule examples

Checked By	Rule	Notes
OWL reasoner	An interface is presented by a component only	The vocabularies assert dozens of such constraints: sub-/superclass relations, property domains and ranges, property cardinalities, disjointness assertions, etc., all of which are checked
	Every assembly belongs to at most one subsystem	
	Every junction joins only interfaces of compatible type	
Query/ test in R	An assembly power mode may map directly to a system power mode or indirectly through a subsystem power mode, but not both	This condition is equivalent to acyclicity of the power mode mapping graph for every system mode. The test uses the R *igraph* library
	Every system power model is mapped from some power mode of every assembly	This condition is equivalent to a reachability condition on the power mode mapping graph for every system mode. The test uses the R *igraph* library
	The duty cycle of a power mode mapping is a real number in the interval [0, 1]	Simple arithmetic comparison
	The duty cycles of all modes mapping to a superior mode sum to 1.0	Simple arithmetic summing and comparison

CAESAR — europa-pel

Spacecraft Power Modes List

The Power Modes List enumerates the power modes available to each type of component on the flight system. Note that this list does not provide information about the configuration and operation of the spacecraft, nor was it intended to provide either. For configuration information, see the block diagrams or the deployment MEL. For some operation information, see the Power Scenarios. Some configuration information can also be seen in the Power Scenarios.

For leaf elements (e.g. cards, sensors, supports), the Maximum Expected Value (MEV) is determined by multiplying the Current Best Estimate (CBE) by a contingency factor.

2002 Telecommunications Subsystem

Assembly	Mode	Power				Waste Heat			Thermal Loop Waste Heat	
		CBE [W]	Contingency [W/W]	MEV [W]	Factor [W/W]	CBE [W]	MEV [W]	Factor [W/W]	CBE [W]	MEV [W]
Frontier Radio A	Ka-Band Transmit	11.46	15%	13.18	1.00	11.46	13.18	1.00	11.46	13.18
	Off		15%		1.00			1.00		
	X + Ka-Band Transmit	15.03	15%	17.28	1.00	15.03	17.28	1.00	15.03	17.28
	X-Band Receive	6.27	15%	7.21	1.00	6.27	7.21	1.00	6.27	7.21
	X-Band Transmit	9.85	15%	11.33	1.00	9.85	11.33	1.00	9.85	11.33
Frontier Radio B	Ka-Band Transmit	11.46	15%	13.18	1.00	11.46	13.18	1.00	11.46	13.18
	Off		15%		1.00			1.00		
	X + Ka-Band Transmit	15.03	15%	17.28	1.00	15.03	17.28	1.00	15.03	17.28
	X-Band Receive	6.27	15%	7.21	1.00	6.27	7.21	1.00	6.27	7.21
	X-Band Transmit	9.85	15%	11.33	1.00	9.85	11.33	1.00	9.85	11.33
Ka-band Traveling Wave Tube A	Off		25%		1.00			1.00		
	On with no RF Applied	40.50	5%	42.52	1.00	40.50	42.52	1.00	40.50	42.52
	On with RF Applied	81.90	5%	86.00	0.50	40.95	43.00	1.00	40.95	43.00

Fig. 4 Example power modes report

CAESAR — europa-pel

Subsystem Power Rollup

2002 Telecommunications Subsystem

Estimate	DOWNLINK_KA	DOWNLINK_X	OFF	UPLINK
CBE [W]	93.36	56.15	0.00	6.27
Contingency [W/W]	0.06	0.07	NA	0.15
MEV [W]	99.17	59.94	0.00	7.21
CBE Waste Heat [W]	52.41	33.00	0.00	6.27
CBE Thermal Loop Waste Heat [W]	52.41	33.00	0.00	6.27
MEV Waste Heat [W]	56.18	35.63	0.00	7.21
MEV Thermal Loop Waste Heat [W]	56.18	35.63	0.00	7.21

Assembly	CBE [W]	Contingency [W/W]	MEV [W]	Waste Heat Factor [W/W]	Thermal Loop WHF [W/W]	Mode	Duty Cycle [s/s]			
Frontier Radio A	11.46	0.15	13.18	1.00	1.00	Ka-Band Transmit	1.00			
Frontier Radio A	0.00	0.15	0.00	1.00	1.00	Off			1.00	
Frontier Radio A	6.27	0.15	7.21	1.00	1.00	X-Band Receive				1.00
Frontier Radio A	9.85	0.15	11.33	1.00	1.00	X-Band Transmit		1.00		
Frontier Radio B	0.00	0.15	0.00	1.00	1.00	Off	1.00	1.00	1.00	1.00
Ka-band Traveling Wave Tube A	0.00	0.25	0.00	1.00	1.00	Off	1.00	1.00	1.00	1.00

Fig. 5 Example subsystem power roll-up

can be produced for use at a later time when it can't. Assessing the balance between demand and availability thus requires simulation over time to integrate the asynchronous contributions. The Europa project has developed a number of simulation capabilities [5] that can integrate both energy and data for this purpose. As a next step, the CAESAR team is looking at developing similar capabilities to use the configuration-managed characterizations of power modes and associated demands, combine that with models of scenarios that describe how those modes are invoked over time, and feed that all into a simulation engine that can perform the integration. The simulations can then integrate relevant trajectory and environmental factors (i.e., solar range and eclipses that affect solar power production) to determine power balance and battery state of charge over time.

Extending the system characterization to include data production/consumption behaviors would enable similar simulation analysis of data balance over time (spacecraft typically store and forward data to the ground as communication links are available and those links are typically asynchronous to the observation events that produce the data of interest).

Engineers have had access to a wide variety of tools and languages for describing and analyzing systems for some time, but they often have to choose between a solution that can directly support the needed mathematical analysis and a solution that can precisely describe the system that the math abstracts. E.g., Mathematica or even Excel can capture mathematical relationships, but they have little means to associate variables or cells to real-world objects in order to help modelers organize and manage the data. Languages like SysML can describe some of the system objects and relations, but cannot directly solve the math. Semantic modeling focuses on describing the system in a way that permits the model to be related to or transformed to other analytical viewpoints. Rather than look for a single solution to every problem, semantic modeling aims to federate diverse models through common vocabulary, modeling patterns, and rules. In this case the analysis is performed in a number of queries and scripts, but the project has also demonstrated the ability to transform from descriptive models into Modelica form to run high-precision simulations.

Most engineering organizations have some combination of tools in place that provide reasonably effective ability to describe parts of the system. Mechanical and electrical CAD tools, for example, have evolved over several decades to become very effective at capturing and analyzing geometric objects and constraints. These tools have already evolved the vocabulary and automation for their disciplines. But connecting those silos of information and capturing aspects of how these design viewpoints combine to make a system are not so well established – at least as commercial products – in part because every system is different.

But the approach still applies to the problem of describing how concepts and relations from one discipline map to those of another, providing the vocabulary needed to compare, for example, architecture with design or mechanical design to thermal design. While it may be possible to connect data from tool to tool by more superficial means, such solutions are often brittle or difficult to maintain as all of the interpretation and transformation logic is hidden in imperative code that is difficult to analyze and review. When the mappings are instead expressed as vocabularies, they can be validated using off-the-shelf reasoners and transformations, queries, and other analysis that can be expressed in simpler, more maintainable forms.

The case study demonstrates that the CAESAR platform provides an effective way to implement discipline-specific model authoring tools, change control the model information in git, and feed this data into a continuous integration process that can validate the essential consistency of the data, provide regular feedback to authors on completeness and consistency with external constraints expressed in other models, and publish reports for consumption by downstream consumers.

Cross-References

▶ Ontological Metamodeling and Analysis Using openCAESAR

Acknowledgment Some of the work described here was performed at the Jet Propulsion Laboratory, California Institute of Technology, under a contract with the National Aeronautics and Space Administration.

References

1. Wagner et al (2020) CAESAR model-based approach to harness design. IEEE Aerospace conference 2020
2. OpenMBEE https://www.openmbee.org/
3. OpenCAESAR Project Ontology https://github.com/opencaesar/jpl-vocabularies/blob/master/src/oml/imce.jpl.nasa.gov/foundation/project.oml
4. Eclipse Modeling Framework https://www.eclipse.org/modeling/emf/
5. Ferguson, Eric W., et al. (2018) Improving spacecraft design and operability for Europa Clipper through high-fidelity, mission-level modeling and simulation. *2018 SpaceOps Conference.*

David A. Wagner manages both the System Modeling and Methodology group as well as product development on the CAESAR project at the Jet Propulsion Laboratory. He holds a bachelor's degree in Aerospace Engineering from the University of Cincinnati, a master's degree in Aerospace from the University of Southern California, and over 40 years of experience in systems engineering large and small projects at JPL.

Mark Chodas is a systems engineer at the Jet Propulsion Laboratory. He is a product owner for the CAESAR project and contributes to other model-based systems engineering tools at JPL. He previously was the Instrument Systems Engineer for the REXIS instrument onboard NASA's OSIRIS-REx asteroid sample return mission. He holds bachelor's, master's, and PhD degrees in Aerospace Engineering from the Massachusetts Institute of Technology.

Maged Elaasar is a senior software architect at the Jet Propulsion Laboratory (California Institute of Technology/NASA), where he technically leads the Integrated Model Centric Engineering program and the CAESAR product. Prior to that, he was a senior software architect at IBM, where he led the Rational Software Architect family of modeling tools. He holds a PhD in Electrical and Computer Engineering and MSc in Computer Science from Carleton University, (2012, 2003) and a BSc in Computer Science from American University in Cairo (1996). He has received 12 US patents and authored over 30 peer-reviewed journal and conference articles. He is a regular contributor to and leader of modeling standards at the Object Management Group like UML and SysML. Maged is also the founder of Modelware Solutions, a software consultancy and training company with international clients. He is also a lecturer in the department of Computer Science at the University of California Los Angeles. His research interests span model-based engineering, semantic web, big data analytics, and cloud computing.

J. Steven Jenkins is a Principal Engineer in the Systems Engineering and Formulation Division, Jet Propulsion Laboratory, California Institute of Technology. Since 2009 he has served as the Chief Engineer of JPL's Integrated Model-Centric Engineering Initiative. His interests include the integration of descriptive and analytical modeling and the application of knowledge representation and formal semantics to systems engineering. He was awarded the NASA Outstanding Leadership Medal in 1999 and was a co-recipient of the NASA Systems Engineering Award in 2012. He holds a BS in Mathematics from Millsaps College, MS in Applied Mathematics, and PhD in Electrical Engineering from University of California, Los Angeles.

Nicolas Rouquette is a Principal Computer Scientist at the Jet Propulsion Laboratory where his pioneering work on comprehensive code generation for JPL's Deep Space One mission paved the way for applying such techniques since then. He made key contributions to several revisions of modeling standards developed by the Object Management Group (OMG) for UML and SysML, including for UML 2.4.1 which became the first revision of UML for which formal verification of consistency was performed by mapping to OWL2 and for SysML 1.4 which became the first revision of SysML for which comprehensive support 13 for the 3rd revision of the Vocabulary of International Metrology (VIM) is available in SysML's ISO 80000 library. He holds a PhD in Computer Science from the University of Southern California where he enjoyed guidance on applied graph algorithms from Pavel Pevzner and then a PostDoc at USC's Math department. He also holds an engineering diploma from the French ESIEE engineering school in Paris.

Modeling Trust and Reputation in Multiagent Systems

39

Michael Sievers

Contents

Introduction 1154
    Problem Statement 1154
    Key Concepts and Terminology 1155
State-of-the-Practice 1157
    Ontologies 1157
Best Practice Approach 1178
Illustrative Example 1181
    JPL's Deep Space Network 1181
    Trust and Reputation Study and Results 1182
Chapter Summary and Future Work 1184
Cross-References 1185
References 1185

Abstract

Understanding agent trustworthiness in a multiagent system is a prerequisite for evaluating network state and determining when corrective actions are needed. Untrustworthy agents may disrupt network operation by sending erroneous data or initiating malicious or inadvertently dangerous transactions. This chapter defines key parameters of trust and reputation and surveys the state-of-practice in trust ontology, modeling, and evaluation. Agent health is a key aspect of trust but is not often included in conventional trust analyses. This chapter describes a means for accommodating health information as a component of trust and

---

Currently employed at Jet Propulsion Laboratory, California Institute of Technology. This work was done as a private venture and not in the author's capacity as an employee of the Jet Propulsion Laboratory, California Institute of Technology.

M. Sievers (✉)
NASA/Jet Propulsion Laboratory, California Institute of Technology, Pasadena, CA, USA
e-mail: msievers@jpl.nasa.gov

© Springer Nature Switzerland AG 2023
A. M. Madni et al. (eds.), *Handbook of Model-Based Systems Engineering*,
https://doi.org/10.1007/978-3-030-93582-5_52

discusses architectures for collecting and using health information. The concluding section demonstrates one model of trust based on an experimental approach developed for a large system of systems.

Keywords

Trust · Reputation · Agent health · Dependability · Ontology · Reinforcement learning

Introduction

An *agent* is an autonomous and continuously operating software component that continuously executes actions for another actor that may be other software components or system users. A *multiagent system* (MAS) is one in which there are multiple interacting agents. Agents are powerful entities that can cause significant damage in an MAS if breached by malicious attack or weakened by disruptive events. An MAS must have means for determining whether individual agents are trustworthy and quarantine those agents deemed untrustworthy.

Trust is an agent's understanding of another agent's capability, honesty, and reliability based on its observations [1]. Reputation reflects the collective view of trust across all agents. Intuitively, trust and reputation might be qualitative terms; however, several practical and theoretical evaluation methods have been developed. While some of these methods are applied in cybersecurity and blockchain contexts, more commonly, though less obvious, quantitative trust and reputation applications evaluate the integrity of product and company ratings.

This chapter refers to a single form of trust that aggregates the following concepts often discussed in the literature:

- Provision trust: An agent's assessment of trust in resources or information from other agents
- Access trust: An agent's assessment of trust in allowing access of its information by other agents
- Delegation trust: An agent's assessment of trust in allowing other agents to perform actions on its behalf
- Identity trust: An agent's assessment of trust that another agent is properly identifying itself
- Context trust: An agent's assessment of trust that the system and infrastructure it lives in support the transactions it needs and provide a trustworthy safety net

Problem Statement

The sophistication of technology and system complexity will continue their explosive growth. Coinciding with that growth is both our decline in understanding how systems work and our willing ceding of essential functions to automation. Ultimately, we trust that our systems do the right thing while also preventing intentional or unintentional actions. Unfortunately, in general, our complacency is only jolted when a high-profile and embarrassing hack is made public. Of course, computer systems compromises are serious. Many dedicated and highly trained experts specialize in preventing computer compromise and clean up after the inevitable damage is done. Missing though in much of those efforts is an understanding that systems can become more vulnerable when disrupted by unexpected internal or external events. More than just an issue of cyberdefense and event postmortems, system trust boils down to a risk assessment that includes consideration of known and potentially novel disruptive events such as described by Avizienis et al. [2]:

- Development errors
- Internal and external faults
- Natural and man-made faults
- Hardware and software faults
- Malicious and nonmalicious faults
- Deliberate and nondeliberate faults
- Accidental and incompetence faults

Key concepts and terminology are discussed next. Section "State-of-the-Practice" examines trust evaluation approaches based on information exchanges and collaboration among system agents. There are also many well-known means for determining system health. However, this chapter focuses on ways in which cybersecurity and health evaluations are combined into an overall trust assessment.

Key Concepts and Terminology

A *dependable* system provides defensibly trustworthy mission services within a specified period despite *disruptive* events and in the presence of potentially *untrustworthy* neighbors. Reputation is the collective measure of trustworthiness (dependability) derived from observations made by peers within a system. A peer's subjective trust is derived from its observations and the observations made by its peers, although psychologists and sociologists distinguish between cognitive (an agent's rational expectation that a peer has the necessary competence, benevolence, and integrity to be relied upon [3]) affected trust (developed from an emotional assessment of security and confidence placed in others), and game-theoretical trust (defined by a subjective probability in which an agent expects that a peer acts in a way that benefits the peer [4–7]). However, in this chapter, we view trust from the perspective of three factors [8]:

- Expectancy: An agent expects certain behaviors from a peer, such as providing valid information or taking mutually cooperative actions.

- Belief: An agent believes that expectancy is valid based on observations made of the information and actions of a peer. Section "Trust Evaluation" discusses an unsupervised learning approach useful in determining belief [9].
- Vulnerability willingness: An agent assesses the risk within a given context of accepting information and actions by a peer.

Evaluating trustworthiness, though, is complicated by several factors such as the following:

- The health of the agent making the evaluation: Disruptions may hamper or nullify an agent's self-defense mechanisms. These disruptions may result from internal fault conditions, maliciously inserted code, design flaws, unexpected and stressing operating conditions, weak internal health and status checks, and so forth.
- The health of agents being evaluated: Peers may not understand their health status and may report conflicting information to their peers, setting up so-called Byzantine fault conditions. A Byzantine fault occurs when a peer collective cannot arrive at a consistent opinion regarding some aspect of system operation.
- Trust is asymmetric: An agent's trust in a second agent may not be the same as the trust the other agent has in the first agent.
- Malicious agent collectives: Collaborating agents subvert reputation evaluations by elevating the trust of the collective and denigrating trust evaluations of otherwise trustworthy agents.
- Low-rate malicious activities: Malicious agents may attack a system at low rates or in short bursts that avoid detection by mechanisms that look for rapid onset and sustained malicious activities. An untrustworthy agent may quietly hide until it attacks, or it may behave nominally most of the time but periodically attempt a malicious activity.
- Identity spoofing: A malicious agent uses the identity of a trustworthy agent for gaining unauthorized system access.
- Code changes: A malicious agent changes the code on other agents to confuse its trust evaluation.
- Normal errors: Agents may take actions that call into question their trustworthiness but are random and harmless. However, persistent errors may indicate an issue with an agent's trustworthiness.
- Data saturation: A system in which reputation is established by asking peers for trust evaluations and then peers of peers followed by peers of peers of peers, and so forth, can quickly get overwhelmed by the quantity of data that must be processed.
- Standards: Although many qualitative and quantitative methods exist (see section "State-of-the-Practice") for evaluating trust and reputation, no single "gold standard" exists. Consequently, system dependability cannot be judged in the absolute, i.e., trust and reputation are fundamentally subjective and strongly affected by the choice of observations made, the frequency of observations, and how those observations are used.

State-of-the-Practice

This section discusses trust and reputation ontologies, trust evaluation methods, and means for estimating belief, i.e., the confidence placed in trust evaluations.

Ontologies

As discussed in other chapters, establishing an ontology is an important part of creating models. An ontology defines the terms, constraints, and relationships for a given domain.

Viljanen [10] establishes important ontological trust relationships:

- Trust is asymmetric: If X trusts Y, it is not always true that Y trusts X.
- Trust is not distributive: If X trusts (Y and Z), it does not necessarily follow that X trusts Y AND X trusts Z.
- Trust is not associative: The expression X trusts (Y) trusts (Z) is not valid.
- Trust may be transitive: If X trusts (Y trusts Z) implies that X trusts Y and Y trusts Z. That X trusts Z may be valid.

A trust model taxonomy is the foundation of ontological semantics:

- Identity aware models: A model includes a notion of target identity. Identity need not be globally understood or unique but minimally must be locally unique and temporally stable. All trust models are identity-aware.
- Action aware models: Trust depends on what action an agent is attempting and for what purpose that agent is being trusted. The set of actions may be open implying no restrictions or closed in which actions or parameters of actions have limitations. For example, an evaluation of a system's belief state may be used to choose an action as discussed in [9] and summarized in section "Best Practice Approach."
- Business value aware models: These models include concepts of risk and reward for taking actions.
- Competence aware models: Evaluate the competence of an agent performing an action. Less competent agents may be associated with a greater risk of harm.
- Capability aware models: Although related to competency, capability in trust models typically refers to whether an agent has proper permission for data access or performing an action.
- Confidence aware models: account for the degree of confidence in computed values of trust and reputation.
- Context-aware models: Trust evaluations may depend on internal system state or external status at a given point in time, e.g., an agent affected by an internal fault condition may be more susceptible to intrusion.

- History aware models: These models evaluate trust and reputation by tracking past behavior. Current behaviors can be compared with past behaviors as a way of finding unusual patterns.
- Third part awareness: Model may use information from external sources in evaluating trust.

Using this taxonomy, the ontology comprises the top-level class, *Principle,* which is a generalization for subclasses *Trustor* and *Trustee.* These two subclasses are associated with the class, *trusts,* which is a composition of subclasses *Context, Confidence, Action, Competence, 3rd Party Information,* and *History.* Subclass, *Action* is a composition of three specializations of the class *Business Value: Risk, Benefit, Importance.* Subclass *3rd Party Information* is a generalization for three specializations: *Reputation, External credential,* and *Recommendation.*

Huang and Fox [7] describe an ontology that is available as an OWL (https://www.w3.org/TR/2012/REC-owl2-primer-20121211) file from the University of Toronto (http://ontology.eil.utoronto.ca/trust.html). The ontology comprises two classes of trust: *trust in belief* (the trust an agent has in its evaluation of belief) and *trust in performance* (an agent's trust in the information or actions taken by a peer). The Web Ontology Language (OWL)-based ontology (see Chapter 4 for an introduction to OWL) refers to the former as Trust_b and the latter as Trust_p which have relationships with concepts in the domain: *trust_Degree* (the trust evaluation), *trustee* (the peer in question), and *trustor* (the agent making the trust evaluation of the peer). The range for this domain comprises between *has_trust* and has_*trusted_certainty_degree_for.*

Huang and Fox develop formal semantics based on *situation calculus* which is a logical construct useful in representing dynamically changing domains. Situation calculus represents system dynamics through a set of *fluents* which are situation-dependent values. Situations change as agents in a system perform actions. Their ontology treats trust and reputation as fluents that are the basis of formal semantics representing a form of predicate logic.

An ontology proposed by Ceolin et al. [11] extends an earlier ontology developed by Alenemr et al. [12] using definitions from O'Hara [13]:

- $Tw < Y, Z, R(A), C>$ (Trustworthiness): An agent, Y, behaves in a way that conforms to a behavior R that benefits a collective, A in a context, C, as determined by an agent Z. Context defines the conditions under which Y is willing, able, and motivated to conform to R.
- $Tr < X, Y, Z, I(R(A), c), Deg, Warr>$ (Trust attitude): An agent, X, believes with confidence, *Deg,* that *Y*'s intentions, capacities, and *motivations* conform to *I(R (A), c)* that agent X believes are consistent with *R(A),* as determined by an authorized agent Z. Warrant is defined as the positive and negative inputs used by X in making its judgment.
- X places trust in Y (Trust action): Placing trust is different than trusting. Trusting requires that *Tr* is true. *X* may begin *placing trust* in Y when *Tr* is *True* by acting in ways that might appear harmful or illogical. X may also trust Y without *placing*

trust in *Y.* However, if *Tr* is *True,* then *X* believes that the action of *placing trust* in *Y* is safe and will conform to *I(R(A), c).*

The extended ontology comprises the following terms:

- Trustor (source): an agent that makes a trust assessment based on available evidence and policies for using that evidence.
- Trustee (target): the agent evaluated by a trustor.
- Trust Attitude Object: a trustor's belief in the trustworthiness of a trustee. Similar to reputation but includes O'Hara's warranty concept.
- Trust Attitude Value: the computed value of Trust Attitude Object as a function of Role, Warranty, Context, and Trust Action Object.
- Trust Action Object: the result when an agent takes the action, *placing of trust.* Since *placing of trust* occurs instantaneously, the trust value is likely either True or False.
- Role, Context, and Warranty: The extended ontology include O'Hara's warranty notion.

For example, a partial semantic web for Trust Attitude Object comprises the following:

In pseudo-OWL notation, *Trust Attitude Object*
hasSource(Trustor)
hasTarget(Trustee)
hasTrustAttitudeValue(Trust Attitude Value)
hasCriteria(Criterion
hasRole(Role)
hasWarranty(Warranty)
hasContext(Context)
CalculatedBy(Computational Algorithm))

More recently, trust and reputation ontologies have concentrated on cybersecurity and cyber resilience. Veloudis et al. [14] describe a metamodel for context-aware security models. A model is associated with three subclasses: *Context Pattern, Permission,* and *Security Context Element.*

Subclass Security *Context Element*
hasHandler (Handler)
isAssociatedWith(Request)
isAttributeOf (Request)
isAssociatedWith(Subject and *Object)*
isAttributeOf(Request)

The class *Subject* represents a requestor, i.e., an agent wanting access to an object or an agent under consideration for allowing object access by another agent. *Object* represents a protected resource. *Handler* processes data related to access control. The

Permission class defines the actions allowable by members of the *Subject* class. *ContextPattern* represents the history of access patterns that is used for judging whether future access by a *Subject* is permitted.

Amaral et al. [15] summarize a long-term research program investigating the ontological concepts of trust for social and organizational modeling. Although this research is not explicitly focused on engineered systems, concepts such as risk, value, preferences, actions, roles, contracts, capabilities, and so forth are readily identified in the ontologies above.

Trust Evaluation

As described by Mui et al. [16], trust, reputation, and reciprocity are linked as shown Fig. 1 in which the arrow shows the direction of influence. Reciprocity involves mutual exchanges, e.g., of services or data; the reputation of an agent is relative to the system in which that agent is a member and represents its perceived behavior by all agents within the system, and trust is the subjective belief that an agent has regarding another agent's future behavior.

If a system comprises n agents ($a_1, a_2, \ldots a_n$), then the following three relationships hold:

- Increasing agent a_i's reputation within the system should increase the trust other agents have in a_i.
- Increasing agent a_j's trust in a_i should increase the believe that a_j reciprocates positively to actions taken by a_i.
- Increasing agent a_i's reciprocity to other agents in the system should increase a_i's reputation.

The paper formulates a trust model as a function of:

- α_i: Actions taken by an agent a_i, $\alpha_i \in A$, and A is the set of all actions.
- $\rho \in [0, 1]$: Reciprocity (a measure of whether cooperative actions are met by cooperative responses).
- $\theta_{ji}(c)$: the reputation of a_i of concern to agent a_j within the context, $c \in C$.
- ε: Encounters: $\varepsilon \in E = \alpha^2 \times C \cap \{\varnothing\}$ where E is the set of all encounters and $\{\varnothing\}$ is the set of no encounters.

Fig. 1 Reinforcing relationships between trust, reciprocity, and reputation

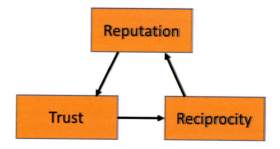

- $D_{ji}(c) = \{E^*\}$: History of the encounters agent a_j has with agent a_i in context, c, and E^* is the set of all finite length strings derived from elements of E (Kleene closure). That is $D_{ji}(c)$ might include other encounters agent a_i has with agents other than a_j.

Using this notation, trust is computed as:

$$\tau(c) = \mathbb{E}[\theta_c | D_c] \tag{1}$$

That is, trust, $\tau(c)$, is the expected value of reputation in context, $\tau(c)$, given the history D_c. Expectation is the belief that some event will occur in the future while belief is an evaluation of truth regardless of supporting or conflicting evidence. The basic formula for expectation is $\mathbb{E} = P(x) * n$, i.e., the probability of an event, x, multiplied by, n, the number of times the event happens. Belief estimation is discussed in section "Bayesian Belief Network."

The paper by Janiszewski [17] discusses the implementation of a trust and reputation management system that chooses a service provider using a five-step process:

1. Observe interactions with other agents, ask other agents for their opinions of peers, and save a history of those interactions.
2. Evaluate trust from the observations made.
3. Choose a service provided using the trust evaluation.
4. Interact with the selected provider and evaluate that interaction.
5. Update trust based on the evaluation in Step 4.

Janiszewski computes trust using a measure of effectiveness defined as the ratio between the sum of positive and negative interaction values and the total number of interactions. A similar construct was described by Kamvar [18] in a seminal paper describing an Eigen Trust algorithm. As described in [18], there are five important agent-agent trust issues:

1. The system must be self-policing and not depend on a central authority. This is important because a central authority can be a single-point failure in that its misbehaviors might bring the rest of the system down.
2. The system should depend on opaque identifiers for agents rather than externally associated identifiers. This allows flexibility in where agents are hosted but also provides some insulation against external spoofing or attack.
3. New agents must successfully complete a number of transactions before earning system trust. Additionally, system rules need to prevent malicious or unhealthy agents from escaping quarantine by changing their opaque identifiers.
4. Trust evaluations must not impart significant overhead, i.e., the storage, computation, and complexity of transactions should not noticeably impact system performance or resources.

5. Most importantly, the system must protect against groups of malicious or unhealthy agents that conspire to disrupt the rest of the system.

The basis for the Eigen Trust algorithm is a network-wide rating of peer-to-peer transactions. A transaction between agents a_i and a_j, tr_{ij}, is assigned a positive value if the transaction is successful and a negative value otherwise:

$$tr_{ij} = \begin{cases} +1, & \text{successful transaction} \\ -1, & \text{unsuccessful transaction} \end{cases} \tag{2}$$

Each peer computes its understanding of local trust value as $lt_{ij} = \sum tr_{ij}$ where the sum is over all interactions between agents a_i and a_j. Local trust is normalized so that malicious collectives cannot bias global trust by assigning high local trust values to other members of the collective and low values to peers not part of a malicious collective. Normalized trust is defined as:

$$c_{ij} = \frac{\max\left(lt_{ij}, 0\right)}{\sum_j \max\left(lt_{ij}, 0\right)} \tag{3}$$

Equation 3 ensures that normalized trust is between 0 and 1, and is set to 0 if $\sum_j \max\left(lt_{ij}, 0\right) = 0$. c_{ij} is a relative value and cannot distinguish between the situation in which agents do not interact and when interacting agents have more unsatisfactory interactions than satisfactory. Moreover, if $c_{ij} = c_{ik}$, then we only know that a_j and a_k have the same trust as viewed by a_i, but we do not know if both a_j and a_k are trustworthy or equally untrustworthy. When a set of peers $p_i \in P$ are known to be trustworthy, then we can define $p_i = 1/|P|$ if $i \in P$ and 0 otherwise. When a system includes trusted agents, then Eq. 3 may be rewritten:

$$c_{ij} = \begin{cases} \dfrac{\max\left(lt_{ij}, 0\right)}{\sum_j \max\left(lt_{ij}, 0\right)}, & \text{if } \sum_j \max\left(lt_{ij}, 0\right) \neq 0 \\ p_j, & \text{otherwise} \end{cases} \tag{4}$$

To an extent, a 0 denominator in Eq. 3 may be avoided using a guaranteed message delivery protocol that assures message delivery even when a direct path between agents is not available [19]. That is, if agent a_i attempts an interaction with a_j, the delivery protocol assures that the interaction occurs. However, it may not be the case that each agent interacts with each other agent as required by Eq. 3, especially in systems comprising heterogeneous agents. Heterogeneous agents are designed for specific purposes and may not share the same inputs, perform the same function, produce the same output, or even interact with each other in the same way. Given these dissimilarities, it may not be clear what messages agents should send for evaluating Eq. 3. A potential mitigation for heterogeneous agents discussed in [19] and in section "Best Practice Approach" is an exchange of health and status among agents. Partitioning a system into cohorts of interacting peers is another option.

39 Modeling Trust and Reputation in Multiagent Systems

Aggregating the normalized local trust is determined by weighting a given peer's normalized trust with the opinions of its peers. The trust peer i places in peer k after asking its neighbors is:

$$t_{ik} = \sum_j c_{ij} c_{jk} \tag{5}$$

In matrix form, let C be matrix $[c_{ij}]$ and $\vec{t_i}$ be the vector of values t_{ik}, then

$$t_{ik} = C^T \vec{c_i} \tag{6}$$

A more complete understanding of trust can be obtained by asking the opinion of peers of peers and so on:

$$t = \left(C^T\right)^n c_i \tag{7}$$

Equation 7 eventually converges to the same vector for each peer producing a consistent global trust evaluation. An iterative algorithm is defined in [18] in which

$$t^{k+1} = \left(C^T\right)^{k+1} c_i \tag{8}$$

is compared to t^k. If the difference exceeds a value δ, then t^{k+2} is computed and compared with t^{k+1}. As an example, in experiments conducted by the author using data from Caltech's Jet Propulsion Laboratory (JPL) Deep Space Network (DSN)), it was found that between 6 and 12 iterations were needed when $\delta = 1.0e^{-4}$[20].

Over time, it should be expected that s_{ij} eventually becomes very large. In that situation, unsuccessful transactions have little impact on normalized local trust in Eq. 3. The implication is that an attack on a peer that had been behaving for a long time is not detected unless the attack persists long enough that it noticeably impacts the results of Eq. 3.

For the DSN experiment, trust was evaluated in fixed-length windows in which earlier trust values were scaled and added to the most recent trust value. Accommodating prior trust with current trust enabled maintaining a history of prior interactions but placed more emphasis on current status as shown in Eq. 9:

$$t_{\text{new}} = t_{\text{window}} * \left(1 - sf\right) + t_{\text{old}} * sf \tag{9}$$

In Eq. 9, t_{new} is the updated global trust, t_{window} is the global trust computed with data from the last time period, t_{old} is the previous global trust, and sf is a scaling matrix in which $sf_{ij} = [0, 1]$. The scaling factor is evaluated during the computation of t_{window}. If a given agent does not participate in transactions with another agent, then the scaling factor associated with those agents is higher than if those agents had interacted. This is done so that lack of interactions does not immediately have an adverse impact on quiet agents. However, Eq. 9 guarantees that the trust between those agents decays so that long-term absence of transactions eventually impacts

Table 1 Typical DSN experimental results

Destination = dst-1	
Source	Global Trust
src-1	0.0154
src-2	0.0111
src-3	0.0246
src-4	0.0101
src-5	0.0100
src-6	0.0239
src-7	0.0116
src-8	0.0136

Table 2 Example anomalies

IP	Anomaly
ip-1	Unusual IP found
ip-2	Unusual IP found
ip-3	Unusual mean trust
ip-4	Unusual mean trust
ip-5	Unusual mean trust
ip-6	Unusual mean trust
ip-7	Unusual IP found

global trust. After some experimentation, it was found that $s_{ij} = 0.8$ provided good results and supported detecting short-term anomalous behavior.

The evaluation process used for the DSN experiment began by collecting statistics over a 1-month period. Several million interactions occurred during the month and assured a good set of initial trust evaluations. In operation, experimental data were collected every m minutes for the prior $m + n$ minutes so that collection windows overlapped by n minutes. Anomalous conditions are detected by evaluating current trust mean and standard deviations from their baseline values.

Table 1 shows typical DSN experiment results for one agent's view of a subset of its peers. Table 2 shows representative anomalous detections in which "Unusual IP found" means that a given IP did not appear in the baseline and "Unusual mean trust" is an indication that the deviation from the baseline of current computed global trust exceeded an acceptable value.

A discussion and analysis of other trust evaluation methods is found in [21].

Bayesian Belief Network

Bayesian Belief Networks, also called Bayesian Networks or Belief Networks, represent the conditional dependencies between random variables through directed acyclic graphs. Bayesian Networks are commonly used for evaluating system trust [22–24].

Fig. 2 Bayesian network example

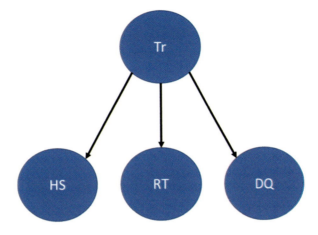

Bayesian Belief Networks support an understanding of the structure of causality relationships within a domain of interest using Bayesian theory of conditional probability. Recall that given two random, dependent, variables, x and y, then

$$P(x|y) = \frac{P(x,y)}{P(y)} \quad (10)$$

$$P(x|y) = \frac{P(y|x)P(x)}{P(y)} \quad (11)$$

If x and y are independent, then

$$P(x|y) = P(x) \quad (12)$$

A belief network then evaluates

$$P(x_1, x_2 \ldots x_n) = \prod_{i=1}^{n} P(x_i | x_1, x_2, \ldots x_{i-1}) \quad (13)$$

As a simple example, suppose we want an evaluation of a service provider's (*SP*) competence and we make observations of its health status (*HS*), response time (*RT*), and data quality (*DQ*). In words, we are after the probability that *SP* can be trusted given observations of *HS*, *RT*, and *DQ*. The Bayesian Belief network is shown in Fig. 2. Suppose we want to know how probable (*Tr* = *1*) given *HS* = good, *RT* = normal, and *DQ* = low. The corresponding equation is:

$$P(Tr = 1 | HS = \text{good}, RT = \text{normal}, DQ = \text{low}) \quad (14)$$

Table 3 Example observations for $P(HS|T, RT, DQ)$

	Tr	RT	DQ	HS = ng	HS = g
1	0	nn	l	0.7	0.3
2	0	nn	h	0.8	0.2
3	0	n	l	0.3	0.7
4	0	n	h	0.5	0.5
5	0	nn	l	0.9	0.1
6	0	nn	h	0.8	0.2
7	0	n	l	0.7	0.3
8	0	n	h	0.6	0.4
9	1	nn	l	0.3	0.7
10	1	nn	h	0.2	0.8
11	1	n	l	0.2	0.8
12	1	n	h	0.5	0.5
13	1	nn	l	0.1	0.9
14	1	nn	h	0.2	0.8
15	1	n	l	0.3	0.7
16	1	n	h	0.5	0.5

From the chain rule in Eq. 9:

$$P(Tr = 1|good, normal, low) =$$

$$\frac{P(good|normal, low, Tr = 1) * P(normal|low, Tr = 1) * P(low|Tr = 1) * P(Tr = 1)}{P(good|normal, low) * P(normal|low) * P(low)} \quad (15)$$

The terms in Eq. 15 come from observations made on system operation. For example, suppose trust, $Tr = 1$, has a Boolean value, i.e., $\{0,1\}$ and the remaining terms in Eq. 15 come from other observations made of system operation. For example, Table 3 shows an arbitrary set of values for $P(DQ|T, HS, RT)$ in which $HS = \{g, ng\}$ (good, not good), $RT = \{n, nn\}$ (normal, not normal), and $DQ = \{h, l\}$ (high, low). From row 11 in Table 2, $P(good|normal, low, Tr = 1)$ is 0.8.

Markov Decision Processes

A Markov Decision Process (MDP)) is a dynamic Bayesian Network in which an agent chooses an action at a time, t, based on the observation of state s_t with the goal of maximizing a reward. An MDP is a Markov Process defined by the tuple $(S, A, \mathbb{P}, \mathbb{R},$ and $\gamma)$ in which:

- S is a finite set of states, $s_i \in S$, and all states are observable.
- A is a finite set of actions that may be taken at each state, $a_t \in A$.
- T is the state transition probability matrix, $T_{ss'}^a = P[s_{t+1} = s'|s_t = s, a_t = a]$ where $P(s_{t+1}|s_t)$ must equal $P(s_{t+1}|s_1, s_2, \ldots s_t)$ (Markov Property).

- $\mathbb{R}$ is a reward function, $R_s^a = [R_{t+1} \mid s_t = s, a_t = a]$.
- γ is the discount factor for taking future actions, $\gamma \in [0.1]$.

The Markov Property requires that a state, s_t, captures the relevant history of all prior states and all prior states can be ignored when evaluating the probability of arriving at state s_{t+1} from state s_t.

Figure 3 is an example of an MDP in which the transitions between states are labeled with probability associated with actions taken. For example, for a given action, the model transitions from state s_1 to state s_2 with the probability 0.3. The sum of all transitions exiting a state must sum to 1.0.

Table 4 shows state transition matrix. $\mathbb{P}$, corresponding to Fig. 3.

The transition matrix after n-steps is $\mathbb{P}^n$, e.g., after two steps:

$$\mathbb{P}^2 = \begin{bmatrix} .1 & .3 & .2 & .4 \\ .3 & .4 & .1 & .2 \\ .2 & .5 & .2 & .1 \\ .4 & .3 & .1 & .2 \end{bmatrix} \begin{bmatrix} .1 & .3 & .2 & .4 \\ .3 & .4 & .1 & .2 \\ .2 & .5 & .2 & .1 \\ .4 & .3 & .1 & .2 \end{bmatrix} = \begin{bmatrix} .30 & .37 & .13 & .20 \\ .25 & .36 & .14 & .25 \\ .25 & .39 & .14 & .22 \\ .23 & .35 & .15 & .27 \end{bmatrix} \quad (16)$$

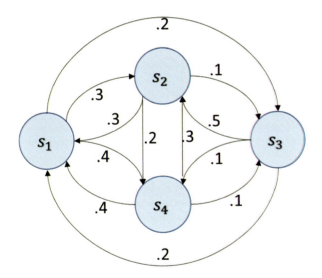

Fig. 3 Example Markov Process

Table 4 State transition matrix for Fig. 3

	s_1	s_2	s_3	s_4
s_1	0.1	0.3	0.2	0.4
s_2	0.3	0.4	0.1	0.2
s_3	0.2	0.5	0.2	0.1
s_4	0.4	0.3	0.1	0.2

A Markov Reward Process (MRP)) is a Markov Process that evaluates the reward function, $\mathbb{R}$, on the basis of immediate reward and discounted reward of future actions. The discount factor, γ, reduces future rewards and represents an understanding in the confidence of future actions. A value of γ near 0 indicates little confidence in the future and gives preference to immediate actions. As γ approaches 1, there is more confidence in future actions and more accumulated reward.

The objective of an MRP is finding the path through a Markov Process that maximizes the *return*, G_t:

$$G_t^* = \max(G_t) = R_{t+1} + \gamma R_{t+2} + \ldots + \gamma^{n-1} R_{t+n} = \sum_{i=0}^{\infty} \gamma^i R_{t+i+1} \tag{17}$$

Let an agent take an action, $a_t \in A_t$, when in state s_t, as dictated by a *policy*, π. In the realm of trust and reputation, a policy might determine whether one agent permits interaction with another or how much faith an agent places in information received from another agent. An agent's policy might be implemented as a static table that represents that agent's history with other agents or a policy may be continually updated as an agent observes the behavior of its peers. For a policy, π, Eq. 17 is updated as the *action value function*, Q:

$$Q^\pi(s,a) = \mathbb{E}_\pi \left\{ \sum_{i=0}^{\infty} \gamma^i R_{t+i+1} | s_t = s, a_t = a \right\} \tag{18}$$

That is, $Q^\pi(s,a)$ is the expected reward when an agent uses policy π in state $s_t = s$ and takes action $a_t = a$.

The maximum value function is

$$Q^*(s,a) = \max_\pi Q^\pi(s,a) \tag{19}$$

Equation 19 is used for finding the optimal policy

$$\pi^*(s) = \underset{a}{\text{argmax}} \, Q^*(s,a) \tag{20}$$

In which the *greedy policy* with respect to Q^* is a practical way for determining $\pi^*(s)$. A greedy policy makes a locally optimum choice at each step with the goal of global optimization. The greedy policy does not guarantee finding the globally optimal solution but finds approximations that are good enough. Note too that argmax(f(x)) is the value of x in which f(x) is the maximum.

Equation 18 may be evaluated iteratively using Bellman's [25] equation which comprises a series of decompositions:

$$\begin{aligned} Q^\pi(s_t,a) &= \mathbb{E}_\pi \left\{ \sum_{i=0}^{\infty} \gamma^i R_{t+i+1} | s_t, a \right\} \\ &= \mathbb{E}_\pi \left\{ \left(R_{t+1} + \gamma \left(R_{t+2} + \gamma R_{t+3} + \gamma^2 R_{t+4} + \ldots \right) \right) | s_t, a \right\} \\ &= \mathbb{E}_\pi \left\{ R_{t+1} + \gamma Q^\pi(s_{t+1}, a) \right\} \end{aligned} \tag{21}$$

39 Modeling Trust and Reputation in Multiagent Systems

Equation 21 is called an *infinite-horizon-discounted model* because the action value function is evaluated over an infinite number of steps. In practice however, iterations of $Q^\pi(s, a)$ reach a limit for $\gamma < 1$ suggesting that iterations may be stopped when the added reward at a step becomes sufficiently small. In a *finite-horizon model*, Eq. 21 is evaluated over a fixed horizon, H. That is, the infinite summation in Eq. 21 is replaced by a summation over the horizon size. *Value iteration* (VI)) and *policy iteration* (PI)) are two commonly used methods for finding optimal policies.

VI starts at an end state and works backward to a beginning state refining an estimate of an optimal policy. Since there is not an actual end state, one is selected arbitrarily and Eq. 21 is repeatedly calculated while keeping track of the best backward path. VI starts by setting $Q_0^*(s)$ to 0 for all states. Then given Q_i^*, calculate for all states using the Bellman Update function:

$$Q_{i+1}^* = \max_a \sum_{s'} T_{ss'}^a \left[\mathbb{R}_{ss'}^a + Q_i^*(s') \right] \tag{22}$$

PI begins by selecting any policy π_k as the starting point and iterates until convergence:

$$Q_{i+1}^{\pi_k}(s) = \sum_{s'} T_{ss'}^{\pi_k} \left[\mathbb{R}_{ss'}^{\pi_k} + \gamma Q_i^{\pi_k}(s') \right] \tag{23}$$

Next step is policy improvement that finds the best action by looking ahead one step:

$$\pi_k(s) = \operatorname*{argmax}_{s'} \sum_{s'} T_{ss'}^a \left[\mathbb{R}_{ss'}^a + \gamma Q^{\pi_k}(s') \right] \tag{24}$$

This two-step iteration repeats until convergence.

Partially Observable Markov Decision Process

In complicated or complex systems, it is likely that system state is only partially observable and state must be inferred from indirect observations, and probabilities. A *Partially Observable Markov Decision Process* (POMDP) [26] is a type of an MDP that has the same state-based architecture, but the underlying state is not observable or there is uncertainty in the outcome of an action. In an MDP, the problem is finding a mapping from states to actions while in a POMDP, the problem is finding a mapping from a probability distribution over the states to actions. A state probability distribution is called a *belief state* that represents probable state history given a set of observations $o_1, o_2, \ldots, o_t$. In belief state, b, the probability of being in state, s, is $b(s)$.

A POMDP may be described as an MDP with the addition of a sensor model, $O(o|s, a) = P(o|s, a)$, which is the probability distribution of observing o when in state s after taking action a. Figure 4 shows an example of a POMDP model. Each state, s, has an associated sensor model, an action policy, and a reward determined by the belief state for taking a selected action.

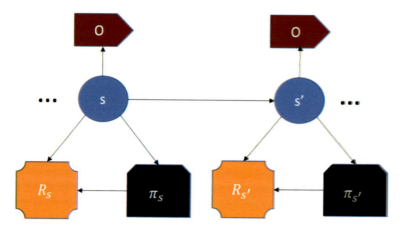

Fig. 4 Example POMDP model

Belief state is computed by initially assigning a belief distribution from knowledge prior to making observations. A new belief state, b', is computed from the current belief state, b, action, a, and the resulting observation, o:

$$b'(s') = P(s'|o, a, b) \qquad (25)$$

Using Bayes Rule, Eq. 25 is rewritten as:

$$b'(s') = \frac{O(o|s', a) \sum_{s \in S} T^a_{ss'} b(s)}{\sum_{s' \in S} O(o|s', a) \sum_{s \in S} T^a_{ss'} b(s)} \qquad (26)$$

The numerator in Eq. 26 enumerates the probabilities of transitioning to s' and observing, o, after taking action, a. The denominator normalizes the numerator by the probability of observing, o, given b and taking action a.

POMDPs are executed in using the following five steps:

1. Initialize the belief state.
2. Choose the action to take from policy $\pi(b)$.
3. Observe the system output.
4. Evaluate the next belief state based on Eq. 25.
5. Go to step 2.

MDP and POMDP Trust Evaluation

The inherent ability of accommodating partial unobservability makes POMPD models particularly useful in trust evaluations. The paper by Wang et al. [27] analyzes human-robot interactions in a dynamic trust environment. The system considered comprises a human-robot team in which the robot searches for weapons in a building and the human must decide whether it is safe to enter the building with or without protective gear. The robot is equipped with multiple weapon sensors, a

39 Modeling Trust and Reputation in Multiagent Systems

camera for detecting armed gunmen, and a microphone that listens for suspicious conversations. Errors are injected into the robot's evaluation by assuming a faulty camera that cannot see armed gunmen and occasionally gives an incorrect "safe" assessment. Trust is evaluated against four precepts:

1. The human's action following their assessment of the robot's trustworthiness. E.g., the human might always ignore the robot's recommendations even when indicating trust in the robot. From a reward perspective, altering actions that increase trust are not worth investigating.
2. Human actions prior to and after a mistake by a robot are related to human trust in the robot. Optimal decisions might be incorrect in hindsight when in an uncertain environment. Errors might reinforce a human's mistrust of the robot or might be quickly set aside as correct recommendations are made.
3. The difference between the times a robotic recommendation is followed or ignored is indicative of human trust in the robot. Reporting and correcting mistakes should result in more trust.
4. The number of corrective decisions made by a human also reflects the human's trust in the robot. A human who understands the strengths and weaknesses of the robot will generally make the right decisions and trust the robot within the scope of its strengths.

Chen et al. [28] describe an MDP model of learning and planning in a human-robot collaboration in which the human is evaluating trust in the robot's actions. The collaboration is described by a finite set of states and a set of human and robot actions. Model dynamics are driven by the state transition matrix that includes both human and robot actions:

$$T_{ss'}^{a^H, a^R} = P\left[s' | s, a^H, a^R\right] \tag{27}$$

in which $a^H \in A^H$, the set of all human actions7 and $a^R \in A^R$, the set of all robot actions. Based on a policy, the human determines a^H at step t by observing the current a^R and h_t which is the history of interactions between the human and robot up to step t.

$$\pi^H\left(a_t^H | s_t, a_t^R, h_t\right) \tag{28}$$

Equation 18 is extended as:

$$Q^{\pi^H, \pi^R}\left(s, a^H, a^R\right) = \mathbb{E}_\pi\left\{\sum_{i=0}^{\infty} \gamma^i R_{t+i+1} | s, a^H, a^R\right\} \tag{29}$$

and the optimal robot policy then is:

$$\pi^{R*}(s) = \operatorname*{argmax}_{a^R} Q^{*\pi^H, \pi^R}\left(s, a^R\right) \tag{30}$$

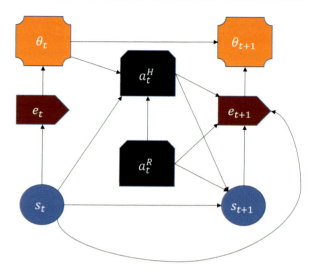

Fig. 5 POMDP trust model

in which $Q^{*\pi^H,\pi^R}(s,a^R)$ is the maximum value of $Q^{\pi^H,\pi^R}(s,a^H,a^R)$ over all policies. Trust is considered a single scalar random variable, θ, which is an approximation of h_t. With that assumption, Eq. 28 becomes:

$$\pi^H\left(a_t^H|s_t,a_t^R,\theta_t\right) = \pi^H\left(a_t^H|s_t,a_t^R,h_t\right) \quad (31)$$

Recognizing the trust changes over time, trust dynamics depend on robot performance, e_t, which is a function f_p of current state, new state after a transition, and human and robot actions:

$$e_{t+1} = f_p\left(s_{t+1},s,a_t^H,a_t^R\right) \quad (32)$$

Equivalent to belief state update, trust dynamics are approximated by Eq. 33. The POMDP model is shown in Fig. 5.

$$\theta_{t+1} = P(\theta_{t+1}|\theta_t,e_{t+1}) \quad (33)$$

Many other trust evaluation formulations have been discussed in the literature such as those in [29–32].

Evaluating Agent Health

Agents may be impacted by known faults with known risks (known-knowns), known faults of uncertain risk (known-unknowns), and unknown faults of unknown risk (unknown-unknowns). In this chapter, the term *error* refers to the difference between an expected outcome or value and the value produced by a system or an agent within a system. A *fault* is defined as the cause of an error. Faults may cause one or more errors depending on system or agent operation, inputs, usage, and environment.

Conventional Error Detection

Conventional fault-tolerance methods are effective in managing known-knows and somewhat effective in managing known-knows. However, conventional fault-tolerance likely is insufficient for trust evaluation when faced with unknown-unknown faults.

Avizienis et al. [2] developed an error detection taxonomy comprising concurrent detection (occurs during system operation) and preemptive detection (occurs when system service is suspended). While preemptive detection mechanisms are valuable, health status is needed currently with operations so that questionable agents are shut down or quarantined quickly.

Mitra and McCluskey's paper [33] describes a number of concurrent hardware detection schemes that depend on some form of redundancy and comparison (Fig. 6). As described in [33–35], there are many implementations of Fig. 6 that fundamentally depend on error codes and comparisons, and time-out. With a few exceptions (e.g., memory protection using error detecting and correcting code), agents do not execute on systems that make extensive use of hardware-based error detection due to the cost of redundant electronics and the added complexity associated with integrating the primary, prediction, and checking functions. Instead, if present, error detection is performed in software. Of course that raises the question of how much trust can we place in software that runs on a potentially faulty computer.

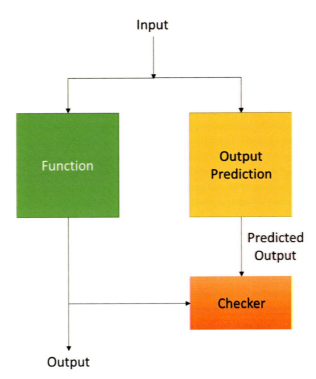

Fig. 6 Generic concurrent hardware-implemented error detection

Software error detection also requires some form of redundancy comprising a check of functional behavior or output against an expected result. Checking may be performed within an agent or by an agent's peers as messages are exchanged [36]. An agent or the computational environment of an agent may check resulting values against limits, timing, attempted access of out-of-bounds memory locations, stack violations, divide-by-zero exceptions, policy violations, value conversion exceptions, inconsistencies, and control flow errors, to name a few typical error checks. Peers may check each other by evaluating the content and timing of message exchanges and observation of unexpected transactions or requests for unapproved access. An interesting construct described in [37] reverses engineers network protocols using program binaries in a shadowing approach that takes advantage of program semantics as well as syntax. Another form of checking entails the exchange of challenge and response messages [38, 39] in which an initiating agent sends a challenge request to another agent. The receiving agent produces a response that is checked by the initiating agent. The initiating agent verifies the correctness of the response and triggers a warning if the response is late or incorrect.

Health Evaluation Using Models and Machine Learning

The current most promising means for managing unknown-unknowns comprise machine-learning (ML) and model-based constructs that create patterns of "nominal operation" and "off-nominal operation." When contrasted with observed system behavior, these patterns detect unexpected operation that can trigger diagnostic and response mechanisms.

There are a number of fundamental challenges to ML and model-based health assessments such as the following: defining the parameters and metrics for the training sets and models, biases; incorrect assumptions regarding operational correctness, completeness, and consistency; and confusing correlation with causality, to name a few. In spite of these challenges, ML and modeling have proven useful as discussed below.

Model-Based Health Assessment

Model-based health evaluations use a physics-, mathematical-, or knowledge-based model of a system. These methods are capable of detecting known errors as well as unknown-unknowns. A comprehensive discussion of model-based health diagnosis is found in Isermann [40]. The general process consists of the following three components:

- Monitoring: processes that evaluate system observables for consistency with nominal values.
- Automatic protection: Dangerous conditions detected by the monitoring function trigger safing events.
- Supervision: System behavior and nominal values are evaluated that produce error syndromes for fault diagnosis.

Figure 6 shows the components of model-based error detection. The fault model generates residuals, parameter estimates, and state estimates. In analytical redundancy applications, a comparison is made between multiple, disparate, analytical methods that produce equivalent results when no faults are present. A residual is the difference in the output of these methods and indicates a fault condition when the disparity exceeds a threshold. The change detection function normal behavior and features with those produced by Feature Generation and creates error syndromes which are signatures that point to the likely underlying fault condition.

An example model-based health monitoring described by Kolcio et al. is a physics-based approach called Model-Based Off-Nominal State Identification and Detection (MONSID) [41]. MONSID uses a diagnostic engine and nominal model of a physical system. Inputs and commands that drive system operation are also routed to MONSID's system model. The diagnostic engine evaluates the consistency of the model and physical system outputs. Inconsistencies indicate off-nominal conditions symptomatic of a system health problem.

MONSID implements analytical redundancy by evaluating forward and reverse constraints. A forward constraint relates system inputs to its outputs. A reverse constraint relates system outputs to system inputs. The two constraints create at least two values as each system interface that should be equal when no faults are present (Fig. 7).

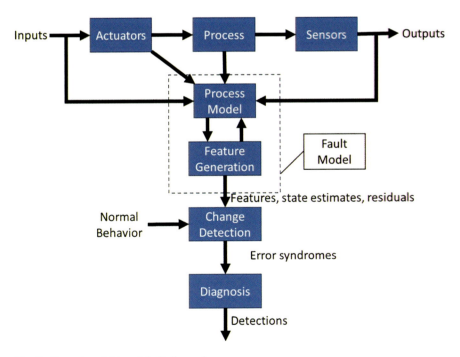

Fig. 7 General model-based fault diagnosis

As previously discussed, policies developed for the actions agents take are based on statistical inference developed through evaluation of the action value function of Eq. 18 and the optimal policy function in Eq. 20. However, attaining confidence that the failure rate of a policy is below ε requires at least $1/\varepsilon$ trials. Obviously obtaining that many trials may be infeasible for very small values of ε. Representative machine learning concepts that have found utility in agent health evaluations are discussed next.

ML-Health Evaluations

Uesato et al. [42] address the issue of ensuring that agents do not make catastrophic mistakes using an adversarial evaluation. The approach filters out situations that are not problematic and focuses on those that are. Recognizing that problematic situations are rare and often unpredictable (unknown-unknowns), the paper discusses a *continuation approach* for learning a *failure probability predictor* (AVF)). That approach develops the failure signatures by using data from agents that fail more often. Two algorithms are looked at for creating failure signatures: *failure search* and *risk estimation*.

The goal of this work is assessing the reliability of trained agents. The assessment is based on executing an experiment with the trained agent given a set of initial conditions. Over all expected randomness related to agent operation and environment, the experiment creates a binary-valued failure indicator in which "1" indicates a catastrophic failure.

Failure search looks for catastrophic agent failures using search algorithms called *adversaries*. Adversary algorithms set up initial conditions, execute the agent, and stop when a catastrophic condition is found. Agent history leading to the catastrophic failure is used for training an agent failure model. Risk evaluation evaluates a probability distribution of catastrophic failure over a set of initial conditions and agent probability distribution.

This work determined that standard reinforcement learning (RL) approaches are inefficient in detecting rare, catastrophic failures. A result is a false sense of system safety. The process of injecting catastrophic faults enables creating characterization of history and symptoms that can be used for monitoring system health.

Of course this approach cannot guarantee that untried conditions will not affect an agent or that the outcome of disruptions is not timing, state, or data dependent. However, it does provide a basis for creating useful failure signatures. Moreover, this approach might not detect noncatastrophic agent health problems such as those that might occur when the agent is under cyberattack or when it suffers nonfatal faults.

Alford et al. [43] examine an approach for monitoring the status of multiple cyber (robotic) agents. The approach looks at physical, sensory, and cognitive aspects of a human-cyber interface. For example, a cognitive feature for a human might be "logic," while the cyber equivalent might be an associative memory database. Cognitive capabilities of cyber entities are provided by *self-agents* while physical and sensory aspects are allocated to multiple, atomic agents. Importantly, self-agents aggregate failure information from the atomic agents and maintain overall health status which is called System Status Evaluation (SSE).

The basis for SSE are data flow patterns and timing relationships important for maintaining correct operation. SSE analyzes agent communication delay statistics similar to commonly used watchdog or heartbeat timers. Using the cumulative distribution function (CDF), agents evaluate the health of their peers by comparing an observed delay to a predetermined threshold.

Two forms of communication are used in the system. In one form, a source agent provides an observing agent with a steady stream of data. Here the time delay, $\delta(t)$, is simply the difference between two successive messages. The second form comprises a controlling agent that commands execution of operations in other agents. A receiving agent sends an acknowledgment back to the controller when the operation has completed. In this second case, $\delta(t)$ for each action is the difference between the receipt of the completion message and the transmission of the operation command.

After gathering statistics for both message types, the CDF determines the probability that $\delta(t)$ is less than or equal to a threshold value. This approach is similar to the fault-tolerance coverage concept described in [44] that uses CDF for comparing an observed recovery time to the time-to-critical-effect (TTCE) for a given fault. TTCE defines that time at which an unmanaged fault results in system failure. A key difference though is that the TTCE evaluation looks at the probability of system recovery while Alford focuses on error detection.

From an implementation perspective, the approach is relatively simple since it requires only collecting $\delta(t)$ values during normal operation. The CDF at a point, x, can be estimated by dividing the number of interaction times that occur up to x by the total number of interaction times collected. Alternatively, an arguably more accurate evaluation is achieved by computing the CDF for a specified probability distribution. However, a fundamental issue with heartbeat-type error checks is that these provide little insight into whether a system is reliably executing its expected functions. From a trust perspective, an unusual delay is suspicious, but a delay within acceptable bounds does not a guarantee trustworthiness.

Additional Resources
- Marginalized importance sampling for off-policy optimization [45]
- Fault detection by cotraining of semisupervised machine learning [46]
- Support vector machine fault detection in wireless networks [47]
- Flow-based intrusion detection [48]
- Cybersecurity Guide [49, 50]
- Network Fault-Tolerance [51]
- Byzantine Generals [52]
- Eigen Trust Byzantine Fault-Tolerance [53]
- Byzantine Fault-Tolerant Network [54]
- Swarm Resilience [55]
- CybersecurityManagement [56].

Best Practice Approach

Evaluating trust and reputation in partially observable environments in which agents may suffer malicious attacks, internal faults, and external events requires a combination of the tools discussed previously. Primarily missing from the formulation leading to Eq. 8 is an evaluation of agent health. Agent health could be incorporated into Eq. 4 if agents exchanged their health information. Health information can come from agent self-evaluations but also from agent-agent challenges. In a challenge transaction, an agent sends another agent a problem. In theory, if the receiving agent is healthy, then it returns the expected solution. Conversely, an unhealthy agent either does not return a solution or produces an incorrect result. Unfortunately, there are several considerations that complicate incorporating simple "good" or "bad" health exchanges into trust models.

There are two general trust architectures: centralized and distributed trust management. An example of a centralized trust architecture is shown in Fig. 8. Agents A, B, C, and D interact with each other and report the success or failure of those interactions to the Central Reputation Analysis (CRA) function. Agents A and D have no interaction history, so prior to accepting a transaction, each asks the CRA the other's reputation score. The CRA computes reputation as described in section "Trust Evaluation." Conceptually, a centralized architecture is simple and reduces the computational burden on agents. However, the CRA function is a prime candidate for malicious attack and moreover can become a bottleneck when accommodating a large number of agents.

The distributed architecture, implied by section "Trust Evaluation," is the preferred trust management architecture. This architecture is more difficult to disrupt because each agent is responsible for assessing the trust of its peers. Figure 9 shows an example of distributed trust management. As shown, agents interact with each other and keep track of the success or failure of those interactions. In this example, agents A and B do not have a prior transaction history and therefore cannot

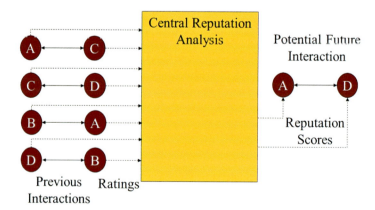

Fig. 8 Centralized trust management

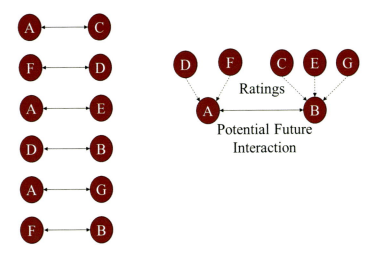

Fig. 9 Distributed trust management

determine the other's trustworthiness without help. Since agent B has interacted with agents D and F, agent A asks those agents for their assessment of agent B. Similarly, agents C, E, and G have transaction history with agent A agent B uses for assessing the trustworthiness of agent A.

Interaction history is necessary for evaluating trust but is not sufficient. Agents may also suffer from transient or permanent fault conditions that may result in fatal or nonfatal service errors. Fatal errors are usually the easiest to detect and manage because the affected agent is removed from service. Conversely, nonfatal or undetected errors may not significantly interfere with the operation of an agent and may or may not require further action. Unfortunately, nonfatal or undetected faults could hide more severe behaviors or weaknesses that may not immediately appear. Consequently, an agent's health is an essential component of an agent's trustworthiness.

Many agents perform self-health evaluations and use exception handlers that report error(s) and reset or terminate the agent. Honest agents will inform their peers when a health problem is detected. However, self-health checks and exception handlers can miss health problems. A potentially worse situation occurs when an agent's self-health mechanisms are compromised and thwarts honest evaluations and reporting.

An approach successfully used in the fault-tolerance community creates a variant of the centralized architecture shown in Fig. 8 in which the Reputation Analysis function is replaced by a so-called system *hardcore*. The hardcore comprises functions that must correctly manage disruptive events even when those functions are challenged by faults or malicious attacks. A network hardcore may consist of one or more trusted health agents that are protected by combinations of commercially available cyber defense products and processes [49, 50] and fault-tolerance methods [2, 51]. These agents evaluate client compliance with a network health policy such as the following:

- Verifying that clients have current health policy software
- Verifying that clients have current threat definitions
- Verifying that clients properly communicate with the network
- Verifying point-to-point transaction timing
- Verifying critical network utilization
- Triggering client self-diagnostic routines
- Receiving an error or exception message from a client
- Verifying challenge responses (more commonly used for thwarting spam)

A distributed hardcore as illustrated in Fig. 9 has a number of advantages over a centralized implementation such as simpler scalability, better efficiency, and greater fault-tolerance and reliability. However, data access and security can be managed relatively easily in a centralized implementation but is more complicated in distributed architectures.

Among the potential problems in distributed architectures are Byzantine [52] conditions in which a fault or attacked agent not only gives a different assessment of its own health to its peers, but also sends confusing information about its peers to other peers. An Eigen-Trust algorithm described by Gao et al. [53] is conceptually similar to Kamvar's Eigen Trust algorithm [18]. Their algorithm constructs a blockchain consensus group by iteratively evaluating local trust and reputation as in [18]. Agent trust is sorted by reputation, and agents with reputation greater than a threshold are included into a consensus group. A related consensus algorithm described by ElDefrawy [54] uses the protocol originally described in [52] and achieves f-Byzantine fault tolerance using transactions from $3f + 1$ agents. Another approach discussed in [55] adjusts belief state (Eq. 26) by weighting the influence of an agent's observations based on the agent's reputation.

Whether a centralized or distributed architecture is used, the response taken for health policy noncompliance depends on the severity of the policy infraction. In the most severe case, a noncompliant client may be disconnected from the network. Less severe options could involve sending a compliance violation warning to the affected client and marking the client a potential risk.

A modification of Eq. 2 enables including health assessment in a trust evaluation. A compliant health evaluation can be considered a successful transaction while noncompliant health assessments are considered unsuccessful transactions. However, the penalty for serious, noncompliant health assessments should carry a more severe penalty than a minor transaction glitches. Conversely, a serious transaction failure must be given more weight than a minor health policy infraction. Table 5 shows an example of penalty scaling based on the severity of noncompliant health and transaction malfunctions and transaction malfunctions.

Table 5 Example penalties for noncompliant health and unsuccessful transactions

Transaction failure	Penalty	Health noncompliance	Penalty
Message check code error	−1	Old threat definitions	−1
Message length error	−1	Correctable transient memory error	−1
Message semantic error	−2	Message time-out	−2
Unexpected request	−3	Application exception	−3
Access violation	−4	Challenge-response failure	−4

Illustrative Example

JPL's Deep Space Network

JPL's Deep Space Network (DSN)) is an international network of large radio antennas that provides two-way communication for guidance and control of interplanetary and Earth-orbiting spacecraft missions (https://www.nasa.gov/directorates/heo/scan/services/networks/deep_space_network/about). The DSN also supports radar and radio astronomy studies of objects within and external to our galaxy.

The DNS comprises three equidistant facilities in California, Spain (Madrid), and Australia (Canberra) that are located approximately $120°$ apart in longitude. The facilities are placed so that there is constant communication with spacecraft as the Earth rotates. All three sites directly communicate with the Space Flight Operations Facility (SFOF) located at the Jet Propulsion Laboratory in Pasadena, California. NASA spacecraft mission data and health telemetry are received at the SFOF for processing and made available to researchers and mission operators. Commands uplinked to spacecraft for controlling operation and mitigating anomalies also originate at the SFOF.

The critical nature of the DSN requires continuous vigilance and very high levels of cybersecurity against attacks that come from externally facing systems and applications. The needed protection is provided by multiple layers of sophisticated cyber protection and a team of highly trained cybersecurity experts who constantly monitor for novel attacks and update the cyber protection.

Unfortunately, there have been successful attacks on the DSN that resulted in significant data loss. In 2011, intruders gained access to 18 DSN servers that supported key NASA missions and stole 87 gigabytes of data. In 2018, an external user's account was compromised allowing an intruder to steal 500 megabytes of data from a major NASA mission. As a result of these attacks, NASA conducted an audit of JPL's cybersecurity [56]. The audit found several weaknesses including the following: keeping track of physical resources and applications that have DSN access, limited connected device visibility, lack of partner network segmentation and isolation, and a backlog of unresolved network issues. These deficiencies have been addressed, and the network is now significantly more secure than prior to the audit. However, an important lesson learned as a result of the audit is the importance of knowing each agent's current state of trustworthiness.

Trust and Reputation Study and Results

A study performed in 2020 investigated the availability of information and methodologies for evaluating agent trust. The study found that transaction information for an Eigen trust analysis was available, and moreover, all needed information was readily available at JPL. Although distributed trust evaluation can provide additional protection, having all needed data centrally available simplified data collection enabling a focus on algorithm experimentation.

DSN cybersecurity applications record essential data from all network transaction. These applications monitor a large number of transaction parameters; however, for our initial study, we used only transaction date and time of every transaction, source and destination addresses and ports, and a course-grained interpretation of transaction success or failure. It was understood that finer granularity monitors could be included by adding penalties to Table 5 and using the modified version of Eq. 2 as discussed in section "Best Practice Approach."

Two Python scripts are needed for the study: one script accessed DSN transaction data through a Splunk interface (https://www.splunk.com/) while the other imported JSON-formatted transaction data and performed the trust analysis. Because we were evaluating whether trust could be used in the DSN cybersecurity arsenal, Splunk data are filtered so only transactions accepted by the DSN cybersecurity were passed to the trust analysis script.

The trust analysis script imports the following seven utilities:

- json (parses json data)
- splunkReader (a JPL-developed script that connects to splunk and reads transaction data)
- numpy (Python numerical analyses)
- datetime (accesses system date and time information)
- sleep (suspends script operation for a specified time period)
- getpass (reads user passwords)
- sys (sets Python runtime environment)

The trust analysis script uses the four following primary functions:

- readJason: reads a JSON file and returns transaction data sorted by time
- localTrust: creates the local trust matrix
- getTime: reads current time
- writeOutput: creates an output report consisting of trust results for each address and statistical data associated with potential problems

The main program begins by obtaining current system time and reading a trusted peer list which is used for resolving situations in which peers have very few or possibly no transaction history. The script comprises two primary phases: baseline formation and network monitoring.

Baseline formation collects the previous month's history of successful transactions. There are typically more than 500,000 transactions in the history. Based on known network performance, the baseline duration was considered sufficiently long that observing transactions from all valid addresses was highly likely. Baseline reputation is evaluated using Eq. 8; however, experimentation showed that computing t^8 converged quicker than iteration and produced results that are sufficient for the study. The baseline history is also used for creating a list of observed client and server addresses.

After completing baselining functions, the script transitions to network monitoring in which it waits τ minutes between Splunk collection requests that cover transactions for the past $\tau + \delta$ minutes. The strategy creates an overlap of δ minutes so that unusual events at window borders are not lost. Experimentation indicated that $\tau = 55$ min and $\delta = 5$ min provided a reasonable compromise between oversampling and unmanageably large data collections for each 1-h window. Table 6 shows typical baseline global trust evaluated for one destination.

The following two checks are performed against the baseline:

- The first check looks for client or server addresses that are not part of the baseline. This is a simple check comprising comparing the addresses in the latest search window to the list of baseline addresses. Although looking for novel addresses is not a perfect indication of an anomaly, novelty was determined sufficient justification for triggering additional investigation by a security expert.

Table 6 Example baseline global trust

Destination = Dest-1	
Source	Global trust
src-1	0.0178
src-2	0.0121
src-3	0.0136
src-4	0.0117
src-5	0.0146
src-6	0.0133
src-7	0.0390
src-8	0.0133
src-9	0.0133
src-10	0.0133
src-11	0.0104
src-12	0.0403
src-13	0.0136
src-14	0.0118

- The second check evaluates the number of standard deviations between the current and baseline reputation means. An address is flagged for additional scrutiny if there are more than three standard deviations between the baseline and the current transaction sample.

Several options were investigated for evaluating statistical differences between acceptable and anomalous global trust. While none of the approaches tried were perfect, a candidate that produced reasonable results finds the number of standard deviations between the baseline and current global means:

$$\|baseMean_i - meanTrust_i\| / stdD_i \tag{34}$$

Where $baseMean_i$ is the mean global trust for the *i-th* address in the baseline address list, $meanTrust_i$ is the mean global trust for the *i-th* address in most recent search, and $stdD_i$ is the standard deviation of the global trust for the *i-th* address in the most recent search. As noted above, an address is flagged if Eq. (34) exceeds 3. Typical results are shown in Tables 1 and 2.

The script executes continuously until interrupted by an operator. Over a period of several weeks, multiple anomalies in the form of novel addresses and unusual statistics were detected. Fortunately, all were due to known maintenance operations and were not malicious attacks. Although malicious attacks did not occur during the study, the algorithm showed promise by detecting unusual behavior in transactions that were otherwise considered nominal.

Chapter Summary and Future Work

Computer networks are constantly threatened by internal and external cyberattacks. Protective cyber defenses may themselves fall victim to attack or become compromised by fault conditions that impact the platform and agents on which they depend. Moreover, coordinated attacks may create Byzantine situations that fool and confuse cyber defenses.

Although researchers are still struggling with means for staying ahead of cyberattacks, real-world exigencies make more near-term solutions an imperative. This chapter has summarized approaches that are in use or have been proposed for use for detecting and managing cyberattacks and potentially fault-weakened defenses. Some of the more promising of these use of ML concepts that learn nominal network behavior by observing network exchanges and status reports. These data are used for evaluating individual agent trust and overall network reputation.

Large networks will produce huge numbers of exchanges every second, and the overhead for collecting and processing all of these by a centralized sentinel is neither practicable. Moreover, a centralized sentinel becomes an obvious target for cyberattacks. In this chapter, we describe the experimental implementation of a modified version of the Eigen Trust Algorithm that combines windowing for reducing the chances of missing rare events and agent health monitoring that looks for potential fault-induced weaknesses.

The experiment collected real-time transaction data in week-long bursts over a period of several weeks. Each burst began by establishing a baseline reputation as well as a profile of network connections. Transaction samples within windows are used for reputation updates as well as for comparison with the baseline connectivity profile. The algorithm successfully found unusual behaviors that were traced to maintenance operations.

Although the initial results were promising, the efficacy of the approach in real attacks was not tested. A long-term study is needed in which malicious attacks occur that are not detected by the extremely robust cyber defenses already in place. Moreover, more work is needed for validating statistical models and the heuristics that resulted from those models. In particular, the approach must be tested by exposing it to recorded actual attack scenarios as well as white hat attacks. However, the key contribution of the approach described is in integrating multiple factors that impact trust and reputation with the ultimate goal of creating cyber-resilient systems.

Cross-References

▶ Cybersecurity Systems Modeling: An Automotive System Case Study
▶ Semantics, Metamodels, and Ontologies

References

1. Wang Y and Vassileva J (2003) "Trust and Reputation Model in Peer-to-Peer Networks," in *Proc. Third International Conference on Peer-to-Peer Computing (P2P2003)*, Linkoping, Sweden, pp. 150–157
2. Avizienis A, Laprie J.-C, Randell, B, and Landwehr C (2004) "Basic concepts and taxonomy of dependable and secure computing," IEEE Transactions on Dependable Computing, vol. 1, no. 1, pp. 11–33, https://doi.org/10.1109/tdsc.2004.2.
3. Komiak SX and Benbasat I (2004) "A Process Tracing Study on Trust Formation in Recommendation Agents," Proceedings of the Third Annual Workshop on HCI Research in MIS, Washington, D.C.
4. Guzmán R, Harrison R, Abarca N, and Villena MG (2020) "A game-theoretic model of reciprocity and trust that incorporates personality traits," Journal of Behavioral and Experimental Economics, Volume 84
5. Chin SH (2009) "On application of game theory for understanding trust in networks," *2009 International Symposium on Collaborative Technologies and Systems*, Baltimore, MD, pp. 106–110, https://doi.org/10.1109/CTS.2009.5067469.
6. Abusitta A, Bellaiche M, and Dagenais M (2018) "A trust-based game theoretical model for cooperative intrusion detection in multi-cloud environments," 2018 21st Conference on Innovation in Clouds, Internet and Networks and Workshops (ICIN), Paris, pp. 1–8, https://doi.org/10.1109/ICIN.2018.8401625.
7. Sabater J and Sierra C (2005) "Review on Computational Trust and Reputation Models," Artificial Intelligence Review 24, pp. 33–60, 2005, https://doi.org/10.1007/s10462-004-0041-5

8. Huang J, and Fox MS (2006) "An Ontology of Trust – Formal Semantics and Transitivity," Proceedings of the International Conference on Electronic Commerce, Association of Computing Machinery, pp. 259–270
9. Sievers M, Madni AM, and Pouya P (2019) "Trust and Reputation in Multi-Agent Resilient Systems," 2019 IEEE International Conference on Systems, Man and Cybernetics (SMC)
10. Viljanen L (2005) Towards an ontology of trust. In: Katsikas, S, López, J, Pernul, G (eds.), TrustBus 2005. LNCS, 3592, pp. 175–184. Springer, Heidelberg
11. Ceolin D, Nottamkandath A, Fokkink W, and Maccatrozzo M (2014) "Toward the Definition of an Ontology for Trust in (Web) Data," CEUR Workshop Proceedings. 1259
12. Alnemr R, Paschke A, and Meinel C (2010) "Enabling reputation interoperability through semantic technologies," I-SEMANTICS, pp. 1–9. ACM
13. O'Hara K (2012) "A General Definition of Trust," Technical Report, University of Southampton
14. Veloudis S, Paraskakis I, Petsos C, Verginadis Y, Patiniotakis I, Gouvas P, and Mentzas G (2019) "Achieving security-by-design through ontology-driven attribute-based access control in cloud environments," Future Generation Computer Systems, 93, 373–391, ISSN 0167-739X, https://doi.org/10.1016/j.future.2018.08.042
15. Amaral G, Prince Sales T, Guizzardi G, and Porello D (2019) "Towards a Reference Ontology of Trust," https://doi.org/10.1007/978-3-030-33246-4_1
16. Mui L, Mohtashem M, and Halberstadt A (2002) "A Computational Model of Trust and Reputation," Proceedings of the 35th Hawaii International Conference on System Sciences
17. Janiszewski M (2017) "Towards an Evaluation Model of Trust and Reputation Management Systems," Intl Journal of Electronics and Telecommunications, vol. 63, no. 4, pp. 411–416
18. Kamvar SD, Schlosser MT, and Garcia-Molina H (2003) "The Eigen Trust Algorithm for Reputation Management in P2P Networks," in *Proc. of the Twelfth International World Wide Web Conference*
19. Sievers M, Madni AM, and Pouya P (2018) "Assuring Spacecraft Swarm Resilience," in *Proc. AIAA Scitech*, San Diego
20. Michael Sievers, Arun Viswanathan (2021) An Experimental Reputation Algorithm for Detecting Anomalous Peers in Large Communication Networks AIAA ASCEND, pp. 11
21. Schlosser A, Voss M, and Brückner L (2004) "Comparing and Evaluating Metrics for Reputation Systems by Simulation," Proceedings of the IEEE Workshop on Reputation in Agent Societies
22. International Conference on Web Intelligence (WI 2003), Halifax, NS, Canada, 2003, pp. 372–378, https://doi.org/10.1109/WI.2003.1241218.
23. D'Angelo G, Rampone S, & Palmieri F (2007) "Developing a trust model for pervasive computing based on A priori association rules learning and Bayesian classification," *Soft Comput* **21**, 6297–6315, https://doi.org/10.1007/s00500-016-2183-1
24. Sardana N, Cohen R, Zhang J, and Chen S (2018) "A Bayesian Multiagent Trust Model for Social Networks," in IEEE Transactions on Computational Social Systems, vol. 5, no. 4, pp. 995–1008, https://doi.org/10.1109/TCSS.2018.2879510.
25. Bellman R (1956) "Dynamic Programming and Lagrange Multipliers," Proc Natl Acad Sci USA, vol 42, no. 10, 767–769
26. Kaelbling LP, Littman ML, and Cassandra AR (1998) "Planning and Acting in Partially Observable Stochastic Domains," Artificial Intelligence 101, pp. 99–134
27. Wang N, Pynadath D, Hill S, and Merchant C (2017) "The Dymanics of Human-Agent Trust with POMDP-Generated Explanations," In: Beskow J., Peters C., Castellano G., O'Sullivan C., Leite I., Kopp S. (eds) Intelligent Virtual Agents, Lecture Notes in Computer Science, vol 10498. Springer, Cham. https://doi.org/10.1007/978-3-319-67401-8_58: Beskow J., Peters C., Castellano G., O'Sullivan C., Leite I., Kopp S. (eds) Intelligent Virtual Agents. IVA 2017. Lecture Notes in Computer Science, vol 10498. Springer, Cham. https://doi.org/10.1007/978-3-319-67401-8_58

28. Chen M, Nikolaidis S, Soh H, Hsu D, and Siddhartha S (2018) "Trust-Aware Decision Making for Human-Robot Collaboration: Model Learning and Planning," In: Proceedings of the 2018 ACM/IEEE international conference on human–robot interaction, pp 307–315

29. Okamura K and Yamada S (2020) "Empirical Evaluations of Framework for Adaptive Trust Calibration in Human-AI Cooperation," *IEEE Access*, vol. 8, pp. 220335–220351, https://doi.org/10.1109/ACCESS.2020.3042556

30. Sun YL, Han Z, Yu W, and Liu K.J, "Attacks on Trust Evaluation in Distributed Networks," 2006 40th Annual Conference on Information Sciences and Systems, Princeton, NJ, 2006, pp. 1461–1466, https://doi.org/10.1109/CISS.2006.286695

31. Koppel A, Warnell G, Stump E, Stone P, and Ribeiro A, "Policy Evaluation in Continuous MDPs with Efficient Kernelized Gradient Temporal Difference," IEEE Transactions on Automatic Control, https://doi.org/10.1109/TAC.2020.3029315, 2020

32. Sardana N, Cohen R, Zhang J, and Chen S (2018) "A Bayesian Multiagent Trust Model for Social Networks," IEEE Transactions on Computational Social Systems, vol. 5, no. 4, pp. 995–1008, https://doi.org/10.1109/TCSS.2018.2879510

33. Mitra S and McCluskey E (2000) "Which Concurrent Error Detection Scheme to Choose," *Proceedings International Test Conference 2000 (IEEE Cat. No.00CH37159)*, Atlantic City, NJ, USA, pp. 985–994, https://doi.org/10.1109/TEST.2000.894311.

34. Mahmood A and McCluskey E (1998) "Concurrent error detection using watchdog processors-a survey," in *IEEE Transactions on Computers*, vol. 37, no. 2, pp. 160–174, https://doi.org/10.1109/12.2145.

35. Gizopoulos D, et al. (2011) "Architectures for online error detection and recovery in multicore processors," *2011 Design, Automation & Test in Europe*, Grenoble, pp. 1–6, https://doi.org/10.1109/DATE.2011.5763096.

36. Yu Y, Li X, Leng X, Song L, Bu K, Chen Y, Yang J, Zhang L, Cheng K, and Xiao X (2019) "Fault Management in Software Defined Networking: A Survey, IEEE Communications Surveys and Tutorials, vol 21, no. 1, pp. 349–392

37. Caballero J, Yin H, Liang Z, and Song D, (2007) "Polyglot: Automatic Extraction of Protocol Message Format Using Binary Analysis," CCS'07, Alexandria, Virginia, USA

38. Alharbi S, Rodriguez P, Maharaja R, Iyer P, Subaschandrabose N, and Ye Z (2017) "Secure the Internet of Things with Challenge Response Authentication in Fog Computing," 2017 IEEE 36th International Performance Computing and Communications Conference (IPCCC), San Diego, CA, pp. 1–2, 2017, https://doi.org/10.1109/PCCC.2017.8280489.

39. Kushwaha P, Sonkar H, Altaf F, Maity S (2021) "A Brief Survey of Challenge–Response Authentication Mechanisms." In: Fong S., Dey N., Joshi A. (eds) ICT Analysis and Applications. Lecture Notes in Networks and Systems, vol 154. Springer, Singapore, https://doi.org/10.1007/978-981-15-8354-4_57

40. Isermann R (2005) "Model-Based Fault Detection and Diagnosis – Status and Applications," Annual Reviews in Control, no. 29, pp. 71–85

41. Kocio K, Fesq L, and Mackey R (2017) "Model-Based Approach to Rover Health Assessment for Increased Productivity," *2017 IEEE Aerospace Conference*, Big Sky, MT, pp. 1–13, https://doi.org/10.1109/AERO.2017.7943835

42. Uesato J, Kumar A, Szepesvari C, Erex T, Ruderman A, Anderson K, Dvijotham K, Heess, N, and Kohli P (2018) "Rigorous Agent Evaluation: An Adversarial Approach to Uncover Catastrophic Failures," arXiv:1812.01647

43. Alford A, Wilkes M, and Kawamura K (2000) "System Status Evaluation: Monitoring the State of Agents in a Humanoid System," *SMC 2000 conference proceedings. 2000 IEEE International Conference on Systems, Man and Cybernetics, vol. 2,* pp. 943–948, Nashville, TN, https://doi.org/10.1109/ICSMC.2000.885971

44. Sievers M and Madni AM (2015) "Defining Credible Faults, A Risk-Based Approach," AIAA Space

45. Xie T, Ma Y, and Wang Y-X (2019) "Towards Optimal Off-Policy Evaluation for Reinforcement Learning with Marginalized Importance Sampling, "33rd Conference on Neural Information Processing Systems (NeurIPS 2019), Vancouver, Canada
46. Abdelgayed T, Morsi W, and Sidhu T (2018) "Fault Detection and Classification Based on Co-training of Semisupervised Machine Learning," in IEEE Transactions on Industrial Electronics, vol. 65, no. 2, pp. 1595–1605, https://doi.org/10.1109/TIE.2017.2726961
47. Zidi S, Moulahi T, and Alaya B (2018) "Fault Detection in Wireless Sensor Networks Through SVM Classifier," in IEEE Sensors Journal, vol. 18, no. 1, pp. 340–347, https://doi.org/10.1109/JSEN.2017.2771226
48. Umer M, Sher M, and Bi Y (2017) "Flow-Based Intrusion Detection: Techniques and Challenges," Computer and Security, vol 70, pp. 238–254
49. U.S. Department of Homeland Security, "Cyber Security Division Technology Guide," vol. 1 https://www.dhs.gov/sites/default/files/publications/CSD%20Tech%20Guide-.pdf
50. U.S. Department of Homeland Security, "Cyber Security Division Technology Guide," vol. 2 https://www.dhs.gov/sites/default/files/publications/csd-ttp-technology-guide-volume-2.pdf
51. Peón, PG, Steiner, W, and Uhlemann, E, "Network Fault Tolerance by Means of Diverse Physical Layers," 2020 25th IEEE International Conference on Emerging Technologies and Factory Automation (ETFA), Vienna, Austria, 2020, pp. 1697–1704, https://doi.org/10.1109/ETFA46521.2020.9212131.
52. Lamport L, Shostak R, and Pease M (1982) "The Byzantine Generals Problem," *ACM Transactions on Programming Languages and Systems*, vol. 4, no. 3, pp. 382–401
53. Gao, S, Yu, T, Zhu, J, and Cai, W, "T-PBFT: An EigenTrust-based practical Byzantine fault tolerance consensus algorithm," in China Communications, vol. 16, no. 12, pp. 111–123, Dec. 2019, https://doi.org/10.23919/JCC.2019.12.008
54. ElDefrawy, K, and Kaczmarek, T, "Byzantine Fault Tolerant Software-Defined Networking (SDN) Controllers," 2016 IEEE 40th Annual Computer Software and Applications Conference (COMPSAC), Atlanta, GA, USA, 2016, pp. 208–213, https://doi.org/10.1109/COMPSAC.2016.76
55. Sievers, M, Madni, AM, Pouya, P, "Assuring Spacecraft Swarm Byzantine Resilience," AIAA Scitech, 2019
56. NASA, Office of Inspector General, "Cybersecurity Management and Oversight at the Jet Propulsion Laboratory, Report No. IG-19-022, June, 18, 2019

Michael Sievers is Senior Systems Engineer/Systems at the California Institute of Technology, Jet Propulsion Laboratory (JPL), Pasadena, CA, and an adjunct lecturer in Systems Architecting and Engineering Department, University of Southern California (USC), Los Angeles, CA. He is responsible for the design and analysis of spacecraft avionics, software, end-to-end communication, and fault protection as well as ground system modeling and mission concepts of operation. He has conducted research at JPL and USC in the areas of trust and resiliency and performed a DSN study that investigated dynamic trust assessments. At USC, he teaches classes in systems and systems of systems architecture, resilience, and model-based systems engineering. Dr. Sievers is an INCOSE Fellow, AIAA Associate Fellow, and IEEE Senior Member.

Modeling and Simulation Through the Metamodeling Perspective: The Case of the Discrete Event System Specification

40

María J. Blas and Silvio Gonnet

Contents

Introduction	1190
What Is *Modeling and Simulation*?	1192
State of the Art	1194
Computational Representation: The DEVS Metamodel	1201
Best Practice Approach	1211
Defining a DEVS Simulation Model Through the DEVS Computational Representation: Software Tool and Illustrative Examples	1211
Limitations of the DEVS Community That Metamodeling Overcomes	1218
Chapter Summary and Conclusion	1221
Cross-References	1222
Appendix	1222
References	1225

Abstract

The Discrete Event System Specification (DEVS) is a modeling formalism that supports a general methodology for describing discrete event systems with the capability to represent both continuous and discrete systems due to its system theoretic basis. This chapter addresses the use of metamodeling and the role of related technologies in the Modeling and Simulation (M&S) field, mainly devoted to implementing DEVS models. The main aim is to answer the following questions: What is the significance of model-based engineering in the M&S field? Can a computational representation of DEVS defined following the principles of metamodeling improve the modeling task? How can DEVS models obtained through metamodeling integrate with simulators? This chapter discusses using metamodels as the foundation to define formal DEVS models from a computational point of view. A metamodel-based computational representation of DEVS formalism is described. From such a computational representation, how current

M. J. Blas (✉) · S. Gonnet
Instituto de Desarrollo y Diseño INGAR (UTN-CONICET), Santa Fe, Argentina
e-mail: mariajuliablas@santafe-conicet.gov.ar; sgonnet@santafe-conicet.gov.ar

© Springer Nature Switzerland AG 2023
A. M. Madni et al. (eds.), *Handbook of Model-Based Systems Engineering*,
https://doi.org/10.1007/978-3-030-93582-5_86

technologies can be used to support DEVS metamodel in existing M&S software tools is discussed. The chapter shows how concrete DEVS models are used in existing M&S software tools to get executable simulation models through meta-model instantiation. Illustrative examples show how the DEVS metamodel improves the simulation model specification. The chapter concludes with a discussion of the current state of the metamodeling approach used in the M&S field and the remaining problems that need to be addressed in the future.

Keywords

Computational representation · Concrete model · Formal model · General-purpose language · Metamodel · Simulation software

Introduction

Any field that contributes to the systems engineering of complex systems is highly valued. Successful Systems Engineering (SE) activities are directly related to the methods by which models are developed and used. Over the past few years, some Modeling and Simulation (M&S) problems of complex systems have been addressed through joint advances in scientific and engineering communities [1]. For example, the Systems Modeling Language (SysML) [2] is a standard from the Object Management Group (OMG) that supports the design, analysis, and verification of complex systems, including software and hardware components [3]. SysML enables the modeling of a wide variety of systems from different perspectives such as behavior, structure, or requirement as a weak-semantic graphical language useful in creating models as part of a Model-Based Systems Engineering (MBSE) process. Combining SE with M&S approaches through Model-Driven Architecture (MDA) concepts, in [4], a transformation of SysML models into executable simulation models is presented. The authors propose a metamodel for implementing discrete-event simulation models that provide a standard representation for the simulation-specific domain. Following the metamodeling perspective, in [5] a concrete textual notation based on metamodeling and a general-purpose simulation engine for abstract state machines specifications is presented. Recently, a metamodel-based Monte Carlo simulation (MCS) method has been developed to accurately capture the dynamic, stochastic behavior of a manufacturing system and allow real-time evaluation of a release plan's performance metrics [6].

The Discrete-Event System Specification (DEVS) formalism [7] is a modeling formalism based on systems theory that provides a general methodology for the hierarchical construction of reusable models in a modular way. Over the years, DEVS has been applied to the M&S of distinct types of real-world systems due to the ease of model composition, reuse, and hierarchical coupling. These features are enabled by the two types of models defined in DEVS: atomic and coupled models. The atomic model is used to specify behavior, while the coupled model denotes structure among other DEVS models. Modelers can easily combine both types of models to define new models starting from existing ones.

Furthermore, it has been shown that DEVS is equivalent to the Iterative System Specification (ISS) [8]. The ISS is an intermediate specification between DEVS at the lower (computational) level and general system behavior at the higher level [8]. The ISS, as defined, is generic enough to be applicable to most (if not all) dynamic computational models. Hence, DEVS can be used to specify general system structure and compute its behavior as a computational implementation of the ISS.

Numerous applications of DEVS are found in the literature. For example, DEVS models have been used to design mobile application behaviors [9] and to compare the performance of distinct network architectures [10]. In [11, 12], a combination of micro- and macro-views based on DEVS is presented for biological systems. A strategy to generate DEVS and Linear Programming models semiautomatically from industry-scale relational databases is proposed in [13]. Given that constructing a real battlefield situation and verifying the performance of military devices is costly, in [14] DEVS formalism is used to define a robotic vehicle model to conduct tests through the simulation of several scenarios. Hence, DEVS is a useful formalism for modeling complex dynamic systems using a discrete-event abstraction.

One of the main advantages of DEVS compared to other discrete-event formalisms is its rigorous formal definition based on set theory and systems theory [15]. Such a mathematical definition with the system basis allows for building hierarchical models using sets and functions. A DEVS formal model is defined using DEVS formal specification. A DEVS concrete model is an implementation of a formal model.

Due to the separation between behavior and structure (i.e., atomic and coupled models, respectively), DEVS is most naturally implemented in Object-Oriented (OO) frameworks. That implies that DEVS modelers need to understand OO concepts. Therefore, building an OO implementation of DEVS formal models (i.e., a computational concrete model) in a way that ensures DEVS formal specification (i.e., a mathematical definition) is not easy. As shown later in this chapter, the dependence between models and simulators with the deployment of simulator engines through OO implementations leads to the need to develop both entities (i.e., model and simulator) using the same programming language. That is a hard requirement since DEVS software tools are not deployed as black box simulation tools. The DEVS simulator must interact with the DEVS model during the algorithm steps through message passing. Then, both components need to be defined over the same basis. However, when concrete DEVS models are developed using programming languages, it is hard to ensure they conform to their formal model [16].

In this regard, using metamodeling can provide a basis to define a computational representation of the DEVS mathematical definition that allows generating concrete model implementations for OO simulation tools through a translation process. DEVS formal models are developed from a metamodel that provides a basis for creating DEVS concrete models as platform-independent formulations (i.e., computational instances of the metamodel). Hence, modelers and systems engineers can create DEVS computational models (i.e., concrete models realized in OO frameworks) without knowing OO concepts through metamodeling instantiation. Such computational models align with the formal implications of DEVS specification. In this way, metamodeling enables the use of M&S to analyze complex systems in new domains.

This chapter discusses a behavioral and structural metamodeling approach based on the DEVS formalism and introduces its support in current M&S software tools. It is focused on matching DEVS formal models, their concrete implementations, and the advantages of a general-purpose computational representation. This chapter discusses an application-oriented compilation of three of the most theoretical and practical topics in M&S through metamodeling (i) the use of DEVS for M&S of systems, (ii) the role of metamodels in capturing DEVS formal specification, and (iii) metamodel-enabled foundation for existing M&S software tools that achieve accurate executable models from formal specifications. These topics together provide a comprehensive, unified system-engineering approach with modeling expressiveness and good simulation performance for DEVS models.

The key topics covered by this chapter are:

- DEVS as the formalism for defining simulation models from a system specification point of view
- Metamodels that support the DEVS formal specification for developing concrete models expressed through a general-purpose computational representation of the formalism
- How a DEVS computational representation combined with existing M&S software tools promotes the collaboration between M&S and SE fields

What Is *Modeling and Simulation*?

Modeling and *Simulation* are distinct activities [17]. *Modeling* is the process of developing and using abstractions to simplify the real world for focused analysis. On the other hand, *Simulation* subsumes modeling activities and focuses on model execution in a simulator. The simulator can be an algorithm executed on paper or software that runs in a computational environment. For practitioners, simulation activities enable model experimentation (e.g., for SE practitioners, these activities are often related to system evaluation through models).

The *M&S progression* [7] is a sequence of three steps: abstraction, formalization, and implementation. Abstraction creates a conceptual system model by extracting only those system elements needed for addressing a stakeholder's concerns. Formalization creates an unambiguous representation of the abstraction using a controlled vocabulary. Finally, an implementation provides a concrete realization of the abstraction through formalization. Such an implementation is often called "reduction to concrete form" [7]. If the implementation is written in a programming language, the concrete model is materialized as a computational model. Once a concrete realization is defined, experimentation can be performed at execution time. For computational environments, such an experimentation requires the setup (e.g., the initial state definition, time boundaries, and so on) of the computational model (i.e., an executable (a simulation model is executable if there is at least one computational simulator with the capabilities to run its instructions) implementation).

As the section "State-of-the-Art" details, a simulation model is linked to a simulator that executes its instructions. Simulation models may be expressed in different ways during the M&S lifecycle (e.g., on paper, as a mathematical formulation, or in a computational environment). The nature of the representation determines the simulator needed for model execution. All representations used across the lifecycle are equally valid. The main difference between them is the foundation used during their conceptualization. Considering the *M&S progression*, a M&S model evolves during its lifecycle through at least four representations [18]: (i) an *abstract model* (nonexecutable representation of a real-world phenomenon), (ii) a *formalization model* (nonexecutable representation defined over a formal basis such as set theory and mathematical functions), (iii) a *computational implementation model* (concrete model materialized as an incomplete executable representation in a programming language), and (iv) an *executable model* (executable representation seeded with simulation parameters and ready for execution through an algorithm implemented as a computational simulator). These models together define the complete simulation model from a broad perspective. Table 1 summarizes the model types that arise from the M&S progression when computational environments are used to perform simulations.

When formal specifications are used to design a *formal model*, the final *executable model* (and all intermediate representations developed during the progression) should be aligned with the formal basis. However, when the formal basis and implementation tools use different paradigms, such an alignment between formal and executable models may be fuzzy. This chapter describes a computational representation of DEVS through a metamodeling-based approach to show how an alignment between formal and computational representations can be performed in M&S software tools.

DEVS formal models are built in a mathematical form over a rigorous common basis of set theory and systems theory. Therefore, DEVS is an abstract formalism for specifying simulation models independent of any implementation [19]. But DEVS is most naturally implemented in a computational form in an OO framework [7]. This

Table 1 Distinct types of models involved in M&S progression

Step	Model	Definition
Formalization	Formal	A model that is defined over the basis of a formal specification
Implementation	Concrete	An implementation of a formal model
	Computational	A concrete model that is realized using a programming language
Execution	Executable	A computational model with a valid setup that can be executed in a computational environment (computational environment: simulator or M&S software tool)

implies that DEVS modelers need to understand OO concepts. Such concepts are embodied in the implemented code (e.g., Java or C++). Hence, when concrete DEVS models are developed using programming languages, it is difficult to ensure they conform to their formal model [16].

State of the Art

The DEVS formalism is an M&S formalism developed in the late 1970s [20] as a specification for describing discrete event systems. In a DEVS specification, a timed sequence of pertinent "events" input to a system causes instantaneous changes to the system's state [21]. These events are generated externally (i.e., by another model) or internally (i.e., by the model itself due to time-outs). External events take place when the state of the model is stopped by an external action. Internal events occur when the time assigned to the current state expires. In both cases, the next state of the system is defined based on the previous state of the system and the event. Between events, the state of the system does not change (i.e., state transitions do not occur between events).

Several authors have improved the formalism specification through DEVS extensions to solve new problems. Starting from Classic DEVS [7], some of the available extensions include (i) the notion of ports (Classic DEVS with ports [7]), (ii) use of cell spaces (Cell-DEVS [22]), (iii) setup for dynamic structures (Dynamic Structure DEVS [23]), (iv) parallel event processing (Parallel DEVS [24]), (v) real-time execution of models (Real-Time DEVS [25]), and (vi) event routes definition (Routed DEVS [26]). These extensions are variations of DEVS formalism that expand the classes of systems modeled with DEVS. Hence, DEVS can be applied to several types of systems.

The core of DEVS includes a M&S framework [27] composed of four entities (Fig. 1): (i) the *source system* as the real/virtual environment to be modeled, (ii) the *model* that describes the set of instructions for generating data comparable to the data observable in the system, (iii) the *simulator* that refers to the computational system that executes the set of instructions specified in the *model*, and (iv) the *experimental frame* that represents the conditions under which the *source system* is observed or exercised. The entities interact through two relationships: the *modeling* relation

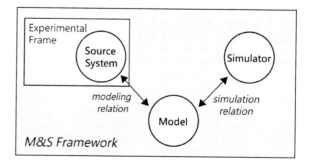

Fig. 1 M&S framework

determines when a *model* is a valid representation of the *source system* within an *experimental frame* and the *simulation* relation specifies what constitutes a correct simulation of a *model* by a *simulator*. Hence, the framework acts as a conceptual model that shows how M&S activities are related. Moreover, it suggests how the main M&S entities are interpreted. For example, it emphasizes the notion of *model* and *simulator* as two independent entities linked when a *model* is executed on a computational environment (*simulator*).

DEVS models are formalized with set theory and systems theory to define behavioral and structural descriptions. Behavior is described using an atomic DEVS model. Eq. (1) presents the atomic model definition using the formalization called *Classic DEVS with ports* (*classic DEVS with ports* is the most widely used extension of DEVS. It simplifies the modeling task by adding couplings between output/input ports of different components. The use of ports allows passing messages from one component to another [28]) [7]. An input port is the site of inputs. An output port is the site of outputs. Each port has a name and a set of input/output values. Events determine values appearing on ports. Two types of events are allowed (i) external events accepted on input ports (i.e., events received by the model), and (ii) internal events on output ports (i.e., events transmitted to other models).

$$DEVS = (X, Y, S, \delta_{ext}, \delta_{int}, \lambda, ta) \tag{1}$$

where
$X = \{(p,v) \mid p \in InPorts, v \in X_p\}$ is the set of inputs, with

InPorts as the set of input ports,
p as the p-th input port that belongs to InPorts,
X_p as the set of input values for the p-th input port, and
v as an element of the input values defined for the p-th input port.

$Y = \{(p,v) \mid p \in OutPorts, v \in Y_p\}$ is the set of outputs, with

OutPorts as the set of output ports,
p as the p-th output port that belongs to OutPorts,
Y_p as the set of output values for the p-th output port, and
v as an element of the output values defined for the p-th output port.

S is the set of sequential states;
$\delta_{ext}: Q \times X \to S$ is the external state transition function, with $Q := \{(s,e) \mid s \in S, 0 \le e \le ta(s)\}$;
$\delta_{int}: S \to S$ is the internal state transition function;
$\lambda: S \to Y$ is the output function;
$ta: S \to \mathbb{R}_0^+ \cup \{\infty\}$ is the time advance function.

Equation (2) presents the Classic DEVS with ports definition of the model known as the "switch" [7]. A switch is modeled as a DEVS atomic model with two input ports (named *in* and *in1*) and two output ports (named *out* and *out1*). External events are received through the set *{in, in1}*. External events are sent through the set *{out, out1}*. When the switch is in the default position (i.e., when the variable named as Sw is true), events arriving on the port *in* are sent out on port *out*, and similarly for ports *in1* and *out1*. The input-to-output links are reversed when the switch is in the opposite setting (i.e., when Sw is false).

$$\text{SWITCH} = (X, Y, S, \delta_{ext}, \delta_{int}, \lambda, ta) \tag{2}$$

where

$X = \{(p,v) \mid p \in \text{InPorts}, v \in X_p\}$ with $\text{InPorts} = \{in, in1\}$ and $X_{in} = X_{in1} = V$ (an arbitrary set);

$Y = \{(p,v) \mid p \in \text{OutPorts}, v \in Y_p\}$ $\text{OutPorts} = \{out, out1\}$ and $Y_{out} = Y_{out1} = V$ (an arbitrary set);

$S = \{passive, active\} \times \mathbb{R}_0^+ \times \{in, in1\} \times V \times \{true, false\}$;

$\delta_{ext}((phase,\sigma,inport,store,Sw),e,(p,v)) =$

(active,processing_time,p,v,!Sw) if phase $=$ passive and $p \in \{in, in1\}$,

(phase, σ-e, inport, store, Sw) in other case;

$\delta_{int}(phase,\sigma,inport,store,Sw) = (passive, \infty, inport, store, Sw)$;

$\lambda(phase,\sigma,inport,store,Sw) =$

(out, store) if phase $=$ active, Sw $=$ true, and inport $=$ in,

(out1, store) if phase $=$ active, Sw $=$ true, and inport $=$ in1,

(out1, store) if phase $=$ active, Sw $=$ false, and inport $=$ in,

(out, store) if phase $=$ active, Sw $=$ false, and inport $=$ in1;

ta(phase,σ,inport,store,Sw) $= \sigma$.

The formal definition detailed in Eq. (2) includes five state variables named *phase*, σ, *inport*, *store*, and *Sw*. Each variable stores a value as part of the model state: (i) *phase* and σ are state variables that denote the state name (*phase*) and time (σ) to spend in it if no external event takes place (i.e., in the absence of external events the model stays in the current *phase* for the time given by σ), (ii) *inport* and *store* are used to retain the input port and value of the external event, and (iii) *Sw* is used to configure the setting of the switch (in the model, the setting is toggled between true and false at each input). Then, the set of sequential states S is the cross product of the sets used to define each one of the state variables (e.g., *phase* $\in$ {passive, active} and $\sigma \in \mathbb{R}_0^+$). The time-advance function *ta()* controls the timing of internal transitions. In the switch model, *ta()* returns the value of σ. The internal transition function $\delta_{int}()$ specifies the next state the model transitions to after the time given by *ta()*. In Eq. (2), the model will transit to a new state with *phase* $=$ passive and $\sigma = \infty$ (the values of other state variables remain as in the current state). The output function $\lambda()$ generates an external event just before an internal transition occurs. For example, the switch model details that if *phase* $=$ active, *Sw* $=$ true, and *inport* $=$ in, the external output will be the value of the state variable *store* at the output port *out*. Finally, the external transition function $\delta_{ext}()$ specifies how the system changes state when an external event v is received at the input port p and the time elapsed in the current state is e. For the switch, such a transition function specifies that, for example, if *phase* = passive and $p \in \{in, in1\}$, the next state should be set as *phase* = active, σ = *processing_time* (i.e., a parameter value), *inport* = p, *store* = v, and *Sw* =!*Sw*. This means that when an input external event breaks a state with *passive* phase, the next state will be in *phase* active for the time detailed by the *processing_time* parameter retaining the port p and value v as *inport* and *store* and changing setting the other way around.

Figure 2 illustrates a black-box representation of the switch as a DEVS atomic model with input and output ports.

In addition to the behavioral specification through an atomic DEVS model, a DEVS structure is described using a coupled DEVS model. Following the *Classic DEVS with ports* formalization, Eq. (3) presents its definition including the external interface (i.e., input and output ports and values), the components (which must be DEVS models), and the coupling relations. The coupled model indicates how to connect several component models to form a new model. This new model can be employed as a larger coupled DEVS model component enabling hierarchical model construction [7].

$$N = (X, Y, D, M_d| \, d \in D, EIC, EOC, IC, Select) \qquad (3)$$

where
$X = \{(p,v) \mid p \in \text{IPorts}, v \in X_p\}$ is the set of inputs, with

IPorts as the set of input ports,
p as the p-th input port that belongs to IPorts,
X_p as the set of input values for the p-th input port, and
v as an element of the input values defined for the p-th input port.

$Y = \{(p,v) \mid p \in \text{OPorts}, v \in Y_p\}$ is the set of outputs, with

OPorts as the set of output ports,
p as the p-th output port that belongs to OPorts,

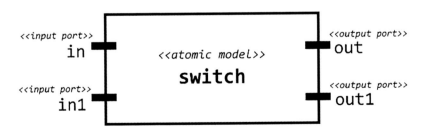

Fig. 2 Representation of the switch model. Stereotypes are used only to illustrate element types

Y_p as the set of output values for the p-th output port, and
v as an element of the output values defined for the p-th output port.

D is the set of the component names (i.e., the names used to identify component models).

For each $d \in D$, $M_d = (X_d, Y_d, S, \delta_{ext}, \delta_{int}, \lambda, ta)$ is a DEVS, with $X_d = \{(p,v) \mid p \in IPorts_d, v \in X_p\}$ and $Y_d = \{(p,v) \mid p \in OPorts_d, v \in Y_p\}$.

EIC $\subseteq \{((N,ip_N), (d,ip_d)) \mid ip_N \in IPorts, d \in D, ip_d \in IPorts_d\}$ is the set of external input couplings that connect external inputs to component inputs.

EOC $\subseteq \{((d,op_d),(N,op_N)) \mid op_N \in OPorts, d \in D, op_d \in OPorts_d\}$ is the set of external output couplings that connect component outputs to external outputs.

IC $\subseteq \{((a,op_a), (b,ip_b)) \mid \{a,b\} \in D, op_a \in OPorts_a, ip_b \in IPorts_b\}$ is the set of internal couplings that connect component outputs to component inputs.

Select: $2^D \rightarrow D$, the tie-breaking function used for selecting one event to process out of a set of contending simultaneous events (i.e., it is used for tie-breaking in case of equal next event times in more than one component).

Abstractions are formalized in a DEVS model using the mathematical specifications in Eq. (1) and Eq. (3). To be useful, DEVS formal models should be implemented in computational environments. Several M&S software tools and libraries are currently available to implement concrete DEVS models' realizations [29]. Most used tools are implemented in Python [30], Java [31, 32], and C++ [33, 34]. However, these tools often require OO knowledge. Due to dependency between model and simulator (illustrated in Fig. 1), computational models are attached to the programming language used in the software tool chosen by the modeler. Executable models are not compiled or binary versions of the model that can be executed on any computer. Instead, they should be understood as a set of instructions handled by a simulation engine through an algorithm based on message exchange. That means executable models should be written as counterparts of the simulator used for their execution. Hence, when engineers want to simulate DEVS models, they need to program them in the input language of a concrete simulator [35].

As the authors of [36] detail, "simulationists must represent a DEVS model as programming code using predefined libraries offered by each simulator." The authors refer to "a representation" of the DEVS model meaning writing code in a general-purpose-programming language to denote the formal model in a computational form. Here, the reference to "general-purpose programming languages" is related to the OO programming languages used to support existent DEVS simulators. Maintaining the correspondence between the DEVS formal model and its implementation during programming is complicated. Therefore, any mechanism that promotes the alignment between DEVS formal and computational models is highly valued. In this regard, a general-purpose computational representation of the DEVS formalism can be the key to getting such an alignment. Figure 3 illustrates how such a computational representation can be linked to the models introduced in Table 1.

40 Modeling and Simulation Through the Metamodeling Perspective: The Case... 1199

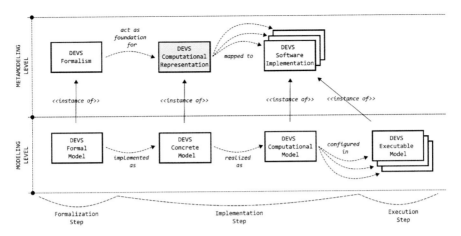

Fig. 3 Understanding of the DEVS M&S progression using a metamodeling-based approach. At the core of the metamodeling level, a computational representation of DEVS (highlighted in gray) allows defining DEVS concrete models without considering the M&S software tool (i.e., in a platform-independent formulation). For platform-dependent formulations (i.e., for specific M&S software tools), a programming language (i.e., the one attached to the simulator) should be used to support the computational model that realizes the concrete model

The approach illustrated in Fig. 3 shows *modeling* and *metamodeling* levels. At the *modeling level*, the definition of a DEVS model comprises: *DEVS formal model*, *DEVS concrete model*, *DEVS computational model*, and *DEVS executable model*. Conceptual relationships are defined between these models to illustrate dependencies between them, such as *implemented as*, *realized as*, and *configured in*. Each model in the *modeling level* is depicted as an instance of a conceptual metamodel defined at the *metamodeling level*. A metamodel makes statements about what can be expressed in the valid models of a certain modeling language [37]. Considering that a metamodel is a generalized model of a model (i.e., a model that structures and constrains the design of other models) that can be instantiated for specific domains, the *metamodeling level* shows three metamodels: (i) *DEVS formalism* metamodel which is the mathematical definition of *DEVS formal models*, (ii) *DEVS computational representation* metamodel that allows instantiating *DEVS concrete models*, and (iii) *DEVS software implementations* metamodel that defines the M&S tools and libraries for defining *DEVS computational models* that are further configured in *DEVS executable models*.

As Fig. 3 illustrates, the *DEVS computational representation* acts as a bridge between *DEVS formalism* and the set of *DEVS software implementations*. The *DEVS formalism* in its mathematical form gives the foundations required to build a *DEVS computational representation*. Then, the *DEVS computational representation* gives itself the foundations required to be mapped into *DEVS software implementations* (i.e., the OO programming languages that support software tools). Hence, such a *DEVS computational representation* should be understood as the core of the alignment between *DEVS formal models* as instances of *DEVS formalism* and *DEVS computational models* as instances of *DEVS software implementations.* Section "Limitations of the DEVS Community That Metamodeling Overcomes" presents the metamodel that supports *DEVS computational representation* for DEVS formalization specified in Eq. (1) and Eq. (3).

DEVS Languages, Ontologies, and Metamodels

By nature, simulation is a technical field focused on the "scientific" elements of the simulation project life cycle (i.e., model development and analysis) [38]. This means that most simulation projects focus their attention on building an "accurate" model (i.e., the model that represents a system or situation). For DEVS projects, this implies dealing with programming activities. As mentioned in section "State-of-the-Art," no common DEVS format is used by all tools, and modelers are tied to their M&S software tool [29]. That is, still, there is no way in DEVS to abstract simulator features, behaviors, and structures so that the DEVS model is agnostic to nuances of the simulator. Hence, the DEVS software implementations can be seen as Domain-Specific Languages (Domain-specific languages are languages tailored to a specific application domain [40]. Conversely, general-purpose languages are broadly applicable across domains) attached to M&S software tools. This means that most approaches are isolated from each other due to the software view that influences their design using the programming language that supports each M&S software tool.

Several model types have been developed to study the DEVS formalism over the years. Most of these models are used to analyze a "view" of DEVS, such as its formalization or implementation through the definition of languages, ontologies, or metamodels. Focused on the formalization view, the models developed have evolved over the years to provide computational representations. Mainly, three representation types were proposed: ontologies, metamodels, and languages. In 2007, as a step toward model interoperability and reuse, an XML schema for representing the structure and the behavior of DEVS-coupled scenarios was proposed in [39]. The XML schema was advocated to create one of the first computational representations of DEVS that was not explicitly founded in a conceptual model of the domain (i.e., a suitable foundation for the generation of platform-independent DEVS models). In 2012, the author of [41] proposed a metamodel of DEVS that provides a pivot format for building simulation models compatible with available software tools. The metamodel defines the specification of DEVS simulation models into two parts: (a) a structural part that deals with state variables, ports, components, and connections, and (b) a behavioral part that describes the temporal evolution of models in terms of the DEVS functions.

From an ontological point of view, in 2013, an ontology-based model representation named DEVSMO (DEVS math ontology) [42] was designed to boost model reuse by employing a network of three ontologies: *DEVS model ontology, model structure ontology,* and *model behavior ontology.* The *DEVS model ontology* describes the classification of DEVS formalisms, such as *Classic DEVS, Parallel DEVS, Dynamic DEVS, Real-Time DEVS,* and so on. The *model structure ontology* provides the terminologies to define the simulation model structure for atomic and coupled models in DEVS with ports. Finally, the *model behavior ontology* specifies the behavior parts of DEVS models, including the specification of functions, actions, mathematical expressions, and control structures. The models represented by DEVSMO cannot be considered DEVS computational models. DEVSMO models become DEVS computational models when they are coding in programming languages.

In 2015, two representation proposals were presented. In [16], a DEVS behavioral metamodel was discussed for developing concrete models from domain-specific metamodels. This metamodel includes a set of abstractions for state transitions in the external and internal transition functions. Abstractions were also defined for output and time advance functions. In [19], a formal modeling language called CML-DEVS (Conceptual Modeling Language for DEVS) is defined. CML-DEVS provides an abstract description of DEVS models using logical and mathematical expressions. By describing DEVS models in their most abstract form, independent of any particular implementation, the language has three main advantages: (i) It improves the communication between practitioners and researchers, (ii) it facilitates both maintenance and modification of the simulation models, and (iii) modelers can define a model without having programming skills.

Even when all the aforementioned representations are useful to define DEVS models, none of them provide a suitable solution to the alignment between formalization (i.e., the definition of DEVS formal models) and concreteness (i.e., the specification of DEVS concrete models), including only concepts related to DEVS specification. That is, neither CML-DEVS nor DEVSMO defines a concrete representation of DEVS simulation models that include all the mathematical entities and relationships defined at the formalization level. The main limitations are the lack of semantic models that support structural relationships between concepts [39], the definition of metamodels that mix metaprogramming elements (e.g., datatypes and operations) with conceptual definitions [16, 41], and the need to learn a new language syntax for representing DEVS models [19].

Computational Representation: The DEVS Metamodel

To overcome the aforementioned issues, this section discusses the recommended DEVS computational representation for instantiating concrete formal models. The representation extends prior work with a set of additional components and relationships:

- The representation is founded in a metamodel as [16, 41]. Practitioners can instantiate the metamodel to get DEVS models without having programming skills or without knowing the particular modeling language of a specific M&S tool. Moreover, they need not learn a new language to define simulation models (as in [19]).
- The core of the metamodel design follows the distinction between behavioral and structural parts as in [16, 41, 42]. It adds a new interaction layer that unifies

structural and behavioral parts to provide all the concepts and relationships required to define DEVS models.
- The entities and their relationships are designed to avoid using programming features (aiming to provide an accurate formal definition). Moreover, the mathematical approach of the formal modeling language proposed in [19] is used as a foundation because it defines the complete formalization of DEVS models (i.e., the modeler truly defines a formal model).

These features combined provide a suitable representation of DEVS formalism through a general-purpose metamodel named *DEVS computational representation*.

The metamodel designed to support DEVS computational representation is defined using Unified Modeling Language (UML) diagrams restricted by Object Constraint Language (OCL) constraints. Figure 4 shows the UML package diagram that illustrates the main structure and dependencies. Here, packages are conceptually placed into three layers. Basic packages are placed at the top, while DEVS-specific packages are placed at the bottom. DEVS-general purpose packages are in the middle. Each package groups classes and relationships according to their scope using a UML class diagram. These diagrams are described in the following subsections. Classes are named according to the formal element represented by them. Roles are used to show how classes are involved in relationships. When class dependencies are needed, the package scope is denoted between brackets.

Classes and relationships are restricted with OCL constraints to ensure consistency between the DEVS formalism and metamodel instances. For example, constraints are added to verify the following: (i) There are no direct feedback loops in coupled models, (ii) correctness of set definitions, and (iii) the correctness of coupling specifications following the sets defined on related ports. Due to the number of OCL constraints attached to the UML models, these constraints are not included in this chapter.

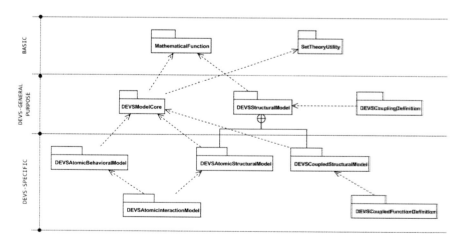

Fig. 4 UML package diagram of the metamodel used to define the DEVS computational representation. Packages are placed in layers to depict their scope

The Basic Layer: "Mathematical Function" and "Set Theory Utility" Packages

The "Basic" layer includes the "Mathematical Function" and "Set Theory" packages that define the general concepts of the DEVS formal specification. The "Mathematical Function" metamodel (Fig. 5) defines the *Function* concept. A *Function* comprises one or more *Specifications* (i.e., a mathematical expression that is used as part of the function definition). A *Specification* may be attached to a *Condition* which is an optional feature (due to the multiplicity 0..0.1). A *Condition* may be an *If* or an *Otherwise* expression.

The "Set Theory Utility" package (Fig. 6) includes the concepts that define the set theory foundation used in DEVS formal models. The metamodel concepts are related to concepts in the middle and bottom package layers (i.e., to sets of input and output ports, state variables, parameters, and so on).

From the *Variable* concept of the "DEVS Structural Model" package, the *Definition* concept is used to abstract a set theory definition. Here, a *Definition* can be either a *Nonempty Set* or a *Structure*. As opposed to the *Nonempty Set*, the *Empty Set* concept is defined.

The *Nonempty Set* concept is specialized in set types as follows:

- The *Boolean Algebra Set* concept represents the set {false, true}.
- The *Mathematical Set* concept generalizes mathematical sets (such as $\mathbb{R}$, $\mathbb{Z}$, and $\mathbb{N}$). Each mathematical set is defined as a specific concept (e.g., $\mathbb{Z}$ is the set of *Integer Numbers*).
- The *New Set* concept provides the capability of defining original sets as a collection of predefined elements. Each element defined as part of a *New Set* is defined as an *Element*. Here, an *Element* can be a *String Element*, a *Mathematical Element* (defined using the specialization of *Real Number*, *Integer Number*, and *Boolean Value*), or a *Structure Element*.

When a *Definition* is based on the *Structure* concept, then structured elements are used to support such a definition. These elements can be either a plain structure (i.e., the *New Structure* concept) or a *Collection*. If a *New Structure* is defined, at least two *Structure Variables* are required. These *Structure Variables* play a role *field* in the composition relationship. By contrast, if a *Collection* is used, it needs to refer to the *Structure* that defines its elements. This means that even when the *Collection* is

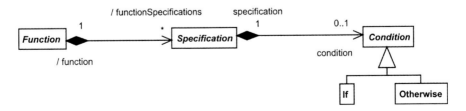

Fig. 5 UML class diagram of the "Mathematical Function" package

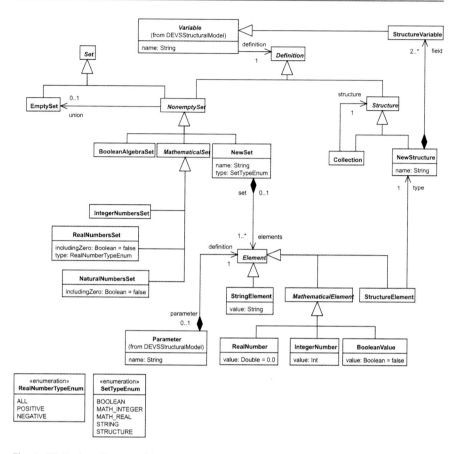

Fig. 6 UML class diagram of the "Set Theory Utility" package

a *Structure* as a kind of *Definition*, the *Collection* refers to another *Structure* that details how the collection elements are composed (i.e., a collection of structured elements).

Finally, following the approach used for the *Variable* definition, the definition of the *Parameter* concept (detailed in the "DEVS Structural Model" package) is based on the *Element* concept. Composition is used to denote that a *Parameter* is composed of an *Element*. As the next section details, the same approach is used in the "DEVS Structural Model" package to denote that *Input* and *Output Definitions* are composed of different kinds of *Variables* (i.e., *Input Variables* and *Output Variables*).

The DEVS-General Purpose Layer: "DEVS Model Core," "DEVS Structural Model," and "DEVS Coupling Definition" Packages

The "DEVS-general purpose" layer includes the set of packages used to define the components shared by both types of DEVS models. The package named "DEVS

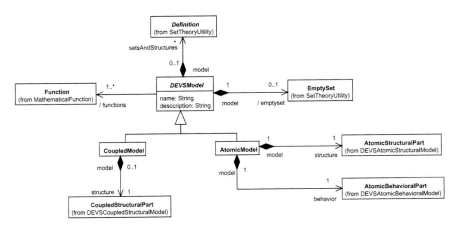

Fig. 7 UML class diagram of the "DEVS Model Core" package

Model Core" (Fig. 7) defines basic concepts used to specify a DEVS Model. Each DEVS Model has a unique name and references to the *Definitions* (i.e., sets and structures) and *Functions* used in its specification. The *Empty Set* (from the "Set Theory Utility" package) provides an individual conceptualization (i.e., multiplicity 1) of the empty set {Ø}. The empty set can be used in the specification of new sets (i.e., it is available for the model specification). For example, the modeler can define a new set with a joined empty set to generate some null events.

A *DEVS Model* can be an *Atomic Model* or a *Coupled Model*. The *Atomic Model* concept acts as a container of two components: (i) the *Atomic Structural Part* (from the "DEVS Atomic Structural Model" package), and (ii) the *Atomic Behavioral Part* (from the "DEVS Atomic Behavioral Model" package). The *Atomic Structural Part* plays the role of *structure*. It defines the inputs and outputs sets along with the state composition. On the other hand, the *Atomic Behavioral Part* plays the role of *behavior*. It specifies state transitions, output, and time advance functions. In both cases (i.e., atomic and coupled), components are mandatory and unique (i.e., multiplicity 1), and the *AtomicModel* concept plays the role of *model*.

Complementary to the *Atomic Model* definition, the *Coupled Model* uses a *Coupled Structural Part* (from the "DEVS Coupled Structural Model" package). The *Coupled Structural Part* plays the role *structure* by defining the inputs and outputs sets, the couplings, and the select function.

The "DEVS Structural Model" package (Fig. 8) is used for the structural parts of DEVS models. The package defines the structure shared by both types of DEVS models (i.e., atomic and coupled). The *Structural Part* concept is a container of (i) a list of *Input Ports* (i.e., has multiplicity *) that plays the role *InPorts*, (ii) a list of *Output Ports* (i.e., has multiplicity *) that plays the role *OutPorts*, (iii) a mandatory input set X, (iv) a list of *Parameters*, and (v) a mandatory output set Y. In all cases, the *Structural Part* plays the role *model* on the container side because it refers to the DEVS model under specification.

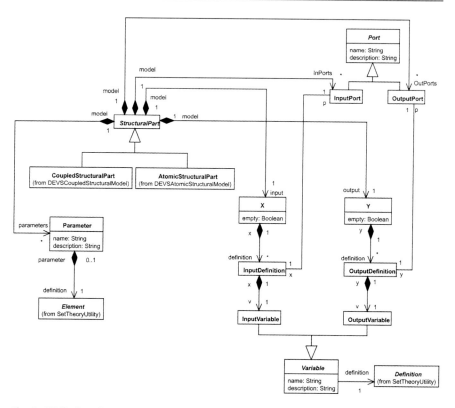

Fig. 8 UML class diagram of the "DEVS Structural Model" package

A *Port* is defined by its name. As mentioned before, each type of port (i.e., *Input Port* and *Output Port*) defines a list. The input set X is defined as a list of *Input Definitions* (i.e., multiplicity *). If the input list is empty, the value of the *empty* attribute is *true*. Otherwise, *empty* is set to false (and there must be at least one instance of *Input Definition* attached to the set). When nonempty list is used, an *Input Definition* represents the pair *(p,v)* of the input set formal specification. Since *Input Port* composes *Structural Part*, an *Input Definition* is linked to the port by a weak association.

Instead, the variable related to the port is detailed as part of the definition (i.e., *Input Definition* is a composition of *Input Variable*). In this way, an *Input Definition* is related to an *Input Variable* (that plays the role *v*) and an *Input Port* (that plays the role *p*). The output set *Y* is defined in the same way using the concepts *Output Definition*, *Output Variable*, and *Output Port*. Both *Input and Output Variables* are defined as a *Variable* with a *Definition* (from the "Set Theory Utility" package).

Finally, a *Parameter* is considered a factor that can be one element of a set that defines the model or a condition for the model operation. Each *Parameter* is defined using an *Element* (from the "Set Theory Utility" package). The list of *Parameters* is added to the *Structural Part* to save the set of parameters used in the model definition.

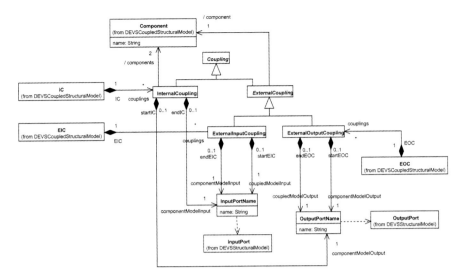

Fig. 9 UML class diagram of the "DEVS Coupling Definition" package

The coupling structure is defined in the "DEVS Coupling Definition" package (Fig. 9). From the *Coupling* concept, *Internal Coupling* and *External Coupling* definitions are included. The *External Coupling* concept is specialized into *External Input Coupling* and *External Output Coupling*. Then, the remaining concepts define how each type of coupling is structured. For example, each *Internal Coupling* is defined based on a pair (*Output Port Name, Input Port Name*). The *Output Port Name* should refer to an *Output Port* of a component model, while the *Input Port Name* should refer to the *Input Port* of a distinct component model. These references are shown in Fig. 9 as dotted associations. Then, over the individual coupling definitions, coupling sets are defined using as containers concepts defined in the "DEVS Coupled Structural Model" package (i.e., *IC*, *EIC*, and *EOC*). It is important to denote that the *Component* concept is used to describe an internal DEVS model of a *CoupledModel*.

The DEVS-Specific Layer: "DEVS Atomic Structural Model," "DEVS Atomic Behavioral Model," "DEVS Atomic Interaction Definition," "DEVS Coupled Structural Model," and "DEVS Coupled Function Definition" Packages

The "DEVS-specific" layer includes the set of packages used to define special components required for the specification of each type of DEVS model. The "DEVS Atomic Structural Model" package (Fig. 10) defines how a DEVS atomic model is structured. It is based on the *Structural Part* concept included in the "DEVS Structural Model" package. The *Atomic Structural Part* is defined by adding the *State* definition as part of the components included in the container. A *State* is a mandatory component of an *Atomic Structural Part*. *State* defines the state composition as a nonempty ordered list of *State Variables* (i.e., multiplicity 1..*). It is not possible to declare a *State* without defining the *State Variables* involved in it.

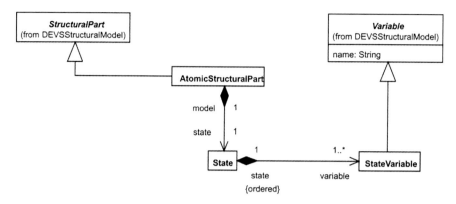

Fig. 10 UML class diagram of the "DEVS Atomic Structural Model" package

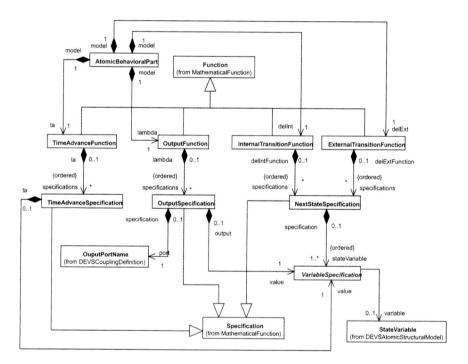

Fig. 11 UML class diagram of the "DEVS Atomic Behavioral Model" package

The "DEVS Atomic Behavioral Model" package (Fig. 11) defines the *Atomic Behavioral Part* concept following the approach of the *Structural Part* (i.e., as a container of several distinct elements). In all cases, the *Atomic Behavioral Part* plays the role *model* because it refers to the behavior of the *Atomic Model* (from the "DEVS Model core" package). All components included in the container are mandatory (i.e., multiplicity 1).

Each component defines a *Function* (from the "Mathematical Function" package). Each function plays a different role in the model as follows: (i) the *External Transition Function* plays the role *delExt*, (ii) the *Internal Transition Function* plays the role *delInt*, (iii) the *Output Function* plays the role *lambda*, and (iv) the *Time Advance Function* plays the role *ta*. Following the *Function* definition, each type of function has its *Specification* type. For example, both *External Transition Function* and *Internal Transition Function* are based on a list of *Next State Specifications*. A *Next State Specification* is related to a collection of *Variable Specifications*. Such a collection cannot be empty (i.e., multiplicity 1..*). Here, the *Variable Specification* plays the role *stateVariable*. Moreover (as Fig. 12 shows), a *Variable Specification* is related to a *State Variable* (from the "DEVS Atomic Structural Model" package). Hence, the *Variable Specification* linked to the *Next State Specification* defines how a *State Variable* evolves during the state transition function.

On the other hand, following the same modeling strategy, the *Output Function* concept is defined over *Output Specifications*, while the *Time Advance Function* uses *Time Advance Specifications*. An *Output Specification* is defined using a *Variable Specification* and an *Output Port Name* (from the "DEVS Coupling Definition" package). Finally, the *Time Advance Specification* only involves a *Variable Specification* that plays the role *value*.

The "DEVS Atomic Interaction Definition" package (Fig. 12) is added to the metamodel to define a set of intermediate elements used for linking the *Atomic Structural Part* and the *Atomic Behavioral Part* of the *Atomic Model* concept. Hence, these concepts abstract the interactions between both parts by modeling components of the behavioral level as elements that depend on the structural level. This allows instantiating the function specification using new independent elements that depend on the model structure.

From the *Variable Specification* (from the "DEVS Atomic Behavioral Model" package) concept, *Value Variable* and *Expression Variable* concepts are defined. The *Value Variable* concept refers to the specification of a variable as a single value. Such a value can be:

- An *Input Port Name* that depends on an *Input Port* (from the "DEVS Structural Model" package)
- An *Input Port Variable* that depends on an *Input Variable* (from the "DEVS Structural Model" package)
- A *Parameter Name* that depends on a *Parameter* (from the "DEVS Structural Model" package)
- A *Previous State Variable* that depends on a *State Variable* (from the "DEVS Structural Model" package)
- A *Value From Set* that depends on a *Set* (from the "Set Theory Utility" package)
- The *Infinity Parameter* or *Elapsed Time Variable* (i.e., concepts frequently used in state transitions)

On the other hand, the *Expression Variable* refers to a variable specification that is defined as an expression (e.g., the case of *If* conditions from the "Mathematical

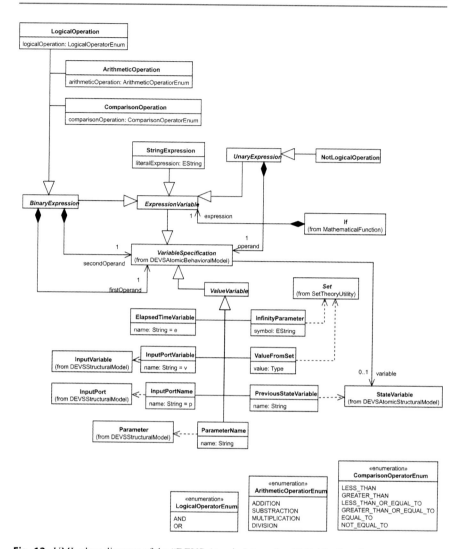

Fig. 12 UML class diagram of the "DEVS Atomic Interaction Definition" package

Function" package). Such an expression can be either a *Unary Expression* or a *Binary Expression* concept. For both types, the operands are defined as *Variable Specifications*. Hence, the model provides expressions designed over the set of elements available in the model definition. Moreover, an expression can be used as an operand in another expression. For each expression type, the model provides a set of predefined operations such as *Not Logical Operation*, *Logical Operation*, *Arithmetic Operation*, and *Comparison Operation*. It also includes the *String Expression* concept to define other types of expressions.

To define DEVS coupled models, two DEVS-specific packages are proposed. The "DEVS Coupled Structural Model" package (Fig. 13) extends the *Structural Part*

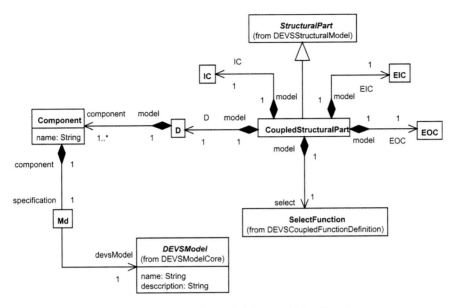

Fig. 13 UML class diagram of the "DEVS Coupled Structural Model" package

definition (from the "DEVS Structural Model" package) to conceptualize the *Coupled Structural Part*. Using the container approach proposed for the *Atomic Structural Part*, the coupled structure is defined as a container of (i) the set *EIC*, (ii) the set *EOC*, (iii) the set *IC*, (iv) the set *D*, and (v) the *Select* function (from the "DEVS Coupled Function Definition" package). Sets defined as *EIC*, *EOC*, and *IC* refer to the coupling definitions. The set defined as *D* contains all the DEVS components linked in the coupled model. For each *Component*, its specification (i.e., *Md*) is mandatory. Such a specification needs to refer to a *DEVS Model* (from the "DEVS Model Core" package).

Finally, the "DEVS Coupled Function Definition" package (Fig. 14) defines the *Select Function* from the *Function* concept (from the "Mathematical Function" package). The *Select Specification* is attached to a set of *Component Names* that refer to *Components* (from the "DEVS Coupled Structural Model" package) as imminent components and the tie-breaking decision.

Best Practice Approach

Defining a DEVS Simulation Model Through the DEVS Computational Representation: Software Tool and Illustrative Examples

As prior sections have shown, the DEVS metamodel includes a set of detailed concepts that a practitioner can use for creating DEVS models (i.e., state, state variables, input/output ports, and so on). In [43], a software tool has been developed

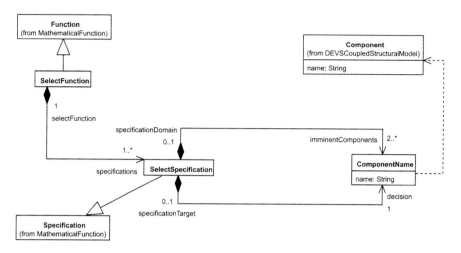

Fig. 14 UML class diagram of the "DEVS Coupled Function Definition" package

to assist modelers during the metamodel instantiation process (i.e., a computer-aided instantiation). This tool is based on Eclipse Modeling Framework (EMF) [44] as a plug-in for Eclipse IDE [45].

The EMF project is a modeling framework and code generation facility for building software tools and other software applications based on a structured data model. Using this technology, the metamodel defined in section "Computational Representation: The DEVS Metamodel" was implemented as an Ecore model. From the Ecore model, the EMF technology allows getting the Java code implementation of the conceptualization. The plug-in developed contains a wizard designed to guide practitioners during the definition (i.e., instantiation) process of the Ecore model. Extra fields were added for model documentation.

Figure 15 shows a UML activity diagram that shows the processing steps used by the tool to instantiate a computational formal model from DEVS computational representation. Once the modeler starts the definition process, the wizard asks for a file name. Such a file will contain the metamodel instance. The metamodel instantiation process starts with the instantiation of the DEVS model type selected by the user. The DEVS model type is defined in the file name through extensions. For example, if the instance refers to an atomic model, then the model will be saved in a file with the extension ".ams" (i.e., an "atomic model specification"). The attribute *name* of the model instance (i.e., *Atomic Model* or *Coupled Model*) is set with the file name. Then (mandatory) instance components are created (e.g., for an instance of *Atomic Model*, *Atomic Structural Part* and *Atomic Behavioral Part* are created).

After creating the model file, the user defines InPorts and OutPorts sets. For each set, the wizard provides a page that allows defining the ports as rows of a table. Figure 16 illustrates the definition of the *InPort* set for the switch model formalized in Eq. (2). Duplicate names are not allowed. With the defined sets, instances of *Input Port* and *Output Port* are created using the names detailed by the user. Such

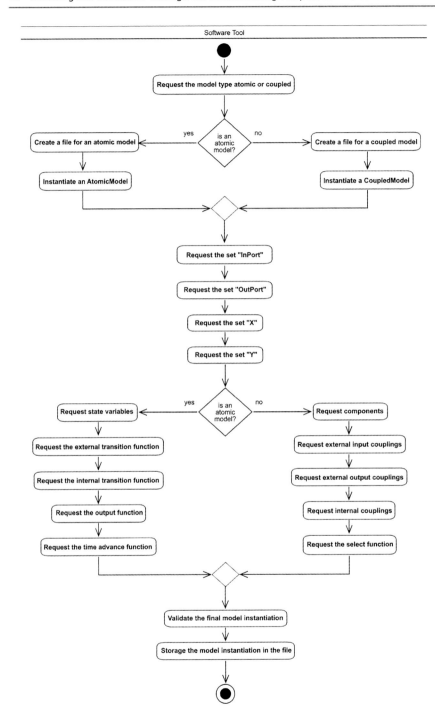

Fig. 15 Steps used to instantiate a valid computational formal model using the metamodel. A different sequence of actions is followed according to the model type. Each action instantiates the related elements

Fig. 16 Screenshot of the step used to define the "InPort" set for the DEVS model specified in Eq. (2)

instances are linked to the model using the proper relationships defined in the metamodel.

If the InPorts set is nonempty, then X should be defined. The same goes for OutPorts and Y. For both cases, if a set is required, the software redirects the modeler into a new page that allows properly defining each set. Over the InPort definition presented in Fig. 16, Fig. 17 shows how the modeler can select the set attached to each input port variable. Using the "New Set" button, the user can define new sets. Once the definition of input and output sets is complete, the wizard instantiates all the elements (i.e., concepts and relationships) required to build ports computational

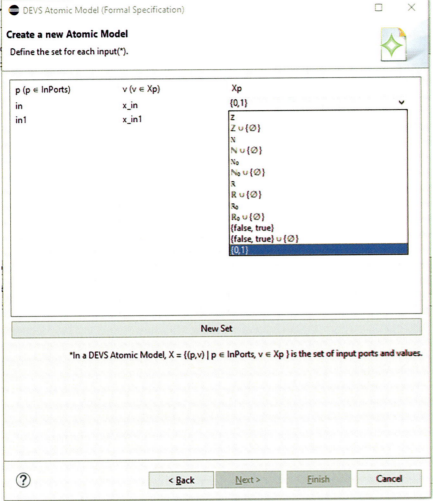

Fig. 17 Screenshot of the step used to define the mathematical sets attached to each input port included in the "InPort" set defined in Fig. 16 for the DEVS atomic model specified in Eq. (2)

representation (i.e., new sets using the concepts included in the "Set Theory Utility" package, the set of *Input Definition* and *Input Variable* attached to the input set X, and the set of *Output Definition* and *Output Variable* attached to the output set Y).

As Fig. 15 shows, the remainder of the instantiation process allows defining each part of the DEVS model following the style of pages presented above. Following the switch example, the state definition is presented in Fig. 18. The definition process is like the one described for input and output sets. Using a table, each one of the state variables is defined (Fig. 18a). Then, the user defines sets for each variable priorly

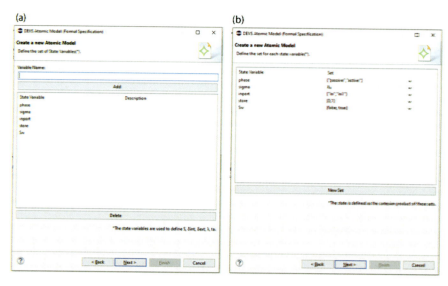

Fig. 18 Screenshot of the step used to define the state for the DEVS atomic model specified in Eq. (2): (**a**) Definition of the state variables; (**b**) definition of the sets attached to each state variable defined in (a)

detailed using a new wizard page (Fig. 18b). Once the definition is complete, the tool creates the *State* instance with the set of *State Variable* required.

For functions definition, the instantiation process is the same. The user defines the function using the set of available elements and instances created during the model definition. Figure 19 presents as an example the specification of the external transition function detailed in Eq. (2). Each specification is defined in a new row. The user uses the "New Specification" button to create a specification. This action redirects the user to a new page that allows detailing how each state variable changes due to the transition. As a result of the function definition, concepts from the "DEVS Atomic Behavioral Model" and "DEVS Atomic Interaction Definition" packages are instantiated and linked to the model.

Once the user ends the model specification, the wizard shows the final definition (Fig. 20). Before ending the process, the tool requires the user validates its specification. Such a verification checks the integrity of all instances created by the wizard following the metamodel definition. When the final instance is not valid, the tool requires the user to go back to the pages with issues and fix them. On the contrary, when the final instance is a valid instantiation of the DEVS metamodel, the creation process is complete. Then, the edition of the file containing the computational representation of the formal model is complete.

Using the computer-aided instantiation detailed, Appendix A presents the metamodel instantiation for the model detailed in Eq. (2). Instead of using the EMF viewer, the model is presented in its XML version. Also, Table 2 summarizes how each part of the formal specification is defined using the concepts of the DEVS computational representation.

40 Modeling and Simulation Through the Metamodeling Perspective: The Case...

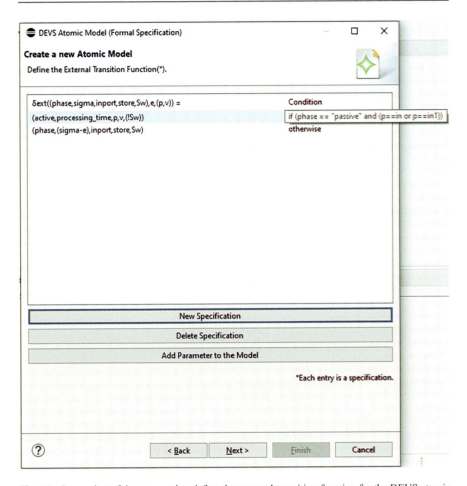

Fig. 19 Screenshot of the step used to define the external transition function for the DEVS atomic model specified in Eq. (2). From this step forward, the modeler can always define new parameters using the "Add Parameter to the Model" button. When adding a new parameter to the model, an instance of the Parameter concept is created and configured into the model following the metamodel definition

As an example of the instantiation detailed in Appendix A, Fig. 21 illustrates the instance of the metamodel for the *State Variable*-named *phase* as part of the *State* instantiation. The object diagram depicted in Fig. 21a uses the concepts of *State*, *StateVariable*, *New Set*, and *String Element* as instances of the metamodel. The instance of *State Variable* (i.e., a concept defined in the "DEVS Atomic Structural Model" package) has the attribute *name* set as "phase." For the definition of such a *State Variable* instance, an instance of *New Set* (i.e., a concept defined in the "Set Theory Utility" package) is detailed. The attribute *type* of the new set is defined as *STRING*. This means that the elements included in the set must be instances of *String Element* (i.e., a concept also included in the "Set Theory Utility" package). Two

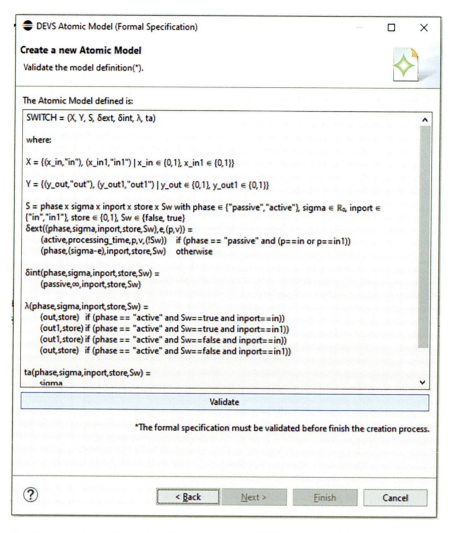

Fig. 20 Screenshot of the final validation step for the DEVS atomic model specified in Eq. (2) considering V = {0,1} and processing_time = 15.85

instances of *String Element* are illustrated. Each instance has a distinct set as part of the attribute *value*. Hence, the set attached to the state variable *phase* is {*passive, active*}.

Limitations of the DEVS Community That Metamodeling Overcomes

DEVS is most applicable for the modeling of systems with component-based modularity. However, it can be used much more generally as a simulation assembly

40 Modeling and Simulation Through the Metamodeling Perspective: The Case... 1219

Table 2 Instances of the DEVS computational representation created for the model specified in Eq. (2)

Model	Instances that represent the model component
SWITCH	An *AtomicModel* with *name* = *"SWITCH"* that contains an *AtomicStructuralPart* and an *AtomicBehavioralPart*
X	A *X* containing two instances of *InputDefinition*: (i) one related to *InputPort* with *name* = *"in"* and *InputVariable* with *name* = *"x_in,"* and (ii) the other related to *InputPort* with *name* = *"in1"* and *InputVariable* with *name* = *"x_in1"*
Y	A *Y* containing two instances of *OutputDefinition*: (i) one related to *OutputPort* with *name* = *"out"* and *OutputVariable* with *name* = *"y_out,"* and (ii) the other related to *OutputPort* with *name* = *"out1"* and *OutputVariable* with *name* = *"y_out1"*
S	A *state* instance that contains five instances of *StateVariable* identified with (i) *name* = *"phase,"* (ii) *name* = *"sigma,"* (iii) *name* = *"inport,"* (iv) *name* = *"store,"* and (v) *name* = *"Sw"*
δ_{ext}	An *ExternalTransitionFunction* instance that contains an ordered set of *NextStateSpecification* (in this case, two). As an example, the first one is defined as an ordered set of *VariableSpecification* that includes the following instances: (i) a *ValueFromSet* with *value* = *"active,"* (ii) a *ParameterName* with *name* = *"processing_time,"* (iii) an *InputPortName*, (iv) an *InputPortVariable*, and (v) a *PreviousStateVariable* with *name* = *"Sw."* Condition instances are defined appropriately for each specification
δ_{int}	An *InternalTransitionFunction* instance that contains a *NextStateSpecification* defined as an ordered set of *VariableSpecification* that includes the following instances: (i) a *ValueFromSet* with *value* = *"passive,"* (ii) an *InfinityParameter*, (iii) a *PreviousStateVariable* with *name* = *"inport,"* (iv) a *PreviousStateVariable* with *name* = *"store,"* and (v) a *PreviousStateVariable* with *name* = *"Sw"*
λ	An *OutputFunction* instance that contains an ordered set of *OutputSpecification* (in this case, four instances). As an example, the first *OutputSpecification* takes as value the *PreviousStateVariable* with *name* = *"store"* and as port the *OutputPortName* with *name* = *"out."* Condition instances are defined appropriately for each specification
ta	A *TimeAdvanceFunction* that contains an ordered set of *TimeAdvanceSpecification*. The single specification included in the set is defined using the *PreviousStateVariable* with *name* = *"sigma"*

language, or as a theoretical foundation for M&S formalisms. Most executable representations are tool-specific, though efforts are underway to define a common standard [46]. In these cases, a well-defined standardized DEVS conceptualization can decrease the effort to develop DEVS simulation models that conform to their formal model depending on the M&S software tool selected by the user.

When a DEVS formal model is detailed using the DEVS computation representation based on the metamodel, the modeler is creating a DEVS formal model in a computational representation. Data is not lost in between. This means that the computational representation acts as a conceptual mechanism to define valid DEVS formal models in computational environments. Figure 21b shows how the instance defined in Fig. 21a for the state variable *phase* using the DEVS computational representation can be mapped to an OO framework. The *State* class represents the model state using an attribute named *phase* of type *Phase Value* to define the OO

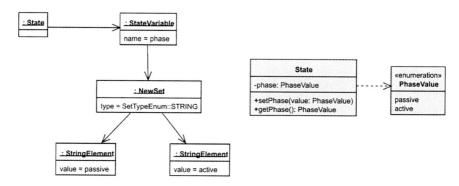

Fig. 21 Conceptualization of the State Variable "phase": (**a**) Instantiation of the metamodel for the State Variable-named "phase" as part of the State instantiation; (**b**) mapping the instance depicted in (a) to an OO framework

composition. This means the *State* class will have as many attributes as *State Variable* instances defined in the model. The *Phase Value* enumeration is obtained considering the set used to specify the formal state variable *phase* (i.e., the string set {*passive, active*}). Each instance of *String Element* is used to define an enumeration literal. Hence, the DEVS computational representation can act as a foundation to build autocoders for M&S software tools that create implementations from metamodel-based instantiations of DEVS formal models.

Using the instantiated model as an Ecore instance, the EMF model can be easily used to produce Java code. Following such a strategy, for example, implementations based on DEVSJAVA [31] and MS4-Me [32] can be obtained. Both software tools are M&S tools based on Java that support the M&S of DEVS models by providing an OO definition for model development and a simulation engine for model's execution. For software tools based on other programming languages, the XML instance (as the structured representation of the formal model) can be used to get computational models that comply with the formal model definition. Hence, a unique formal definition defined as an instance of the metamodel (i.e., a concrete model) can be used to obtain several model implementations for different tools (i.e., a set of computational models). This is the main benefit of using metamodels to support formal specifications. Moreover, as a standard practice in software development, developers can improve the computational model following programming best practices to enhance the performance when executed.

This contribution can be seen as the fundamental part of the metamodeling-based formalization of DEVS extensions. Classic DEVS with ports was employed to define the metamodel. Since most DEVS extensions are based on port formalization, the computational representation of DEVS can be extended to support other formalisms designed over DEVS. This means that extension points can be used to define new extensions of the metamodel related to the conceptual models of other DEVS-based formalisms. The formal definition of such DEVS extensions could be conceptualized from the metamodel by extending the concepts and relationships, if and only if such extensions are based on DEVS atomic models (e.g., Parallel DEVS, Dynamic

Structure DEVS, and Routed DEVS). For example, Parallel DEVS introduces the confluence function to deal with concurrent events. This function can be defined using the proposed metamodel by extending the concept *Function* of the "Mathematical Function" package and following the structure of external and internal transition functions. With an accurate definition of the remaining concepts, DEVS extensions could be modeled using the DEVS computational representation as a foundation. Hence, future developments could include the metamodeling of specific DEVS extensions and the study of how metamodeling can support multiformalism definitions. These open problems are derived from this chapter and need to be addressed in the future.

As a remark, it is interesting to briefly discuss how our proposal is related to the MDA approach. MDA involves the specification of models at three levels: models, metamodel, and metameta models. The capabilities of the proposed metamodel can be improved with a metameta model that acts as a standard definition of formalisms. Hence, such a metameta model will allow bridging the gap between DEVS formal models and other alternative modeling formalisms.

Chapter Summary and Conclusion

Formalisms help describe a system unambiguously. If formal models are implemented without any consistent computational frameworks, however, it is hard to guarantee that there is no semantic gap between formal models and their programming-code implementation. DEVS is an abstract formalism for M&S. Due to DEVS generality, its formal specification based on set theory defines an abstraction for concrete model implementations. However, given the difficulties of getting programming code implementation for such an abstraction, practitioners are devoted to the development of DEVS representations attached to code implementations.

In this chapter, a metamodel-based computational representation of DEVS formal specifications based on Classic DEVS with ports has been presented to show the significance of model-based engineering in the M&S field. The approach presented follows the guidelines of existing conceptualizations of DEVS, showing that building languages, ontologies, and metamodels for DEVS is a highly popular research field. The models instantiated from DEVS computational representation are conceived as computational formal models (i.e., platform-independent concrete models) that can be translated to any M&S software tools when transformation rules are defined (i.e., platform-specific implementation models). Such a feature improves the modeling task by providing practitioners the option of creating DEVS models without knowing OO concepts. The formal implications of DEVS specification are represented in the metamodel, so practitioners can define "accurate" DEVS models (i.e., the complexities of dealing with M&S software tools and libraries implementations are hidden from modelers). In this way, metamodeling can enable the use of M&S to analyze complex systems in new domains for DEVS users that are nonexpert programmers. Moreover, it is expected that the approach motivates practitioners to create other metamodels for M&S.

Since the metamodel has been defined as a conceptual model of DEVS formalism, the DEVS metamodel proposed can be seen as the cornerstone for MBSE approaches applied to the M&S field. We strongly believe that metamodels can help the M&S community to bring standards to model-driven methodologies and model-based engineering practices. The use of these conceptual models could advance the discipline and profession of DEVS M&S with a set of conventions and notations for the formal specification of DEVS models. This approach can enrich other existing approaches where DEVS formal models are used specifically to define domain-to-formal DEVS transformations.

Cross-References

▶ Conceptual Design Support by MBSE: Established Best Practices
▶ Semantics, Metamodels, and Ontologies

Appendix

```
<?xml version="1.0" encoding="UTF-8"?>
<model:AtomicModel xmi:version="2.0" xmlns:xmi="http://www.omg.
org/XMI" xmlns:xsi="http://www.w3.org/2001/XMLSchema-instance"
xmlns:model="http://www.devs.formalization.atomic.org/model"
name="SWITCH">
  <structure>
    <InPorts name="in" x="//@structure/@input/@definition.0"/>
    <InPorts name="in1" x="//@structure/@input/@definition.1"/>
    <OutPorts name="out" y="//@structure/@output/@definition.0"/>
    <OutPorts name="out1" y="//@structure/@output/@definition.1"/>
    <input>
      <definition p="//@structure/@InPorts.0">
        <v name="x_in" definition="NS01" />
      </definition>
      <definition p="//@structure/@InPorts.1">
        <v name="x_in1" definition="NS01" />
      </definition>
    </input>
    <output>
      <definition p="//@structure/@OutPorts.0">
        <v name="y_out" definition="NS01" />
      </definition>
      <definition p="//@structure/@OutPorts.1">
        <v name="y_out1" definition="NS01" />
      </definition>
    </output>
    <state>
      <variable name="phase" definition="NSPA" />
      <variable name="sigma" definition="//@setsAndStructures.8" />
```

```xml
      <variable name="inport" definition="NSIP" />
      <variable name="store" definition="NS01" />
      <variable name="Sw" definition="//@setsAndStructures.10" />
    </state>
    <parameters name="processing_time">
      <definition xsi:type="model:RealNumber" value="15.85"/>
    </parameters>
  </structure>
  <setsAndStructures xsi:type="model:IntegerNumbersSet" />
  <setsAndStructures xsi:type="model:IntegerNumbersSet" union="//
@emptyset"/>
  <setsAndStructures xsi:type="model:NaturalNumbersSet" />
  <setsAndStructures xsi:type="model:NaturalNumbersSet" union="//
@emptyset"/>
  <setsAndStructures xsi:type="model:NaturalNumbersSet"
includingZero="true"/>
  <setsAndStructures xsi:type="model:NaturalNumbersSet" union="//
@emptyset"
  includingZero="true"/>
  <setsAndStructures xsi:type="model:RealNumbersSet" />
  <setsAndStructures xsi:type="model:RealNumbersSet" union="//
@emptyset"/>
  <setsAndStructures xsi:type="model:RealNumbersSet"
includingZero="true"/>
  <setsAndStructures xsi:type="model:RealNumbersSet" union="//
@emptyset"
  includingZero="true"/>
  <setsAndStructures xsi:type="model:BooleanAlgebraSet" />
  <setsAndStructures xsi:type="model:BooleanAlgebraSet" union="//
@emptyset"/>
  <setsAndStructures xsi:type="model:NewSet" name="NS01"
type="MATH_INTEGER">
    <elements xsi:type="model:IntegerNumber" value="0"/>
    <elements xsi:type="model:IntegerNumber" value="1"/>
  </setsAndStructures>
  <setsAndStructures xsi:type="model:NewSet" name="NSPA"
type="STRING">
    <elements xsi:type="model:StringElement" value="passive"/>
    <elements xsi:type="model:StringElement" value="active"/>
  </setsAndStructures>
  <setsAndStructures xsi:type="model:NewSet" name="NSIP"
type="STRING">
    <elements xsi:type="model:StringElement" value="in"/>
    <elements xsi:type="model:StringElement" value="in1"/>
  </setsAndStructures>
  <emptyset />
  <behavior>
    <delExt otherwiseConditionAlreadySet="true">
      <specifications>
        <condition xsi:type="model:If" conditionAsText="if (phase
== "passive"
        and (p==in or p==in1))"/>
```

```xml
        <stateVariable xsi:type="model:ValueFromSet"
        variable="//@structure/@state/@variable.0" value="active"/
>
        <stateVariable xsi:type="model:ParameterName"
        variable="//@structure/@state/@variable.1"
name="processing_time"/>
        <stateVariable xsi:type="model:InputPortName"
        variable="//@structure/@state/@variable.2"/>
        <stateVariable xsi:type="model:InputPortValue"
        variable="//@structure/@state/@variable.3"/>
        <stateVariable xsi:type="model:NotLogicalOperation"
        variable="//@structure/@state/@variable.4">
          <operand xsi:type="model:PreviousStateVariable"
name="Sw"/>
        </stateVariable>
      </specifications>
      <specifications>
        <condition xsi:type="model:Otherwise"
conditionAsText="otherwise"/>
        <stateVariable xsi:type="model:PreviousStateVariable"
        variable="//@structure/@state/@variable.0" name="phase"/>
        <stateVariable xsi:type="model:ArithmeticOperation"
        variable="//@structure/@state/@variable.1"
arithmeticOperation="-">
          <firstOperand xsi:type="model:PreviousStateVariable"
name="sigma"/>
          <secondOperand xsi:type="model:ElapsedTimeVariable"/>
        </stateVariable>
        <stateVariable xsi:type="model:PreviousStateVariable"
       variable="//@structure/@state/@variable.2" name="inport"/>
        <stateVariable xsi:type="model:PreviousStateVariable"
       variable="//@structure/@state/@variable.3" name="store"/>
        <stateVariable xsi:type="model:PreviousStateVariable"
     variable="//@structure/@state/@variable.4" name="Sw"/>
      </specifications>
    </delExt>
    <delInt>
      <specifications>
        <stateVariable xsi:type="model:ValueFromSet"
        variable="//@structure/@state/@variable.0"
value="passive"/>
        <stateVariable xsi:type="model:InfinityParameter"
        variable="//@structure/@state/@variable.1" symbol="∞"/>
        <stateVariable xsi:type="model:PreviousStateVariable"
       variable="//@structure/@state/@variable.2" name="inport"/>
        <stateVariable xsi:type="model:PreviousStateVariable"
        variable="//@structure/@state/@variable.3" name="store"/>
        <stateVariable xsi:type="model:PreviousStateVariable"
        variable="//@structure/@state/@variable.4" name="Sw"/>
      </specifications>
    </delInt>
    <lambda>
      <specifications>
```

```
            <condition xsi:type="model:If" conditionAsText="if (phase
== "active"
        and Sw==true and inport==in))"/>
        <value xsi:type="model:PreviousStateVariable"
name="store"/>
        <port name="out"/>
      </specifications>
      <specifications>
        <condition xsi:type="model:If" conditionAsText="if (phase
== "active"
        and Sw==true and inport==in1))"/>
        <value xsi:type="model:PreviousStateVariable"
name="store"/>
        <port name="out1"/>
      </specifications>
      <specifications">
        <condition xsi:type="model:If" conditionAsText="if (phase
== "active"
        and Sw==false and inport==in))"/>
        <value xsi:type="model:PreviousStateVariable"
name="store"/>
        <port name="out1"/>
      </specifications>
      <specifications>
        <condition xsi:type="model:If" conditionAsText="if (phase
== "active"
        and Sw==false and inport==in1))"/>
        <value xsi:type="model:PreviousStateVariable"
name="store"/>
        <port name="out"/>
      </specifications>
    </lambda>
    <ta>
      <specifications>
        <value xsi:type="model:PreviousStateVariable"
name="sigma"/>
      </specifications>
    </ta>
  </behavior>
</model:AtomicModel>
```

References

1. D. Gianni, A. D'Ambrogio, and A. Tolk, Eds., *Modeling and Simulation-Based Systems Engineering Handbook*. Boca Raton, FL, USA: CRC Press, 2015.
2. J. Holt, and S. Perry, *SysML for systems engineering*. London, United Kingdom: IET, 2008.

3. S. Wolny, A. Mazak, C. Carpella, V. Geist, and M. Wimmer, "Thirteen years of SysML: a systematic mapping study," *Softw. and Syst. Model.*, vol. 19, pp. 111–169, 2020, https://doi.org/10.1007/s10270-019-00735-y.

4. G. Kapos, V. Dalakas, A. Tsadimas, M. Nikolaidou and, D. Anagnostopoulos, "Model-based system engineering using SysML: Deriving executable simulation models with QVT," in *Proc. 2014 IEEE Int. Syst. Conf.*, 2014, pp. 531–538, https://doi.org/10.1109/SysCon.2014.6819307.

5. A. Gargantini, E. Riccobene, and P. Scandurra, "A metamodel-based language and a simulation engine for abstract state machines," *J. Univers. Comput. Sci.*, vol. 14., no. 12, pp. 1949–1983, 2008, https://doi.org/10.3217/jucs-014-12-1949.

6. M. Li, F. Yang., R. Uzsoy, and J. Xu, "A metamodel-based Monte Carlo simulation approach for responsive production planning of manufacturing systems," *J. of Manuf. Syst.*, vol. 38, pp. 114–133, 2016, https://doi.org/10.1016/j.jmsy.2015.11.004.

7. B. P. Zeigler, A. Muzy, and E. Kofman, *Theory of modeling and simulation: discrete event & iterative system computational foundations*, 3rd ed. London: Academic Press, 2018.

8. A. Muzy, B. P. Zeigler, and F. Grammont, "Iterative specification as a modeling and simulation formalism for I/O general systems," *IEEE Syst. J.*, vol. 12, no. 3, pp. 2982–2993, Sept. 2018, https://doi.org/10.1109/JSYST.2017.2728861.

9. Y. J. Kim, J. Y. Yang, Y. M. Kim, J. Lee, and C. Choi, "Modeling Behavior of Mobile Application Using Discrete Event System Formalism," in *Proc. 2016 Asian Simul. Conf.*, 2016, pp. 40–48, https://doi.org/10.1007/978-981-10-2158-9_4.

10. G. Wainer, M. Etemad, and B. Kazi, "Modeling Coordinated Multipoint with a dynamic Coordination Station in LTE-A mobile networks," in *Proc. 2017 IEEE Int. Conf. on Netw., Sens. and Control*, 2017, pp. 807–812, https://doi.org/10.1109/ICNSC.2017.8000194.

11. A. Uhrmacher, D. Degenring, B. Zeigler, "MultiLevel discrete event modeling in systems biology," *Trans. Comput. Syst. Biol. I*, pp. 66–89, 2004, https://doi.org/10.1007/978-3-540-32126-2_6.

12. A. Uhrmacher, R. Ewald, M. John, C. Maus, M. Jeschke, and S. Biermann, "Combining micro and macro-modeling in DEVS for computational biology," in *Proc. 2007 Winter Simul. Conf.*, 2007, pp. 871–880, https://doi.org/10.1109/WSC.2007.4419683.

13. S. Gholami, H. Sarjoughian, G. Godding, D. Peters, and V. Chang, "Developing Composed Simulation and Optimization Models using Actual Supply-Demand Network Datasets," in *Proc. 2014 Winter Simul. Conf.*, 2014, pp. 2510–2521, https://doi.org/10.1109/WSC.2014.7020095.

14. M. Moallemi, G. Wainer, A. Awad, and D.A. Tall, "Application of RTDEVS in Military," in *Proc. 2010 Spring Simul. Multiconf.*, 2010, pp. 29–36, https://doi.org/10.1145/1878537.1878568.

15. P. Carreira, V. Amaral, and H. Vangheluwe, Eds., *Foundations of multi-paradigm modelling for cyber-physical systems*. Cham, Switzerland: Springer Nature, 2020.

16. H. S. Sarjoughian, A. Alshareef, and Y. Lei. "Behavioral DEVS metamodeling," in *Proc. 2015 Winter Simul. Conf.*, 2015, pp. 2788–2799, https://doi.org/10.1109/WSC.2015.7408384.

17. T. Ören, S. Mittal, U. Durak. "A Shift from Model-Based to Simulation-Based Paradigm: Timeliness and Usefulness for Many Disciplines," *Int. J. Comput. Softw. Eng.*, vol. 3, 2018, Art no. 126, https://doi.org/10.15344/2456-4451/2018/126.

18. M. J. Blas, S. Gonnet, and B. P. Zeigler, "Towards a Universal Representation of DEVS: A Metamodel-Based Definition of DEVS Formal Specification," in *2021 Annu. Model. and Simul. Conf.*, 2021, https://doi.org/10.23919/ANNSIM52504.2021.9552162.

19. D. A. Hollmann, M. Cristia, and C. Frydman, "CML-DEVS: A specification language for DEVS conceptual models," *Simul. Model. Pract. Theory*, vol. 57, pp. 100–117, Sept. 2015, https://doi.org/10.1016/j.simpat.2015.06.007.

20. B. P. Zeigler, *Theory of modeling and simulation*, New York, NY, USA: John Wiley & Sons Inc., 1976.

21. Y. Van Tendeloo and H. Vangheluwe, "Classic DEVS modelling and simulation," in *Proc. 2017 Winter Simul. Conf.*, 2017, pp. 644–658, https://doi.org/10.1109/WSC.2017.8247822.
22. G. Wainer and N. Giambiasi, "N-dimensional Cell-DEVS models," *Discret. Event Dyn. Syst.*, vol. 12, pp. 135–157, 2002, https://doi.org/10.1023/A:1014536803451.
23. F. J. Barros, "Modeling formalisms for dynamic structure systems," *ACM Trans. Model. Comput. Simul.*, vol. 7, no. 4, pp. 501–515, 1997, https://doi.org/10.1145/268403.268423.
24. A. C. Chow and B. Zeigler, "Parallel DEVS: a Parallel, Hierarchical, Modular Modeling Formalism," in *Proc. 1994 Winter Simul. Conf.*, 1994, pp. 716–722, https://doi.org/10.1109/WSC.1994.717419.
25. J. S. Hong, H. S. Song, T. G. Kim, and K. H. Park, "A real-time discrete event system specification formalism for seamless real-time software development," *Discret. Event Dyn. Syst.*, vol. 7, pp. 355–375, 1997, https://doi.org/10.1023/A:1008262409521.
26. M. J. Blas, S. Gonnet, and H. Leone, "Routing Structure Over Discrete Event System Specification: A DEVS Adaptation To Develop Smart Routing In Simulation Models," in *Proc. 2017 Winter Simul. Conf.*, 2017, pp. 774–785, https://doi.org/10.1109/WSC.2017.8247831.
27. B. P. Zeigler and J. J. Nutaro, "Towards a framework for more robust validation and verification of simulation models for systems of systems," *The J. of Defense Model. and Simul.*, vol. 13, no. 1, pp. 3–16, 2016, https://doi.org/10.1177/1548512914568657.
28. X. Hu, B. P. Zeigler, and S. Mittal, "Variable structure in DEVS component-based modeling and simulation," *Simulation*, vol. 81, pp. 91–102, 2005, https://doi.org/10.1177/0037549705052227.
29. Y. Van Tendeloo and H. Vangheluwe, "An evaluation of DEVS simulation tools," *Simulation*, vol. 93, pp. 103–121, 2017, https://doi.org/10.1177/0037549716678330.
30. Y. Van Tendeloo and H. Vangheluwe, "The modular architecture of the Python(P)DEVS simulation kernel," in *Proc. 2014 Spring Simul. Multiconf.*, 2014, pp. 387–392.
31. H. S. Sarjoughian and B. Zeigler, "Devsjava: Basis for a devs-based collaborative m&s environment," *Simul. Series*, vol. 30, pp. 29–36, 1998.
32. MS4 Systems. *MS4 me simulator.* (3.0). Accessed: June 27, 2022. [Online]. Available: http://ms4systems.com/pages/ms4me.php
33. G. Wainer, "CD++: a toolkit to develop DEVS models,", *Softw.: Pract. and Experience*, vol. 32, no. 13, pp. 1261–1306, Nov 2002, https://doi.org/10.1002/spe.482.
34. C. Ruiz Martin, G. Wainer. *Cadmiun.* Accessed: June 27, 2022. [Online]. Available: https://github.com/SimulationEverywhere/cadmium
35. M. Cristia, D. A. Hollmann, and C. Frydman, "A multi-target compiler for CML-DEVS," *Simulation*, vol 95, pp. 11–29, 2019, https://doi.org/10.1177/0037549718765080.
36. M. Nikolaidou, V. Dalakas, L. Mitsi, G. Kapos, and D. Anagnostopoulos, "A SysML Profile for Classical DEVS Simulators," in *3rd Int. Conf. on Softw. Eng. Adv.*, 2008, pp. 445–450, https://doi.org/10.1109/ICSEA.2008.24.
37. E. Seidewitz, "What models mean," *IEEE Softw.*, vol. 20, no. 5, pp. 26–32, Sept.–Oct. 2003, https://doi.org/10.1109/MS.2003.1231147.
38. S. Robinson, "Conceptual Modelling for Simulation: Progress and Grand Challenges," *J. of Simul.*, vol. 14, no. 1, pp. 1–20, 2020, https://doi.org/10.1080/17477778.2019.1604466.
39. J. L. Risco Martin, S. Mittal, M. A. López-Peña, and J. M. de la Cruz, "A W3C XML schema for DEVS scenarios," in *Proc. 2007 Spring Simul. Multiconf.*, San Diego, CA, USA, pp. 279–286, 2007.
40. M. Mernik, J. Heering, and A. M. Sloane, "When and how to develop domain-specific languages," *ACM Comput. Surv.*, vol. 37, no. 4, pp. 316–344, Dec. 2005, https://doi.org/10.1145/1118890.1118892.
41. L. Touraille, "Application of Model-driven Engineering and Metaprogramming to DEVS Modeling & Simulation". Ph.D. dissertation, Université d'Auvergne, France, 2013.
42. Y. Hu, J. Xiao, H. Zhao, and G. Rong, "DEVSMO: An Ontology of DEVS Model Representation for Model Reuse," in *Proc. 2013 Winter Simul. Conf.*, 2013, pp. 4002–4003.

43. M. J. Blas, and S. Gonnet, "Metamodel-based formalization of DEVS atomic models," *Simulation*, 2021, https://doi.org/10.1177/00375497211045628.
44. The Eclipse Foundation. *Eclipse modeling project. Eclipse modeling framework*. Accessed: June 26, 2022. [Online]. Available: https://www.eclipse.org/modeling/emf/
45. The Eclipse Foundation. *Eclipse*. Accessed: June 26, 2022. [Online]. Available: https://www.eclipse.org/
46. G. Wainer, K. Al-Zoubi, and D. R. C. Hill, "An introduction to DEVS standardization," in *Discrete-event modeling and simulation: theory and applications*, Boca Raton, FL: CRC Press, 2010, pp. 393–425.

María Julia Blas is an Assistant Researcher at the Consejo Nacional de Investigaciones Científicas y Técnicas (CONICET) and an Assistant Professor at Universidad Tecnológica Nacional – Facultad Regional Santa Fe (UTN-FRSF). She received her PhD degree in Engineering from Universidad Tecnológica Nacional in 2019. Her research interests include discrete-event modeling and simulation and metamodeling approaches to support the Discrete Event System Specification (DEVS) formalism.

Silvio Gonnet received his PhD degree in Engineering from Universidad Nacional del Litoral in 2003. He currently holds a Researcher position at the Consejo Nacional de Investigaciones Científicas y Técnicas (CONICET). His research interests include conceptual modeling, ontology engineering, and discrete-event modeling and simulation.

Part VIII
Future Outlook

Exploiting Transdisciplinarity in MBSE to Enhance Stakeholder Participation and Increase System Life Cycle Coverage

41

Azad M. Madni

Contents

Introduction	1232
MBSE and Decision Analysis	1233
Preferences, Values, and Utility	1233
Quantifying Uncertainty	1235
Updating Belief	1236
Valuing Information and Experimentation	1236
Biases in Human Decision-Making	1237
Institutional Barriers	1238
MBSE and Digital Engineering	1238
MBSE and Digital Twins	1239
MBSE and Digital Thread	1239
Industry 4.0	1239
Internet of Things	1240
MBSE and AI/ML	1241
MBSE and Social Networks	1244
Crowdsourcing	1244
MBSE and Entertainment Arts	1245
Transdisciplinarity: Key to Exploiting Convergence	1248
Implications for MBSE Curriculum	1248
Conclusions	1249
Cross-References	1250
References	1250

Abstract

As MBSE continues to extend its reach into the later phases of the systems life cycle, it is becoming apparent that MBSE has much to gain by leveraging concepts from other disciplines such as decision analysis, digital engineering, AI/machine learning, social networks, and entertainment arts. This chapter

---

A. M. Madni (✉)
Intelligent Systems Technology, Inc., Los Angeles, CA, USA
e-mail: amadni@intelsystech.com

© Springer Nature Switzerland AG 2023
A. M. Madni et al. (eds.), *Handbook of Model-Based Systems Engineering*,
https://doi.org/10.1007/978-3-030-93582-5_69

illuminates the synergy between MBSE and these other disciplines along multiple dimensions including the modeling of humans and groups, quantifying uncertainty, capturing human preferences, reasoning about the problem space, aggregating information, employing digital twins to increase system life cycle coverage, leveraging crowdsourcing to facilitate access to expertise, and exploiting storytelling to increase stakeholder participation in collaborative planning and decision-making. Recommended changes and additions to existing systems engineering curricula and MBSE courses are also presented.

Keywords

MBSE · Decision analysis · Digital engineering · AI/machine learning · Social networks · Entertainment arts

Introduction

Systems engineering (SE) is undergoing a welcome transformation fueled by advances in model-based systems engineering (MBSE) and the advent of Industry 4.0 and Internet of Things (IoT). As MBSE continues to extend its reach into the later phases of the systems life cycle, it is becoming apparent that MBSE can leverage concepts from other disciplines such as decision analysis, digital engineering, AI/ML, social networks, and entertainment arts. Exploiting synergy of systems engineering with other disciplines is at the heart of transdisciplinary systems engineering which seeks to exploit the convergence of engineering with other disciplines to address sociotechnical systems problems that appear intractable when viewed solely through an engineering lens [1].

With this new mindset, systems engineers can begin to formulate solution approaches for complex sociotechnical problems (e.g., NAE Grand Challenges) using key concepts from other engineering disciplines (e.g., decision analysis, digital engineering, AI/ML), as well as from humanities (e.g., social networks, entertainment arts).

Decision analysis (DA) offers a rigorous approach to ranking and valuing decision alternatives. It has been applied successfully in diverse industries, e.g., pharmaceuticals, oil and gas, manufacturing, and large enterprises including acquisition organizations within the Federal government. DA is used to (i) identify preferences, values, and levels of risk tolerance, (ii) quantify uncertainties and provide methods for representing and updating belief in the light of new information using appropriate methods to frame the decision, (iii) methods to explore new and feasible alternatives, (iv) methods to frame the decision, (v) method to incorporate the human element, and (vi) methods to analyze the situation using sound logic [2] and [3]. DA also provides a means to address cognitive biases in decision-making.

Digital engineering (DE) is an integrated digital approach that uses authoritative sources of systems data and models within a continuum across disciplines to support lifecycle activities from concept to disposal. Digital twin and digital thread are two important concepts in DE that can enable MBSE cover the later stages of the system life cycle such as verification and validation testing. The concept of digital twins is already being leveraged in MBSE [4, 5].

AI/machine learning (AI/ML) represents a significant advance in computer science and smart data processing that is rapidly transforming several industries. AI generally refers to processes and algorithms that simulate human by mimicking cognitive functions such as perception, learning, and problem-solving. ML and deep learning (DL) are subsets of AI. AI/ML have the potential to extend MBSE to modeling complex systems with partially known state space operating in uncertain, dynamic environments.

Social networks (SN) refer to the use of Internet-enabled social media sites that allow people to stay connected to other people (e.g., friends, family, colleagues, customers, clients), socialize, or conduct business. Examples include Facebook, Twitter, LinkedIn, and Instagram. Social networks can complement traditional collaboration platforms and thereby enable increase access to qualified participants not part of the original collaboration team.

Entertainment arts (EA) encompass transmedia storytelling, visualization and animation, concept design, video games, and digital movies. In recent years, EA concepts such as storytelling and video games are being increasingly employed in engineering to inform, engage, entertain, and persuade stakeholders engaged in collaborative decision-making [6].

This chapter presents the synergy between MBSE and these other disciplines and provides examples of how transdisciplinary thinking can be exploited to incorporate key concepts from these other disciplines into SE curricula and MBSE courses [7].

MBSE and Decision Analysis

The systems engineering community is beginning to recognize that individual and organizational decision-making in systems engineering can potentially benefit from incorporating certain concepts from decision analysis (DA). Specifically, upfront collaborative systems engineering activities stand to benefit from a structured human-centered framework for evaluation of alternatives. Figure 1 presents the potential synergy between MBSE and DA that can be exploited in upfront engineering. As shown in this figure, MBSE is used to frame the decision problem and problem context including informational and physical constraints. DA is used to structure the problem situation, generate/access and evaluate alternatives, and make recommendations that consider both individual and organizational preferences, values, and utility.

Preferences, Values, and Utility

DA has three important concepts that can be exploited potentially in systems engineering models to aid decision-making. DA helps identify a decision maker's

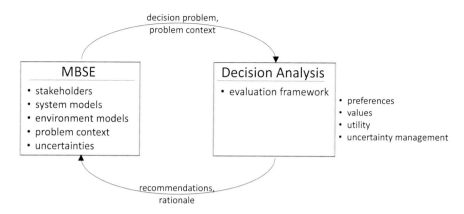

Fig. 1 Exploiting synergy between MBSE and decision analysis

preferences for prospects (i.e., likely outcomes) of decisions as well as the *value* for each prospect. It also defines a *utility* function that captures the decision maker's risk attitude [8]. These concepts apply to decision-making in systems engineering (e.g., [9, 10–12]).

Specifically, in systems engineering, it is important to understand stakeholder preferences, values, and utility for engineering decision and be able to distinguish between attributes that provide *direct value* and those that are a *means to an end*. For example, for a profit maximizing firm, total *shareholder return* would be considered an attribute of direct value, while *demand* or *product quality* would be considered means to maximizing *profit*, an attribute of direct value. In this example, profit is the end objective. In other instances, there may be more than one attribute of direct value such as safety and climate effects. The purpose of design is to maximize value as identified by the attributes of direct value [13].

Systems engineering relies on customer provided and/or elicited requirements and targets to meet specific goals and to distribute decision-making among collaborative teams. Often, requirements acquired in this fashion can lead to suboptimal designs unless the trade-offs among the feasible design attributes correspond to the contours of constant value [14].

One of the first attempts to introduce DA into systems engineering was to harness the concept of value in design. Subsequently, the concept of value-driven design has been successfully introduced into specific aspects of systems engineering [15–17]. However, systems engineering can further benefit from a broader understanding of inconsistencies that target-setting might inadvertently introduce into the actual preferences of an organization. In this regard, it is important to consider the effects of requirements on the ultimate design and how setting requirements may in fact lead to suboptimal products that deviate from the value maximizing objective of design. As important, systems engineering curricula and MBSE courses stand to benefit from courses in decision analysis that cover preference, value, and utility concepts.

Quantifying Uncertainty

Even though uncertainty is inherent in the design and implementation of any large-scale system, uncertainty characterization and management are given short shrift in most engineering courses. Today most probability courses in engineering curricula tend to focus exclusively on tools and models to compute likelihoods, durations, and expectation of events when some initial probabilities are provided. For example, students may be given probabilities of the Dow Jones Industrial Average and the S&P 500 increasing at a certain rate as well as some measures of association between them (e.g., conditional probability, correlation coefficient), and then asked to provide the joint probability of both the Dow and the S&P 500 increasing. Based on the material taught in probability courses, students can make this computation. However, when students are required to assign a probability to an event (e.g., the Dow increasing in a week, rain tomorrow), students find themselves at a loss. This is not surprising because students have been taught probability computation, not probability assignment. In other words, they are provided probabilities and asked to make further computations using them. To prevent making this problem too obvious, uncertainties are often represented as a "fair coin" or a "fair die" or a "perfectly shuffled deck of cards." Students can then assign probabilities (e.g., 1/2, 1/6, 1/52) from symmetric or uniform distributions. Today with the advent of big data, there is a growing trend to calculate limit values and rely on ratios for probabilities. For example, "probability of rain tomorrow" can be computed by reviewing data over the last 10 years to identify when it rained on that particular day in the past 10 years and compute the ratio from that (a frequency measure). In this case, when such data is unavailable, students once again find themselves at a loss on how to proceed.

While there is clearly a need for such probability computation methods in the material that is currently being taught, there is also a pressing need for structured methods to: (a) elicit probabilities from decision makers and (b) capture their beliefs in real-world problem situations in which problems invariably do not involve a perfectly shuffled deck of cards. For such problems/problem contexts, *probability encoding* is the right techniques to be employed. Surprisingly, while probability encoding is a classic and fundamental skill in DA, it is conspicuously absent in engineering courses taught in engineering schools [3].

It is important to realize that problems such as assigning "probability of rain tomorrow" is a microcosm of much larger and more complex problems in which engineers are required to quantify uncertainty in *engineering* and *operating* complex systems. For example, the probability of a successful launch or a successful landing on another planet is a more complex version of the "probability of rain tomorrow" problem. Unless exposed to this concept of probability encoding, even without specific data, students are likely to revert to methods that fail to incorporate uncertainty or that use simplified (somewhat arbitrary) methods for capturing uncertainty.

Updating Belief

Updating belief, a key DA concept, is especially important for complex engineered systems that operate in uncertain, reactive environments. To appreciate the concept of updating belief, it is worth reviewing Ohm's law, a basic law of engineering/physics which is taught in most undergraduate engineering curricula. Ohm's Law states that Voltage = Current × Resistance (V = IR). However, in the real world, due to noise, measurement errors, and sensor inhomogeneity, a voltmeter reading may deviate from what the formula (i.e., law) dictates. Therefore, it is important to incorporate uncertainty in the model to account for this random deviation. Traditionally, the approach taken to account for uncertainty has been to use *threshold values* and *tolerances*. The issue with using tolerances (as opposed to quantifying uncertainty using probability) is that the use of tolerances does not provide a means for updating belief.

Today Bayes' law (also called Bayes' theorem or Bayes' rule) and other methods for updating belief are taught in probability classes. However, to incorporate this law into systems engineering, there is a need to construct examples and content that reflect real-world problems relevant to systems engineering [18]. In this regard, potential collaborations between DA researchers and systems or manufacturing engineers can potentially yield fruitful new results that could potentially benefit both fields.

Valuing Information and Experimentation

Experimentation, validation, and experimental design are essential components of systems engineering and closely related to the concept of *value of information* and *experimentation* in DA [2]. In the real world, system validation is especially difficult because validation scope is usually constrained by schedule and budget [19]. To address this challenge, the concept of *risk-based validation* has emerged. Popularized in the pharmaceutical industry, this approach has been employed in a somewhat ad hoc fashion. Madni [19] introduced the concept of model-driven risk-based validation to formalize this process within the context of system acquisition and engineering.

The system modeling approach employed depends on the a priori information available on the system and the characteristics of the environment. Thus, the system model can range from deterministic models to probabilistic models to probabilistic learning models [1]. This modeling approach is somewhat analogous to the value of information concept from decision analysis [2]. Using DA concepts of value of information and experimentation, systems engineers can design experiments based on their value to decisions and designs. They can then compare values and costs for different alternatives and make appropriate recommendations. These DA concepts are also applicable to smart diagnostics and internet of things where tests can be made to maximize value (e.g., money). Elements of value can include materials, labor, time, as well as residual uncertainty about faulty system components [20, 21, 22].

Biases in Human Decision-Making

Quantifying uncertainty in decision-making invariably requires human input. In fact, the concept of probability becomes moot without a human decision maker. Probability encoding requires a sound understanding of *cognitive* and *motivational* biases that influence human perception of likelihoods and proper decision-making. It also requires understanding of noise and inconsistencies that result in flaws in decision-making [23]. For complex decisions in uncertain environments, it is often the case that that initial information is elicited from humans. This elicitation becomes especially challenging when humans are required to provide probabilities. It turns out that depending on what probabilities are asked for, humans can be more or less accurate. For example, when it comes to subjective probabilities, humans can provide P (evidence/hypothesis) much more accurately than P (hypothesis/evidence). Simply being aware of this fact can positively impact experiments, elicitation of initial probabilities, interactive problem-solving, and human-systems integration. Errors in human probability assessments arise from cognitive biases, which are systematic errors in thinking that occur when humans process and interpret information from their environment. Cognitive biases affect human judgment and decision-making. Simply explained, cognitive biases invariably result from the human decision maker's attempt to simplify information processing during decision-making. Some of the more important cognitive biases are availability bias, anchoring bias, hindsight bias, and confirmation bias. In quantifying uncertainty, it is important to be mindful of cognitive bias and errors in judgment [23]. These biases need to be addressed in any type of elicitation, including probability elicitation. Given the important of cognitive biases in decision-making, it is important that engineering schools begin teaching cognitive and motivational biases in behavioral decision-making as part of their systems engineering curricula.

Sunk cost bias (also called sunk cost fallacy) is an often-overlooked bias that can impair decision-making. The sunk cost bias manifests itself when irrational decisions are made that lead to suboptimal outcomes. This bias arises from the fact that decision makers focus on past investments instead of present and future costs and benefits, and thereby commit to decisions that are no longer in the best interests of decision makers. An example of sunk cost is the replacement of 3D with ultrahigh-definition TV. The world has moved on from 3D streaming to ultrahigh-definition TV. However, people who have substantial investment in the former tend to cling to it. This example helps create awareness of the problem and the remedy that follows from the recognition of the sunk cost bias. Finally, there are motivational biases which stem from incentive structures employed, and the trade-off between long-term profit and fast short-term return.

There are other human decision-making attributes that need to be considered in systems engineering. For example, humans are better at ordinally ranking than weighting attributes in multi-attribute decision-making. Also, humans tend to confuse objectives and options when framing decision situations and structuring objectives especially in complex systems engineering [24].

Institutional Barriers

Organization structure can be an important enabler or inhibitor of innovation. For example, decomposing a product to match the organization structure is less likely to lead to innovation than decomposing the product using a structure that helps exploit innovation while also maximizing reuse of legacy components. And, finally, there are real-world constraints. A solution that looks elegant in theory may not play out in practice. Such situations can be traced back to unrealistic or unwarranted assumptions. For example, an unrealistic assumption about policy can have a deleterious impact on decision-making.

MBSE and Digital Engineering

Drawing on Industry 4.0 and Internet of Things, and with the strong leadership of the US Department of Defense, Digital Engineering (DE) has advanced rapidly in the past 5 years. Today both the US government and industry are working on exploiting the synergy between MBSE and DE [4, 5]. While both disciplines are concerned with models, data, communication, and integration, their relative emphasis on these different aspects are quite different. MBSE is focused on advancing the state of the art in modeling systems and the environments in which they operate. DE is concerned with communication and integration using the digital thread, linking models to data, and ensuring interoperability. While a few claim that DE subsumes MBSE, this chapter views MBSE and DE as complementary and synergistic technologies (Fig. 2). Against this backdrop, MBSE provides decision problems and problem contexts to digital Engineering (DE), while DE offers the concept of digital twins and digital thread to MBSE. Specifically, the digital twin construct can be exploited in MBSE for systems, products, components, processes, and human agents involved in complex systems. DE

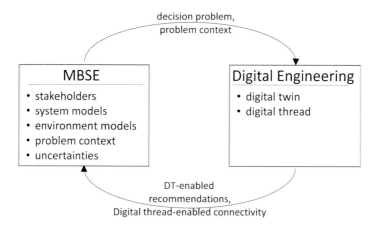

Fig. 2 Exploiting synergy between MBSE and digital engineering

provides digital twin technology and digital thread to link upstream and downstream processes and link digital twins to their real-world physical counterparts. Figure 2 presents this synergy between these two mutually supportive disciplines.

MBSE and Digital Twins

The concept of digital twins, introduced by Grieves in 2002 in the context of product life cycle Management in the manufacturing sector, was first applied to a real-world space problem by NASA in 2010 but without the digital twin label. A *digital twin* is a virtual representation that serves as the real-time digital counterpart of a physical object, process, or human.

In recent years, the concept of digital twin has become ubiquitous in DoD, construction, real estate, manufacturing, aerospace, and automotive industries. More recently, a Digital Twin Consortium has been formed to develop standards for defining digital twins. Today digital twin technology is viewed as a key contributor to MBSE in extending MBSE coverage of the system life cycle [4, 5]. Digital twin today is defined a bit differently as an up-to-date representation of a product, system, process, or humans. It mirrors an organization's machines, controls, workflows, and systems.

MBSE and Digital Thread

A digital thread is defined as a communication network that enables flow of data within connected entities in the network and affords an integrated view of an asset's data across its lifetime through various isolated functional perspectives. Simply stated, a digital thread is a record of a product or systems history from creation to disposal. Specifically, digital threads ensure the traceability of decisions, data uses and usage, data created throughout a product's life cycle. A digital thread links all the data associated with a digital twin. This data includes designs, performance data, product data, supply chain data, software, and all other information that goes into product creation. While useful in the creation and optimization of both physical and digital processes across the product value chain, digital twin and digital thread implementations are expanding into the digital marketplace. Thus, in the autonomous vehicle sector, if a vehicle experiences unexpected acceleration that leads to an accident, the digital thread by virtue of its connectivity has the requisite traceability across the vehicle's life cycle and with access to models from MBSE can identify the source of the problem. Both digital twin and digital thread are central to the ongoing digital transformation in industry.

Industry 4.0

Industry 4.0 is a vision of manufacturing in which smart, interconnected production systems optimize the complete value-added chain to reduce cost and time-to-market.

At the core of Industry 4.0 is the smart factory of the future, whose successful deployment requires solving challenges from many domains. Model-based systems engineering (MBSE) is a key enabler for such complex systems of system.

Industry 4.0 aims to revolutionize manufacturing through integrating software-intensive systems from business plans to manufacturing systems to products and across the complete value-added chain. A prime objective of Industry 4.0 is the smart factory in which interoperability, information transparency, and decentralized decision-making aim to reduce unscheduled downtime to improve resource efficiency. Model-based systems engineering is a key enabler for efficiently engineering the complex automation systems of systems of Industry 4.0.

This "fourth industrial revolution" raises the new challenges for future smart factories driven by four disruptions: (1) data volumes, computational power, and connectivity; (2) the emergence of analytics and business-intelligence capabilities; (3) new forms of human-machine interaction; (4) and improvements in transferring digital instructions to the physical world, such as advanced robotics and 3D printing. The interplay of these four disruptions led to recognizing four particular Industry 4.0 design principles:

- Interoperability: connect production systems, devices, sensors, and people.
- Information transparency: query data and connect digital planning with the runtime data collected from sensors.
- Technical assistance: provide the right abstraction to understand the complexity of Industry 4.0 systems and processes.
- Decentralized decision-making: enable autonomous systems.

For successfully engineering Industry 4.0 systems of systems, the integration of MBSE and digital engineering (DE) plays a crucial role in enabling these design principles. Specifically, digital twin-enabled MBSE can enhance digital system representation, system and process integration, and process definition.

Digital twin-enabled MBSE can contribute to:

1. Informed decision-making through increased transparency and insights resulting from the integration of digital twins and their physical counterparts.
2. Increased understanding of where to introduce flexibility in design.
3. Improved resource utilization leading to increased efficiency.
4. Increased confidence in a system's ability to perform as expected.

Internet of Things

Model-based systems engineering is concerned with the creation and use of a single authoritative source of truth (ASOT) system model which is distributed over multiple repositories and supported by multiple tools. MBSE is a critical enabler in the development of Internet of Things (IoT) products. In MBSE, the disciplines and tools involved in the engineering process are brought to bear in the IoT network.

IoT product development represents a rigorous series of modeling challenges.

- Cyber-physical – As a combination of software, electronic, and mechanical components, IoT products require multidisciplinary approaches, where no single design or analysis tool is sufficient.
- Agile – IoT products are designed to change rapidly, so the development process must be closely coupled to configuration and project management.
- Secure – As network elements, IoT products are vulnerable to outside actors. Building in security, safety, and reliability requires recognition of nonobvious extended connections between features and functions.
- System-of-systems (SoS) – As components of larger networks, IoT models must be easily federated into larger models to evaluate emergent behaviors.

Linking MBSE models to the IoT brings the data needed to understand how a real-world system (e.g., manufacturing assembly line, autonomous vehicles network) behaves and performs in the operational environment. Furthermore, the combination of IoT and system models can enhance preventive maintenance and support analytics/AI-based optimization of the system and operational processes. Acting as a bridge between the system and models, the IoT can deliver operational performance and health status data from the system to the virtual modeling environments. Combining insights from the real-world data with predictive modeling can enhance the ability to make informed decisions that can potentially lead to the creation of effective systems, optimized production operations, and new business models. Multisource/multi-sensor information (e.g., outside temperature, moisture content, production status of current batch) can be delivered to the model along with information from traditional sensors (e.g., SCADA) to facilitate predictive modeling [25].

MBSE and AI/ML

Artificial intelligence (AI) and machine learning (ML) are terms that are used together in the recent literature. In reality, ML is a subset of AI. AI is concerned with performing tasks that require human-like intelligence (e.g., perception, reasoning), while ML is the ability of a machine to ingest, parse, and learn from data. These technologies are especially relevant for problems in which the system state space and the state of the environment are only partially known because of partial observability and uncertainties, and the initial conditions are not fully known or specified. An example of this case is the operation of unmanned aerial vehicles operating behind enemy lines. These vehicles are equipped with onboard sensors and receive updates from external sensors. These updates can be exploited by machine learning algorithms such as reinforcement learning to update the system state and status as well as reduce uncertainty in the knowledge of the environment. AI can be used in multiple ways: (a) create a domain ontology that enable framing of the problem space; (b) map textual descriptions to models; (c) reason within the problem state space

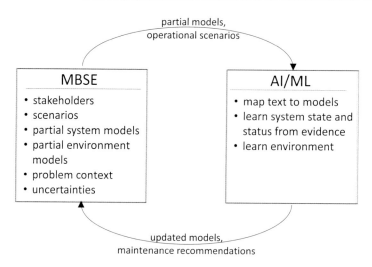

Fig. 3 Exploiting synergy between MBSE and AI/ML

using probabilistic models such as partially observable Markov decision process models; and (d) recommend need to maintenance based on observed conditions. Figure 3 presents the information exchange that can occur between MBSE and AI/ML.

The application of MBSE can produce large amounts of data which can be mined for patterns and trends using appropriate machine learning and data analytics algorithms. In particular, learning agents by themselves or in conjunction with matrix methods can add value to MBSE.

Learning Agents For simple applications, MBSE offers value without having to employ machine learning. Simple applications are characterized by a limited number of variables and an easily discoverable linear relationship between inputs and outputs. However, most real-world systems that contend with multiple data streams stand to benefit from machine learning and analytics to make sense from data. Machine learning, in this context, implies any algorithm that is applied to a data stream to uncover/discover patterns that can be subsequently exploited in a variety of ways. For example, machine learning can automate complex analytical tasks, evaluate data in real-time, adjust behavior with minimal need for supervision, and thereby increase the likelihood of desired outcomes. Machine learning can also contribute to producing actionable insights that can lead to cost savings. Machine learning uses within a MBSE include supervised learning (e.g., using neural network) of operator/user preferences and priorities in a simulation-based, controlled experimentation testbed; unsupervised learning of objects and patterns using, for example, clustering techniques in virtual and real-world environments; and reinforcement learning of system and environment states in uncertain, partially observable environments [25].

Matrix Methods Matrix methods such as dependency structure matrix (DSM) provide a formal means to represent and analyze system architecture models. Since data in matrix form is machine-interpretable, matrix representations lend themselves to machine learning and data analytics. In a matrix representation, a system model is represented as $N \times N$ square matrix in which the rows and columns represent N system entities while the body of the matrix represents the interactions between them. The matrix representation can be effectively used to describe architectures using multiple layers that represent perspectives such as operational, functional, logical, implementation, and organizational. This representation can also be customized to capture interface specifications, temporality, as well as technical, social, and economic characteristics of system entities. For analysis of small sub-matrices, the submatrix can be transformed into graph form for superior human understanding and interaction with the model. Figure 4 presents the two views. The graph view presents architectural entities as blocks and interactions as line segments between blocks. The identical information is presented in the matrix view, where all entities are presented in the first row and first column of the matrix in the same order, and connection between the entities are shown in the matrix body.

Today computational hardware is increasingly focused on matrix manipulations to support AI computations. Also, data representation in matrix form is widely used in machine learning algorithms. Most commercial applications of machine learning use images and sound data which is stored in the form of matrices and arrays. Matrix-based data storage also allows parallel processing of data; hence hardware such as graphics processing units (GPUs) and tensor processing units (TPUs) are

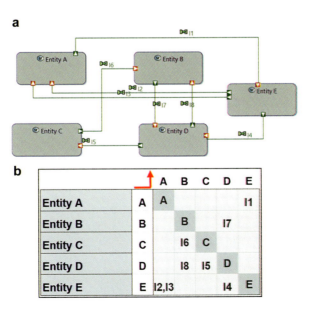

Fig. 4 (**a**) Graph-based view and (**b**) matrix view of architecture

mostly designed for matrix-based data processing. Matrix methods are foundational to data analytics, signal processing, and machine learning. This makes system models in matrix form highly suitable for building AI agents since hardware and algorithms are already available for matrix-based data processing.

MBSE and Social Networks

Distributed collaboration, a cornerstone of systems engineering, is concerned with information sharing and problem-solving in virtually co-located teams comprising stakeholders and experts from different disciplines. For distributed collaboration to be effective, all stakeholders need to participate to share their needs and concerns, and contribute to resolution of issues as they arise. Occasionally, not all disciplines or specialties are represented in these groups so that certain subproblems go unaddressed or are deferred. It is in this situation that social networks with the ability to crowdsource can play an important role. Figure 5 presents the synergy between MBSE and social networks (SN). As shown in this figure, SN can augment MBSE by providing the needed expertise missing in the team as well provide opinions on specific decisions. The SN can accomplish these objectives through crowdsourcing problems and providing recommendations including voting results to collaborative MBSE teams based on feedback from the "crowd."

Crowdsourcing

The term crowdsourcing, which is a combination of crowd and outsourcing, was coined in 2006 by Wired magazine author Jeff Howe in his article "The Rise of Crowdsourcing." Crowdsourcing is concerned with acquiring expertise, information, or opinions from a large or small group of paid freelancers or well-meaning

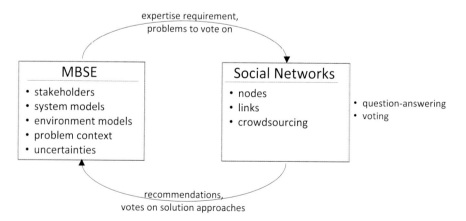

Fig. 5 Exploiting synergy between MBSE and social networks

volunteers who submit their responses via the internet, social media, and smartphone apps. Crowdsourcing in systems engineering can be employed by organizations that do not have organic expertise in specific areas. From a MBSE perspective, crowdsourcing allows an organization to acquire needed information or input into a task or project by enlisting the services of a pool of paid or unpaid experts, usually via the internet or social networks. Importantly, crowdsourcing is a cost-effective strategy to acquire missing expertise within collaborative teams. Interestingly, this approach changes the very definition of "collaboration" and "collaborators" used in traditional distributed collaboration. No longer does all the expertise have to reside in the collaborative team because it can be acquired on demand through crowdsourcing. Also, no longer do collaborators have to be previously known individuals within the collaborative team. They can be complete strangers (e.g., qualified freelancers) who can participate in the collaborative team when needed with the flexibility to enter and exit the social network at any point during the collaboration. Madni [1] calls this "disruptive collaboration."

Crowdsourcing has numerous applications in both system development and manufacturing. Technical subproblems that require specific expertise not available in-house can be crowdsourced and can potentially result in the discovery of one or more experts who have previously encountered and solved that subproblem. Similarly, ample opportunities exist in making manufacturing processes flexible and efficient. The need for process improvement is pervasive because of globalization, ever more demanding customers, and need to accelerate innovation. In fact, in today's global competition, it is prudent to assume that there are better performing global competitors. Thus, the need to continually improve products and processes always exists.

The business case for crowdsourcing is no longer questioned. In today's competitive environment, companies can no longer continue to increase permanent staff given that investors demand profitable growth. Adding permanent staff does not square with this objective. Fortuitously, the labor market is changing with a dramatic increase in qualified freelancers and start-ups. Therefore, it is reasonable to assume that there are smart people out there who can be tapped through crowdsourcing. Finally, it has become impossible to keep up with new and emerging technologies given the rapid pace of technological advance. At the same time, companies are no longer willing to pay for services that do not translate directly into business impact. Against this business landscape and attitudes, crowdsourcing is not only an attractive business proposition, it is an indispensable business option.

MBSE and Entertainment Arts

Today languages such as SysML, UML, and OPM are used to model systems from different perspectives (i.e., requirement, structure, behavior, parametrics) within the MBSE rubric. While suitable for engineers, such languages are unknown to several nontechnical stakeholders who need to have a say especially in upfront systems engineering. As a result, these stakeholders are unable to share their requirements and needs the way they want to. Quite often they don't speak up because of their unfamiliarity with SysML and other such modeling languages. This deficiency shows up in the form of extraneous design iterations and costly rework. Importantly, these results contribute to program and schedule overruns.

In recent years, there is a growing recognition of the increasing convergence of computational methods in engineering with methods from the humanities, arts, and social science [1]. In particular, the convergence of entertainment arts with engineering was successfully defined by Madni [6] in his journal publication and in his chapter on model-based interactive storytelling [1]. The central idea is that *if a picture is worth a thousand words, and a model is worth a thousand pictures, then a story is worth a thousand models*. In this approach, system models and use cases captured in SysML are semiautomatically transformed into stories that can be simulated and interactively executed by stakeholders in virtual worlds [1]. Figure 6 presents the synergy between MBSE and entertainment arts.

As shown in this figure, MBSE provides system models and use cases depicting sunny-day and rainy-day scenarios in SysML or some other modeling language to entertainment arts. A transformation engine helps semiautomatically transform system models and use cases into stories that can be interactively executed in virtual worlds provided by the entertainment arts (Fig. 7).

The implementation of virtual worlds has become both affordable and fast given advances in game engines, virtual reality (VR), and augmented reality (AR) technologies.

Game engines such as Unity 3d and the Unreal Engine provide the capability to create virtual worlds rapidly and cost-effectively. Game engines offer services to build simulations, collect data, connect with the cloud, and connect with multiple agents on the network. These capabilities allow streamlining of various activities in systems engineering life cycle phases.

In addition, game engine technologies allow building virtual environments within which system models can be explored through simulation by stakeholders with

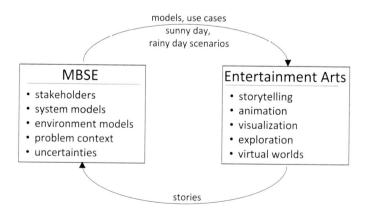

Fig. 6 Exploiting synergy between MBSE and entertainment arts

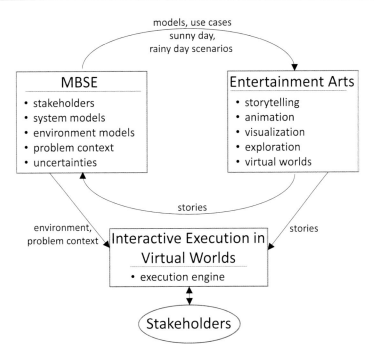

Fig. 7 Interactive story execution in virtual worlds

various what-if assumptions and changes to engineering constraints and policies in the systems engineering process. This capability assures greater participation and as well as timely and opportunistic collaboration. AR and VR technologies can be used in both preliminary design reviews and detailed design reviews.

Virtual worlds can also be used to capture historical cases in story form [6]. Then users can explore how the outcomes of historical cases change with changes in assumptions and constraints. This capability can be best described as "experience accelerator" in that it enriches and accelerates user experiences. Capturing case studies and building experience accelerators is also possible with modern gaming technologies.

Finally, systems development and sustainment programs tend to have long life cycles and generally experience high personnel turnover. Compounding the problem is the fact that departing or retiring program managers and system architects and engineers are unable to transfer their experiences, expertise, and lessons learned to new program managers and the rest of the organization. This is where storytelling comes in. Capturing their knowledge in the form of digital stories can help capture and disseminate their experiential knowledge base. Subsequently, interactive storytelling in virtual worlds can offer an effective means to further explore the stories and gain additional insights.

Table 1 Core concepts from other disciplines useful for MBSE

Discipline	Key concepts	Use in SE/MBSE
Decision analysis	Preferences Values Utilities	Modeling decision maker Reaching consensus in teams
Digital engineering	Digital twin Digital thread Integration of virtual and physical world	Systems V&V testing Condition-based maintenance Linking virtual and real worlds Linking upstream and downstream activities
AI and machine learning	Reinforcement learning Text to model translation Information prefetching	Requirements engineering Closed-loop system modeling Keeping system and environment model current Transforming schedule-driven to condition-based maintenance
Social network	Crowdsourcing Crowd questioning Crowd voting	Information gathering Spontaneous collaboration Conflict resolution
Entertainment arts	Storytelling Animation Visualization	Collaboration Exploration Decision-making Consensus building

Transdisciplinarity: Key to Exploiting Convergence

The preceding discussion clearly shows that convergence of MBSE with other disciplines can potentially enrich not only MBSE, but engineering in general [26]. Table 1 presents the key disciplines, core concepts from those disciplines, and the use of those concepts in MBSE activities. The other disciplines that have synergy with MBSE are shown in the leftmost column in this table. The key concepts from each of these disciplines that can contribute to MBSE are shown in the center column. The specific uses of these concepts in MBSE are shown in the rightmost column.

There are other disciplines that have similar synergy with systems engineering and specifically with MBSE. These include biology which contributes biologically inspired architectures to the systems architecting process employed in MBSE. And there is economics that has contributed the concept of financial options to systems engineering where the analogous concept of real options is created.

Implications for MBSE Curriculum

Table 1 presents the key concepts that should be taught in systems engineering and specifically in MBSE courses. One of the key thrusts in the ongoing systems engineering transformation is to introduce rigor into SE through formal methods such as DA. This opportunity provides the impetus for introducing DA into SE courses and in so doing expose systems engineers to important concepts such as value of information, uncertainty quantification, individual and group preferences, and information aggregation. It also provides an ideal opportunity to enhance the material being taught in engineering economy (e.g., how to deal with internal rate of return (IRR) versus net present value (NPV)). Furthermore, the emphasis of SE on MBSE provides an ideal framework for introducing DA and decision-driven architecting and engineering into systems engineering. For example, courses in systems architecting and model-based systems engineering (MBSE) provide the "low-hanging fruit" for introducing DA concepts into SE.

Digital engineering (DE) is being so tightly coupled to MBSE that it is essential for MBSE courses to cover DE fundamentals including the key concepts of digital twins and digital threads. AI/ML need to be part of systems modeling lectures in MBSE courses. AI/ML effectively transform the concept of system modeling into a closed-loop process thereby allowing more complex system models to be developed with partial initial information [1]. Social networks (SN) and concepts of crowdsourcing should be covered in systems engineering curricula, and specifically in MBSE courses. Entertainment arts (EA) concepts that include storytelling, visualization, and animation should be covered in the topic on systems simulation within MBSE courses and more broadly with systems engineering curriculum. There are other important topics that should be covered in systems engineering curriculum. These include concepts from biology (e.g., use of immune system models to create artificial immune systems) and concepts adapted from economics (e.g., real options) that should be incorporated in the courses on systems architecting.

The primary challenge to accomplishing this curriculum transformation/enhancement is getting the right faculty involved who are interested in pursuing interdisciplinary/transdisciplinary approaches research and are willing to reflect these advances in their engineering courses.

Conclusions

Leveraging transdisciplinarity is one of the more promising advances in systems engineering [1, 27]. Transdisciplinary thinking can potentially lead to methods that exploit the growing convergence of SE with other engineering disciplines such as DA, DE, AI/ML, SN, and EA. These disciplines can contribute key concepts to MBSE. For example, DA provides concepts such as "preference" and "value" in realizing value-driven systems engineering. DE contributes the concepts of "digital twins" and "digital thread" to MBSE and thereby enable model-based system verification and validation and facilitate transformation of schedule-driven maintenance into condition-based maintenance – the keys to saving costs and increasing system availability. AI/ML contribute the ability to model complex systems with only partial initial information and transform system modeling into a closed-loop process, as well as afford the ability to map textual descriptions of systems to models. SN contribute the concept of crowdsourcing to MBSE, redefining the concept of "collaboration" and "collaborators" in the process. Specifically, SN through crowdsourcing can bring on-demand expertise to collaborative engineering teams thereby accelerating schedule by eliminating extraneous design iterations and costly rework. EA contribute the concepts of storytelling, role-playing, animation, visualization, and interactive story execution in virtual worlds. As presented in this chapter, MBSE can harness the concept of storytelling to transform models into stories that can be interactively executed and visualized in virtual worlds from the perspectives of the different stakeholders.

In conclusion, this chapter has focused on how transdisciplinary thinking can be exploited to identify and incorporate key concepts from other disciplines into MBSE Specifically, it presented how the growing convergence of systems engineering with other disciplines can be exploited in practice. This chapter is intended to stimulate interest within the systems engineering practice and education communities to come together and define new best practices and create new courses that exploit the growing synergy of MBSE with other disciplines.

Cross-References

▶ Model-Based System Architecting and Decision-Making
▶ Overarching Process for Systems Engineering and Design

References

1. Madni, A.M., *Transdisciplinary Systems Engineering: Exploiting Convergence in a Hyperconnected World,* (foreword by Norm Augustine), Springer, September 2018.
2. Howard, R. A., and A. E. Abbas. 2015. Foundations of Decision Analysis, Pearson, NY.
3. Spetzler, C.S. and Van Holstein, C.S. S. Probability Encoding in Decision Analysis, Management Science, 1975, vol. 22, issue 3, 340–358.
4. Madni, A.M., Madni, C.C., and Lucero, D.S. Leveraging Digital Twin Technology in Model-Based Systems Engineering, MDPI *Systems,* special issue on *"Model-Based Systems Engineering,"* Feb 2019a.
5. Madni, A.M., Erwin, D. and Madni, A. Exploiting Digital Twin Technology to Teach Engineering Fundamentals and Afford Real-World Learning Opportunities, 2019 ASEE 126th Annual Conference and Exposition, Tampa, FL, June 15–19, 2019b.
6. Madni, A.M. "Expanding Stakeholder Participation in Upfront System Engineering Through Storytelling in Virtual Worlds," *Systems Engineering,* Vol. 18, No. 1, pp. 16–27, Jan. 2015.
7. Madni, A.M. "Transdisciplinarity: Reaching Beyond Disciplines to Find Connections," *Journal of Integrated Design and Process Science*, Vol. 11, No. 1, March 2007, pp. 1–11.
8. Arrow, K. J. (1971). "2". Essays in the Theory of Risk Bearing, Exposition of the Theory of Choice under Uncertainty. Markham Publishing, Chicago. ISBN 978-0-444-10693-3. Retrieved 2009-05-25
9. Keeney, R and Howard Raiffa. 1993. Decisions with multiple objectives. Cambridge University Press.
10. Hazelrigg, 2012, FUNDAMENTALS OF DECISION MAKING: For Engineering Design and Systems Engineering, Neils Corporation, January 2012
11. Hazelrigg, 1998. A Framework for Decision-Based Engineering Design, Journal of Mechanical Design, Vol. 120, 653–658

12. Howard, R. A., Matheson, J.E., North, D.W., The Decision to Seed Hurricanes, Science, 16 June 1972, Volume 176, pp. 1191–1202
13. Keeney, R. 1996. Value-focused Thinking. Harvard University Press.
14. Abbas, A. E and J. Matheson. 2010. Normative Decision Making with Multiattribute Performance Targets. Journal of Multicriteria Decision Analysis, 16 (3, 4), 67–78.
15. Collopy, P. (2001). "Economic-Based Distributed Optimal Design". American Institute of Aeronautics and Astronautics, Reston, VA. Retrieved 2009-05-24.
16. Collopy, P.; Horton, R. (2002). "Value Modeling for Technology Evaluation". American Institute of Aeronautics and Astronautics. Retrieved 2009-05-25.
17. Castagne, S.; Curran, R.; Collopy, P. (2009). "Implementation of value-driven optimisation for the design of aircraft fuselage panels". International Journal of Production Economics. 117 (2): 381–388
18. Sun Z, Hupman A, Ritchey H, and A.E. Abbas. 2016. Bayesian Updating of the Price Elasticity of Uncertain Demand. IEEE Systems Journal Vol 10 issue 1, 136–146.
19. Madni, A.M. Risk-Based Validation Research for System Acquisition and Development, USCSAE, 2021.
20. Hupman, A., Abbas, A., and Schmitz, T., 2015a, Incentives Versus Value in Manufacturing Systems: An Application to High-Speed Milling, Journal of Manufacturing Systems, 36:20–26.
21. Abbas, A. E. 2013. Normative Perspectives on Engineering Systems Design. IEEE Systems Conference, SYSCON 2013. April 15th–18th, 2013. Orlando, FL.
22. Hupman, A, A. E. Abbas, B. Tibor, H. Kannan, C. Bloebaum and B. Mesmer. 2015b. Calculating Value Gaps Induced by Independent Requirements, Deterministic Modeling, and Fixed Targets. Proceedings of the 56th AIAA/ASCE/AHS/ASC Structures, Structural Dynamics, and Materials Conference, Huntsville, Alabama
23. Kahneman, D., Sibony, O., and Sustein, C.R. *Noise: A Flaw in Human Judgment,* Little Brown Spark, 2021.
24. Madni, A.M. "Generating Novel Options During Systems Architecting: Psychological Principles, Systems Thinking, and Computer-Based Aiding," *Systems Engineering*, Volume 17, Number 1, pp. 1–9, 2014.
25. Madni, A.M. Transdisciplinary Systems Engineering: Exploiting Disciplinary Convergence to Address Grand Challenges, *IEEE SMC Magazine*, Vol. 5, Issue 2, pp. 6–11, April 2019.
26. Madni, A.M. and Sievers, M. Model-Based Systems Engineering: Motivation, Current Status, and Research Opportunities, *Systems Engineering*, Special 20th Anniversary Issue, Vol. 21, Issue 3, 2018.
27. Chami, M., Zoghbi, C., and Bruel, J-M., A First Step toward AI for MBSE: Generating a part of SysML Models from Text Using AI, INCOSE AI for Systems Engineering, 2019 Conference Proceedings (pp. 123–136), 1st Edition, October 24, 2019.

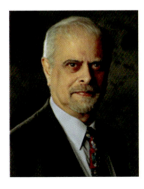

Dr. Azad M. Madni is the founder, CEO and Chief Technologist of Intelligent Systems Technology, Inc. He is a University Professor and the Northrop Grumman Foundation Fred O'Green Chair in Engineering Professor of Astronautics and of Aerospace and Mechanical Engineering in the University of Southern California. He also has a joint appointment in the Sonny Astani Department of Civil and Environmental Engineering. He is the Executive Director of USC's Systems Architecting and Engineering Program and the founding Director of USC's Distributed Autonomy and Intelligent Systems Laboratory. He is the co-founder and chair of IEEE Systems, Man and Cybernetics Society's Model-based Systems Engineering Technical Committee. He is also the Chief Systems Engineering Advisor to The Aerospace Corporation. He defined the field of transdisciplinary systems engineering in his highly acclaimed book on the subject. He received his B.S., M.S., and Ph. D. degrees in Engineering from UCLA. He is a graduate of the Executive Institute at Stanford University. He is a Life Fellow or Fellow of nine professional societies including AAAS, AIAA, IEEE, IISE, INCOSE, and WAS, and an elected member of Omega Alpha systems engineering honor society. He is internationally known for his pioneering contributions to transdisciplinary systems engineering including interactive storytelling in virtual worlds, and virtual reality-enabled distributed simulation-based training. He is also a leading researcher in formal and probabilistic modeling in the planning and decision-making in autonomous systems and system-of-systems. In 2022, he received the IEEE AESS Industrial Innovation Award. In 2021, he received the IEEE AESS Judith A. Resnik Space Award for excellence in Space Engineering. He is the recipient of the *Pioneer Award* from both IEEE Aerospace and Electronics Systems Society and INCOSE. In 2020, he received the *IEEE Systems Man and Cybernetics Norbert Wiener Outstanding Research Award* and National Defense Industrial Association's *Lt. Gen. Thomas R. Ferguson Award* for Excellence in Systems Engineering.

Toward an Engineering 3.0

42

Norman Augustine

Contents

Introduction 1254
Complex Systems Challenges 1254
Two Complications for Engineering 3.0 1255
Further Challenges 1255
Limitations of Modeling and Simulation 1256
Beyond Established Equations 1257
Chapter Summary 1257
References 1258

Abstract

Systems complexity continues to increase as global communications, massive software, and complicated hardware – and people – are increasingly linked together. Hidden in these systems of systems lie immeasurable opportunities for failure – including some that may even be intentionally caused. The system designer needs new tools that can contend with such immensely important but fragile entities, with MBSE being one of the most important of those tools.

Keywords

Systems · Complexity · MBSE · Unintended consequences

Norman Augustine is Retired.

N. Augustine (✉)
Lockheed Martin Corporation, Bethesda, MD, USA

© Springer Nature Switzerland AG 2023
A. M. Madni et al. (eds.), *Handbook of Model-Based Systems Engineering*,
https://doi.org/10.1007/978-3-030-93582-5_71

Introduction

The history of major engineering projects traces back at least 5000 years. It began in earnest with the construction of large stationary structures: pyramids, walls, roads, bridges, and aqueducts – what became known as civil engineering. The need to construct objects whose parts move relative to one another gave identity to the practice of mechanical engineering. Then the ability to separate, combine, and capitalize on the elements of matter enabled chemical engineering. The eventual ability to control the behavior of electrons defined the province of electrical engineering.

A second era followed in which needs and capabilities arose that did not neatly fit the traditionally defined engineering categories. As a result, new, more specialized engineering disciplines were born: petroleum, aerospace, biomedical, computer, entertainment, and many more. Given the profusion of knowledge in each of the traditional engineering disciplines, the teaching of engineering over the generations became highly compartmentalized to the point of occasional disconnection with society and reality. The "stovepiping" trend was exacerbated by such forces as the academic accreditation process, which encouraged this narrower focus. Similarly, industries largely structured themselves around highly specialized disciplines.

Then dawned the era that might be termed Engineering 3.0, a pursuit to better engage with complex systems of systems demanding great breadth as well as depth of knowledge. The engineering profession, well founded in methodology, nonetheless found itself ill prepared to deal with the interconnectedness and enormity of connected but uncoordinated systems and their consequences.

Complex Systems Challenges

It is clear that complex systems require far more than traditional engineering. For example, how can the natural environment be preserved when any solution demands that literally dozens of autonomous geopolitical entities work in concert? How can America's public pre-K-12 education system, with its 14,000 independent school districts, be made to produce students who are uniformly competitive on a global scale? How can health care be provided to over 300 million people without bankrupting the nation? And how can congestion and gridlock on the nation's 8 million miles of roads serving 247 million motor vehicles be eliminated?

Any objective assessment of the current state of the art in engineering such complex systems would likely conclude that there is a great deal of room for improvement and unifiability – "a target-rich environment," as they say in the Pentagon.

Beginning with climate change, the carbon concentration in the atmosphere has now risen to well over 400 ppm for the first time in at least 900,000 years. In education, US 15-year-olds finish in 25th place on international tests in combined reading, science, and mathematics scores – even as this country spends more per student than any other nation but one. America now devotes 7 more percentage points of its GDP to medical care than the next highest spending nation, yet has a declining life expectancy and fails to excel across many other health indices. The average adult American wastes 54 hours a year in traffic delays, and 36,000 Americans die in automobile accidents each year.

Two Complications for Engineering 3.0

Two particular complications confront the opportunities for Engineering 3.0. The first is that it involves humans. Many systems include people who appear not only as individuals but also collectively as society.

Humans can be not only inconsistent but also notoriously irrational. They may refuse to take vaccines that are known to save lives. They are more frightened of shark attacks than bee stings although the latter kill 60 times more people in the United States each year. They may oppose the prospect of nuclear fusion energy because there have been accidents in nuclear fission plants and because of a fear of nuclear weapons. As research in behavioral sciences has repeatedly shown, people implausibly value something they have more highly than the identical thing they do not have. Attempts to model the behavior of the stock market or project election outcomes provide classic examples of systems tortured by such idiosyncrasies.

The second emergent complexity multiplier concerns a relatively recently discovered colorless, odorless, weightless substance called software. It flourishes in complex systems, but the accidental omission of a single bar among many thousands of lines of code can cause a spacecraft mission to Venus to fail (see Mariner 1). Further, adding a few lines of code to a major system is usually not very costly on the margin but has led to the adage among some engineers that "If it isn't broken, it doesn't have enough functions yet." A modern automobile contains around 100 million lines of code, about a thousand times the number of lines in the Apollo spacecraft. It is a software app on wheels – and driverless cars are still in the future.

Further Challenges

Systems of systems involve feedback, interconnectedness, instabilities, nonlinearities, and discontinuities. Philosophers and metaphysicists through the generations have puzzled over Lorenz's conundrum that asks whether a butterfly flapping its wings in, say, New York, can cause a hurricane in China. We now know the answer: A microbe in China can shut down New York. Similarly, the assassination of an archduke in Sarajevo can trigger a world war. Or an argument between a street vendor and a police officer in Tunis can spark an "Arab Spring" throughout much of the Middle East when connectivity is provided through widespread availability of cell phones. And a tree branch in Ohio can trigger a cascade of events that shuts off electric power for over 50 million people in the northeastern United States and part of Canada for up to 4 days (see 2003 blackout).

Further, the challenge of designing and analyzing interdisciplinary systems usually requires accommodating legacy components of existing systems while maintaining operability as change is introduced: the classic problem of rebuilding an airplane in flight or restructuring a national healthcare system or introducing resilience into the nation's existing electric grid.

Friedrich Wiekhorst of the Max Planck Institute derived the equation that describes the number of states in which a system of n elements can exist, assuming each element can affect each other element in the simplest of possible manners, a binary connection. A system of two elements thus has four possible states. But a system of just seven elements has a number of possible states that approximates the number of stars in our galaxy. While in most actual systems every element is not directly connected to every other element, the magnitude of the number of theoretical possibilities does suggest, among other things, why many failure modes are not caught in testing.

The pace of technological change intensifies the challenges faced by the modern systems engineer when a system can be out of date by the time it is deployed: The number of transistors on a chip has increased by a factor of about 10 million in just 50 years; the cost of gene sequencing has declined by over 6 orders of magnitude in less than 20 years; the number of smartphones in use has grown from 0 to 3.5 billion (half the world's population) in 13 years.

Further, complex unifiable systems are often adaptive, as is particularly true of biological systems. Engineering such systems often entails compromises and trade-offs of unlike qualities.

Limitations of Modeling and Simulation

The rigorous practice of modeling and simulation as part of systems engineering can offer important insights into the design and analysis of complex systems of systems sometimes aptly referred to as wicked problems. But even with these tools, challenges abound. When it comes to systems of systems, the optimum of the whole rarely equals the sum of the optima of its parts. Contrary to ritual, the best way to eat this kind of elephant is not one piece at a time.

If a model is too encompassing, it may defy analysis. But if it is too narrow, it may omit critical aspects of a system's behavior. Unfortunately, it is not uncommon for system failures to be caused by elements that did not rise to the level of adequate concern by system designers. It was, for example, not one of the 25,000 tiles that received so much attention on the Space Shuttle's thermal protection system that caused the failure of the Challenger. It was an O-ring on a solid rocket booster on a very cold day.

A regional telephone company performed an extensive analysis of what would be needed in order to recover from a major hurricane in its operating area. It stockpiled wire, telephone poles, vehicles, and more. But when the hurricane struck, the bottleneck that emerged was absent from the models: It was daycare centers for children. With schools closed, employees' families with two working parents had to have one parent remain at home to care for the children, just at the time a full workforce was critically needed.

So fundamental an issue as identifying figures of merit can be ambiguous in complex systems. There is, for example, the tension between controlling system cost and ensuring system resilience, e.g., just-in-time inventory versus "just-in-case" inventory. Is it better to be efficient or resilient with regard to stockpiling empty beds in a hospital?

Beyond Established Equations

Evaluating systems involving humans may require somehow placing a value on a human life, a year of human life, a quality-adjusted year of human life, or some other such measure. Should a new highway be constructed through the middle of a city that will save thousands of travelers many hours but will create a barrier to community life in the affected neighborhood? What is the exchange rate between tons of carbon emitted into the atmosphere and its social cost? Is it appropriate to put millions of people out of work, many of them into poverty, in order to save thousands of lives in a pandemic? Engineering complex systems not uncommonly finds itself more engulfed in the field of ethics than engineering – confronting issues that have no standard equations for their solution.

When it comes to engineering complex unifiable systems, both the profession and the practice may be better at reflecting on questions and offering insights than in delivering absolute solutions.

Finally, a critical factor that the construction of many complex systems often fails to adequately address is their vulnerability to external interference, intentional or otherwise. The design of the World Wide Web does not appear to have adequately accounted for the impact of malevolent individuals or nations or even of nature itself disrupting the intended functioning of the system.

America's electric grid is a canonical example of this problem. With 7300 power plants and 160,000 miles of high-voltage line, the latter is owned by some 500 independent firms. The US grid possesses substantial vulnerabilities, and a massive failure of the system could prevail for months, creating disruption of a magnitude even beyond that of covid-19. Communications would be curtailed, pumps in filling stations would not operate, refrigerators storing food would fail, and entire regions would go dark. A near-term task is to take hostile threats into consideration when designing self-driving cars that will be used on connected highways.

Chapter Summary

Systems complexity, like entropy, always seems to increase. The advent of the Internet of Things will one day likely represent a canonical example of this trend as it connects massive software, global scale communications, and new types of hardware. Entities ranging from one's iPhone to the thermostat in one's home; from one's door lock to one's medical records; and from one's bank account to one's safety in trains, airplanes, and automobiles will be dependent upon the detailed design of systems of heretofore unimagined scope and scale. Such will be the role of Engineering 3.0 and MBSE.

As ordained in the variously attributed euphemism, "Every system is perfectly designed to get the result it gets."

References

Augustine, Norman. "Engineering 3.0." "The Bridge," National Academy of Engineering, Winter 2020

Norman Augustine, holds a BSE and MSE in Aeronautical Engineering from Princeton University. He has served as chairman and CEO of Lockheed Martin Corporation, Under Secretary of the U.S. Army, and Lecturer with the Rank of Professor in the Princeton School of Engineering and Applied Science. He served 16 years on the President's Council of Advisors on Science and Technology, has chaired the National Academy of Engineering, and has been awarded the Presidential Medal of Technology. He holds honorary degrees from 35 universities.

Category Theory

43

S. Breiner, E. Subrahmanian, and R. D. Sriram

Contents

Introduction	1260
Composition and Context	1263
A Model Is a Mapping	1266
Isomorphism and Identity	1271
Picturing Processes	1276
Nondeterminism	1280
Possibility	1281
Probability	1282
State	1284
Visual Reasoning	1288
Duality	1290
Further Study	1292
Conclusion	1294
References	1296

Abstract

Category theory (CT) is a branch of mathematics concerned with the representation and composition of structured relationships. Recent interest in systems engineering (SE) stems from the possibility that CT might provide a principled mathematical foundation that SE currently lacks. The case is bolstered by a broad

---

S. Breiner (✉) · R. D. Sriram
Information Technology Lab, National Institute of Standards and Technology, Gaithersburg, MD, USA
e-mail: spencer.breiner@nist.gov; ram.sriram@nist.gov

E. Subrahmanian
Information Technology Lab, National Institute of Standards and Technology, Gaithersburg, MD, USA

Engineering Research Accelerator/Engineering and Public Policy, Carnegie Mellon University, Pittsburgh, PA, USA
e-mail: sub@cmu.edu

© Springer Nature Switzerland AG 2023
A. M. Madni et al. (eds.), *Handbook of Model-Based Systems Engineering*,
https://doi.org/10.1007/978-3-030-93582-5_85

array of applications in probability, computing, data, and dynamics, as well as a track record of unification in science and mathematics. However, the tools and methodology for applying CT within engineering are mostly prototypes and proofs-of-concept. This chapter introduces the key ideas and terminology needed to engage with this emerging research area.

Keywords

Category theory · Systems engineering · Mathematical modeling · Composition

Introduction

This chapter gives an informal introduction to some of the core ideas and methods from category theory (CT), a branch of mathematics concerned with the representation of *compositional systems*. It offers an extension of traditional set-based mathematics that emphasizes structural relationships (arrows $X \to Y$) rather than internal structure (elements $x \in X$).

CT is a topic of growing interest within systems engineering (SE) because it offers the possibility of a principled foundation that would justify and sharpen SE approaches in the same way that Newton and Maxwell's equations underwrite mechanical and electrical engineering. Despite early recognition of the potential [1], serious consideration of categorical foundations has long required a doctorate in mathematics or a related field. However, recent years have seen a substantial effort to increase CT's accessibility through friendlier introductions and domain-focused use cases and examples. This has revealed that "abstract" categorical ideas are often quite intuitive, at least once they are specialized to a familiar context.

For the practicing engineer, CT offers an extremely powerful modeling toolkit for managing structured information of all kinds, with precise, expressive mechanisms for specification, transformation, composition, and abstraction (generalization). Composition encourages us to separate point solutions into reusable components and relationships, leaving behind a computational infrastructure that can be reused, audited, and extended to solve new problems in the future. As we deepen our understanding of the system, CT guides the model evolution process, making it easier to substitute and recombine these pieces into new forms. Such an approach is potentially relevant at every stage of the SE process, from requirements and design through operation and retirement.

Categories provide a universal framework to organize and evolve the diverse collection of models and methods that go into answering important questions about cost, risk, safety, and other critical concerns. In her popular science book *How to Bake π* [2], Eugenia Cheng describes CT as "the mathematics of mathematics," and "[w]hatever mathematics does for the world, category theory does it for mathematics." CT describes all manner of mathematical phenomena, from geometry and dynamics to data and computation, in terms of a small collection of abstract concepts, which we can compose in different ways to represent different types of systems. Crucially, deep connections to computing, data science, probability, and physics can help engineers manage the ongoing transition to pervasive computing, sensing, and actuation.

However, significant work is needed to establish the methodology and develop the tools needed to create such a vision. For most of its history, research in CT has focused on developments in mathematics, theoretical physics, and computer science. The last decade, though, has seen the rise of a new community of researchers devoted to Applied Category Theory (ACT), including more substantial interaction with engineering and related fields. Although reduced, the barriers to entry for CT remain substantial; much of the literature is extremely technical, and even "basic" examples may assume background knowledge (topology, group theory) that engineers lack. Our goal in this chapter is to provide a "travel guide" that will help systems engineers who are interested in exploring this emerging field to engage with the technical literature.

Practical applications of CT (see section "Further Study") usually involve an interaction between several technical concepts, and the resulting presentation is often inscrutable or top-heavy, depending on whether the author merely cites the necessary definitions and results or takes the time to explain. Of course, any application would feel top-heavy if it required the introduction of basic concepts like matrix algebra or differential equations. These methods are *generic* – context independent – and their reuse throughout science and engineering entails substantial savings in cognitive overhead and computer implementation. CT is similarly generic, but the value of reuse is invisible when the concepts are unfamiliar.

In this chapter, we have opted for readability over breadth, putting practical SE examples beyond our reach. The systems we do consider are very simple: labeled graphs to introduce categorical data structures, stepping and counting for process representation, and resistor equations to illustrate CT's visual logic. The methods themselves apply much more broadly – any database schema, Bayes net, or matrix derivation can be modeled with the tools we introduce – but the added complexity of a sophisticated example would obscure the method itself and provide an additional barrier for readers who lack that context. Instead, we encourage the reader to follow along with parallel examples from their own domain of expertise. (Understanding CT requires active reading, and the reader should be prepared with pen and paper to write down additional diagrams, calculations, and examples in parallel with the text.)

To situate CT within the SE landscape, we can triangulate against existing technologies. We will position CT as a general-purpose modeling language, comparable to something like the Systems Modeling Language (SysML) or Object Process Methodology (OPM) [3]. These are themselves rather different, and the comparisons highlight different features of the categorical approach. The profusion of diagram types and elements found in SysML emphasizes CT's comparative parsimony, covering the same breadth of application by composing and recomposing the same small set of core concepts.

On the other hand, OPM shares this parsimonious worldview and has a similar set of core concepts (objects and processes). Since OPM models and categories look broadly similar, this comparison points out differences at the level of metamodels.

CT is self-referential – there is a category of categories – and this allows us to structure and manage our modeling activities themselves. Indeed, this supports a multimodel perspective that OPM lacks. Moreover, the object-level features inside a model and the metalevel relationships between them interact, especially when we shift the context of analysis. CT provides a language to understand these subtleties.

As a modeling language, CT's general-purpose usage is complemented by extensive applications in formal methods, suggesting further comparison to a range of domain-specific modeling languages. When we build system models from component descriptions in Simulink or Modelica, the composition takes place in a category of dynamical systems [4]. When we analyze statistical trends in SAS or estimate probabilities from a Bayes net, we compose probabilistic relationships [5]. Graphs [6], Petri nets [7], temporal logic [8], gradient descent [9], and fuzzy logic [10] can all be profitably analyzed in terms of categorical structure, and the "shared DNA" of categorical structure helps to fit all these different methods together, especially when they are used to analyze different facets of the same system.

The remainder of the chapter is organized around three extended examples in sections "Composition and Context, A Model Is a Mapping, Isomorphism and Identity, Picturing Processes, Nondeterminism, State, Visual Reasoning and Duality," summarized below, followed by a brief review of the ACT literature and a guide for further study in section "Further Study."

We will begin in section "Composition and Context" with a brief introduction to the language of categories, using labeled graphs to motivate the concept of categorical composition. Section "A Model Is a Mapping" looks at a more explicit representation for the graphs in the previous section, using categorical mappings called functors for data specification and transformation. Because functors compose, CT is strongly self-referential – there is a category of categories – and this allows us to distinguish internal structure (relationships inside a category) and external structure (relationships between categories). Section "Isomorphism and Identity" introduces two categorical concepts, an internal notion of equivalence (isomorphism) and an external relationship between functors (natural transformation), and puts them together to consider the way that model equivalence changes as we transform between the representations introduced in section "A Model Is a Mapping."

Much of the interest in ACT over the past decade has been driven by applications of string diagrams, a graphical syntax that relates to process models in the same way that schemas relate to data structures. Section "Picturing Processes" introduces the diagrammatic method and uses it to model the interaction between two simpler processes, a stepper, and a counter. Section "Nondeterminism" introduces nondeterminism into the previous example, including both possibility and probability, and shows how composition spreads this nondeterminism throughout the system, even to components that are deterministic when viewed in isolation. Section "State" layers on a further complication by introducing internal (hidden) state for the components. Here we reuse the machinery introduced in sections "Composition and Context" and "A Model Is a Mapping" for a different purpose, using functors and natural transformations to transform component representations from one semantic context to another.

43 Category Theory

Where informal diagrams support only intuitions, formal syntax supports rigorous analysis. Section "Visual Reasoning" introduces the equational logic of string diagrams, which allows us to reason about process equivalence using picture proofs. We introduce a process-theoretic interpretation of matrix algebra and use it to derive the classical equation for serial resistance. Section "Duality" introduces the concept of network duality, which allows us to "reverse" systems of spatial networks, in a way that does not make sense for temporal processes. Using this, we interpret Ohm's laws and derive the formula for parallel resistance, as well.

Notation We use ***bold italics*** to indicate the introduction of a new technical term. For mathematical variables, we use italicized font for elements inside a category (objects and arrows X, Y, f), and bold font for categorical structures (categories, functors, and natural transformations **C**, **f**, **α**). We also use (upper/lower) case to distinguish objects (X, Y, Z) from arrows (f, g, h), as well as categories (**C**, **D**, **L**) from functors (**f**, **g**, **h**). Natural transformations are written in bold and distinguished by lower-case Greek letters (**α**, **β**, **γ**).

Another set of conventions governs the use of arrow notation. The most important is a distinction between type-level relationships, indicated by an ordinary arrow $\rightarrow$, and element-level relationships, which use a tailed arrow $\mapsto$. For example, squaring defines a function $\mathbb{R} \rightarrow \mathbb{R}^{+}$, which maps the set of all decimal numbers $\mathbb{R}$ into the set of positive numbers $\mathbb{R}^{+}$. At the level of elements, the squaring function relates $2 \mapsto 4$ and $-3 \mapsto 9$.

Composition and Context

One road into category theory is via graphs. Like graphs, categories focus on the relationships between entities as much as the entities themselves. In graphs, we call the entities and relationships vertices and edges; in a category, the entities are ***objects*** (Readers with an object-oriented background (e.g., Java, SysML, etc.) should be aware that "objects" in CT do not represent individuals, as in object-oriented terminology. They are more like classes, i.e., collections of individuals.) or types and the (directed) relationships are called ***arrows***, ***maps***, or ***morphisms***.

Arrows in a category are like edges in a directed graph. Every arrow has a ***source object*** X and a ***target object*** Y, specified by writing $h: X \rightarrow Y$. The key difference between categories and graphs is a ***composition*** operation (There are conflicting notations for composition. We prefer the diagrammatic order of composition $h = f \cdot g$ (also written as $f; g$) in this chapter. However, composition is traditionally written in applicative order, so that $h = g \circ f$ (note the reversed order). Applicative order arises from the case where f and g are functions, and the composite function is defined by the formula $h(x) = g(f(x))$.) $h \cdot k$ which allows us to combine a sequence of arrows $X \rightarrow Y \rightarrow Z$ into a single map $X \rightarrow Z$.

An example helps to fix ideas, so consider the directed, labeled graph **g** shown in Fig. 1a, with six vertices and 11 edges. Although **g** has no direct relationship $A \rightarrow C$, it does contain a path that connects the two vertices. Paths compose by concatenation;

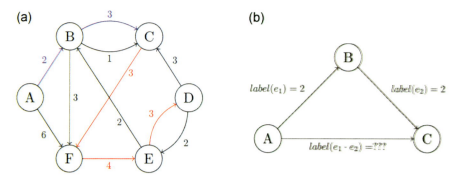

Fig. 1 (a) A directed graph with labeled edges. (b) Composition of labeled edges must specify a composition operation for labels

if h is the two-step blue path $A{\to}B{\to}C$ in Fig. 1a and k is the three-step red path $C{\to}F{\to}E{\to}D$, then the composite $h \cdot k$ is a five-step path $A{\to}D$.

Thus, every graph **g** defines an associated category of paths called the *free category* generated by **g**. All the features of **g** (e.g., connectivity) are reflected in the free category, so we lose nothing in the move from graphs to categories. Without labels, this would be the end of the story; these add an extra layer of complexity that allows us to encode some additional context from the domain.

When we compose edges in a *labeled* graph, we are led to ask how labels compose. Write e_1 and e_2 for the blue edges in Fig. 1a, labeled by 2 and 3, respectively, and h for the two-step path $e_1 \cdot e_2$. Given that $label(e_1) = 2$ and $label(e_2) = 3$, how should we assign $label(h)$? Labeling is a semantic question because its answer depends on what the labels *mean*. Suppose the vertices of **g** are locations, and the edges are routes between them. If the labels represent distances (say, in kilometers), then the labels should combine by addition: The distance from $A{\to}C$ (along the given path) is

$$dist(h) = dist(e_1) + dist(e_2) = 2\text{ km} + 3\text{ km} = 5\text{ km}.$$

Alternatively, the labels might represent a clearance height (in meters) along the route, in which case we should take the minimum:

$$hgt(h) = \min\{hgt(e_1), hgt(e_2)\} = \min\{2\text{ m}, 3\text{ m}\} = 2\text{ m}.$$

Similarly, in electrical networks, series resistance combines by addition, but capacitance requires a more complicated formula

$$res(h) = res(e_1) + res(e_2) = 2\,\Omega + 3\,\Omega = 5\,\Omega,$$

$$cap(h) = \left(cap(e_1)^{-1} + cap(e_2)^{-1}\right)^{-1} = \left((2\,\mu F)^{-1} + (3\,\mu F)^{-1}\right)^{-1} = 6/5\,\mu F.$$

These examples show how composition forces us to confront the semantics of our representations. The same considerations apply to more complicated semantic labels, for example, we might label a link in a communications network with a vector incorporating speed, distance, latency, and noise, and these quantities might interact in the composite if, e.g., increased noise along a single link might lead to increased latency for multiple hops as the system waits for repeated messages. However, if we do not know the appropriate semantics, there is also a "free" option, where we combine the labels into a list, just like we combined edges into paths

$$label(h) = \langle label(e_1), label(e_2) \rangle = \langle 2, 3 \rangle.$$

It is important to recognize that CT does not proscribe the composition operation; this is at the user's discretion. Instead, CT establishes two rules that such a composition must satisfy. The first and most important is **associativity**:

$$(h \cdot k) \cdot l = h \cdot (k \cdot l)$$

Associativity is important computationally because it replaces a nested composite like $(h_1 \cdot h_2) \cdot (h_3 \cdot (h_4 \cdot h_5))$ with the simpler term $h_1 \cdot h_2 \cdot h_3 \cdot h_4 \cdot h_5$. This exchanges a tree representation for a list, so only the order matters. Concatenation of paths in a free category is automatically associative, as are the composition operations given above for distance, height, resistance, and capacitance.

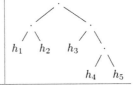

The second requirement for a category is that every object X must have an arrow $id_X: X \to X$ called the **identity** on X, which has no effect under composition: for any $f: X \to Y$,

$$id_X \cdot f = f = f \cdot id_Y.$$

We usually just write id, omitting the subscript, since the objects X and Y are already known from f.

The identities in a free category are the paths of length zero. For labels, we have no choice but to assign id to the unit for the composition operation. For distance, this is the expected result: a path with no steps has length zero:

$$dist(f) = dist(f \cdot id) = dist(f) + dist(id) \Rightarrow dist(id) = 0 \text{ km}$$

Similarly, an identity in a resistor network has $res(id) = 0\ \Omega$, corresponding to a short circuit (ideal wire). Clearance height and capacitance also have unit elements, as long as we allow infinite values. An infinite clearance height represents "no restriction," while an infinite capacitor again approximates an ideal wire (at least for AC circuits).

While serial composition allows us to consider individual paths, we need something more to analyze the network as a whole. For instance, we are usually interested in the *shortest* route or the *maximum* clearance between two locations. To model the impact of multiple paths, we introduce a second operation $h \times k$ to combine paths with the same source and target. Just like serial composition, this operation is determined from the semantics.

For example, when the path labels represent distances, we usually want to find the minimum distance between nodes. In fact, $\mathbf{g}$ contains a second path $k = e_1 \cdot e_3$ that is shorter than h since $dist(e_3) = 1$ km. Hence,

$$dist(A, C) = dist(h) * dist(k) * dist(A \rightarrow F \rightarrow E \rightarrow D \rightarrow C) * \ldots$$

$$= \min \{2 + 3 \text{ km}, \underline{2 + 1 \text{ km}}, 6 + 4 + 3 + 3 \text{ km}, \ldots\} = 3 \text{ km}$$

There are other paths, but none shorter than k, so the distance from A to C is 3 km.

Alternative clearances, on the other hand, combine with max rather than min: If we want to move a large item, it only needs to clear one of the available paths. In this case, the four-step path $A{\rightarrow}F{\rightarrow}E{\rightarrow}D{\rightarrow}C$ is important because of the restrictive height constraint on e_1: $A{\rightarrow}B$.

$$hgt(A, C) = hgt(h) * hgt(k) * hgt(A \rightarrow F \rightarrow E \rightarrow D \rightarrow C) * \ldots$$

$$= \max \left\{ \min \{2, \ 3\}, \ \min \{2, \ 1\}, \ \underline{\min \{6, \ 4, \ 3, \ 3\}}, \ \ldots \right\} m = 3 \text{ m}$$

Parallel paths in electrical network have a different character than the transportation examples above. When we are interested in a "best" alternative, we join paths with order-theoretic operations like max and min, but analyzing concurrent flows requires algebra rather than ordering. For resistors and capacitors, these come from the familiar operations of parallel composition:

$$res(h) * res(k) = \left(res(h)^{-1} + res(k)^{-1} \right)^{-1}$$

$$cap(h) * cap(k) = cap(h) + cap(k)$$

Adding algebra makes this case somewhat more complicated, and we use it to motivate the rest of our discussion. We will return to the resistor equations in section "Visual Reasoning," but for now we look at CT's approach to combinatorial data structures like directed and undirected graphs, and the relationships between them. With this, we can transform the data presented in Fig. 1a into a more appropriate format for resistor networks.

A Model Is a Mapping

This section discusses the categorical approach to data and semantic modeling as one might typically encounter in formal ontologies, relational databases, or object-oriented class diagrams. In particular, we focus on the way that CT represents

43 Category Theory

relationships between different data structures, with directed and undirected graphs as our motivating example.

One unique feature of categorical modeling is an explicit separation of "syntactic" and "semantic" model elements. In slogan form, "a model is a mapping,"

$$\textbf{model : Syntax} \rightarrow \textbf{Semantics}.$$

As indicated by the bold type, the source and target are categories, and the mapping is a categorical relationship called a *functor*. There is a category **Cat** where the objects and arrows are categories and functors. The category of categories introduces an element self-reference into CT, a powerful feature of categorical analysis.

In modeling terms, one should think of the elements of this mapping as follows:

- Syntax specifies the numbers and types of system components, and their arrangement.
- Semantics define explicit representations for component structures or behaviors, and algorithms for composing them.
- Functors assign semantic representations to syntactic components, and calculate system semantics according to the specified algorithms.

Rather than implementing a model "by hand" in the target technology, we can provide a high-level description of the system in terms of its components and their interactions, and then evaluate the semantic functor to construct the desired implementation.

For combinatorial data structures like graphs, the target semantics is **Set**, the category of sets and functions. The objects are sets X, Y; $\mathbb{R}h$: $X \rightarrow Y$ is a function assigning every element $x \in X$ to a unique element $y = h(x) \in Y$. Semantic categories are usually defined a priori, based on preexisting formal methods. Other important classes of semantics include categories of relations **(Rel)**, matrices **(Vect)**, probability distributions **(Prob)**, and dynamical systems **(Dyn)**. (In fact, each has many variants, e.g., finite versus continuous probabilities or dynamics.)

For now, though, our object of focus is the syntax of the model, which we call a *schema*. We start with a simple schema containing only two objects and two arrows

$$\textbf{D} = \left\{ \ E \ \underset{t}{\overset{s}{\rightrightarrows}} \ V \ \right\}$$

The symbols in the schema are mnemonic for *s*ource, *t*arget, *E*dge, and *V*ertex, and the schema's name stands for **D**irected graph.

The objects and arrows in the schema represent placeholders or variables that range over sets and functions. To model the graph **g** drawn in Fig. 1a, we would substitute $\{A, B, C, D, E, F\}$ for the object V, and a set $\{e_1, \ldots, e_{11}\}$ for E. For example, if e_7 is the bottom edge of the graph **g**, labeled 4, then s: $e_7 \mapsto F$ and t: $e_7 \mapsto E$. The full assignment of source and target for a smaller graph is shown in Fig. 2.

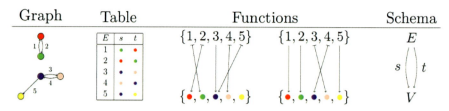

Fig. 2 The representation of a directed graph (left) with five numbered edges and five colored vertices. We can encode this data as a pair of functions or, equivalently, a database table with two columns

We formalize this substitution process as a mapping **D**→**Set**. In general, an arrow **f**: **C**→**D** between categories is called a *functor*. Every functor involves two functions, $\mathbf{f}^{ob}$ and $\mathbf{f}^{ar}$, that map objects to objects and arrows to arrows, respectively. These mappings must preserve connectivity, just like a graph homomorphism: Given an arrow h: $X \to Y$ in **C**, the image $\mathbf{f}^{ar}(h)$ is an arrow $\mathbf{f}^{ob}(X) \to \mathbf{f}^{ob}(Y)$ in **D**. In practice, we omit the decorations and rely on context to distinguish applications $\mathbf{f}(X)$ and $\mathbf{f}(h)$.

Now we can represent a specific directed graph as a functor **g**: **D**→**Set**, which is called an *instance* of the schema **D**. More importantly, the same style of representation generalizes to any database or formal ontology. In a relational database, the tables are objects, the arrows are columns, and a functor maps each table to its set of rows. In logic, objects are formulas, arrows are proofs, and a functor sends each formula to the set of elements that satisfy a property. Part of the value of the categorical approach is to unify superficially different representations.

Next, we need to add labels to our graphs, and more generally data attributes to the objects of our schemas. In this case, we add a new arrow l: $E \to R$ (mnemonics: *l*abel, *R*eal number):

$$\mathbf{L} = \left\{ R \xleftarrow{l} E \xrightarrow[t]{s} V \right\}$$

The only distinction between the attribute l and any other arrow like s or t is that we always assign the target object R to the same set; every model **L**→**Set** sends $R \mapsto \mathbb{R}$. Consequently, the function $\mathbf{g}(l)$ will always assign numeric labels. If E and V act like variables over sets, R acts like a constant.

We now have two schemas **D** and **L**, and their relationship defines a second functor **i**: **D**→**L**. Functors are built from functions, and they compose in the same way – $\mathbf{f} \cdot \mathbf{g}$ sends X to $\mathbf{g}(\mathbf{f}(X))$ – so we can define a category **Cat**, where the objects are categories, and the arrows are functors.

Functors between schemas create transformations between the associated instances. The simplest takes a labeled graph and "forgets" the labels, returning the underlying directed graph. In categorical terms, this is just the composite functor **i** · **g**: **D**→**L**→**Set**. This situation occurs frequently in programming, whenever a class **C** extends an abstract interface **A**. The class **C** "inherits" any method $f \in \mathbf{A}$ by first expanding *f* in terms of **C** and then expanding again according to the implementation of **C**, corresponding to a pair of functors **A**→**C**→**PrLang**.

Alternatively, we may start from an unlabeled graph **g** and ask for a *lift* of **g** along **i**, as shown in the diagram on the right. In this case, each lift corresponds is a different assignment of labels for the same underlying graph. Lifting problems are often underdetermined; there is no way to guess the "right" labels for **g**. Nonetheless, CT provides a "free approximation" that lifts without any additional data. The *left Kan extension*, denoted Σ_i, extends the unlabeled graph by inserting dummy variables (labeled nulls) for any unknown information. Even if we do not know their values, these variables can still carry logical constraints and inferences.

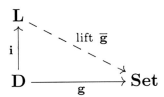

As categories, the schemas **D** and **L** are a bit lacking: They involve no composition. To remedy this, we introduce the schema for *un*directed graphs. The idea is to model an undirected edge $X — Y$ as a pair of directed edges $X \leftrightarrows Y$. Here is the schema:

$$\mathbf{U} = \left\{ r \circlearrowright E \xrightarrow[t]{s} V \;\middle|\; r \cdot s = t,\; r \cdot r = id \right\}$$

There are several elements to note. First, we added a loop $r: E \to E$, allowing for composition with s and t, and of r with itself. Next, we have two *path equations* that formalize semantic constraints of the representation. The first says that the source of the reverse is the target. The second says that reversing twice is the same as doing nothing at all. We can use ordinary equational logic to derive further consequences: The target of the reverse is the source

$$r \cdot t = r \cdot (r \cdot s) = (r \cdot r) \cdot s = id \cdot s = s$$

In addition to the schema itself, we also get a second schema functor **j**: **D**→**U**. Just like before, we can compose with **j** to get the "underlying" directed graph of **u**: **U**→**Set**, noting that **j** · **u** has two directed edges for each undirected edge in **u**. Since most directed graphs are not of this form, the free lift Σ_j "fixes" this by doubling *every* edge $X \to Y$ to a pair $X \leftrightarrows Y$. For example, the graph **g** from Fig. 1a has 11 edges, $\Sigma_j(\mathbf{g})$ would have 22.

CT also provides formal mechanisms for integrating schemas that are developed independently. The initial data is a pair of functors **L** ⟵ **D**→**U** with the same arrangement as **i** and **j**. **L** and **U** are the schemas to be integrated, while **D** defines their conceptual overlap. (CT neither provides nor restricts the mechanisms used to identify the overlap. This is a hard problem, in part because it may depend on the

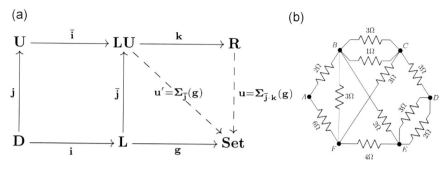

Fig. 3 (a) Transforming a labeled directed graph into an undirected resistor network; (b) the resistor network $\Sigma_{\bar{j}\cdot k}(g)$

intent of the schemas' authors.) Given this data, the *pushout* is a new schema **LU** along with two functors **L**→**LU** ⟵ **U** that complete the left-hand square in Fig. 3a. Intuitively, we produce **LU** by "gluing" **L** to **U** along **D**. The two functors show how the original schemas sit inside the joint context, and when we restrict these embeddings to **D**, they agree.

Now we can use the functor **L**→**LU** to transform the labeled, directed graph **g**: **D**→**Set** into a labeled, undirected graph **u'**: **LU**→**Set**. The transformation doubles each edge, just like $\Sigma_{\bar{j}}$, and the labels in **L** come along for the ride. However, we are not quite done because we need to set the label values for the reversed edges.

It frequently happens that the pushout schema is incomplete, in the sense that there are natural semantic constraints between **L** and **U** that cannot be inferred from the schemas themselves. In this case, we know that resistance labels (coming from **L**) should be symmetric under reversal (coming from **U**). This symmetry corresponds to an unstated equational constraint $r \cdot l = l$. (Note that these constraints are semantically motivated and cannot be derived from the schemas themselves. We could use the same schemas to model a network of batteries, but, in that case, we should set $r \cdot l = -l$ to reflect the fact that voltage is oriented.) Adding this constraint to the pushout creates a sixth schema **R** as well as a comparison functor **k**: **LU**→**R**.

In the final step of the transformation, we apply Σ_k to **u'**. This adds in the new constraint and closes the data structure under logical inference. This will automatically fill the reverse resistances needed to complete our network.

Let us review what we have done. We started by representing data schemas as categories, and the associated data structure as **Set**-valued functors. This provides three important mechanisms for formal data manipulation:

1. Functors express relationships between schemas.
2. Functors between schemas create data transformations using composition and lifting.
3. Pushouts combine schemas based on conceptual overlap.

43 Category Theory

We started with a directed, labeled graph $g: L \to Set$. First, we related L to a second schema U based on a shared subschema D. Next, we pushed out this relationship to construct a joint schema LU. Then we extended LU with additional semantic constraints (resistance values are symmetric) to obtain a final schema R. With all the schemas in place, we then lifted g twice, first from $L \to LU$ to double the edges, then from $LU \to R$ to fill in the missing resistance values.

The categorical treatment is a bit top-heavy for directed graphs; the effort to establish the mathematical machinery out-weighs its value when applied to such tiny schemas. Amortizing the up-front cost over a larger project improves that cost-benefit comparison. Our solution also leaves behind a compositional infrastructure of schemas and functors that can be reused and extended for future problems. Additionally, functorial transformations track *how* a model is built, rather than what is inside, and we can view this as a very explicit form of traceable documentation.

Isomorphism and Identity

In the last section, we spent some effort converting a directed graph of resistor values into an undirected graph. Why? What makes one representation better than another for a given purpose? In this section, we consider one important answer that is both intuitive and historically important: The choice of representation helps us understand whether two things are "the same." CT can help us to give precise answers to fuzzy questions like this, often by sharpening the concepts in play in a way that distinguishes conflicting interpretations.

What does it mean for two things to be the same? This is a problem not only for philosophers, but also for programmers, who must decide whether to compare two data structures by their location in memory (reference equality) or by the data stored there (structural/value equality). Identity turns out to be a surprisingly slippery question.

It is also an important question in practice: If we want to optimize over all structures of a given type or test against all possible failures of a certain kind, we must be able to recognize whether two of these are the same. Failure to recognize an equivalence leads to extra work; the resistor network in the previous section has 2048 different representations as a directed graph, and analyzing the resistance of all these would be a waste of time. Even worse, asserting a false or unjustified equivalence may lead us to ignore cases that are meaningfully different, invalidating our attempted optimization or assurance.

To get a sense of the problem, let us ask which of the following three graphs are the same?

$$g_1 = \left\{ A \xrightarrow{2} B \right\} \quad g_2 = \left\{ B \xrightarrow{2} A \right\} \quad g_3 = \left\{ A \xleftarrow{2} B \right\}$$

If we ask the printer, the answer is none, since all three look different on the page. On the other hand, we usually want to think of graphs as combinatorial structures, so that positioning on the page is irrelevant, and in that case, g_1 and g_2 are identical.

For g_3, the answer is even less clear. We need to ask whether the labels "A" and "B" are semantically meaningful or "just names." If the labels are meaningful, then g_1 and g_3 are meaningfully distinct, and otherwise, all three graphs are equivalent.

To make sense of this, CT combines two formal constructions: isomorphism and natural transformation. The first is an **internal** concept, in the sense that it refers to relationships (between objects) *inside* a category **C**. The second is **external**, because it involves relations *between* categories (inside **Cat**). Self-reference allows us to combine these to define the concept of natural isomorphism, which provides a context-relative definition of sameness for any schema **C**. As we will see, the interplay between internal and external concepts provides a rich language for expressing subtle concepts and intuitions.

An **isomorphism** (**iso**) in a category **C** is an arrow that can be reversed. Isomorphisms come in pairs, so that any iso $i: X \to Y$ is matched with an **inverse** $j = i^{-1}: Y \to X$, and together these satisfy

$$i \cdot j = id_X \quad j \cdot i = id_Y.$$

We can "undo" the composition with i by composing with j, and vice versa.

An iso in **Set** is a function $h: X \to Y$ that is **bijective**, meaning that for every y there is exactly one x such that $h(x) = y$. The function $exp(x) = e^x$. is a bijection $\mathbb{R} \to (0, \infty)$, and it is an isomorphism because there is an inverse function $log: (0, \infty) \to \mathbb{R}$. It is also an algebraic isomorphism between plus and times, because

$$e^{x+y} = e^x \cdot e^y \qquad log\,(x \cdot y) = log\,(x) + log\,(y).$$

However, it is not a metric isomorphism because it does not preserve distance

$$dist(x, y) \neq dist(e^x, e^y).$$

We say objects X and Y are **isomorphic** and write $X \cong Y$ if there is at least one isomorphism relating the two objects. Two sets are isomorphic if they have the same number of elements, in which case we can construct a bijection by pairing them off one by one. Two vector spaces are isomorphic if they have the same dimension, and an isomorphism $\mathbb{R}^n \to \mathbb{R}^n$ is an invertible $n \times n$ matrix.

Isomorphism is an **equivalence relation**, meaning that it satisfies all the usual rules for equality. The reflexive and transitive laws follow directly from identity and composition:

$X = X$	$X = Y \Rightarrow Y = X$	$X = Y \,\&\, Y = Z \Rightarrow X = Z$
$X \cong X$	$X \cong Y \Rightarrow Y \cong X$	$X \cong Y \,\&\, Y \cong Z \Rightarrow X \cong Z$
id	$h \mapsto h^{-1}$	$(h, k) \mapsto h \cdot k$

Functions preserve equality – if $x = y$, then $f(x) = f(y)$ – by the substitution property of equals for equals. In much the same way, a functor $\mathbf{f}\colon \mathbf{C}\to\mathbf{D}$ preserves inverses and isomorphism:

$$h : X \cong Y \in \mathbf{C} \Rightarrow \mathbf{f}(h)^{-1} = \mathbf{f}(h^{-1})$$

$$\mathbf{f}(h) \cdot \mathbf{f}(h^{-1}) = \mathbf{f}(h \cdot h^{-1}) = \mathbf{f}(id_X) = id_{\mathbf{f}(X)}$$

Now we have a well-behaved notion of structural equality inside of any category.

We would like to use isomorphisms to analyze the data structures from the last section. Since instances are mappings (functors), the relationships between them, called natural transformations, are mappings between mappings. In diagrams, they are represented by two-dimensional cells, as in Fig. 4 (top), and we also use a double-shafted arrow (Not to be confused with implication arrow $\Rightarrow$ used above.) $\Rightarrow$ to help distinguish natural transformations from other kinds of arrows.

Given functors $\mathbf{f}$ and $\mathbf{g}$ as shown in Fig. 4, a *natural transformation* $\alpha\colon \mathbf{f} \Rightarrow \mathbf{g}$ is a family of arrows α_C from $\mathbf{D}$ indexed by objects $C \in \mathbf{C}$. Natural transformations raise dimension: Each object $X \in \mathbf{C}$ is sent to an arrow $\alpha_X\colon \mathbf{f}(X)\to\mathbf{g}(X)$ in $\mathbf{D}$, called the *component* of α at X, and every arrow $h\colon X\to Y$ is assigned an equational constraint called a *naturality square*: $\mathbf{f}(h) \cdot \alpha_Y = \alpha_X \cdot \mathbf{g}(h)$, as shown on the bottom right of Fig. 4.

Specializing to directed graphs $\mathbf{f}, \mathbf{g}\colon \mathbf{D}\to\mathbf{Set}$, a natural transformation $\alpha\colon \mathbf{f} \Rightarrow \mathbf{g}$ is just a graph homomorphism. Because the schema has two objects V and E, the transformation involves two functions: α_V sends vertices to vertices, and α_E sends edges to edges. The naturality squares ensure that the homomorphism preserves incidence in the graph: If α_E sends $e \mapsto e'$, then α_V should send $s(e) \mapsto s(e')$ and $t(e) \mapsto t(e')$.

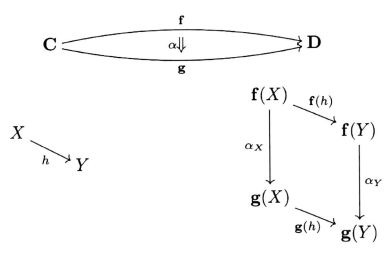

Fig. 4 A natural transformation α between two functors $\mathbf{f}, \mathbf{g}\colon \mathbf{C}\to\mathbf{D}$

Now suppose **f** and **g** are labeled graphs **L**→**Set**. The interpretation of the label set $R \in \mathbf{L}$ is the same for every graph: $\mathbf{f}(R) = \mathbf{g}(R) = \mathbb{R}$. In this case, the naturality square reduces to a triangle, with the resulting equation

$$\mathbf{f}(E) \xrightarrow{\ \alpha_E\ } \mathbf{g}(E)$$
$$\mathbf{f}(l) \searrow \quad \swarrow \mathbf{g}(l)$$
$$\mathbf{f}(R) = \mathbf{g}(R)$$

$$\mathbf{g}(l)(\alpha_E(e)) = \mathbf{f}(l)(e)$$

This is not very easy to parse. When working with natural transformations, we often simplify notation by omitting the indexing object. This is usually clear from context; if v is a vertex, then $\alpha(v)$ is an application of α_V, not α_E. We can also drop **f** and **g** from the notation, and simply write l for any of the label functions $\mathbf{f}(l)$, $\mathbf{g}(l)$, …. We already used this convention above, when we wrote $s(e) \mapsto s(e')$ rather than $\mathbf{f}(s)(e) \mapsto \mathbf{g}(s)(e')$. With lighter notation, the new equation is much easier to read

$$l(\alpha(e)) = l(e).$$

It says that α cannot change the labels on the edges and, more generally, natural transformations cannot modify attribute values.

Putting these two ideas together, we obtain the definition of a **_natural isomorphism_**: an invertible natural transformation. Equivalently, a natural transformation α where each component α_D is an isomorphism in the target category. By this definition, a graph isomorphism is a homomorphism that is bijective on vertices and on edges, the same as the usual definition. For labeled graphs, the isomorphism is not allowed to modify edge labels.

If we model the graphs above as functors **L**→**Set**, then all three are isomorphic. The first isomorphism $\mathbf{g_1} \cong \mathbf{g_2}$ is just the identity, since the graphs have the same vertices, edges, sources, targets, and edge labels. On the other hand, the iso $\mathbf{g_1} \cong \mathbf{g_3}$ has a nontrivial vertex function that sends $A \mapsto B$ and $B \mapsto A$.

Suppose we extend **L** by adding a new attribute $n: V \to Str$. This generates a new schema **L**' as well as a schema inclusion functor $\ell: \mathbf{L} \to \mathbf{L}'$. Once n internalizes the names A and B into the schema, an isomorphism is no longer allowed to permute them: As **L**'-instances, we have $\mathbf{g_1} \cong \mathbf{g_2}$, but $\mathbf{g_1} \ncong \mathbf{g_3}$. Similarly, we could attach coordinates x, $y: V \to R$ to each vertex in order to distinguish $\mathbf{g_1}$ from $\mathbf{g_2}$. Since schemas carry equations, we can also attach relative constraints to distinguish the two, even when we do not know their exact values:

$$x(A) = x(B) - 1 \text{ cm} \in \mathbf{g_1}, \mathbf{g_3} \qquad x(A) = x(B) + 1 \text{ cm} \in \mathbf{g_2}$$

Whatever we include in the schema is regarded as semantically meaningful and must be preserved by all natural transformations; anything left out, we can scramble.

To ask whether two data structures are the same, we should first ask "For what?" The context of the problem determines which features are semantically relevant.

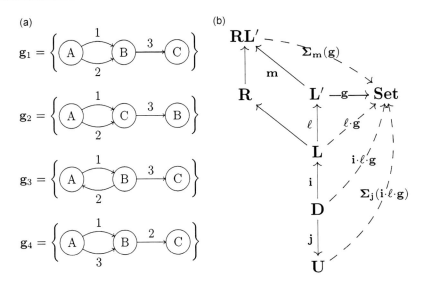

Fig. 5 (a) Four similar graphs that are distinct in $\mathbf{L'}$, but isomorphic in other contexts; (b) the schemas and functors involved in our comparison

Once we know the right question, we can use functorial transformations to recontextualize our structures before asking about equivalence.

As an example, consider the four edge- and node-labels graphs shown in Fig. 5. Regarded as functors $\mathbf{L'} \to \mathbf{Set}$, none of them are isomorphic. If we compose with ℓ, this "forgets" the node labeling, introducing a new isomorphism

$$\ell \cdot \mathbf{g}_1 \cong \ell \cdot \mathbf{g}_2.$$

If we ignore the edge labels as well (e.g., reachability analysis), we pick up another iso

$$\mathbf{i} \cdot \ell \cdot \mathbf{g}_1 \cong \mathbf{i} \cdot \ell \cdot \mathbf{g}_2 \cong \mathbf{i} \cdot \ell \cdot \mathbf{g}_4.$$

Finally, we can ignore direction by lift along $\mathbf{j}$. This makes all four of the graphs isomorphic

$$\Sigma_j(\mathbf{i} \cdot \ell \cdot \mathbf{g}_1) \cong \Sigma_j(\mathbf{i} \cdot \ell \cdot \mathbf{g}_2) \cong \Sigma_j(\mathbf{i} \cdot \ell \cdot \mathbf{g}_3) \cong \Sigma_j(\mathbf{i} \cdot \ell \cdot \mathbf{g}_4).$$

Alternatively, we can imagine that the node labels *are* semantically meaningful. In that case, we can integrate $\mathbf{L'}$ with the resistor schema $\mathbf{R}$ by pushing out the semantic overlap $\mathbf{R} \leftarrow \mathbf{L} \to \mathbf{L'}$. This produces a new schema $\mathbf{RL'}$ as well as a functor $\mathbf{m}: \mathbf{L'} \to \mathbf{RL'}$. Lifting a graph along $\mathbf{m}$ symmetrizes the edges and edge labels, so in this case only $\mathbf{g}_4$ is distinct

$$\Sigma_{\mathbf{m}}(\mathbf{g}_1) \cong \Sigma_{\mathbf{m}}(\mathbf{g}_2) \cong \Sigma_{\mathbf{m}}(\mathbf{g}_3).$$

We begin to see some benefits of a principled compositional approach. We could have easily defined (directed) graph isomorphism directly, but then we would have needed another definition for labeled graphs, another for undirected graphs, and another for undirected, labeled graphs, and each of these would have been trivial on its own, but the need for constant tweaking, refactoring, and updating existing methods is a burden that grows with a code base.

Instead, we defined a general notion of equivalence phrased internally in the language of objects and arrows. Natural transformations, maps between functors, allowed us to use it. Both concepts are independently useful, but we can mix them to generate new and more refined concepts. Critically, we exploit the uniform description of schemas and functors to define these relationships all at once, rather than one data structure at a time. Then we were able to reuse the schemas and functors from the previous section to explore this concept in a specific example.

Picturing Processes

Section "A Model Is a Mapping" introduced the idea that a model is a mapping and showed how we could use this approach to represent and transform combinatorial structures like graphs. Now we want to generalize this story to other classes of models from probability, dynamics, and more. The syntax and semantics of these models are called string diagrams and process categories, respectively.

String diagrams are a formal picture language describing networks of interacting processes. Such diagrams are ubiquitous throughout engineering, ranging from informal flow charts to fully formal computational models like circuit diagrams or Bayes nets. String diagrams provide a uniform approach to a wide range of analytic techniques, helping to link models of different kinds through common reference. The very simple diagram $\mathfrak{D}$ that will guide the discussion is shown in Fig. 6.

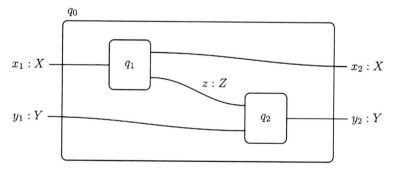

Fig. 6 A string diagram $\mathfrak{D}$ represents a composite process. The outer box q_0 represents the environmental boundary, while q_1 and q_2 are the interfaces for (black-box) subprocesses

43 Category Theory

A ***process interface*** q is a pair of sets $(A; B)$. We call the elements of A and B **ports** and think of them as the inputs and outputs of some process. In many cases, we have different types of ports (e.g., electrical vs. mechanical), corresponding to a pair of functions $(A{\rightarrow}T; B{\rightarrow}T)$ for some set of interaction types T.

A string diagram is specified by

- One ***external interface*** q_0 (the ***environment***)
- Any number of ***internal interfaces*** $q_1, q_2, \ldots, q_k$ (the ***components***)
- A set of interactions (the ***strings*** or ***wires***) matching compatible ports to one another

Explicitly, the diagram above has types $T = \{X, Y, Z\}$, external interface $q_0 = (\{x_1, y_1\}; \{x_2, y_2\})$, and internal interfaces $q_1 = (\{x_1\}; \{x_2, z\})$ and $q_2 = (\{y_1, z\}; \{y_2\})$. The combinatorial representation of the interactions is sensitive to our assumptions about the problem: Does it makes sense for strings to split and merge? Is feedback allowed? There are many flavors of string diagrams tuned to support different answers to these and other questions. See [11] for an extensive survey.

A model for a string diagram starts with the choice of a target category **S** that encodes the semantics of the model. Arrows in **S** model individual processes; we might use tensors in **Vect**, stochastic process in **Prob**, or smooth flows in **Dyn**, among many other alternatives. We can also construct tailor-made semantic categories, like the data structures and natural transformations in sections "A Model Is a Mapping" and "Isomorphism and Identity."

Once the target category is selected, we assign an object to every string and an arrow to every interface. For this to work, we must make sense of multiple inputs and outputs, so we assume that **S** carries a second operation $X{\otimes}Y$ that represents parallel (noninteracting) composition. (The use of tensor notation in CT does not depend on other uses in mathematics and physics, though it does apply to those cases.) With this, we can model the internal interfaces as arrows $q_1: X{\rightarrow}X{\otimes}Z$ and $q_2: Z{\otimes}Y{\rightarrow}Y$.

Now we use the diagram $\mathfrak{D}$ as a recipe to construct a new arrow $q_0: X{\otimes}Y{\rightarrow}X{\otimes}Y$, noting that the construction requires parallel composition for arrows in addition to objects (e.g., $q{\otimes}q'$). We use the term (The usual term is symmetric monoidal category, but the concept is too important for such dense jargon.) ***process category*** for a category together with a chosen operation of parallel composition. Unpacking the abstract definition of q_0 depends on which category **S** we have chosen for the target semantics. We will spend some time in the next few sections working out the details in a few concrete cases.

The canonical process category is **Set**, using the Cartesian product $X{\times}Y$ for parallel composition. To interpret the diagram $\mathfrak{D}$, we start by assigning each type in $T = \{X, Y, Z\}$ to a set, and each component q_i to a function with the appropriate interface. For our first model, we set

$X \mapsto \mathbb{R}$	$(q_1: X \to X \otimes Z) \mapsto (f_1: \mathbb{R} \to \mathbb{R} \times \mathbb{B})$
	$f_1(x: \mathbb{R}) = (x + 0.1, \cos(x) > .5)$
$Y \mapsto \mathbb{N}$	
	$(q_2: Z \otimes Y \to Y) \mapsto (f_2: \mathbb{B} \times \mathbb{N} \to \mathbb{N})$
$Z \mapsto \mathbb{B} = \{T, F\}$	$f_2(b: \mathbb{B}, n: \mathbb{N}) := $ if b then $n+1$ else n

In this model, f_1 executes some very simple dynamics on X and sends a Boolean measurement of the input to f_2, which simply increments a running count of the positive outcomes. The composite process is a function

$$f_0 : \mathbb{R} \times \mathbb{N} \to \mathbb{R} \times \mathbb{N}$$

$$f_0(x: \mathbb{R}, n: \mathbb{N}) := (x + 0.1, \text{if } (\cos(x) > .5) \text{ then } n+1 \text{ else } n) : \mathbb{R} \times \mathbb{N}$$

We can iterate f_0, since it has the same inputs and outputs, and when we do (Fig. 7, top), we see that x rises steadily while n starts and stops as x enters and leaves the measured region.

Of course, if we choose different functions, we get different dynamics. The string diagram places no constraints on how wild the component functions might be. What we can see, though, is that the dynamics on n will never influence the dynamics on x, because there is no channel from the n input to the x output. And, if we replace the assignment $q_2 \mapsto f_2$ with some more complicated dynamics, (Here m % k is the modulus operation, or the remainder of m when divided by k. The equation n % 3 = 0 says that n is divisible by 3.) we can see that this is the case (Fig. 7, bottom):

$$f_3(b: \mathbb{B}, n: \mathbb{N}) = \text{if } (b \text{ OR } n\%3 = 0) \text{ then } n+1 \text{ else } n-2$$

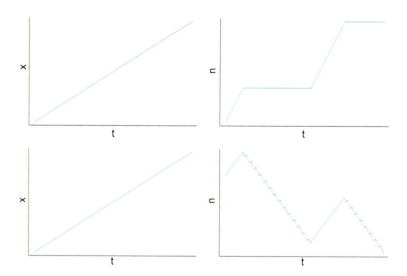

Fig. 7 Iterated dynamics defined by two process functors $\mathbf{f_1}$ (top) and $\mathbf{f_2}$ (bottom)

We would like to think of this construction in light of the earlier principle that "a model is a mapping." The first step is to construct $\mathbf{P} = \mathbf{P}(\mathfrak{D})$, the *free process category* associated with the diagram $\mathfrak{D}$. An arrow in $\mathbf{P}$ is a string diagram that involves only the components in $\mathfrak{D}$, called *generators*, noting that a particular component might appear several times within the same diagram, or not at all. Serial composition $p \cdot p'$ matches the output strings of p to the input strings of p', where we require the types on the strings agree. For parallel composition $p \otimes p'$, we just place the two diagrams side by side.

Rather than a representation of our system, P is a context for representation. We should think of P as a universe of possibility; it describes all possible interactions between the components of $\mathfrak{D}$, without regard to how they are arranged. In particular, $\mathbf{P}$ contains a special arrow $q_0 \colon X \otimes Y \to X \otimes Y$ that represents the diagram $\mathfrak{D}$ itself in terms of an explicit sequence of serial and parallel composition:

$$q_0 = (q_1 \otimes id_Y) \cdot (id_X \otimes q_2).$$

When we defined serial composition, we required two properties: associativity and identity. These are also requirements for parallel composition, but now we weaken the restrictions to isomorphisms. For example, in **Set** we have two different products

$$X \times (Y \times Z) \neq (X \times Y) \times Z$$

because elements on the left have the form $(\bullet, (\bullet, \bullet))$, while those on the right look like $((\bullet, \bullet), \bullet)$. However, there is an obvious bijection $(x, (y, z)) \leftrightarrow ((x, y), z)$. More generally, in any process category we have an isomorphism

$$X \otimes (Y \otimes Z) \cong (X \otimes Y) \otimes Z.$$

In practice, this means we can drop the parentheses, just like we did for serial composition, though the justification for this claim is more involved [12].

The identity for parallel composition involves a special object I, called the *unit object*, which satisfies

$$I \otimes X \cong X \cong X \otimes I$$

In **Set**, the unit object is a chosen singleton $1 = \{*\}$, corresponding to bijections $(*, x) \leftrightarrow x \leftrightarrow (x, *)$. More generally, it represents "no objects" in the same way that $X \otimes Y \otimes Z$ represents "three objects." In a formal sense, it represents the white space behind the string diagram and can be generalized to allow diagrams with colored regions [13].

Unit objects are important for defining *states*, arrows with no inputs, and effects, which have no outputs. We usually emphasize this in string diagrams by drawing states and effects as triangles.

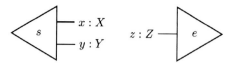

In **Set**, a state $x: 1 \to X$ picks out a single element $x_0 = x(*) \in X$, and once we associate states with elements, function application is just a special case of composition: $f(x_0) = x \cdot f$. We can often think of states in an arbitrary process category as some sort of "element" in the target object.

A functor $\mathbf{f}: (\mathbf{P}, \otimes) \to (\mathbf{Set}, \times)$ is called a ***process functor*** if it preserves parallel composition (in addition to identity and serial composition):

$$\mathbf{f}(id_X) = id_{\mathbf{f}(X)} \qquad \mathbf{f}(p \cdot p') = \mathbf{f}(p) \cdot \mathbf{f}(p')$$

$$\mathbf{f}(p \otimes p') = \mathbf{f}(p) \times \mathbf{f}(p').$$

To define $\mathbf{f}$, all we need is a set $\mathbf{f}(X)$ for each type $X \in T$ and a function $\mathbf{f}(q_i)$ for each component. This is the same information we provided for the models above, so we have already defined two process functors: $\mathbf{f_1}: q_1 \mapsto f_1, q_2 \mapsto f_2$ and $\mathbf{f_2}: q_1 \mapsto f_1, q_2 \mapsto f_3$.

Once the component functions are specified, we can use them as building blocks to construct a new function $\mathbf{f}(p)$ for *any* diagram $p \in \mathbf{P}$. In particular, we can compose a model for the system of interest by applying $\mathbf{f}$ to the distinguished arrow q_0 that represents $\mathfrak{D}$. This provides the recipe needed to construct the functions displayed in Fig. 7:

$$\mathbf{f_1}(p_0) = \left(f_1 \times id_{\mathbf{f}(Y)}\right) \cdot \left(id_{\mathbf{f}(X)} \times f_2\right)$$

$$\mathbf{f_2}(p_0) = \left(f_1 \times id_{\mathbf{f}(Y)}\right) \cdot \left(id_{\mathbf{f}(X)} \times f_3\right).$$

The principal limitation of functions is that they are deterministic – once we know the inputs, the outputs are fully determined – and this is not always how life works. Fortunately, the syntax of string diagrams is flexible enough to interpret in other contexts. In the next section, we show how to model nondeterministic processes using the same framework.

Nondeterminism

In this section, we introduce two more process models illustrating different types of nondeterminism: possibility and probability. String diagrams let us describe and reason about our processes at a high level, while process functors attach these to concrete computations and analyses. In all the examples we consider here, the objects will be sets, emphasizing the importance of relationships in defining the semantic context.

Possibility

A *relation* (We use a special arrow $\rightarrowtail$ to denote relations. We also violate our usual convention of using lower-case names for arrows because a relation is itself a set.) R: $X \rightarrowtail Y$ is a subset $R \subseteq X \times Y$. We can visualize a relation as a bipartite graph, with vertices X and Y and an edge x —y whenever $(x, y) \in R$. We can also think of a relation as a truth function $R: X \times Y \rightarrow \mathbb{B}$, and we write R(x, y) $\in$ {T, F} for the associated truth value. Relations compose serially using an existential quantifier and in parallel using Cartesian product

$(x, z) \in R \cdot S$	$\Leftrightarrow$	$\exists y. (x, y) \in R \,\&\, (y, z) \in S$
$(x_1, x_2, y_1, y_2) \in R \times S$	$\Leftrightarrow$	$(x_1, y_1) \in R \,\&\, (x_2, y_2) \in S$

Unlike functions, relations can also be flipped around. Every $R: X \rightarrowtail Y$ has a *dual* relation $R^*: Y \rightarrowtail X$ defined by $R^*(y, x) \Leftrightarrow R(x, y)$.

A *network category* (As before, we prefer this to the more common term compact closed category.) is a process category with duals. The arrows in a process categories are inherently directed, often along the flow of time, whereas arrows in a network category can be composed either front-to-back or back-to-front. This often gives network categories a spatial, rather than temporal, orientation. The definitions above assemble sets and relations into a network category **Rel**. We will return to the topic of network duality in section "Duality."

However, we can also think of a relation $R: X \rightarrowtail Y$ as a nondeterministic process. Rather than a single output, R associates each input $x \in X$ with a *set* of possible outputs $R(x) \subseteq Y$. This allows us to represent relations as ordinary functions (in **Set**) that map into the *power set* $\mathcal{P}$, i.e., the set of subsets:

$R: X \rightarrow \mathcal{P}(Y)$	$R^*: Y \rightarrow \mathcal{P}(X)$
$R(x) = \{y \mid (x, y) \in R\} \subseteq Y$	$R^*(y) = \{x \mid (x, y) \in R\} \subseteq X$

Viewed as nondeterministic functions, composition of relations $R: X \rightarrowtail Y$ and $S: Y \rightarrowtail Z$ is given by a union over intermediate y

$$R \cdot S : x \in X \mapsto \cup \{S(y) \mid y \in R(x)\}$$

By "non-deterministic" process, we actually mean "not necessarily deterministic" process, because the deterministic processes – functions – are relations, too. For any function $f: X \rightarrow Y$, we can define a relation $\{(x, y) \mid f(x) = y\}$ called the **graph** of f. (This sense of "graph" is unrelated to the earlier discussion of combinatorial graphs. You may remember the "vertical line test" from high school algebra that distinguishes this type of graph from other relations.) This representation preserves serial and parallel composition, so it defines a process functor rel: **Set**→**Rel**.

Now we are ready to attach a relational model **R**: **P**→**Rel** to the string diagram $\mathfrak{D}$ from the previous section. We will not change the types, the Boolean test applied to x, or the counting function $f_2: \mathbb{B} \times \mathbb{N} \rightarrow \mathbb{N}$. However, we will modify the dynamics on

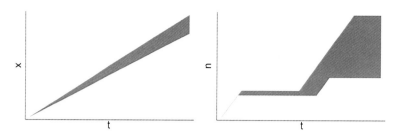

Fig. 8 Relational dynamics of a process functor **R**: **P**→**Rel**

x to include a small tolerance. Since there is now a set of possible outputs, this defines a relation

$$R_1(x : \mathbb{R}) = (x + .1 \pm .01, \ \cos(x) > .5) \subseteq \mathbb{R} \times \mathbb{B}$$

Our model **R** sends $q_1 \mapsto R_1$ and $q_2 \mapsto \mathbf{rel}(f_2)$. When we run the dynamics, we see (Fig. 8, right) that the nondeterminism on x feeds into nondeterminism on n through the Boolean channel, even though the relation operating on n is deterministic.

As we can see in the Fig. 8, the states in **Rel** are subsets rather than values. We did not change the parallel product, so 1 is still the unit object, but now an arrow $1 \nrightarrow X$ is given by $R \subseteq 1 \times X \cong X$. We can still see an echo of the original dynamics in the shape on the right, but the gap between best- and worst-case scenarios compounds more for n than for x.

Probability

Next, we turn to probability. We can model a stochastic process that transforms X into Y as an ordinary function $p: X \to \mathcal{D}(Y)$, where $\mathcal{D}(Y)$ is the set of probability distributions over Y. The distribution $p(x)$ is a function $Y \to [0, 1]$, and its values are conditional probabilities; we emphasize this by writing $p(y \mid x)$ rather than the double application $p(x)(y)$.

Probabilistic functions, which we write $X \rightsquigarrow Y$, form a process category **Prob**. We continue with the Cartesian product for parallel composition, so 1 is the unit. Consequently, a state $x: 1 \rightsquigarrow X$ is just a probability distribution over X, which we think of as a random variable of type X. Parallel composition of arrows $p \otimes p': X \times X' \to Y \times Y'$ uses the product distribution to encode probabilistic independence

$$p \otimes p'(y, y' \mid x, x') = p(y \mid x) \cdot p'(y' \mid x').$$

In particular, a product of states $x \otimes x$ represents a pair of independent, identically distributed (i.i.d.) random variables.

For the serial composition, we compose two arrows $p: X \rightsquigarrow Y$ and $q: Y \rightsquigarrow Z$ by marginalization, summing over the intermediate y's

$$(p \cdot q)(z \mid x) = \sum_y p(y \mid x) \cdot q(z \mid y).$$

This is similar to the definition of relational composition given above, with sums replacing unions. There is also a corresponding functor **prob**: **Set**→**Prob** that associates each ordinary function with a probabilistic function that has zero variance. Both the inclusion and the composition arise from a more general categorical pattern called a ***monad***, a core concept in functional programming that helps to manage interaction logic in the presence of nondeterminism. Monads isolate the key features of aggregation ($\bigcup, \sum$) and inclusion (**rel**, **prob**) that are needed to define well-behaved compositional systems.

Now we are ready to build our model **p**: **P**→**Prob**. Just like before, we leave the types, the counting dynamics and the Boolean test unchanged. This time, we add a Gaussian error term $\epsilon \sim N(0,0.03)$ to the dynamics on x, rather than a fixed tolerance:

$$p_1(x : \mathbb{R}) = (x + .1 + \epsilon, \cos(x) > .5) \in \mathcal{D}(\mathbb{R} \times \mathbb{B})$$

Then we define **p** by sending $q_1 \mapsto p_1$ and $q_2 \mapsto \mathbf{prob}(f_2)$. This yields $\mathbf{p}(p_0)$: $\mathbb{R} \times \mathbb{N} \rightsquigarrow \mathbb{R} \times \mathbb{N}$, and we can examine the resulting dynamics using a Monte Carlo simulation, as shown in Fig. 9.

Even though the standard deviation in **p** is significantly larger than the relational tolerance in **R**, these trajectories are much more tightly bunched because probabilistic errors can cancel out in a way that possibilistic error cannot. This corresponds to the fact that the relational tolerance grows $\propto t$, whereas the standard deviation grows $\propto t^{1'}$ [2]. As before, we can also see that randomness on x feeds into variability on n, even though the process operating on n is deterministic.

Stepping back from the details of the example, what can we say about the role of CT? Worst-case and Monte Carlo analyses are nothing new, so what does the mathematics do for us? String diagrams formalize the sorts of pictures that engineers already like to draw. They enable high-level reasoning about a system, but they also crystalize that intuitive understanding into a combinatorial structure that can be used to parameterize more concrete analyses. In doing so, the diagrams knit these models together through shared reference to the underlying abstract system (Fig. 10), improving traceability and supporting many day-to-day engineering activities such as change propagation and model comparison.

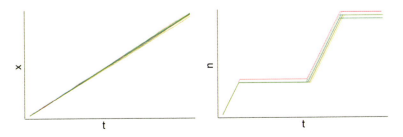

Fig. 9 Monte Carlo dynamics of a process functor **p**: **P**→**Prob**

Fig. 10 A summary of categories, process models (vertical), and semantic relationships (horizontal) involved in our discussion

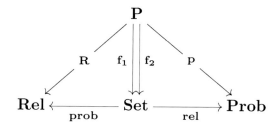

State

In the last two sections, we considered categorical representations of deterministic and nondeterministic processes. Stateful behavior lies somewhere between the two, since it can appear nondeterministic "from the outside" even if it is perfectly predictable "on the inside" with access to the hidden state.

For this example, we equip our counter f_2 with a hidden state variable u with two states: up and down. At each time step, the counter will increment n either up or down according to the state, and the state itself will evolve according to the following finite state machine.

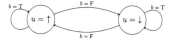

A finite-state machine that takes input and produces output is called a **Mealy machine**, and these machines form the arrows in a category **State**. To define a Mealy machine $s: X \nrightarrow Y$, we must provide three elements: a **state space** U and two functions, an **update** u, and an **output** o

$$u : X \times U \to U \qquad o : X \times U \to Y$$

Equivalently, we can provide a single function $s: X \times U \to Y \times U$.

The description above corresponds to a Mealy machine $s_2: \mathbb{B} \times \mathbb{N} \nrightarrow \mathbb{N}$ with state space $U = \{\uparrow, \downarrow\}$. Explicitly, the update and output functions are given by

$$u(b : \mathbb{B}, n : \mathbb{N}, u : U) = \text{if } b \text{ then } u \text{ else } switch(u)$$

$$o(b : \mathbb{B}, n : \mathbb{N}, u : U) = \text{if } u =\uparrow \text{ then } n + 1 \text{ else } n - 1$$

We can think of a Mealy machine as a stream transformer. Given an initial state u_0 and a sequence of input values $x_0, x_1, x_2, \ldots$, we get a new sequence of output values as follows

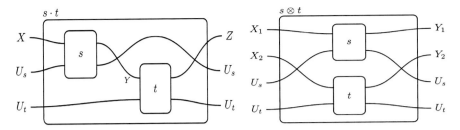

Fig. 11 String diagrams and traditional notation for stateful composition

$$
\begin{array}{ccc}
x_0 = (1, \text{T}), & u_0 = \uparrow & \longmapsto & y_0 = 2 \\
& \downarrow & & \\
x_1 = (2, \text{F}), & u_1 = \uparrow & \longmapsto & y_1 = 3 \\
& \downarrow & & \\
x_2 = (3, \text{T}), & u_2 = \downarrow & \longmapsto & y_2 = 2 \\
& \downarrow & & \\
& \vdots & &
\end{array}
$$

State spaces for separate subprocesses do not interact directly, so the joint state space of a composite process (serial or parallel) is the product of the component state spaces. For the composite operations themselves, we can write down explicit formulas, but we find a string-diagrammatic definition easier to read (Fig. 11).

Now we are ready to build a stateful process model. For the q_1 process, we go back to our original dynamics f_1. This uses the fact that any function is a Mealy machine with a trivial (singleton) state space $U = \{*\}$. As we have seen before, this kind of inclusion is modeled by a functor **state**: **Set**→**State**. Then we define our process functor **s**: **P**→**State** by mapping $q_1 \mapsto$ **state**(f_1) and $q_2 \mapsto s_2$, the Mealy machine defined above.

When we compute the dynamics for the composite process $\mathbf{s}(p_0)$ (Fig. 12, top), the result is essentially the same as the original plot. We can see the state oscillation on the plateaus, but the behavior is broadly similar. This remains true for over a thousand timesteps, but if we zoom out far enough, we begin to see new regimes where the down state takes over, with dramatic departures from the previous behavior (Fig. 12, bottom). Hidden interactions like this are one of the central challenges in testing and verification of modern systems.

The variable behavior would be easier to see if we could compose s_0 with the nondeterministic dynamics from the previous section. This would avoid the unlucky tuning that causes problems for the deterministic analysis.

In order to compose our Mealy machine $s_2 \in$ **State** with a relation $R_1 \in$ **Rel** or a probabilistic function $p_1 \in$ **Prob**, we need to transform them into a common context. To do so, we observe that the construction of the **State** category is essentially diagrammatic. For any process category **P**, a *stateful arrow* s: $X \nrightarrow Y$ in

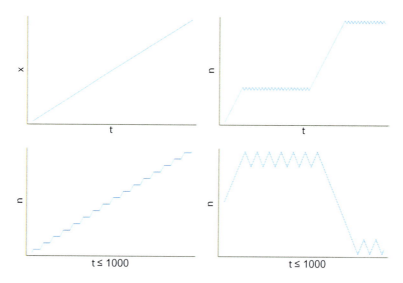

Fig. 12 Iterated dynamics for a stateful process functor **s**: **P**→**State**. Long-term dynamics (bottom right) reveal new interactions with markedly different behavior

just an ordinary arrow $s: U \otimes X \to U \otimes Y$. Serial and parallel composition are defined by the same diagrams given above. This defines a new process category **State(P)**.

Every function is a relation; because of this, every stateful function is a stateful relation.

$s: U \times X \to U \times Y$	$\mapsto$	$\mathbf{rel}(s): U \times X \twoheadrightarrow U \times Y$
$s(u_1, x) = (u_2, y)$	$\leftrightarrow$	$(u_1: U, x: X, u_2: U, y: Y) \in \mathbf{State}(s)$

State sends categories to categories and functors to functors, preserving composition. In other words, this is a functor at the meta-metalevel, relating **PCat**, the category of process categories and functors, with itself.

Similarly, we observed earlier that every function can be regarded as a Mealy machine with a trivial state space. There is a similar construction in any process category **P** using the unit object I for trivial state

$$f: X \to Y \mapsto id \otimes f: I \otimes X \to I \otimes Y$$

This defines a functor **state$_P$**: **P**→**State(P)**, and all these components assemble into a natural transformation **state**: **id** $\Rightarrow$ **State**, again at the meta-metalevel. The naturality square (You, reader, should draw it right now.) for **rel** says that relations and stateful functions "mean the same thing" when they talk about ordinary functions.

Figure 13a shows how to put these constructions together to compose our stateful counter s_2 with the nondeterministic dynamics R_1 and p_1 from the previous section. The individual components correspond to functors with domains ⊣☐⊢ and ⊣☐⊢, representing single-component string diagrams (free process categories). The top

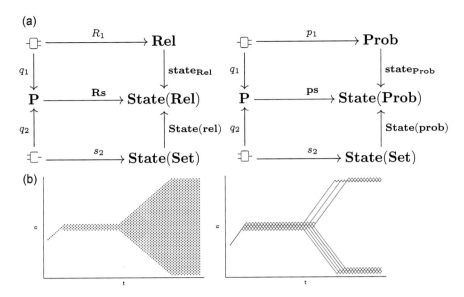

Fig. 13 (**a**): Functors (bottom) and natural transformations (top), let us put together stateful functions with relations (left), and probabilistic functions (right); (**b**): the iterated dynamics of the composed processes

squares import R_1 and p_1 using the **state** transformation, while the bottom squares import s_2 using the **State** functor. Once we transport them to a common context, we can put the components together using the ambient composition defined in Fig. 11. When we plot the iterated dynamics of **Rs**(p_0) or **ps**(p_0) in Fig. 13b, the variable behavior of the counter is much easier to see.

In sections "A Model Is a Mapping" and "Isomorphism and Identity," we represented models as mappings, and used functors between schemas to define model transformations. In the last few sections, we have taken the opposite approach, leaving the syntax of the system unchanged and using semantic functors to relate different modeling contexts. Thus, let us compose components modeled in different formalisms by placing them into a common category.

In this section, we also saw an interaction between the "lower level" categorical structure that describes the processes we are modeling, and the "higher level" categorical structure that describes categories and functors themselves. By framing the **State** construction in purely categorical terms, we could apply the same definition in **Set**, in **Rel**, and in **Prob**, without further modification.

Functoriality and naturality of the construction encode two different types of "simple" stateful processes. These are not complicated constructions. Indeed, layered abstractions like these are often quite intuitive. Categorical language helps us to untangle these intuitions by framing them in a rich vocabulary of explicit formal relationships, and self-reference lets us use the same toolbox at all levels of abstraction.

Visual Reasoning

In this section and the next, we return to the resistor networks from earlier, with an eye toward their semantic representation. The linear algebra that sits underneath electrodynamics has a process interpretation in terms of the interaction between adding, scaling, and copying. For the string diagram shown in Fig. 5, the components q_1 and q_2 represented arbitrary processes, but here we have a very precise meaning for each component. These can be described axiomatically, in terms of diagrammatic equations, allowing for rigorous picture proofs and calculations.

Our starting point is the matrix equation that governs a resistor $R = R\,\Omega$, which translates directly into a string diagram

Each end of the resistor is characterized by a current and a voltage, viewed as objects (We use the traditional symbol I for current, although it conflicts with our notation for the unit object.) $I \cong V \cong \mathbb{R}$. Even though these objects are "the same" in some sense, distinction in naming and color helps to keep things in order. The blue dot is a copy, the red bull's-eye is an add, and the orange R is a scaling operation (plus units). The entire diagram represents a linear function $I \times V \to I \times V$.

Linear algebra is a mechanism to understand the interaction of the four atomic operations shown in Fig. 14a: copy, delete, plus, and zero. These come in sets, with copy matched to delete (●) and plus with zero (⊕). The two are mirror images of one another, satisfying a set of dual equations shown in Fig. 14b, which we summarize by saying that zero and plus form a **commutative monoid**, while copy and delete define a **cocommutative comonoid**. (More generally the prefix "co-" always indicates reversal of direction (cf. limits and colimits).)

The (co)commutative (co)monoid rules let us "comb" any tree that involves only ● or only ⊕ into a **normal form**, which is entirely determined by the number of leaves in the tree representation. Since any two diagrams with the same normal form are equal, this licenses various diagrammatic substitutions that we will use in our arguments.

The following diagrams show what happens when ● and ⊕ meet. If we copy and then add, we double the original value: $x + x = 2x$. Similarly, we can construct a scaling operator $n\colon \mathbb{R} \to \mathbb{R}$ for each integer $n \geq 0$. Operator addition has a similar flavor: We copy the inputs, scale both sides, and then add the results. More generally, any diagram with ●s on the left and ⊕s corresponds to a $k \times \ell$ matrix of nonnegative integers, where k and ℓ are the number of output and input strings, respectively. The entries in the matrix count the number of paths connecting each input-output pair.

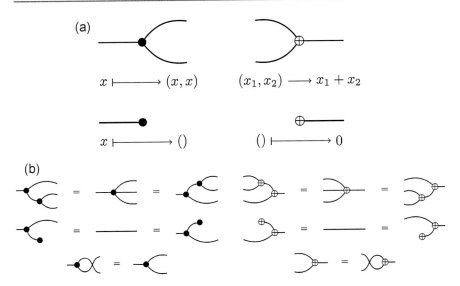

Fig. 14 (a): The generators of linear algebra: copy, delete, plus, and zero; (b): the dual axioms of a (co)commutative (co)monoid

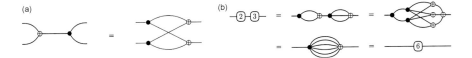

Fig. 15 (a) The Hopf rule pushes copy through add by inserting a complete bipartite graph; (b) scaling operations compose by multiplication

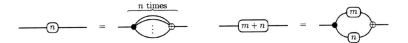

Units like Ohms (Ω) are similar to scaling operators, except that they connect quantities of different types.

Finally, we need to consider addition on the left and copy on the right. The **bialgebra rule**, shown in Fig. 15a, says that we can replace this pattern with a complete bipartite graph connecting all inputs to all outputs. The Hopf rule allows us to "push addition through copy," noting that this leaves the path count between inputs and outputs unchanged. Since we can always move ●s to the left and ⊕s to the right, this is enough to get a matrix normal form for any diagram involving ● and ⊕. Using the Hopf rule, we can show that scaling composes by multiplication, as shown in Fig. 15b, and that arbitrary diagrams compose by matrix multiplication.

With all that machinery in place, let us revisit the resistor diagram from above. The absence of a connection v_0—i_1 corresponds to the matrix entry 0, and the

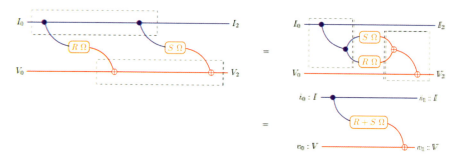

Fig. 16 The (co)monoid laws allow us to derive the formula for serial resistance

unmarked lines across correspond to the 1 s on the diagonal. The orange box $R\,\Omega$ incorporates the resistance as a scaling factor and a unit.

We can use the visual logic to analyze a composite system. For the serial composite of resistors $R\,\Omega$ and $S\,\Omega$ (Fig. 16), we first rearrange the red and blue wires to equivalent normal forms. The rearranged diagram exhibits the composition as a sum, so the serial resistance is $R + S\,\Omega$. This is a special case of composition as matrix multiplication.

Of course, calculating serial resistance is not an impressive result. The main point is that string diagrams support an equational logic based on graph matching and transformation. We can attach rigorous arguments to pictures that allow us to prove and calculate entirely diagrammatically. We close out this line of argument in the next section with a consideration of parallel resistance, but for this, we need one more ingredient: the network structure of reversibility.

Duality

The first step in understanding parallel resistance is to recognize that this is *not* a parallel composition in the sense of process categories. The parallel composition $f \otimes g$ should map $X \otimes X' \to Y \otimes Y'$, with separate inputs and outputs for both components. On the other hand, when we compose resistors in parallel, we split the inputs and merge the outputs.

Kirchoff's laws govern current flows at junctions. These are rendered diagrammatically as split and merge operations on $I \times V$, as shown in Fig. 17a; in slogan form, "currents add, voltages copy." To make this work, we need a way to reverse copy and plus, so we view the functions as relations and take their dual, as described in section "Nondeterminism."

None of the reversed generators are functions. The duals of addition and deletion are multivalued, as indicated by the set braces {} and the introduction of new variables (degrees of freedom) in Fig. 17b. The duals of copy and zero are *partial functions*; the result is deterministic, but the process may fail to return a value (degree reduction).

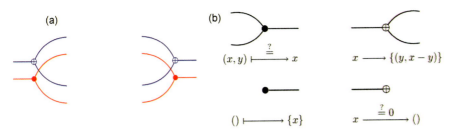

Fig. 17 (a) Split and merge operations on $I \times V$; (b) reversed addition and deletion are multi-valued. Reversed copy and zero are partial

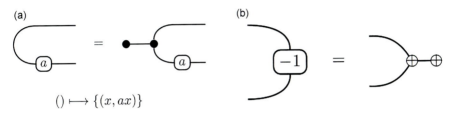

Fig. 18 (a) Composing with a cup transforms a scaling operator into a linear subspace; (b) caps and cups built from plus and zero introduce negative scaling

Reversing the generators allows them to interact in new ways. **Caps** and **cups**, (The names make more sense when diagrams are drawn vertically.) which are defined by plugging, delete into copy; let us turn inputs into outputs and vice versa. For example, flipping the input of a scaling operator defines a linear subspace in two dimensions (Fig. 18a). Playing the same game with plus and zero introduces negative numbers, since two elements that sum to zero must be equal and opposite (Fig. 18b).

Even though the new generators are not functions themselves, we can still use them to build composites that are. For example, if we reverse a scaling operator n, the result is another scaling, this time by $1/n$. This is a function because there is exactly one way to divide a quantity into n equal pieces. Nondeterminism and partiality cancel out, leaving a single value. Consequently, the equational theory now supports fractional scalings in any of our diagrams.

Figure 19 puts these elements together to derive the formula for parallel resistance. The first diagram unpacks the diagrammatic form of the parallel circuit. The first isomorphism uses caps and cups to move currents to the left of the diagram and voltages to the right. After factoring the negatives, we apply the Hopf rule on each side to group the scaling factors. On the next line, we reverse the entire diagram; this inverts the scaling factors and introduces a sum in the center. Finally, we rearrange the diagram back into normal form, including another inversion and the insertion of some negatives, which cancel to yield the result.

Of course, graph transformations are nothing new to electrical engineers; the YΔ-transform is a graph transformation on resistor networks that eliminates an

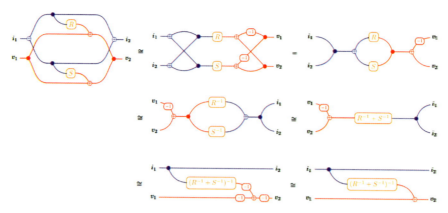

Fig. 19 A diagrammatic derivation of parallel resistance

"internal" node (the center of the Y), without changing resistance through the rest of the network. What CT provides is a general framework for expressing arguments like these in any domain, and for attaching arguments from different domains to the same system. By specifying "the rules of the game," CT provides a rigorous framework to guide our exploration through the space of formal models.

Further Study

In this section, we close with a short guide to additional resources. We organize the discussion by topic: computer science, natural science, formal methods, applications, and resources for learners.

By far the most well-developed application of CT is in *functional programming*. A large and growing community of software engineers explicitly use categorical structures like monads to structure and simplify the design and analysis of large software projects [14], and especially in programming language design [15]. The nondeterministic process categories presented in section "Nondeterminism" are based on these methods. For engineers with programming experience, functional programming can be an excellent introduction CT because it provides a sandbox to play with the mathematical structures. We recommend Milewski's *Programming Cafe* [16] as an introduction to this circle of ideas, or a search for "Functional programming in..." your favorite language.

There are also a number of applications in other areas of computer science, though these are less developed. Our discussion in sections "Composition and Context" and "A Model Is a Mapping," where we presented schemas as categories and data structures as functors, draws from the literature on *categorical databases* [17]. For a more detailed discussion of these ideas, along with working code, see the AlgebraicJulia blog [18]. AlgebraicJulia [6] is an open-source project developing CT tools for scientific computing, including a core library (Catlab) as well as a variety of domain-specific extensions in relational algebra, Petri nets, multiphysics, and more.

Another area of research, closely related to CT but somewhat distinct, is the area of *topological data analysis* [19], which provides methods to extract robust combinatorial structures from data clouds in continuous space. The analysis turns on the fact that the "shape" of a data set depends on the scale; the point-cloud on the right may appear zero-dimensional, one-dimensional, or two-dimensional depending on the level of resolution.

Other applications of CT in computer science include a formal analysis of machine learning via gradient descent in [20], techniques for extracting structure from high-dimensional data [10], concept mining, and methods for integrating grammar and statistics in natural language processing [21].

There is also an established community of practice applying CT to mathematical physics, especially in quantum mechanics. Many of the ideas in our discussion of process and network categories were developed to understand the relationship between quantum theory and computation. *Picturing Quantum Processes* [22] is an excellent introduction to these methods, even for those who are not interested in quantum mechanics per se. Baez and Stay's *Rosetta Stone* paper [23] is also a nice introduction to these ideas, and their relationship with other areas of mathematics.

Resistor networks have been an important motivating example for the use of functorial semantics in the analysis of other physical phenomena. Baez and collaborators developed the theory of *structured cospans* (The original formulation of the idea as decorated cospans [25] had a technical defect that was later corrected in [73].) [24] to analyze the behavior of open systems, networks whose dynamics depend on boundary states and flows. The methods described here are easily extended to other linear circuit components by introducing a time derivative [25], and related research considers similar phenomena for Markov processes [26, 27], reaction networks [28, 29], Petri nets [30], and control theory [31]. More recently, these ideas have been used to construct software for epidemiological modeling [32].

In addition to modeling scientific phenomena directly, CT also helps us manage other math. The material presented in sections "Duality" and "Further Study" is part of a broader effort to develop diagrammatic interpretations for many classes of formal methods. For a gentle but thorough introduction to this approach, see the *Graphical Linear Algebra* blog [33]. The graphical approach can also be applied to study the relational [34, 35] and probabilistic [5, 36, 37] nondeterminism from section "Nondeterminism" as well as the dynamical systems from section "State" (and their continuous counterparts) [4, 38]. Other topics include vector calculus [39], delay and feedback [40], and the theory of computation [41–43].

Today, CT for engineering is a niche topic, but new applications and use cases have led to growing interest. The theory of *codesign* [44–47] models system components as open optimization problems, which can be composed (with feedback) to study global optima. *Operads* are structures to represent and analyze the nesting of components and subsystems in hierarchical structures [48–50]. Robotics researchers have used CT to construct new stability analyses for hybrid dynamical systems [51], among other applications that were presented (and recorded) at a recent workshop at the International Conference on Robotics and Automation [52]. **Bidirectional transformation** (Bx) studies the problem of maintaining consistency in distributed data systems, often using categorical structures called *lenses* and *optics* that model bidirectional information flow [53, 54].

Those who want to learn more will benefit from many excellent resources for further study. In section "Introduction," we quoted from Eugenia Cheng's *How to Bake π* [2], a popular nonfiction book on modern mathematics, in general, and CT, specifically. We also recommend her keynote address at the LambdaWorld conference [55]. Other recommended short works include *What is Applied Category Theory* [56] and the *Rosetta Stone* [23] paper mentioned above.

For a more systematic introduction, we recommend several textbooks which take an applied perspective and try to avoid traditional mathematical prerequisites (e.g., topology, abstract algebra). Spivak's *Category theory for the Sciences* [57] and Perrone's *Notes on Category Theory* [58] both introduce the "standard sequence" of introductory CT topics: categories, functors and natural transformations, limits and colimits, adjunctions, and monads. Spivak and Fong's *Invitation to Applied Category Theory* (*7 Sketches*) [59] uses applications like categorical databases and codesign to introduce more advanced topics like enriched categories and profunctors.

Alongside these formal resources, CT benefits from a vibrant online community. *The n-Category Cafe* [60] is a group blog covering many CT topics in a (relatively) friendly and informal manner; its archive and comments supply deep dives into many topics in pure and applied CT. Other good online resources include *Math3ma* [61] and the *Graphical Linear Algebra* blog mentioned above [33]. The AlgebraicJulia project also hosts a blog [6] to discuss methods and use cases for their CT-based software.

Alongside written sources, there are also a growing number of video resources available for both learning and advanced study. These include recordings of Milewski [62] and Fong and Spivak [63] lecturing on the texts mentioned above, as well as online archive of a course at ETH Zurich on *Applied Compositional Thinking for Engineers* [64]. For more advanced topics, one can consult the proceedings of the last few conferences on Applied Category Theory (ACT) [65, 66], many of which include video recordings due to the pandemic [67]. The Topos Institute, a CT-focused nonprofit, also hosts an online colloquium series [68] as well as other meetings and events [69].

Conclusion

Our foray into CT has touched on many topics briefly, favoring breadth over depth. We introduced categories *vis-a-vis* labeled graphs, asking what it might mean to compose the edges, and found that our answer depended on the semantic context. From there, we moved on to the explicit representation of combinatorial data structures as functors, and transformations of that data using composition and lifting.

Then we introduced isomorphisms and used this to show how two data structures might be "the same" in one context, but different in another.

The next set of topics focused on process categories, particularly the use of string diagrams to represent interacting processes. We constructed a sequence of models based on functions, relations, probability, and state and saw how nondeterminism leaks into deterministic functions via composition. We defined state in purely diagrammatic terms, allowing us to compose stateful processes with relations and probabilistic functions.

Finally, we looked at the two-dimensional logic of string-diagrammatic equations, where proofs are pictures based on graph matching and transformation. We looked at the equations that generate matrix algebra from the interaction of copy and plus, and we used them to generate the classical equations for serial and parallel resistors.

What is lacking, so far, are the tools and tacit knowledge necessary to apply these ideas out in the world. Mathematicians, computer scientists, and physicists have developed a well-honed toolbox of formal techniques for modeling systems of various kinds, but outside of functional programming, we do not yet know how to match these to the appropriate engineering applications. Research in this area is essentially greenfield, which readers should take as both a warning and an invitation.

The first deficit is in methodology. SE is a broad subject, with many processes, practices, and concerns; CT is similarly broad, providing many tools, constructions, and methods. Mapping the two fields requires us to identify which mathematical tools are relevant for each engineering analysis. Nor is this relationship one-to-one: A given engineering problem will often require a mix of several CT constructions. This is particularly difficult to navigate for new learners, so it may be helpful to partner engineers with knowledge of the problems with theoreticians that are able to recognize the abstract patterns that sit beneath.

Tool support is another substantial obstacle. It is still difficult to construct and manipulate categorical models on a computer. The exception here is functional programming, which has a relatively well-developed ecosystem of tooling and support. We have already mentioned AlgebraicJulia [18], an open-source project developing CT software for scientific computing, with applications in several areas. There are also some tools for categorical data [70] and string diagrams [71]. See [72] for a list of other software projects. However, there is currently nothing that supports intuitive style of dynamic argument that we used in our discussion: combing, pushing, flipping, and mixing.

Learning resources for engineers are also in need of improvement, although the situation has improved dramatically in the last decade. However, most texts are focused on explicit knowledge – definition and proof – rather than the tacit knowledge needed to recognize and take advantage of these structures. Engineers require less proof and more demonstration, based on examples that are more than toys. Better tool support would help significantly, providing new users an opportunity to learn about these structures dynamically, through direct interaction, rather than fixed on the page. Better resources for producing diagrammatic syntax would speed up the production of expository materials by an order of magnitude.

In our vision, the model of a system is a system of models. Each model is unique, designed to understand different properties of different systems, leading to different methods, different assumptions, and different levels of fidelity. But models are also tied together. Models for the same system share an architecture and parameters; changes to system design impact them all. On the other hand, models for different systems rely on the same tools and formal methods, and improvements should benefit all of them. Any time we use the same architecture to parameterize a range of analyses, or apply the same analysis to study multiple architectures, we reap the benefits of a compositional infrastructure, and doubly so for changes to an existing system.

Systems engineers interact with categories, functors, and natural transformations on a daily basis, though they might not know it. CT offers an organizing principle that can help us understand a system better – any kind of system – by helping to systematically express the relationships between different views of the system, and further relationships between those. Recognizing the shared calculus of information that sits beneath the tremendous heterogeneity of SE will unify existing methods, provide the basis for new approaches, and help to manage the ongoing transition to a more deeply connected society.

To develop such a vision will require a deep conversation between mathematicians and engineers, between academics and industry, and between theory and practice. It is not only a grandiose vision, but also a unifying one, with CT playing the role of a lingua franca, binding together the myriad threads of systems engineering, across disciplines, across phases, and across companies, helping to organize and connect the individuals, the infrastructure, and the data that drive the modern world.

Disclaimer Commercial products are identified in this chapter to adequately specify the material. This does not imply recommendation or endorsement by the National Institute of Standards and Technology, nor does it imply the materials identified are necessarily the best available for the purpose.

References

1. A. Wayne Wymore, "The tricotyledon theory of system design," 1975, pp. 224–230. https://doi.org/10.1007/3-540-07142-3_87.
2. Eugenia Cheng, *How to Bake Pi*. Basic Books, 2015.
3. D. Dori, *Model-based systems engineering with OPM and SysML*. 2016. https://doi.org/10.1007/978-1-4939-3295-5.
4. David Jaz Myers, "Categorical Systems Theory," *http://davidjaz.com/Papers/DynamicalBook.pdf*, 2022.
5. Bart Jacobs, "Structured Probabilistic Reasoning," *http://www.cs.ru.nl/B.Jacobs/PAPERS/ProbabilisticReasoning.pdf*.
6. AlgebraicJulia, "AlgebraicJulia Blog," *https://www.algebraicjulia.org/blog*.
7. J. C. Baez, F. Genovese, J. Master, and M. Shulman, "Categories of Nets," in *Proceedings - Symposium on Logic in Computer Science*, 2021, vol. 2021-June. https://doi.org/10.1109/LICS52264.2021.9470566.
8. B. Fong, A. Speranzon, and D. I. Spivak, "Temporal Landscapes: A Graphical Temporal Logic for Reasoning," Apr. 2019.

43 Category Theory

9. G. S. H. Cruttwell, B. Gavranović, N. Ghani, P. Wilson, and F. Zanasi, "Categorical Foundations of Gradient-Based Learning," 2022. https://doi.org/10.1007/978-3-030-99336-8_1.

10. L. McInnes, J. Healy, and J. Melville, "UMAP: Uniform Manifold Approximation and Projection for Dimension Reduction," Feb. 2018.

11. P. Selinger, "A Survey of Graphical Languages for Monoidal Categories," 2010, pp. 289–355. https://doi.org/10.1007/978-3-642-12821-9_4.

12. nLab, "Coherence theorem for monoidal categories," *https://ncatlab.org/nlab/show/coherence+theorem+for+monoidal+categories*.

13. M. Stay and J. Vicary, "Bicategorical Semantics for Nondeterministic Computation," *Electronic Notes in Theoretical Computer Science*, vol. 298, pp. 367–382, Nov. 2013, https://doi.org/10.1016/j.entcs.2013.09.022.

14. Haskell.org, "Haskell," *https://www.haskell.org/*.

15. Wikipedia, "Comparison of functional programming languages," *https://en.wikipedia.org/wiki/Comparison_of_functional_programming_languages*.

16. Bartosz Milewski, "Category Theory for Programmers," *https://bartoszmilewski.com/2014/10/28/category-theory-for-programmers-the-preface/*.

17. P. Schultz, D. I. Spivak, C. Vasilakopoulou, and R. Wisnesky, "Algebraic databases," *Theory Appl. Categ.*, vol. 32, pp. 547–619, Paper No. 16, 2017.

18. AlgebraicJulia, "Catlab.jl," *https://www.algebraicjulia.org/*.

19. Robert Ghrist, *Elementary Applied Topology*. CreateSpace, 2014.

20. G. S. H. Cruttwell, B. Gavranović, N. Ghani, P. Wilson, and F. Zanasi, "Categorical Foundations of Gradient-Based Learning," Mar. 2021.

21. B. Coecke, M. Sadrzadeh, and S. Clark, "Mathematical Foundations for a Compositional Distributional Model of Meaning," Mar. 2010.

22. B. Coecke and A. Kissinger, *Picturing quantum processes*. Cambridge University Press, 2017.

23. J. Baez and M. Stay, "Physics, topology, logic and computation: A Rosetta Stone," *Lecture Notes in Physics*, vol. 813, 2011, https://doi.org/10.1007/978-3-642-12821-9_2.

24. J. C. Baez and K. Courser, "Structured Cospans," Nov. 2019.

25. B. Fong, "The Algebra of Open and Interconnected Systems," Sep. 2016.

26. J. C. Baez and K. Courser, "Coarse-Graining Open Markov Processes," Oct. 2017.

27. B. Pollard, "Open Markov Processes: A Compositional Perspective on Non-Equilibrium Steady States in Biology," *Entropy*, vol. 18, no. 4, p. 140, Apr. 2016, https://doi.org/10.3390/e18040140.

28. J. C. Baez and B. S. Pollard, "A compositional framework for reaction networks," *Reviews in Mathematical Physics*, vol. 29, no. 09, p. 1750028, Oct. 2017, https://doi.org/10.1142/S0129055X17500283.

29. J. C. Baez, B. S. Pollard, J. Lorand, and M. Sarazola, "Biochemical Coupling Through Emergent Conservation Laws," Jun. 2018.

30. J. C. Baez and J. Master, "Open Petri nets," *Mathematical Structures in Computer Science*, vol. 30, no. 3, pp. 314–341, Mar. 2020, https://doi.org/10.1017/S0960129520000043.

31. J. C. Baez and J. Erbele, "Categories in Control," May 2014.

32. S. Libkind, A. Baas, M. Halter, E. Patterson, and J. Fairbanks, "An Algebraic Framework for Structured Epidemic Modeling," Feb. 2022.

33. Pawel Sobocinski, "Graphical Linear Algebra," *https://graphicallinearalgebra.net/*.

34. F. Bonchi, D. Pavlovic, and P. Sobocinski, "Functorial Semantics for Relational Theories," Nov. 2017.

35. E. Patterson, "Knowledge Representation in Bicategories of Relations," Jun. 2017.

36. T. Fritz and P. Perrone, "Bimonoidal Structure of Probability Monads," *Electronic Notes in Theoretical Computer Science*, vol. 341, pp. 121–149, Dec. 2018, https://doi.org/10.1016/j.entcs.2018.11.007.

37. T. Fritz, T. Gonda, P. Perrone, and E. F. Rischel, "Representable Markov Categories and Comparison of Statistical Experiments in Categorical Probability," Oct. 2020.

38. D. I. Spivak, "The operad of wiring diagrams: Formalizing a graphical language for databases, recursion, and plug-and-play circuits," *arXiv preprint https://arxiv.org/abs/1305.0297arXiv:1305.0297*, 2013.

39. J.-H. Kim, M. S. H. Oh, and K.-Y. Kim, "Boosting vector calculus with the graphical notation," *American Journal of Physics*, vol. 89, no. 2, pp. 200–209, Feb. 2021, https://doi.org/10.1119/10.0002142.

40. E. di Lavore, A. Gianola, M. Román, N. Sabadini, and P. Sobociński, "A Canonical Algebra of Open Transition Systems," 2021, pp. 63–81. https://doi.org/10.1007/978-3-030-90636-8_4.

41. D. Pavlovic, "Monoidal computer I: Basic computability by string diagrams," Aug. 2012.

42. D. Pavlovic, "Monoidal computer II: Normal complexity by string diagrams," Feb. 2014.

43. D. Pavlovic and M. Yahia, "Monoidal computer III: A coalgebraic view of computability and complexity," Apr. 2017.

44. A. Censi, "A Mathematical Theory of Co-Design," Dec. 2015.

45. A. Censi, "A Class of Co-Design Problems With Cyclic Constraints and Their Solution," *IEEE Robotics and Automation Letters*, vol. 2, no. 1, pp. 96–103, Jan. 2017, https://doi.org/10.1109/LRA.2016.2535127.

46. G. Zardini, A. Censi, and E. Frazzoli, "Co-Design of Autonomous Systems: From Hardware Selection to Control Synthesis," in *2021 European Control Conference (ECC)*, Jun. 2021, pp. 682–689. https://doi.org/10.23919/ECC54610.2021.9654960.

47. G. Zardini, N. Lanzetti, M. Salazar, A. Censi, E. Frazzoli, and M. Pavone, "On the Co-Design of AV-Enabled Mobility Systems," in *2020 IEEE 23rd International Conference on Intelligent Transportation Systems (ITSC)*, Sep. 2020, pp. 1–8. https://doi.org/10.1109/ITSC45102.2020.9294499.

48. S. Breiner, O. Marie-Rose, B. Pollard, and E. Subrahmanian, "Modeling Hierarchical Systems with Operads," in *Applied Category Theory 2019*, 2020.

49. J. D. Foley, S. Breiner, E. Subrahmanian, and J. M. Dusel, "Operads for complex system design specification, analysis and synthesis," *Proceedings of the Royal Society A: Mathematical, Physical and Engineering Sciences*, vol. 477, no. 2250, p. 20210099, Jun. 2021, https://doi.org/10.1098/rspa.2021.0099.

50. S. Libkind, A. Baas, E. Patterson, and J. Fairbanks, "Operadic Modeling of Dynamical Systems: Mathematics and Computation," May 2021.

51. A. D. Ames, P. Tabuada, and S. Sastry, "On the Stability of Zeno Equilibria," 2006, pp. 34–48. https://doi.org/10.1007/11730637_6.

52. ICRA2021, "Compositional Robotics: Mathematics and Tools," *https://idsc.ethz.ch/research-frazzoli/workshops/compositional robotics.html*, 2021.

53. F. Abou-Saleh, J. Cheney, J. Gibbons, J. McKinna, and P. Stevens, "Introduction to Bidirectional Transformations," 2018, pp. 1–28. https://doi.org/10.1007/978-3-319-79108-1_1.

54. Mario Román, "Composing Optics," 2020.

55. Eugenia Cheng, "Category Theory in Life," *https://www.youtube.com/watch?v=ho7oagHeqNc*, 2017.

56. T.-D. Bradley, "What is Applied Category Theory?," Sep. 2018.

57. D. I. Spivak, *Category theory for the sciences*. MIT Press, Cambridge, MA, 2014.

58. P. Perrone, "Notes on Category Theory with examples from basic mathematics," Dec. 2019.

59. B. Fong and D. I. Spivak, "Seven Sketches in Compositionality: An Invitation to Applied Category Theory," Mar. 2018.

60. Multiple authors, "The n-Category Cafe," *https://golem.ph.utexas.edu/*.

61. Tae-Danae Bradley, "Math3ma," *https://www.math3ma.com/*.

62. Bartosz Milewski, "Category Theory," *https://www.youtube.com/user/DrBartosz/playlists*.

63. David I. Spivak and Brendan Fong, "Applied Category Theory," *https://ocw.mit.edu/courses/18-s097-applied-category-theory-january-iap-2019/pages/lecture-videos-and-readings/*.

64. Andrea Censi, Jonathan Lorand, and Gioele Zardini, "Applied Compositional Thinking for Engineers," *https://applied-compositional-thinking.engineering/*.

65. D. I. Spivak and J. Vicary, Eds., "Applied Category Theory 2020," in *https://act2020.mit.edu/*.

66. K. Kishida, Ed., "Applied Category Theory 2021," in *https://www.cl.cam.ac.uk/events/act2021/*, 2021.

67. Conference recording, "Applied Category Theory," *https://www.youtube.com/channel/UC1Kxtc6DOexi4JT-t57Ey9g/playlists*.

68. "The Topos Institute Colloquium," *https://topos.site/topos-colloquium/*.

43 Category Theory

69. "Topos Institute YouTube Playlists," *https://www.youtube.com/c/ToposInstitute/playlists*.
70. Ryan Wisnesky and David I. Spivak, "Categorical Databases," *https://www.categoricaldata.net/*.
71. et al. Jamie Vicary, "Homotopy.io," *https://homotopy.io/*.
72. S. Breiner, B. Pollard, and E. Subrahmanian, "Workshop on applied category theory:," Gaithersburg, MD, Feb. 2020. https://doi.org/10.6028/NIST.SP.1249.
73. J. C. Baez, K. Courser, and C. Vasilakopoulou, "Structured versus Decorated Cospans," Jan. 2021.

Dr. Spencer Breiner is a mathematician at the US National Institute for Standards and Technology, working in the Software & Systems Division of the Information Technology Lab. His research focuses on applications of category theory to problems in systems modeling and interoperability. Dr. Breiner received his Ph.D. from Carnegie Mellon University in 2013 before joining NIST in 2015.

Dr. Eswaran Subrahmanian is a research professor at the Engineering Research Accelerator and Engineering and Public Policy at Carnegie Mellon University. He is also an associate at the National Institute of Standards and Technology. He has held visiting professorships at the Faculty of Technology and Policy Management at TU-Delft (Netherlands), the University of Lyon II, and the International Institute of Information Technology, Bangalore. His research is in the areas of socio-technical systems design, decision support systems, engineering informatics, design theory and methods, and engineering design education. He has worked on designing design processes and collaborative work support systems for Westinghouse, ABB, Alcoa, Bombardier, Boeing, and Robert Bosch. He has been a consultant to a number of organizations, including ABB, Bosch, and Lytix, and is a co-founder of a Bangalore-based non-profit simulation and gaming startup called Fields of View. He has published extensively in several disciplines; co-edited three books on Empirical Studies in Engineering Design, Knowledge Management, and Design Engineering; and co-authored a book on ICT for Development. He is the co-author of the book 'We are not Users: Dialogues, Diversity, and Design,' MIT Press. He was awarded the Steven Fenves Award for contributions to Systems Engineering at CMU. He is a Distinguished Scientist of the ACM and a Fellow of the American Association of Advancement of Science.

Dr. D. Sriram is currently a division chief of the Software and Systems Division with the National Institute of Standards and Technology, Gaithersburg, MD, USA. Prior to joining NIST, he was on the engineering faculty (1986–1994) at the Massachusetts Institute of Technology (MIT) and was instrumental in setting up the Intelligent Engineering Systems Laboratory. He is a distinguished alumnus of the Indian Institute of Technology and Carnegie Mellon University, Pittsburgh, PA, USA. He has co-authored or authored nearly 250 papers, reports, and several books. His current research interests include developing knowledge-based expert systems, natural language interfaces, machine learning, object-oriented software development, life-cycle product and process models, geometrical modelers, object-oriented databases for industrial applications, healthcare informatics, bioinformatics, and bioimaging. Dr. Sriram was a recipient of the NSF's Presidential Young Investigator Award in 1989, the ASME Design Automation Award in 2011, the ASME CIE Distinguished Service Award in 2014, and the Washington Academy of Sciences' Distinguished Career in Engineering Sciences Award in 2015. Sriram is a Fellow of the American Society of Mechanical Engineers (ASME), the American Association for the Advancement of Science (AAAS), the Institute of Electrical and Electronics Engineers (IEEE), the Solid Modeling Association (SMA), International Council on Systems Engineering (INCOSE), and the Washington Academy of Sciences (WAS). He is also a Distinguished Member (life) of the Association for Computing Machinery (ACM) and a Senior Member (life) of the Association for the Advancement of Artificial Intelligence (AAAI). He is the President-Elect of the Washington Academy of Sciences and federal representative to ONC's Health IT Advisory Committee.

Perspectives on SE, MBSE, and Digital Engineering: Road to a Digital Enterprise

44

H. Stoewer

Contents

Introducing the Digital Enterprise (DE)	1302
Historical Perspective on Systems Engineering (SE)	1304
The Broader Context of Today's Systems	1304
Implications of a Circular Economy on SE	1305
State of the Practice	1305
System Boundaries	1306
End-to-End Approach and SE Focus	1306
System Architecting	1308
Assumption and Sensitivity Analyses	1309
SE and Project Management Overlap	1309
Engineering Simplicity	1310
System Engineering Failures	1311
Risk Management	1312
Model-Based Systems Engineering	1313
Concurrent System Design Facilities	1314
Lessons from CAD/CAE	1316
Current SE Issues Affecting MBSE	1318
From Fixed to Living Systems	1318
Project Handovers	1319
Models Needed for MBSE	1320
Relationship of MBSE to MBE	1321
US DoD Digital Engineering Strategy	1322
The MBSE Landscape	1324
Implications of the Emerging Digital World for SE	1325
The DT Building Block for SE and Digital Enterprises	1325
The Advancing Simulation and "Virtualization" World	1328

H. Stoewer (✉)
TU Delft, Delft, Netherlands
e-mail: heinzstoewer@spaceassociates.net

© Springer Nature Switzerland AG 2023
A. M. Madni et al. (eds.), *Handbook of Model-Based Systems Engineering*,
https://doi.org/10.1007/978-3-030-93582-5_83

1301

Best Practice Approach for Combining MBSE and DE	1331
Toward a System-Driven Digital Enterprise (DE)	1331
Interrelationships Between SE Tools and Their DE Contributions	1332
MBSE Can Help Shape the Digital Enterprise	1333
How Does MBSE Support Digital Enterprises?	1334
Chapter Summary	1335
References	1336

Abstract

Model-based Systems Engineering (MBSE) is a promising advance in systems engineering. Systems Engineering (SE) has a new mindset and a new set of tools. Models and algorithms partially replace empirical intuition, disconnected models, and stove-piped SE tools. Of course, anyone embarking on MBSE must first become a competent system engineer (i.e., one who understands the fundamentals of SE). Another promising new development is digital engineering. The advent of digital engineering and its potential synergy with MBSE is viewed as potential enablers of the Digital Enterprise (DE). This chapter presents key perspectives on SE, MBSE, and digital engineering and how the synergy between it, other tools, and MBSE can contribute to realizing the Digital Enterprise (DE).

Keywords

Systems engineering · Digital engineering · Model-based systems engineering · Systems architecture · Digital Twin (DT)

Introducing the Digital Enterprise (DE)

This section presents key perspectives on Systems Engineering (SE), Model-Based Systems Engineering (MBSE), and Digital Engineering. Digital Enterprises (DEs) are viewed differently by different people. To some, they are "uncontrollable monsters" endangering employment. To others, they are "simply digital" organizations. Today, most enterprises are partially digital, but very few are entirely digital. Start-ups, for example, are often digital from day 1 with virtually nonexistent paper documents. Whether an enterprise can survive without digitalization is similar to asking whether "horse-based transport" enterprises could survive without automobiles. Non- DEs cannot survive the continued growth in system interconnections, interactions, and interdependence.

There is uncertainty when asking which functions of an enterprise (Fig. 1) could or should become digital. While the functions in Fig. 1 depend on human involvement, digitization can significantly enhance each one's efficiency and effectiveness. The appropriate question is not which enterprise function should become digital but which cannot be digitized or fully digitized.

A suitable IT infrastructure is a prerequisite for the functioning of a digital enterprise. DE will need an inventory of tools to address the various functions of the enterprise in ways allowing seamless interconnection and data exchange between

Fig. 1 Generic enterprise and its functional organizational units

multiple models, tools, and data [1]. A DE function needs individualized tools compatible with staff competencies, the function, and its interoperability with other functions. Moreover, data must allow aggregation for higher-level analyses, including strategic decision-making. Creating a DE necessitates systems thinking [2, 3, 4], which is a holistic approach that examines how a system's components interoperate over time within the context of the larger system.

The construction of a DE begins by establishing its goals and its vision. Common goals include: increasing collaboration, increasing agility, improving service, increasing enterprise efficiency, process optimization, improving data security, and as an overall objective retaining or gaining a competitive advantage. The vision should reflect customer needs and an understanding of how emerging technologies apply. For example, an enterprise might aspire to become the first to enable its employees, customers, and suppliers to be digitally connected enabling reduced product delivery times while not incurring unnecessary purchasing or manufacturing costs. Visions usually span 5–15 (or more) years.

Strategic plans become the blueprint for operational plans and DE implementation. Strategic plans outline the actions needed to achieve the goals and enterprise vision. Strategic plan timescales typically cover 3–6 years while operational plans span 1–2 years. According to McKinsey & Company (http://dln.jaipuria.ac.in:8080/jspui/bitstream/123456789/2218/1/Six%20building%20blocks%20for%20creating%20a%20high-performing%20digital%20enterprise.pdf), there are six essential components of a DE strategic plan:

- Strategy and innovation: Focuses on future value
- Customer decision journey: Analyses that examine how and why customers make decisions
- Process automation: Reinventing processes and how customers interact with the enterprise
- Organization: Changes that increase enterprise agility, collaboration, and capabilities
- Technology: Needed IT in support of core functions
- Data and Analytics: Identifying essential metrics and the means for their capturing and analysis

Historical Perspective on Systems Engineering (SE)

The advent of SE is often associated with the early missile and the Apollo programs in the 1950s and 1960s. While undoubtedly true in the narrow sense of formalized SE definitions and documentation, SE has been practiced for centuries in the broader sense. The complicated constructions of ancient architectural marvels, such as the Pyramids, the Roman Kolosseum, water supply systems traversing hundreds of kilometers, the impressive Mayan cities, medieval sailing ships, and early industrial factories employing steam engines are all examples of creative SE. (This chapter uses the term, "complex" in reference to emergent system behaviors, i.e., unanticipated behaviors resulting from interaction of system components and its environment that are not explicitly designed. The term "complicated" applies to systems consisting of large numbers of interacting components that make analyses difficult.) Furthermore, early aircraft, rail systems, and automobiles were all created before SE was formalized. However, these developments would not have been possible without systems thinking and system engineering practices.

Since the 1960s, slide rules and pocket calculators have given way to computers, heavily interlinked networks, and cloud-based computing [5, 6]. Today, SE increasingly benefits from applying mathematics, physics, AI, machine learning, IT, and many other disciplines. At the same time, SE computing support tools continue to advance rapidly.

The complexity of modern systems has continually increased and skyrocketed over the past two decades. Interfaces and interdependencies between system elements have become crucial SE drivers. Unfortunately, our understanding and ability to deal with complexity have not kept pace with this development.

SE has become increasingly popular and a quasi-formalized discipline during the past decade. Its focus has evolved from mechanical and electrical systems to software-intensive mechatronic systems. At the same time, software-intensive systems have evolved into network-intensive systems, cyber-physical systems, systems of systems, and enterprise systems. This evolution has been accompanied by increasingly uncertain operational environments, the advent of non-deterministic systems, and increased risk.

The Broader Context of Today's Systems

Today's complicated and complex systems increasingly need to respond to societal and environmental factors. The United Nation's long-term sustainability goals for humankind (https://www.un.org/sustainabledevelopment/development-agenda/) are an excellent example of the multifaceted challenges (e.g., technical, economic, social) facing systems engineering. The complicated interconnected infrastructure and industrial products of the twenty-first-century demand "smart" solutions, which invariably require extensive transdisciplinary cooperation [7, 8, 9]. One conclusion drawn from the UN goals is that fundamental human needs translate into societal needs, which must be satisfied by policy and system solutions.

Fig. 2 Cruise ships and automobiles, examples of substantially reduced development times and technical progress

Systems engineering today goes beyond "faster, better, cheaper," to include agile, resilient, and secure. Today this imperative is omnipresent in industries worldwide and, to some extent in government organizations. With the help of SE, even highly complicated systems such as cruise ships (whose predecessors have sometimes taken decades to build) are built within a couple of years. Today the automotive industry is producing new models at breathtaking speeds (Fig. 2).

Implications of a Circular Economy on SE

A global trend profoundly influencing how we conceive future systems, use and re-use resources, maintain/update/upgrade existing systems, and more is called the "Circular Economy." The Circular Economy concept recognizes that global material resources are limited and that our "linear throw-away" economy is unsustainable. The implications of a circular economy are profound for SE in that it adds significantly to systems thinking considerations. SE must respond to circular economy challenges by introducing new technologies and holistic solutions emphasizing reuse rather than disposal.

State of the Practice

This subsection discusses key SE fundamentals to orient aspiring system engineers (SEs) and provide a bridge to MBSE. At the outset, a couple of prerequisites to SE and MBSE are important:

- An aspiring system engineer first needs to become a good engineer
- Before embarking on MBSE, a system engineer must master the fundamentals of SE.

In a world of ever-growing complexity, system engineers must understand the basics of SE and strive for simplicity, some call it elegance [10], in both processes and solutions. Often simple solutions prove to be the most valuable. Furthermore, simple processes have the added advantage of being more readily understood, changed, and adopted within teams. While "real-world" SE can be complicated or even complex, system engineers benefit from a sound understanding of essential SE fundamentals. With such understanding, SEs can more easily maintain perspective and adhere to structure and simplicity in the face of complexity. After all, "seeing the forest from the trees" is an essential competency of SEs.

System Boundaries

An SE needs a good understanding of their system boundaries (domain), the context (environment) in which their system exists, and the form of interactions necessary within and with its environment. There are three common forms of interaction:

- Compatible: Systems or system components do not interfere with each other's functionality. There are no assumptions about exchanging data or services.
- Interoperable: Systems can function independently but lose functionality if external connections are unavailable.
- Integrated: Often applies to subsystems within a system in which components depend on each other and the system cannot perform its intended function unless all components are present and operational.

Importantly, these interactions are not mutually exclusive within a system and may change with time as system boundaries and definitions change.

End-to-End Approach and SE Focus

The SE process spans the lifecycle of a product. It contributes to and enables successive incremental decisions as the system matures. SE iteratively addresses systems analysis, architecting, design, implementation, testing, manufacturing, deployment, operation, upgrade, and retirement. The NASA Systems Engineering Handbook [11] describes a typical SE lifecycle comprising:

- Pre-Formulation (Pre-Phase A):
  - Purpose: concept studies that explore system goals, needed capabilities, the concept of operation, scope, constraints, draft system-level requirements, technology needs, and assess performance, cost, and schedule feasibility
  - Outcome: Feasible system concepts in the form of simulations, analyses, study reports, models, and mock-ups
  - Review: Mission Concept Review

- Formulation
  - Phase A (Concept and Technology Development)
    - Purpose: Determines the feasibility and desirability of the proposed system, develops finalized mission concepts, system-level requirements, needed technology development, and program and technical management plans
    - Outcome: System concept definition in the form of simulations, analyses, models, mock-ups, and needed trade studies
    - SE Role: Supports the development of mission concepts, requirements, technology needs, and technical management plan
    - Review: System Requirements Review and Mission Definition Review
  - Phase B (Preliminary Design and Technology Completion)
    - Purpose: Defines the system in sufficient detail to establish an initial baseline capable of achieving mission goals. Develop system structure and end product requirements and preliminary design
    - Outcome: End products in the form of mock-ups, trade studies, specifications, interface documents, and prototypes
    - SE Role: Baseline design, trade studies, specifications, and interface documents
    - Review: Preliminary Design Review
- Implementation
  - Phase C: Final Design and Fabrication
    - Purpose: Complete detailed design of systems and subsystems, fabricate hardware and code software, develop verification and validation plans
    - Outcome: End product design details, end product component fabrication, and software development
    - SE Role: Supports development of system-level validation and integration and test plans
    - Review: Critical Design Review, System Integration Review
  - Phase D: System Assembly, Integration, Test, and Deployment
    - Purpose: Assemble, integrate, and test hardware and software in the expected operational environment
    - Outcome: Operations-ready system with supporting and enabling products
    - SE Role: Review and approve system validation results. Support readiness and post-deployment reviews
    - Review: Operational Readiness Review, Flight Readiness Review, Post-Deployment Assessment Review
  - Phase E: Operations and Sustainment
    - Purpose: Conduct the mission, validate performance and behavior, and when necessary and possible, make system updates
    - Outcome: Desired operational system
    - SE Role: Support reviews and anomaly resolution as needed
    - Review: Critical Events Readiness Review, Post-Flight Assessment Review, Decommissioning Review
  - Phase F: Closeout

Purpose: System decommissioning/disposal
Outcome: System Closeout
SE Role: Support closeout operations and reviews
Review: Disposal Readiness Reviews

System Architecting

System architecting (SA) is the foundational discipline for developing system concepts and realizing systems with the required quality attributes. SA is a component of SE that begins in the pre-formulation phase and continues into the formulation phase. SA determines component allocation levels and validates derived requirements.

The system architecture describes logical functions and relationships in pre-formation and early formulation phases. A logical architecture usually defines functions and functional relationships without physical component allocations. References to physical features may be present in logical architectures when the architecture includes legacy or customer-provided components. Physical or "concrete" architectures are developed in the latter formulation in which functions and functional interfaces are allocated to physical components. Physical architectures are decomposed into lower-level components consistent with the requirement decomposition level. Figure 3 shows how SA fits into the lifecycle and the depth of analysis it generally reaches.

SA occasionally makes use of architecture frameworks. An architecture framework is a metamodel-like construct consisting of generalized hierarchies,

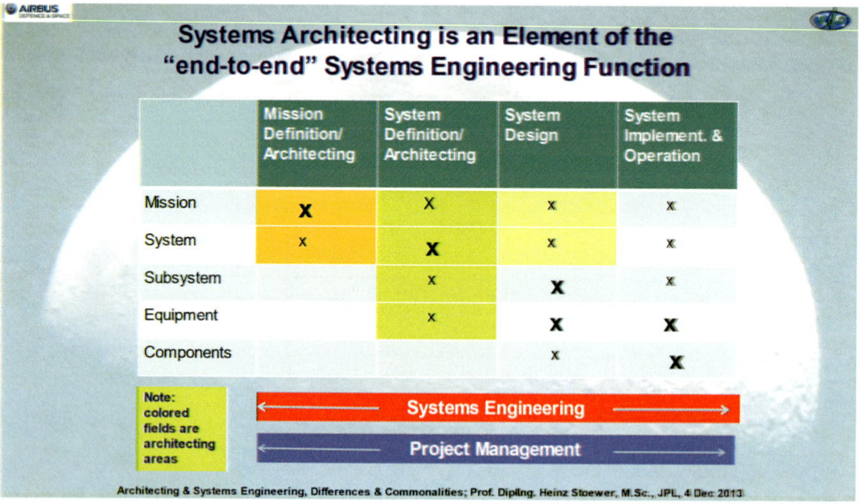

Fig. 3 Systems architecting in the SE lifecycle

components, and relationships [12]. Frameworks facilitate assessing and analyzing large-scale systems. A key benefit of architecture frameworks is that they promote an understanding of system constraints and implementation options at the beginning of a project. Frameworks can also define usage scenarios, reference use cases, and general technical, business, and schedule options.

Assumption and Sensitivity Analyses

As we progress through the lifecycle, we sometimes forget that the basis of our conceptual design are initial assumptions about technology readiness, vendor availability, the cost of procuring a component, delays due to sending or hosting data in the cloud, and so forth. We make assumptions for parameters we cannot fully describe or understand in the early system development and definition phases. System engineers must systematically record their early assumptions and periodically revisit them, verifying whether they are still valid or need revision.

Another practice is occasionally conducting sensitivity analyses of a design and programmatic assumptions. Sensitivity analysis entails varying one or more system parameters, such as data rates, mass, or battery lifetime, and assessing the effects of such variations on the preferred solution. These analyses may lead a system engineer to a design or technology change and provide feedback on the robustness of a solution.

SE and Project Management Overlap

SE and program management (PM) disciplines are complementary and intertwined. While each may have different educational paths and professional societies, they are highly interdependent. Both also depend on other engineering, business, marketing and sales, and manufacturing disciplines. The interaction between SE and PM starts during the establishment of initial system concepts (Fig. 4).

Examples of issues affecting SE and PM include:

- Who translates stakeholder needs into requirements and specifications?
- Who conducts early architecture trade-offs?
- Who negotiates with suppliers and accepts their inputs?
- Who identifies schedule estimates and on which design or definition baseline?
- Who drives the selection of the most suitable technologies?
- Who ensures proper processes are in place to report test campaign progress and performance achievements?
- Who identifies risks and evaluates mitigation strategies?

Although SE and PM have their primary competence and authority, most of the tasks in Fig. 4 need close cooperation. With SE advice and recommendations, PM usually makes the final decisions for a project. The PM also reports directly to top

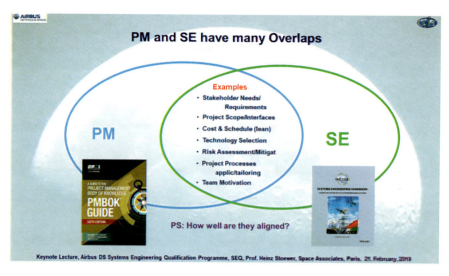

Fig. 4 PM and SE overlap

management. Conversely, the SE and Product Assurance leaders report project and quality engineering status to their line management.

Often the question arises regarding whether MBSE could support SE-communication. MBSE can be helpful if it creates better project transparency and greater consistency across engineering specialties, development, and production. MBSE and a compatible project lifecycle management (PLM) tool can simplify SE and PM interactions and help arrive at a deeper understanding of relevant product realization issues.

Engineering Simplicity

A good system engineer must strive for simple solutions, however hard they may be to come by. Engineering simplicity is not minimalism. Minimalist elegance thrives on simplicity, while engineering elegance thrives on complexity by simplifying, hiding, and explaining complexity. Paraphrasing Albert Einstein: "Everything should be made as simple as possible, but not simpler." (Einsteins actual quote: "It can scarcely be denied that the supreme goal of all theory is to make the irreducible basic elements as simple and as few as possible without having to surrender the adequate representation of a single datum of experience.")

The story of Jon illustrates the point of simplicity (Fig. 5). Jon entered a competition organized by a marine outboard engine company. Entrants were each provided an identical outboard motor and nothing else. The challenge was to be the first to cross a river. Some entrants raced to nearby hardware stores to buy wooden planks, nails, and more to build a floating device. Others rushed to the woods to come back with trunks of trees to build a float for their motor. Jon went to the judges

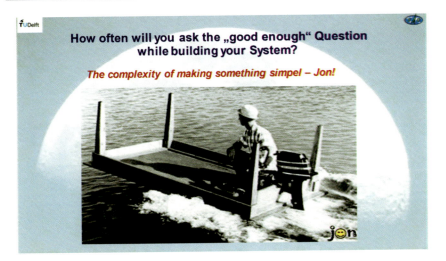

Fig. 5 Story of Jon

and asked them politely to remove their papers and prizes, grabbed their table, and went off across the river before the other entrants even had a chance to design their floating devices. The moral of this story: simple solutions can sometimes suffice.

Another example is called the "luxury of numbering games." When setting up new projects, PMs and SEs generally create a work breakdown structure for organizing the project's tasks. PMs, SEs, and subsystem product managers sometimes make a product breakdown structure that identifies the end item deliveries. PMs may also develop organizational schemes for their teams that show internal and external relationships. Additionally, financial managers develop a financial reporting and accounting scheme.

Each of these organizational structures often have a different numbering system. These numbering systems are diligently updated, configuration controlled as a project proceeds, and changes are made throughout the project lifecycle. There are often additional numbering systems for administrative and budgetary purposes, procurements, materials bills, and more. In reality, the different numbering systems create substantial non-value added efforts in correlating tasks and responsibilities, controlling and accounting comparisons, and aligning change processes. Having one project structure and one numbering system for all functions and tasks would be much more efficient.

System Engineering Failures

Modern industrial systems have reached a level of complexity, making SE failure more likely. A system SE failure occurs when a product fails to meet its goals or behaves incorrectly and possibly dangerously due to errors and oversights committed during SE processes. A few examples of SE failure causes include:

- Errors in assumptions regarding performance and availability of new technology
- Lack of communication within and external to the SE team
- Missing or incorrect requirements
- Missing or incorrect use cases
- Poor understanding of the operational environment
- Errors and omissions in interface specifications
- Errors in legacy use
- Poor understanding of how the system will be used and controlled
- Incorrect performance and fault analyses
- Inattention to detail
- Malicious actors during design and in operation
- Ignoring fault cases deemed "not credible"
- Poor project planning and misjudgment of cost and schedule
- Poorly planned verification and validation
- Lack of top-level management support

The theoretical basis for most of the above causes of SE failure is incomplete, inaccurate, or unreliable. A key reason is that the disciplines involved rely partly on anecdotal and statistical data. The point is that systems engineering relies as much as possible on physics and mathematics-based engineering methods but also on qualitative methods which are dependent on partial theories, empirical databases, and non-calculable, so-called "soft" parameters. Fundamentally, SE is failure-prone because it needs to "marry" hard and "soft" disciplines in meaningful ways. The way to deal with the soft factors is by acknowledging uncertainties in certain variables, assigning technical and programmatic margins, assessing and mitigating risk, and incorporating reserves for project development.

Risk Management

The objective of risk management is risk avoidance. A vital system engineering task is the timely identification and mitigation of risks. There are risks in requirements, technology selection, design, supplier agreements, business models, integration and test, validation and verification, production, and maintenance plans. Risk management should occur sufficiently early to implement risk circumvention or mitigation measures.

Standard industry practice creates special teams that implement risk management using traffic light indicators or risk matrices. Importantly risk management frequently fails without analyses that anticipate and prevent risks.

Attempting to uncover potential risks in early designs is inherently difficult because it requires a deep understanding of system design and functionality and a mindset focused on potential problems. Risk assessment is quantitative where possible and often somewhat qualitative, based on intuition. While not entirely ignoring low risks, risk management should focus on high likelihood and high consequence risks rather than trying to address all risks. An important aspect of

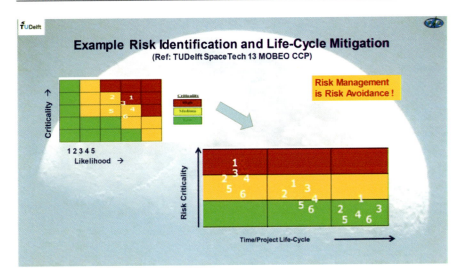

Fig. 6 Risk matrix example

risk management is maintaining constant awareness of priority risks and mitigations at all project levels.

Figure 6 shows risks inserted into a risk matrix (upper left) and how risk levels are reduced from red to orange to green during a project (lower right). The figure also alludes to management reserves needed at any given project phase. For example, if risk is low and stays low, fewer resources are required than if risks remain high toward at later stages of the project.

Model-Based Systems Engineering

MBSE is SE supported by engineering models and computer tools. MBSE is normal and potentially good SE, employing modern databases, tools, and IT. MBSE is also a transition from document-based to digital SE. Significantly, languages such as SyML are not, by themselves, MBSE but rather provide a means for naming and combining SE entities and relationships just as the parts of a natural language are not a manuscript.

First, models are representations of reality that approximate existing or anticipated reality. However, models are seldom a perfect representation of reality but become better as a project proceeds. Final details are fleshed out after calibration and validation in real-world testing.

Like most engineers, systems engineers are modelers but do not (yet) have a set of models that address all systems engineering issues. Nor do they have the means for seamless data transfer between heterogeneous tools needed for difficult SE design and analysis. Although some MBSE tool providers claim to support seamless interoperability with external tools, the reality is that interoperability is limited and may only apply to portions of a model. In some cases, it is necessary to export model content to an intermediate file format, e.g., a spreadsheet, to make it available to another tool. However, efforts such as OpenMBEE (https://www.openmbee.org) are actively working to improve tool integration and means for including the participation of non-modeler subject matter experts in the MBSE process.

Today's SE challenges have grown more complex with the increasing complexity of our technologies and societies. SE today must also cope with intangible aspects and soft judgments, which some call elements of art. These include system elements/tasks/functions that are neither physics-based nor mathematics-based but rely partly on human judgment and incomplete databases (e.g., in the business, cost, schedule, or risk analysis fields). Intangibles are based on or derived from engineering baselines, based on human-made assumptions and specifications of uncertainties. Hence model results require interpretation, analogies for comparisons, and experience-based judgments. The combination of engineering intangibles and the lack of a unifying fundamental SE theory currently prevents creating a seamless SE toolset that addresses all SE issues.

However, systems engineering has come a long way, and there is a lot to be gained from models that allow system engineers to acquire otherwise impossible insights. Also, fully integrated MBSE will not come about through a "big bang." Instead, it will progress with incremental improvements in both tools, databases, and skill sets. Findings from these models and tools can replace or augment assumptions about characteristics such as mass, energy, dynamic behavior, and cost estimates. They can also substantiate analytical data and deeper insights into sensitivities of system design parameters to various changes and associated technology maturity assessments. The comparison between the "big bang" MBSE machine and the iterative automated SE approach is shown in Fig. 7.

Concurrent System Design Facilities

There are many examples of early MBSE successes. These have all followed evolutionary paths, often spanning decades. For example, different concurrent system design methods include those employed by Team X at NASA's Jet Propulsion Laboratory (JPL) in Pasadena, California, the European Space Agency's (ESA) Concurrent Design Facility, and various industry System Design Offices (SDOs).

Concurrent design facilities enable design teams to interact with each other quasi-concurrently through quick data exchanges from different disciplines. For example, assessing thermal impacts as a function of different energy generation means or the impact on a car's architecture and components as a function of a new requirement for a shorter braking distance.

These facilities are possible because their underlying data are brought together in one place and one repository. Even today, the IT means for facilitating needed interactions are often spreadsheets, PCs, and a host computer.

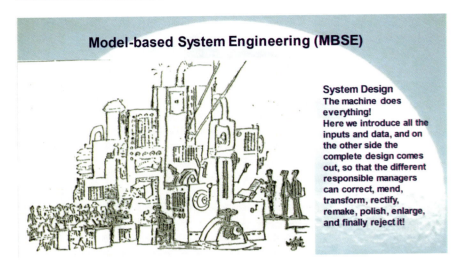

Fig. 7 MBSE approaches

Fig. 8 Early MBSE in a concurrent conceptual design facility

These design facilities and tools have become increasingly more elaborate and comprehensive, replacing spreadsheets with digital tools and providing more automatic, near real-time system assessment and design interactions. Figure 8 shows ESA's Concurrent Design Facility.

Concurrent design facilities have revolutionized early space system study phases and led to an order of magnitude increase in efficiencies. An early metric from JPL

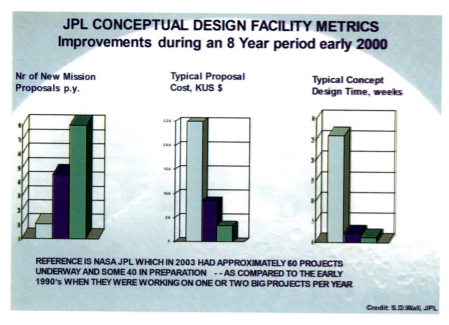

Fig. 9 Conceptual design facility metrics

(Fig. 9) summarizes achievements for the years 1995 through 2003. It shows that in the first seven or so years since the introduction of JPL's Team X conceptual design facility, the number of study proposals has multiplied by more than 8, while cost per study has come down to less than 15%. Study durations have reduced from some 26 weeks to about 1 week.

Concurrent design facilities have, until now, primarily addressed early phase conceptual developments. However, most engineering disciplines comprising Model Based Engineering (MBE) have adequate tools and data sets in their application domains for conceptual and often for the later project implementation phases. Today, concurrent design facilities are moving toward better integration with MBE tools.

Lessons from CAD/CAE

There are parallels between the development and application of MBSE and early computer-aided design (CAD) and engineering tools (CAE). It has taken decades from initial CAD/CAE ideas to their use in broad industrial applications. Similarly, the foundations of MBSE began in 1990s but only recently has it become somewhat accepted practice in many industries such as aerospace, automotive, defense, telecommunications, marine systems, rail, and manufacturing. There is an expectation that MBSE tools, integration, and sophistication will follow a development path similar to CAD/CAM.

CAD/CAE examples from the aircraft, automobile, and machine building industries demonstrate that the underlying models and toolsets have matured well (Fig. 10). Their capabilities are continuously improving and today support many analyses and design tasks.

JPL's Curiosity and Perseverance Mars rovers and their entry, descent, and landing subsystem designs are an excellent example of a high payoff CAD/CAE application (Fig. 11). Without CAD/CAE, designing the many elements that make

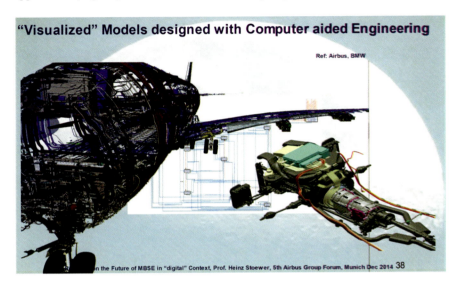

Fig. 10 Computer-aided engineering

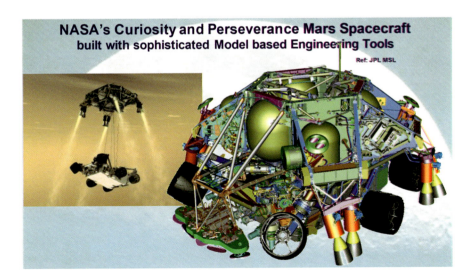

Fig. 11 NASA/JPL Mars Rovers

up the spacecraft (surface rover, descent module, protective shell, parachutes, transfer propulsion, and more) that fit into a confined space with millimeter tolerances under dynamic and thermally stressed conditions would have been impossible. The construction industry has also made substantial progress in digital design tools. Their BIM (Building Information Model) is advanced and widely used by architects and the global building industry.

Current SE Issues Affecting MBSE

Before discussing selected MBSE pilot projects and methodologies, it is helpful to identify specific SE issues that MBSE must address. MBSE cannot just be mechanization or computerization of current SE practices. And, of course, MBSE methods and practices need to be based on sound SE.

INCOSE Vision 2025 (https://www.incose.org/products-and-publications/se-vision-2025) notes three areas of SE challenges that impact the future of MBSE:

- Multiple specialized tools for each discipline: although correlations are natural between many engineering disciplines, interoperability between specialty tools do not always exist.
- Proprietary data formats: Many tools, even those used by the same specialty, have proprietary data formats impeding information sharing and interoperability.
- Limited standardization: There are many incompatible frameworks for system representation which inhibits shared understanding as well as tool interoperability. Additionally, weak semantic MBSE tools such as SysML enable great flexibility in defining custom components and relationships. However, that flexibility is often tool-specific, reducing common understanding and tool interoperability.

From Fixed to Living Systems

As we confront the circular economy era and try to capitalize on investments over more extended periods, questions of technology obsolescence arise for end-to-end SE. Ships, defense systems, aircraft, hydrogen production facilities, smart cities, and hundreds of similar systems will need life-long maintenance to keep them current, inject new technologies, and upgrade them at various points of their useful life. This commitment requires, amongst others, detailed system knowledge and a systematic change management process, to cover several decades. In some sense, we can say that our systems evolve from a state of fixed to living systems in the future (Fig. 12).

MBSE can surface lifecycle issues and create an environment for addressing them rationally and promptly. With Digital Twins (DTs), MBSE will support simulating upgrades by digital insertion into operational systems well before deployment.

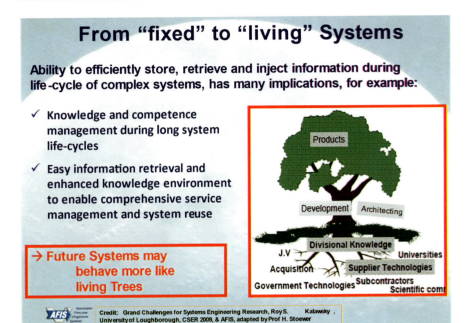

Fig. 12 Dynamic systems evolution

Digital insertion and simulation will save cost while minimizing outages incurred during an upgrade transition.

Carrying complex system designs over into a new, similar, or derivative product becomes substantially easier and allows the re-use of many system artifacts of previous projects. A digital database, or a DT, can facilitate and mitigate life-long system maintenance issues [13]. For example, the International Space Station (ISS), in orbit for more than 20 years, has proven that extending its useful life can be accomplished through anticipating an evolution during the design phase and by systematic preventive maintenance and technology upgrades. Admittedly, this is an exceptional example. But if system evolution is anticipated during early design phases, e.g., for an energy plant, a cruise ship, an aircraft, and if a digital replica supports it, it can be realized faster and with less effort.

Project Handovers

Quite frequently we witness the deployment of different teams to support the acquisition, development, production, or the operation/servicing phases. With hopes of winning a contract, acquisition teams occasionally make promises to customers and/or their management that the design and development teams cannot keep later. Development teams sometimes generate designs that the test and production teams cannot easily, if at all, realize without costly adjustments. The process of

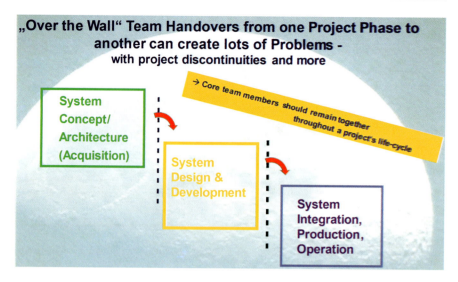

Fig. 13 Project handovers

"distributed" responsibilities and "over the wall" (Fig. 13) handovers is at the root of a many over cost and schedule projects.

MBSE can help with handover problems by creating transition continuity and early transparency. Specifically, MSBE supports preserving all relevant data, e.g., specifications, designs, cost and schedule estimates, and supply chain information in an authoritative (possibly federated) database. These data are then accessible and used by all subsequent teams ensuring effective continuity.

Soliciting all related inputs and integrating them into a coherent project plan and database becomes much easier with inputs from all stakeholders in the early stages of the system lifecycle. This approach, supported by MBSE, is likely to mitigate handover issues, provide transparency throughout the lifecycle and can help to reduce negative hand-over effects. Moreover, MBSE can facilitate the assembly of all staff involved in a project's early and later phases during the conceptual phase, assuring that all stakeholders are present in upfront engineering.

Models Needed for MBSE

As stated earlier, system engineers have always modeled. They currently use multiple models in their day-to-day jobs. These models address, for example, requirements, architecture, design, analysis, and system trade-offs.

Models employed in MBSE need to address the full spectrum of system characteristics, such as physical and functional, stakeholder needs and requirements, hardware and software, test and validation/verification, and support simulations of the entire system, or at least major parts of it. Steve Prusha et al. from JPL sketched the types of models needed for MBSE along with the types of questions they are

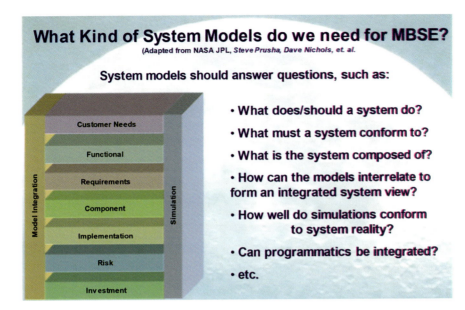

Fig. 14 System models

supposed to answer in Fig. 14. This figure provides a good characterization of the necessary physics-based models for MBSE.

This list is by no means comprehensive or conclusive. Additional programmatic, architectural, or visualization/virtualization models, to name just a few softer examples, need to be also included in the list.

Our current systems model inventory is fragmented. While tool providers are growing in numbers, their tool suites do not yet span and entirely seamlessly address MBSE needs. Therefore, many user companies are creating proprietary system interface layers and/or model entry portals.

MBSE is about SE connections and interdependencies as shown in Fig. 15. The figure shows several MBSE parameters and the need for connectivity with discipline-specific models. It also shows that MBSE can be embedded within other digital enterprise functions and models.

Another MBSE concern is trust in the models and cybersecurity associated with the supporting IT infrastructure. These concerns are prevalent as soon as questions of granting access to tools and data come up, especially when contemplating extensions of model-based techniques throughout the supply chain. In light of the preceding, MBSE can be expected to see a steady stream of iterations and evolution over the coming years.

Relationship of MBSE to MBE

MBSE is not evolving independently. The evolution of Model-based Engineering (MBE) strongly influences MBSE and some view MBSE as part of MBE.

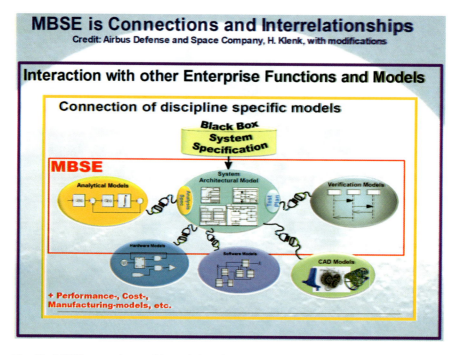

Fig. 15 MBSE connections and interrelationships

MBE and MBSE (Fig. 16) deal with analyses, simulation, and design of solutions to engineering problems. Both are continuously evolving toward better models and simulations. Most engineering disciplines, even more so than MBSE, have sound empirical models, databases, and decades of applications that have proven to represent reality closely.

MBSE depends crucially upon MBE, and vice versa. MBSE needs to integrate constraints, requirements, detail design solutions and more coming from other disciplines, such as thermal or structural analyses, data management schemes, along with risk, schedule, and cost estimations, change and configuration control, customer demands, emerging technologies, and the like into holistic solutions for which much of the analysis and design elements come from MBE.

US DoD Digital Engineering Strategy

Many digital engineering, MBE, and MBSE strategies and plans have been published in recent years [14–16]. Not all are noteworthy because many failed to address important issues. Most attempted to simply expand current methods and practices into the digital world – clearly an inadequate approach that runs the danger of failure. To succeed, one needs to anticipate the future digital world in which

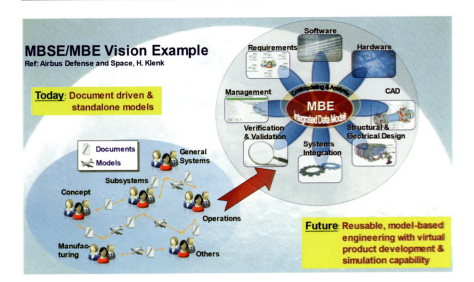

Fig. 16 MBSE vision

business models, enterprises, organizations, and working methods are fundamentally different.

In June 2018, the US DoD released a Digital Engineering Strategy that has the potential to help catapult the DoD and its services into the new MBE and, eventually, MBSE and PLM digital worlds [16] (Fig. 17).

The five goals of this strategy are well-formulated and encompassing. The primary objective is to transform engineering culture and workforce while introducing new methods, models, tools, and practices. Expected benefits are enormous, especially since system development and fielding are likely to become shorter through this initiative. Moreover, decision-making can be based on more dependable information, such as a single authoritative source of truth, which will be more accurate and timelier, provided it is updated in not too long intervals.

Under this engineering transformation, it is anticipated that the DoD acquisition process can also be migrated toward models. Both industry and DoD customer organizations will then discuss models, not physical documents. An MBSE migration will save reams of paper and should increase mutual understanding and collaboration among the parties. Accuracy and comprehension of system proposals and ongoing projects in industry and the customer side are likely to be substantially improved. Continuously updated and adjusted models will represent the project throughout the entire lifecycle of projects. However, well recognized, such a transformation takes time and careful attention to models, simulations, standards, supporting IT, workforce training, and cultural change issues. It will also change how public projects are managed and the roles of customer and contractor teams are defined. A more efficient interaction should be the result.

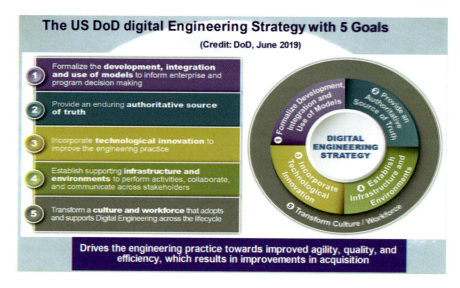

Fig. 17 DoD digital engineering strategy [16]

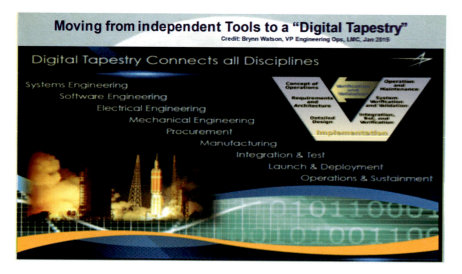

Fig. 18 Digital tapestry

The MBSE Landscape

One of the best illustrations of the future MBSE landscape comes from Brynn Watson of Lockheed Martin (Fig. 18). This figure captures the interconnections and interdependencies of an entire tool/model chain across the lifecycle of a project. She coined the term "digital tapestry" to communicate that all disciplines and attendant tools are part of an overall concept and belong together, i.e., they need to interact/interchange results and data seamlessly.

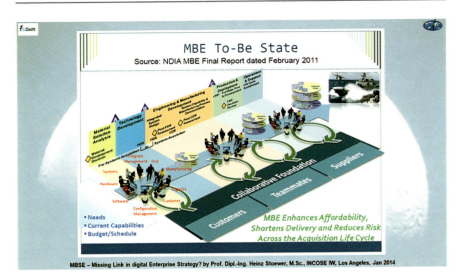

Fig. 19 MBE (and MBSE) "to-be" state

Another excellent visualization of the MBE/MBSE role(s) in an increasingly digital lifecycle scenario has been developed by the USA's National Defense Industries Association (NDIA) (Fig. 19). It depicts the interactive flow of teams, tools, hardware, software, test, configuration control, needs, budgets, and capabilities along a product's lifecycle. It stresses in particular, the need for continuous team interactions along the lifecycle of a (complex) project or product.

Implications of the Emerging Digital World for SE

There is little doubt that a comprehensive digital revolution is in the making today. This revolution impacts engineering and business processes globally. Modeling, simulation, concurrent engineering, manufacturing, and virtualization will advance rapidly everywhere. In particular, the emerging "virtual world" is likely to affect SE more than other past developments and technologies. SE through MBSE will likely become an even stronger competitivity discriminator for enterprises trying to succeed in digital transformation [17].

The next subsections attempt to shed light on the building blocks of the digital environment and how these impact the digital enterprise.

The DT Building Block for SE and Digital Enterprises

A Digital Twin (DT) is a representation of reality or an approximation of reality. But models are also representations and approximations of reality. The difference and relationship between the two is explained in Fig. 20.

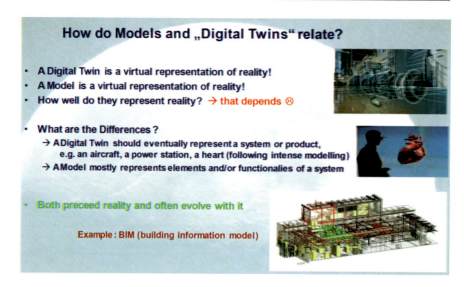

Fig. 20 Relation between models and DTs

A digital representation of a system is achieved by combining multiple models from different fields and disciplines. Examples include architecting, requirements analysis, conceptual design, functional behavior diagrams, energy diagrams, structural designs, etc. Each of these models provides a partial answer for a system representation and can eventually be captured by MBSE. Each of these models produces data describing system elements, functions, or characteristics for which they have been developed. Together, the resulting database represents the extent to which the system has been modeled and its digital representation.

All of these data items taken together eventually lead to, or constitute, a DT. The DT is a system representation to the extent that the underlying models enable or contribute to system representation. A DT depends on the underlying models and is an approximation of a system's reality. Its accuracy follows the depth of the SE lifecycle evolution when a product has completed manufacturing. Reality and the DT are intertwined and interrelated throughout the development and manufacturing process (Fig. 21).

In terms of Industry 4.0, a German industry initiative, the DT is a digital cyber-physical representation of the product to be built, manufactured, operated, serviced, and more. It is truly at the heart of Industry 4.0 enabling an enormous spectrum of applications during the different project phases from product design and development, the manufacturing process, and the operational market introduction (Fig. 22).

In principle, there are no limits to what a DT can represent as long as the underlying data and model base enable this. However, the DT will always be a model-based representation, an approximation of reality. Hence, no matter how powerful the computational assets which support such DTs are, they are limited by the precision of the underlying models and databases from which they have been derived. However, the closer a development approaches the manufacturing stage, the closer to 100% a DT will represent the manufactured product.

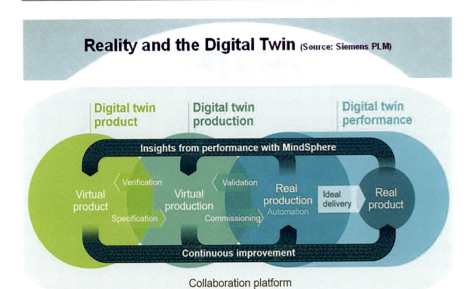

Fig. 21 Reality and DT

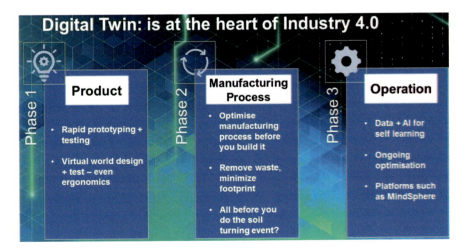

Fig. 22 DT is at the heart of Industry 4.0

In the future, customers will receive a physical product as before, e.g., a textile-producing machine or a skyscraper. In addition, tons of paper documentation will be replaced by a digital file describing the "as manufactured" physical product. This file will enable maintenance and future modifications based on the information captured in the DT. Therefore, they will have a complete digital system description that remains up to date until the eventual discharge of their product, a substantial advantage. An airline can thus maintain full cognizance of the "actual" state of each aircraft in its inventory. Maintenance, repair, and more will be simplified and

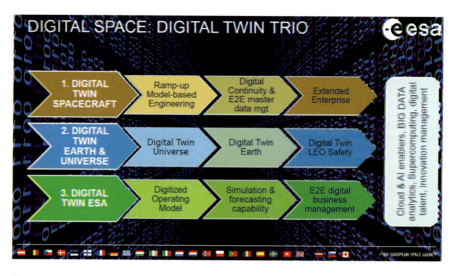

Fig. 23 DT Trio

more accurate since the DT is up to date for every piece of equipment, component, harness, or connector. DTs can represent simple products, such as an integrated starter/generator/clutch assembly for an electric car, a medical object, or an aircraft. For emergent service-oriented enterprises this means business simplification and competitive advantage.

DTs have, in principle, no inherent limitation. The European Space Agency promulgates the vision to eventually represent not only a spacecraft but planet Earth as a DT, or for that matter any complex enterprise (Fig. 23).

The Advancing Simulation and "Virtualization" World

In the past, we created systems on a drawing board and later with the help of models and computers. Future systems will primarily be designed in immersive "Virtualization Labs" long before they are turned into hardware or software in a manufactured system.

Ideas, analyses, designs, and more will become 3D+ visualizations representing system concepts in virtual environments. A virtual environment allows us to "see and feel" emerging products in several visual dimensions. They enable real-time visual feedback and let us "virtually" compose and modify designs along with their many artifacts and functional characteristics until we are satisfied that development, production, and fielding are promising.

Immersive virtualization labs/facilities can contain all system parameters and represent/approximate reality long before realization in the real-world is initiated. If virtualization maturity could be advanced to the point that a DT can be created, this would enable and accelerate the transfer to manufacturing, in a way that CAD does today, for components, and system elements.

Fig. 24 Concurrent virtualization will impact our engineering world

Fig. 25 Examples of virtualizations

The above examples (Figs. 24 and 25) from different industrial sectors show the enormous potential that advanced simulation and virtualizations have for the design and operation of systems. Their potential quality improvements, and cost and schedule gains, when compared to classical development approaches, are enormous. But so is the effort one needs to expend in transforming the design environment into a useful and representative virtual environment. Figure 26 shows different stages of Virtualization an extended Realities continuum.

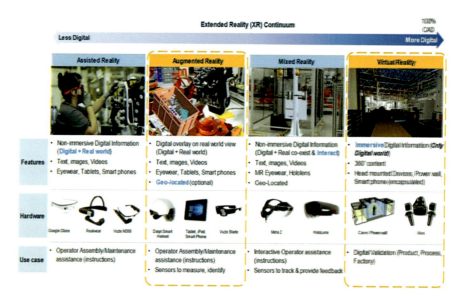

Fig. 26 Extended realities continuum. (Credit: Chris Davey, Ford Co., and INCOSE Vision 2035 draft; Reality-Virtuality-Continuum, source: trekk.com)

Immersive virtual reality (iVR), stage 4, will not come about as a "big bang." Depending on the product, it will go through or take various stages, from assisted reality to augmented reality, mixed and fully immersive virtual reality. Assisted and augmented reality methods have advanced in many enterprises and may suffice for some applications for a long time. Hybrid and fully digital iVR environments are advancing rapidly and have the potential to change the way we conceive and implement future systems dramatically. When advanced space-based cloud storage and quantum processors become available and affordable, we can expect a step-function in the realization of comprehensive and merging MBSE and iVR capabilities (Fig. 27).

The immersive VR transformation is a key, albeit not an easy step, toward realizing a digital enterprise! The DT will be part of this trajectory as one important output of virtualization efforts. AI and machine learning capabilities will be an integral part of this trajectory. Provided this "iVR transformation" is approached systematically and in well-planned and measured steps, a fundamental breakthrough in product development and realization efficiency can be achieved. An enterprise that has mastered this trajectory can be expected to reap enormous added value on the trajectory toward a Digital Enterprise. Clearly, a lot can go wrong during realization of this capability.

For iVR to become the projected asset for industrial or operations-oriented enterprises, substantial investments in models, IT infrastructure, standardization, employee training, and more are needed for successful deployment through a well-thought-out multi-year strategic plan.

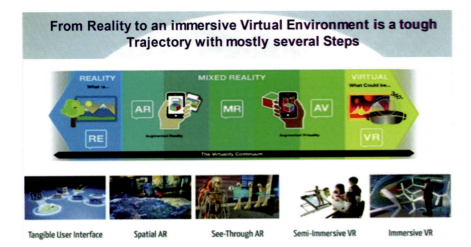

Fig. 27 Virtual environments

Best Practice Approach for Combining MBSE and DE

This section discusses how systems engineering can help digital engineering and how MBSE can help digital engineering. The naïve notion that a digital enterprise would employ a single tool that can facilitate and "integrate" all functions and data is likely to remain a "fata morgana."

The challenge is instead that a digital enterprise will need an inventory of tools to address every function and value stream of the enterprise and allow seamless interconnection and data exchange between multiple models, tools, and data. This involves models and tools for diverse purposes, e.g., to categorize staff competencies, market projections, R&D, engineering, and manufacturing. They need to be "architected" so that they can "relate to each other" seamlessly. Data must allow aggregation for higher-level analyses, even strategic considerations, and decision making, so that digital governance from top to bottom and vice versa is enabled by information flow as a standard practice. This capability is a prerequisite for the functioning of a digital enterprise.

Toward a System-Driven Digital Enterprise (DE)

A system-driven Digital Enterprise is one of the contributions that MBSE can make to a DE. The answer primarily lies in holistic systems thinking, functional analysis, and systematic architecting/engineering of an enterprise's functions. The organizational setup must respond to an analysis and understanding of related enterprise functions. The grouping of these functions, e.g., in a matrix and associated decision

Fig. 28 Digital enterprise layers

diagrams, leads eventually to organizational structures which follow a logical derivation of the process [2].

The message of Fig. 28 is that a digital enterprise can have as many different organizational layers as a "traditional" one, the big difference being that all the layers are "digitally connected." MBE and MBSE are essential pillars in realizing such interconnections, transparency, and competitive advantages.

Interrelationships Between SE Tools and Their DE Contributions

So far, various methods and tools contributing to SE advancements have been discussed. Figure 29 shows their evolution and relationships in a simplified manner. It also indicates that engineering advancements are evolutionary and interdependent. For example, all those tools noted contribute to Digital Twin DTs and ensure that a DT captures engineering characteristics and data from wherever relevant information is available to ensure engineering quality and depth. These various inputs assure the maximum representation of a DT at any given state of an engineering project.

Figure 29 also relate the tools' maturity to the timescales of their introduction and general use by the engineering and SE communities. It also notes that the "digital thread" and the "authoritative source of truth" draw from all the tools engaged. All of these ultimately have a very close relationship with integrating DT, the mainstay and pivot point for future engineering-based product development, manufacturing, and servicing activities. In this sense, the DT will be a foundational element of Digital Enterprises featuring engineering products.

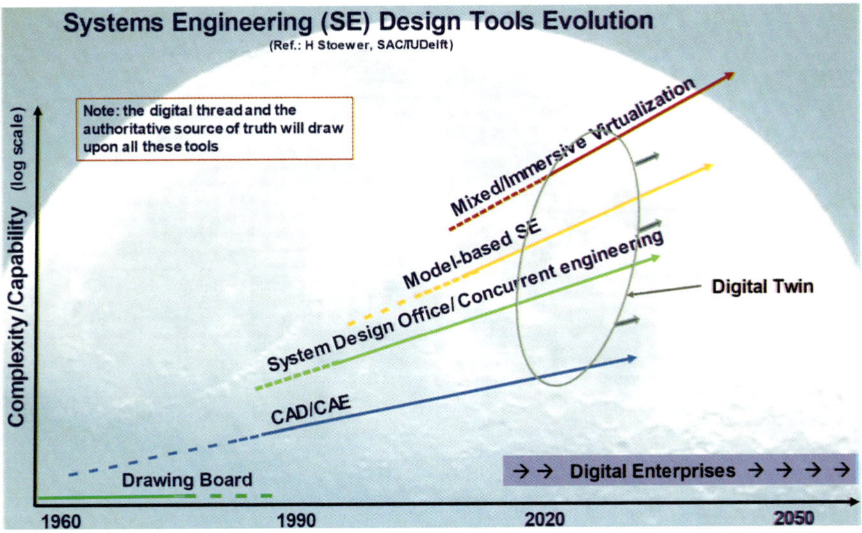

44 Perspectives on SE, MBSE, and Digital Engineering: Road to a Digital Enterprise 1333

Fig. 29 The evolution and relationships between SE tools

MBSE Can Help Shape the Digital Enterprise

The more complex a product portfolio and the associated programs/projects of an enterprise and the faster the business environment changes, the more it can benefit from a holistic systems approach to products and organization. Setting up, updating, and changing an organization while evaluating associated options and forcing decisions for the best functional allocations and organizational structures can thus be assured. Data analytics, AI, and machine learning can contribute substantially to such evaluation processes.

An MBSE setup can help to analyze, evaluate, and rationalize complex organizational situations and prepare decision options based upon competent business models, data, and tools. But what kind of MBSE setup will deliver analyses and results for such a variety of complex tasks? The analogy to Google Earth could be helpful in this regard.

The Google Earth Example. A dream model for MBSE is represented as an analogy to "Google Earth" (GE). GE enables users to:

- Quasi seamlessly transition from the global Earth System to local features without losing the "system" context. This capability corresponds to seamlessly transferring from one "view" and dataset to another, i.e., from the entire system to system elements, equipment, and piece parts/components
- View local streets, electricity networks, or rivers. GE enables us to pass from one highway to another or from one river to another without "being lost" since we do not lose the (geographic) system context. This capability corresponds in our

engineering design world to analyzing circuit diagrams, or power and data lines, cross over from one electrical diagram to another in search of a failure or anomaly.

What is missing in this MBSE analogy with Google Earth is the ability to change anything, i.e., to assess the effect of a change at one place of a system upon another. Clearly, this is where the analogy with Google Earth does not quite hold. Can we nonetheless imagine MBSE to follow the GE example? Not only to navigate between system components but also to change our design and realize the effects upon the other system components. Eventually, it takes a "big" MBSE capability to arrive at such a function. Elements for this already exist! Algorithms and visualizations from "electronic" games and "virtualization advances" can be leveraged. "Immersive Virtual Reality" could sometime come close to this dream model.

How Does MBSE Support Digital Enterprises?

As for the future, your task is not to foresee it, but to enable it. – Antoine de Saint Exupéry, French writer

Nothing on Earth is impossible, you only need to find the means to overcome the hurdles. – Hermann Oberth, German Space Pioneer

Digital devices and networks capture and guide our personal worlds at breathtaking speeds. The Covid pandemic has further accelerated this trend. Engineering analysis and design, concurrent engineering, smart manufacturing, transport, and logistics of huge supply chains are examples of rapid advances in digital worlds. Automation, software, artificial intelligence with data analytics, and machine learning are other elements characterizing our emerging digital world.

Germany's Industry 4.0 initiated and characterized these new systems and cyber-physical/social world as our current industrial phase (Fig. 29). Profound and rapid digitaltransformations beyond robotics, IT, electronics, and software give way to cyber-physical systems with significant social, economic, and ecological implications (Fig. 30).

The transition from the IT, electronics, and robotics industrial environments toward "cyber-physical" environments occurring at multiple places, is swift and a dramatic industry change in most domains. Industry 4.0 addresses many technology trends and new approaches for industrial product development and realization. The "intelligent" integration of many different technologies, methods, processes, and cyber parameters is a huge challenge for years ahead as depicted in Fig. 31.

This ongoing industrial transformation intensely affects and drives our "systems world." Every function within an enterprise undergoes change, be that in engineering, sales, finance, production, or logistics. Systems engineering (SE) plays an important role in this transformation. This is in part so because our modern increasingly complex systems have become more interconnected and interdependent, and because they show elaborate interfaces between system elements and the supporting IT infrastructure. The need for "seamless" SE across the entire value chain of modern enterprises has never been greater.

44 Perspectives on SE, MBSE, and Digital Engineering: Road to a Digital Enterprise 1335

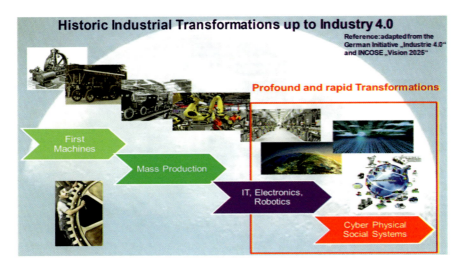

Fig. 30 Industry transformation

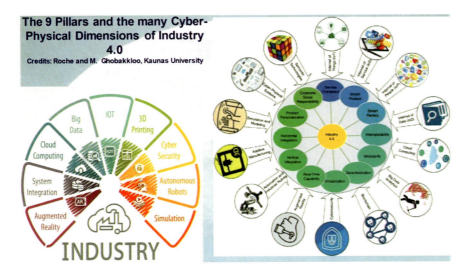

Fig. 31 Multiple dimensions of Industry 4.0

Chapter Summary

This chapter has provided perspectives on SE, MBSE, and digital engineering and how these provide the foundation of the digital enterprise. MBSE has made significant strides in the last decade but still has a long way to go before it fully achieves its potential. However, it has enormous potential to advance both SE and the DE. This potential is slowly being understood and realized through several pilot applications worldwide. At the same time, realizing comprehensive MBSE implementation remains elusive because of the lack of common standards and seamless toolsets. Once these shortcomings are resolved, widescale adoption and advancement will become possible.

Sound MBSE depends on sound engineering practices and a holistic "end-to-end" SE approach. While tackling the physics-based and mathematics-based MBSE challenges is hard, they are relatively well-understood. Also, the need to align MBSE with PLM is well-recognized [18, 19]. However, incorporating economics and social science-based "soft" disciplines into the system development and operation remains a significant challenge [20].

Technology advances in AI, machine learning, data analytics, visualization, virtualization, digital movie, and gaming, along with a myriad of related other IT technologies, need to be harnessed and adopted for MBSE. Once there, MBSE can achieve breakthroughs that promise major long-term value and efficiency increases for digital enterprises in the future.

Seamless interconnections and interaction of inventories of PM, SE, engineering, marketing, and administrative toolsets will eventually create transparency and leverage for moving forward with a digital enterprise's product portfolio and bring superior quality to corporate decision-making. MBSE (in line with a PLM process) can be expected to become a central capability and authoritative source of truth to create the platform for real-time digital interactions with customers, supply lines, manufacturing, and services organizations. Altogether, such an MBSE capability will provide a decisive edge, resulting in a sustainable competitive advantage for a digital enterprise!

References

1. Madni, A.M. and Sievers, M. Model-Based Systems Engineering: Motivation, Current Status, and Research Opportunities, *Systems Engineering*, Special 20th Anniversary Issue, Vol. 21, Issue 3, 2018.
2. Purohit, S. and Madni, A.M. A Model-Based Systems Architecting and Integration Approach Using Interlevel and Intra-level Dependency Matrix, in *IEEE Systems Journal*, 2021, https://doi.org/10.1109/JSYST.2021.3077351.
3. Checkland, Peter. "Systems thinking." Rethinking management information systems (1999): 45–56.
4. Monat, Jamie P., and Thomas F. Gannon. "What is systems thinking? A review of selected literature plus recommendations." American Journal of Systems Science 4.1 (2015): 11–26.
5. Sahlin, J. (2012). 'Cloud Computing: Past, Present, and Future.' In X. Yang, & L. Liu (Eds.), Principles, Methodologies, and Service-Oriented Approaches for Cloud Computing
6. Sahlin, J. (2014). 'Federal Government Application of the Cloud Computing Application Integration Model', In Khosrow-Pour, M. (Ed.). Encyclopedia of Information Science and Technology (3rd Ed.)
7. Sahlin, J. P. (Ed.). (2015). 'Social media and the transformation of interaction in society'. Hershey, PA: IGI Global. ISBN13: 9781466685567
8. Sahlin, J. P., Lobera, K. (2016). 'Cloud Computing as a Catalyst for Change in STEM Education', In Chao, L. (Ed.) Handbook of research on cloud-based STEM education for improved learning outcomes

44 Perspectives on SE, MBSE, and Digital Engineering: Road to a Digital Enterprise

9. A. M. Madni, Transdisciplinary Systems Engineering: Exploiting Convergence in a Hyper-Connected World. Cham: Springer International Publishing AG, 2017.
10. Madni, A. M., "Elegant Systems Design: Creative Fusion of Simplicity and Power," INCOSE Systems Engineering, vol. 15, no. 3, 2012, pp. 347–354
11. NASA Systems Engineering Handbook, Revision 2, 2020
12. Urbaczewski, Lise, and Stevan Mrdalj. "A comparison of enterprise architecture frameworks." Issues in information systems 7, no. 2 (2006): 18–23.
13. Zweber, J. and et. al., Digital Thread and Twin for Systems Engineering: Requirements to Design, 55th AIAA Meeting, AIAA SciTech Forum, (AIAA 2017-0875)
14. US Department of Defense 2018, Digital Engineering Strategy, Office of the Deputy Assistant Secretary of Defense for Systems Engineering, Washington, DC (US).
15. Space and Naval Warfare System Command (SPAWAR) 2018, Model-Based Systems Engineering, SPAWARINST 5401.6, Department of the Navy, San Diego, CA (US).
16. Zimmerman, P. "DoD digital engineering strategy," Proceedings of the 20th Annual National Defense Industrial Association (NDIA) Systems Engineering Conference, Springfield, VA, October 23-26, 2017. OUSD R&E, DoD Digital Engineering Strategy, (June 2018).
17. Madni, A.M. MBSE Testbed for Rapid, Cost-Effective Prototyping and Evaluation of System Modeling Approaches, Special Issue "Model-Based Systems Engineering: Rigorous Foundations for Digital Transformations in Science and Engineering," *Applied Sciences*, 2021, 11(5), 2321.
18. Hans, P. and et. al. 'PLM System Support for Modular Product Development', Computers in Industry, Vol 67, Feb. 2015 pp 97–111, Elsevier.
19. Kossiakoff, A., Sweet, W., Seymour S. & Biemerthe, S., Engineering Principles and Practice, 2nd Ed., John Wiley & Sons, 2011
20. Madni, A.M. and Purohit, S. Economic Analysis of Model Based Systems Engineering, MDPI *Systems,* special issue on *"Model-Based Systems Engineering,"* Feb 2019

Prof. Heinz Stoewer, Degrees in Technical Physics, Business, Systems Management, Germany/USA. Airbus, Germany, Project Engineer European launcher upper stage; Head Programmatics Unit. Boeing, USA, System Engineer space launchers, Project Manager "post Apollo" and NASA Space Tug/USAF Orbit to Orbit Shuttle. European Space Agency, first Programme Manager Spacelab, Europe's first human space project; later Founder of ESA's Systems Engineering and Programmatics Dept. Delft University of Technology, Netherlands, first Professor/Chair Space Systems Engineering, Founder international MSci programme in Systems Engineering; all part-time. Managing Director Science and Application programmes, German Space Agency; Member ESA Council, European Union Space Advisory Group, et al. Founder/President Space Associates GmbH. Member Supervisory Council of OHB SE, Europe's third largest Space Prime contractor. Advisor Systems Engineering Airbus, until 2021. Member Advisory Board Dutch National Space Agency (NSO), until 2018. Fellow, Life Member, Past President INCOSE. Distinguished Visiting Scientist NASA JPL. Member International Academy of Astronautics (IAA); emeritus member Board of Trustees and chair of Engineering Section. Lecturer/guest professor/keynote speaker universities, industry, governments. Greater than 90 publications; multiple national and international honours, e.g., German Parliament, NASA, ESA, IEEE 2018 Simon Ramo Medal.

Index

A

Abstraction, 390, 391, 927–929
  process, 909
Accessible driverless cars, 1085, 1086
Acquire information function, 491
Acquisition, MBSE
  acquisition life cycle, 730–735, 748–749
  benefits to acquirer enterprises, 735–736
  contracting and procurement, 732–733, 746–748
  development, 734
  digital engineering and implications, 736–738
  enterprise gap assessment, 730–731
  materiel solution analysis, 731–732, 743–746
  operations and support, 735
  production and deployment, 734
  technology maturation and risk reduction, 733–734
  unique nature of, 729–730
Action value function, 1168
Activity diagram, 353, 501, 503, 507, 508, 514, 556, 902
ActivityFinal, 906
Actual system performance data, 646
Adapted for Integration, 845
Adaptive Digital Twin, 644
Advanced Digital Twin, 381, 398–400, 410, 412
Advanced System Model, 385, 393–398
Adversaries, 1176
Aeronautics, 594
Aerospace Corporation's studies, 875
Agent Based Simulation (ABS), 603, 606
Agile development process, 109
AI-driven systems, 6
Aileron command, 579, 582
Aimless modeling, 336, 337

AirCADia tool, 598
Aircraft braking system, 668
Aircraft design, 603
Aircraft on-board systems, 603
Air Traffic Management (ATM), 483
Air Transport Association (ATA), 592
AlgebraicJulia project, 1294
Amazon Web Services (AWS), 763
4A Method, 796
Analyses models, 199
Analysis of alternatives (AoA), 732, 746, 747, 848
Analysis Server, 135
Analysis techniques, 865
Analytic Hierarchy Process (AHP), 900
Analytic system architecting, 308
Analytic Target Cascading (ATC), 599
Ant Colony Optimization (ACO) algorithm, 605, 606
AnyLogic™, 428
Apollo Program, 316
Apparent design space, 614
Application programming interfaces (API), 98, 879
Application Service Layer, 835
Applied category theory (ACT), 1261
Architecting, 1018
Architectural decision-making, 291
Architectural Frameworks, 162, 164
Architecture Analysis and Design Language (AADL) framework, 504, 554
Architecture Decision Graph (ADG), 601, 602
Architecture design space elements
  component characterization, 607, 609
  component connections, 607
  components, 608
  concept, 609
  function decompositions, 609
  function fulfillment, 607

© Springer Nature Switzerland AG 2023
A. M. Madni et al. (eds.), *Handbook of Model-Based Systems Engineering*,
https://doi.org/10.1007/978-3-030-93582-5

Architecture design space elements (*cont.*)
functions, 608
modeling, 607
ports, 609
semantic representation, 607
Architecture Design Space Graph (ADSG),
607, 611, 619
additional design variables, 612
conditionally active, 614
coupling, 614
decisions, 611
design variables, 613
hierarchical nature, 614
imputation, 614
Architecture Evaluation, 607
Architecture frameworks
DOD Architecture Framework (DODAF),
260, 261
NATO architecture framework (NAF),
259, 260
nature of frameworks, 259
OMG Unified Architecture Framework
(UAF), 262
Architecture Generation, 606
Architecture generator step, 614
Architecture instance, 612
Architecture processes, 265, 266
Architecture proxy-metrics, 616
Army Research Laboratory (ARL), 456
Artificial intelligence, 5
algorithms, 405
machine learning, 1241–1244
ASELCM Innovation Ecosystem Pattern, 173
Assistive technologies (AT)
business cases, 1097, 1098
challenge, 1080
disabilities, 1082
economic modeling, 1090, 1091
economic projections, 1094
health, 1080
human-centered applications, 1080
investment strategies, 1095–1097
market economics model, 1080, 1091–1093
mobility, 1082
model-based approach, 1086, 1088, 1089
older adults, 1082, 1083
policy implications, 1099
population of interest, 1081, 1082
sensory disabilities, 1083
social inclusion, 1083, 1084
tax revenues, 1099
use case models, 1080
Associativity, 1265

Astrium's Flexbus system, 788
Attack trees, 1051–1052
Audible signal scoring function, 228
Authoring power modes, 1145–1146
Authoritative source of truth (ASOT), 658, 795
system model, 1240
Auto-code, 959–961, 964, 975
Automated identification of dependencies, 361
Automated synthesis, 604
Automated testing, 960, 982
Automatic exploration, 591
Automation, 476, 485
Automotive industry, 1305

B

BaConLaws model, 199–201, 203–205
Baseball-bat collision model, 200–202
Base class, 835
Bayesian Belief Networks, 1164–1166
Behavior, 1195
elements, 507
model, 1120, 1121
Belief state, 1169
Beneficiary group, 665
Bialgebra rule, 1289
Bidirectional relationships, 822
Bidirectional transformation, 1294
Big data, 405
Bikeshed, 945
Binary decision variables, 318
Black-box simulation tools, 599
Block definition diagrams (BDDs), 502, 506,
507, 513, 522, 560, 562, 809, 1000
Blue Origin's New Shepard suborbital system,
786
Brain storming techniques, 900
Built-in self-test, 221, 245, 247
Built-in sensors, 404
Business Alternative and Hybrid, 1040
Business Process Model Notation (BPMN), 490
Byzantine fault conditions, 1156

C

CAESAR, 1137–1138
CAESAR PEL
advantages, 1143–1144
application, 1141–1143
authoring power modes, 1145–1146
implementation, 1144
CallBehaviorAction (CBA), 903
Cameo Workbench, 455

Index 1341

Campaign/Enterprise Context stereotype, 804
Candidate solutions, 911
Capabilities Development Document, 872
Capability Maturity Model Integration for
    Development (CMMI-DEV), 961
Capability Views/Strategic Views, 869
Case-based reasoning (CBR), 487
Category theory, 318, 1260
    benefits from online community, 1294
    composition and context, 1263–1266
    to data and semantic modeling,
        1266–1271
    description, 1260
    duality, 1290–1292
    functional programming, 1292
    isomorphism and identity, 1271–1276
    non-determinism, 1280–1283
    picturing processes, 1276–1280
    possibility, 1281–1282
    probability, 1282–1283
    state, 1284–1287
    topological data analysis, 1293
    visual reasoning, 1288–1290
Catia/NoMagic Cameo Systems Modeler, 935
Cellular communication, 657
Center for Systems and Software Engineering
    (CSSE), 835
Centers for Medicare and Medicaid Services
    (CMS), 1091
Central repository, 829
Central Reputation Analysis (CRA)
    function, 1178
Change control board (CCB), 103
Change request matrices (CRMs), 879
Chocolate Chip Cookie Acquisition System
    evaluation criterion from, 226–227
    requirement, evaluation criterion, and risk
        comparison for, 228–229
    requirement from, 224
    risk from, 227
Circular economy, 1305
Climate change, 1254
Closed system, 351
Code scavenging, 835
Codified design knowledge reuse, 791, 792
Cognitive and Physical Structure/Function
    Analysis (CPSFA), 490
Cognitive assistance, 372
Cognitive function analysis (CFA) method, 490
Cohesion, 819, 820, 822
Combinatorial explosion of alternatives, 591
Combined Assessment of Software Safety
    and Security Requirements
    (CHASSIS), 1056

Command and Telemetry Dictionary, 970–972
Comma-separated file (CSV), 827
Comma separated value (CSV), 873
Comment Resolution Matrix, 878
Commercial off-the-shelf (COTS), 959, 962
    elements, 787
Commonality-modularity spectrum, 788
Commutative monoid, 1288
Compatibility matrix, 601
Competitive Advantage (CA), 1098
Complexity handling techniques, 383
Complex systems, 1254
Component architecture, 388, 393,
    397–398
Component diagram Toolbox, 907
Components failures, 557
Composability, 418, 424
Composable design methodology, 790
Composite requirements, 361
Composition, 1260, 1263–1266
    modeling, 670
Computational representation, DEVS, 1202
    DEVS Atomic Behavioral Model package,
        1208
    DEVS Atomic Interaction Definition
        package, 1209
    DEVS Atomic Structural Model package,
        1207
    DEVS Coupled Function Definition
        package, 1211
    DEVS Coupled Structural Model
        package, 1210
    DEVS Coupling Definition package, 1207
    DEVS Model Core package, 1205
    DEVS Structural Model package,
        1205–1206
    general purpose layer, 1204–1207
    Mathematical Function and Set Theory
        packages, 1203–1204
Computer aided design (CAD), 480
    engineering tools, 1316–1317
    tools, 103
Computer Software Component (CSC), 970
Computer Software Units (CSU), 969
Computer-supported cooperative work
    (CSCW), 476
Concept Design Center (CDC), 1034
Concept exploration, 1019–1021, 1026, 1042
Concept Maturity Levels concept, 903
Concept of Operations (CONOPS), 455, 872,
    965, 972–974
    environment, 304
    operational stakeholders, 304
    stakeholder expectations, 303

Concept of Operations (CONOPS) (*cont.*)
  system design process, 304
Conceptual Design (CD), 897, 898, 900–903,
    907–911, 913–916, 918–920
Conceptual interoperability, 425
Conceptualized for Reuse, 843
Conceptual modeling, 656, 660, 661, 668
Conceptual ontology, 434
Conceptual system model, 320
Concrete model, DEVS, 1191, 1192, 1199,
    1201
Concurrent CD Proposal (CCD), 898, 910, 918
Concurrent design facilities, 1314–1316
Concurrent virtualization, 1329
Condition-based/predictive maintenance, 649
Conduct Design, 880
Conduct Mission Design, 876
Conduct Workspace Reach Analysis, 812
Configuration Control Board (CCB), 966, 982
Configuration of a Pattern, 165
Confirmation bias, 237–239
Consistent problem-space mapping, 326
Constellation Synthesis, 1038
Constructive Cost Model (COCOMO), 838
Constructive Systems Engineering Cost Model
    2.0 (COSYSMO 2.0), 800
Constructive Systems Engineering Cost Model
    (COSYSMO), 837, 848, 849, 851,
    853–856
Consultative Committee for Space Data
    Systems (CCSDS), 964
Contingency, 1142
Continuation approach, 1176
Contract Data Requirements List (CDRL), 733
Control Documents and Perform Change
    Control, 879
1553 Controller, 515
Conventional error detection, 1173, 1174
Cookie acquisition system, 216, 220
Cooperation, 488
Copy & paste engineering, 838
Core activity, 865
Core Flight Executive (cFE), 962
Core Flight Services (cFS), 962
Core library, 1292
Core MA Process (CMP), 865
Core-vocabularies, 948
Coronavirus Infectious Disease Ontology
    (CIDO), 1111
Cost baselines, 848
Cost drivers, 849
Cost-effectiveness analysis (CEA), 848
Cost estimating relationship, 847, 852
Cost model, 1026

Coupling, 614, 819, 820, 822
Courses of Action (COA), 258
COVID-19
  disease spread, 1111
  factors, 1107
  multi-faceted threat, 1108
  pandemic models, 249
  US supply chains, 1121
Creativity, 898, 900, 910, 911, 915, 918, 919
Creeper, 1047
Cross-disciplinary base, 400, 401
Crowdsourcing, 1244
CRUD use cases, 824
Cultural change, 401, 412
Cumulative distribution function (CDF), 1177
Curation, 436
Current best estimate (CBE), 1139
Customer Satisfaction Rating (CSR), 899
Customer value, 391, 394, 402, 403
Custom MODA models, 137–140
Cyberattacks, 1047
Cyberphysical systems (CPS), 380, 409
  attack tree, 1051–1052
  CHASSIS method, 1056
  description, 1046
  HSUV security issues, 1063–1071
  MB3SE, 1057–1059
  MBSEsec methodology, 1059–1061
  model-checking tools, 1053
  security approach, 1061–1063
  simulation, 1053–1055
  state-based model, 1054
  state-based models, 1053
  state-of-the-art, 1050–1051
  SysML-Sec method, 1057
  UAF, 1056
Cybersecurity, 1047
  availability, 1048
  confidentiality, 1047
  integrity, 1048
  loss prevention, 1048
  simulation, 1054
Cyberspace, 1047
Cyber threat, 1048

D
Dashboard technology, 650
Data collection, 270
Data-driven Modeling, 640
Data generation, 640
Data Item Descriptions (DID), 733
Data sets, 397
Decision analysis, 121, 144, 145

Index 1343

Decision hierarchy, 613
Decision making, 88, 132–134
  behavior track, 306
  centricity, 320
  cognitive tasks, 294
  combinations, 315, 316
  criterion, 309
  environment, 306
  execution, 307
  front end, 306
  mental ability, 306
  model-centered approach, 307
  morphisms, 319
  operands, 294
  planning, 321
  problem domain, 319
  reasoning, 294, 305, 308, 309
  recognization, 310
  system architecture, 294
  system's performance and cost, 315
  user experience, 306
Decision nodes, 566
Decision science, 122
Decomposition, 388, 389
  layers, 386
Deep learning, 5
Deep Space Network (DSN), 981, 1163
Department of Defense Architecture
     Framework (DoDAF), 303, 869
Dependability, 1155, 1156
Dependency matrix, 112
Derived classes, 835
Description logics (DL), 932, 938
Design Assurance Core MA Practice
  analysis techniques, 877, 879
  branching, 879
  changes, 876
  core activities, 876
  document-based approaches, 876
  document-based SE approaches, 878, 879
  functional flow, 877
  goal, 875
  MBSE MA practitioner, 878
  model management tools, 879
  physical constraints, 876
  traceability, 878
Design Core Activity, 877
Designed for Reuse, 843
Design Efforts reuse, 791
Design Failure Mode and Effects Analysis
    (DFMEA), 646
Design flexibility, 477, 478
Design Implemented, 843

Designing solutions, 910
Design knowledge reuse, 786–789
Design model, 1026
Design Modified, 844
Design pattern, 707
Design process, 591
Design reuse, MBSE
  4A Method, 796
  capabilities, 794
  context, 798
  hierarchy, 794, 795
  high-level conceptual ontology, 806
  implementation, 802
  lunar rover robotic arm system, 807
  methodology dashboard, 808
  reuse logical process, 797
  robotic arms on the MoI and RCM, 810
  space industry (*see* Space industry design
    reuse)
  SysML, 801
  SysML custom tables, 811
  SysML Reuse profile, 803
  technical inheritance, 799
  value assessment process, 800
Design science research methods (DSRM), 450
Design Structure Matrices, 1097
Design thinking, 63
Design to Cost (DTC), 848
Design variables, 611, 613
Develop/Demonstrate Autonomous
    Algorithms, 812
Development for Reuse (DFR), 835–837,
    841–844, 846, 851, 852, 854–856
Development with Reuse (DWR), 835–837,
    839–846, 849, 851–855
Diagram Predictive Frameworks (DPF), 34
Differential Equation Specified System
    (DESS), 428
Digital banking, 8
Digital ecosystem, 103–113
Digital engineering (DE), 4, 121, 634, 635, 657,
    658, 736–738, 834, 863, 864, 1018,
    1238
  community, 651
Digital enterprise
  layers, 1332
  MBSE, 1333–1334
  and SE tools, 1332–1333
  system-driven, 1331
Digital revolution, 657
Digital tapestry, 1324
Digital thread, 9, 635, 744, 864, 865, 1239
Digital transformation, 656

Digital Twin, 9, 398, 411, 478–480, 528, 676, 678, 679, 683, 687, 688, 692, 698, 701, 702, 1239, 1318, 1325–1328
  cost, 648, 649
  data and simulation, 634
  definition, 634
  digital engineering, 634
  enabler and complement, 638
  history, 634
  IoT, 648
  levels, 642
  links virtual and physical environments, 635
  and machine learning, 539–540, 647
  in manufacturing, 649
  in MBSE (*see* Model-based systems engineering (MBSE))
  NASA and USAF researchers, 637
  ontology use in, 535–536
  physical twin, 635
  structural SysML model, 541
  technology, 530–533
  uses, 637, 638
  virtual system model, 634
  vs. traditional CAD/CAE model, 637
Digital twin-enabled MBSE framework, 635
Directed rooted tree, 928
Discount DT (DDT), 480
Discrete event system specification (DEVS), 428, 1190
  advantages of, 1191
  applications of, 1191
  computational representation, 1201–1212
  concrete model, 1199
  formalism, 1194
  formal models, 1198
  limitations, 1218–1221
  modeling and metamodeling progression, 1199
  simulation model, 1211–1220
  switch model, 1195–1197
Distributed hardware environment, 685–687
Distributed Ontology, Model, and Specification Language (DOL), 818
Diversity, 409
3D Modeling, 640
DOD Architecture Framework (DODAF), 260, 261, 873
DoD Digital Engineering Strategy, 1322–1323
DoD's Digital Engineering initiative, 636
Domain-specific approach, 602
Domain-specific language (DSL), 401, 901, 930
Driveline model, 644

Driver behavior tracking system, 313, 317
Dynamic interoperability, 425, 429
Dynamic mapping, 604
Dynamic systems evolution, 1319
Dysfunctional Behaviour Database (DBD), 561

E
Eclipse-based Rosetta workbench, 944
Eclipse modeling framework (EMF) project, 943, 1145, 1212
Economic modeling, 1090, 1091
Economic projections, 1094
E/e architecture, 397
Effective systems engineering, 926
Effort multipliers, 849
Eigen Trust algorithm, 1161, 1162
Electrical engineering, 384
Electro-Hydrostatic Actuator (EHA), 572
Electro-Mechanical Actuator (EMA), 559
Electronic Data Interface Control Document (EDICD), 978
Electronics Architecture and Software Technology-Architecture Description Language (East-ADL), 386
Electronics warfare (EW), 638
Elephant Specification Language (ESL), 599
Embedded device attribute set, 671
Embedded systems, 504, 506
Energy storage system, 708–709
Engineering 3.0
  challenges, 1255, 1256
  complications, 1255
  limitations, 1256
Engineering, 926
  developers, 383
Engineering change request (ECR) control, 101
Engineering Development Units (EDUs), 968
Engineering manufacturing development (EMD), 843
Enhanced Functional Flow Block Diagram (EFFBD), 909
Enterprise architecture model, 742
Enterprise gap assessment, 731
Enterprise systems engineering (ESE), 740, 741
Entertainment arts, 1245–1247
Environment, 49
  sensing, 406
Environmental Control System (ECS), 596
eReq, 849
ESA's Concurrent Design Facility, 1315
Essence, 663
Europa Clipper project, 103

Index 1345

European Meteorological Satellite
  (EUMETSAT) Agency, 1022
Evaluating systems, 1257
Evidence of completion (EOC), 881
Executable digital representation, 644
Executable systems engineering method
  (ESEM), 105
Experimental modeling, 640
Experimentation support, 688–689
Exploitation, 591
Exploratory evolution, 995
Expression variable, 1209
Extended realities continuum, 1330
Extensible Markup Language (XML), 817
External transition function, 1197
Extract, Transform, Load (ETL), 1123

F
Failure Modes and Effects Analysis (FMEA),
  555, 560–562, 567–569
Failure probability predictor (AVF), 1176
Failure Review Boards (FRB), 881
Failure search, 1176
Fast False Start, 337
Fault, 1172
Fault behavior, 553
Fault management (FM), 960
Fault models, 554
Fault Tree Analysis (FTA), 555, 557–560, 566,
  567, 576, 579–581
Feasible design space, 614
Federal Insurance Contributions Act, 1091
Figures of merit (FOM), 805, 898, 908, 909,
  911
Filtering process, 317
Financial investment model, 203
Finite element (FE), 434, 435
Finite-horizon model, 1169
Finite state machine (FSM), 701
Fire-Fighters System (FFSys) system, 905, 906
FireSat, 904, 950–951
Flight control system (FCS), 571, 603
Flow ports, 562
Fly-By-Wire technology, 572
Food Supply Model, 1120
Formal architecture synthesis, 604
Formal model, DEVS, 1191, 1193, 1198, 1203,
  1219
Framework for the Evaluation of MBSE
  Methodologies for Practitioners
  (FEMMP), 73
Framing, 240–242

Free category, 1264
Full-scale development (FSD), 897, 898, 908,
  910, 919
Fully specified evolution, 994
Function, 665
  Architecture, 393–396, 398, 402, 405, 407,
    411, 412
  decomposition, 593
  flows, 566
  list, 389, 391, 394
  misbehavior, 405
  programming, 1292
Functional Analysis System Technique (FAST),
  909
Functional Hazard Analysis (FHA), 555–557,
  564–566, 575, 578
Functional Mock-up Interface (FMI)™, 427
Function-form map, 308
Function-form-structure mapping, 593
Function Mapping Matrix (FMM), 601
Function-oriented approach, 390
Functor, 1268
Fusion Fire-Picture (FP), 905
Fuzzy front end, 290

G
Game engines, 1246
Generalized Reuse Framework (GRF), 835
  COSYSMO, 848, 850
  Development for Reuse (DFR) weights,
    854, 855
  Development with Reuse (DWR) weights,
    853
  GRF-based cost estimating relationship,
    851, 852
  parametric cost estimating, 848
  system definition, 842
  system design, 843, 844
  system implementation, 844
  validated system component, 845
General-purpose programming languages, 1198
Generic forms (GFs), 301
Generic Modeling Environment (GME),
  935–936
GENESYS, 603
Genetic Algorithm (GA), 605, 606
GEOINT Analyst Control System, 461
Geospatial Intelligence (GEOINT) Analyst
  Control Station, 457
Geospatial UAV study, 456
Geostationary Earth Orbit (GEO) Landsat
  Earth-imaging satellite, 788

Geostationary equatorial orbit (GEO)
satellites, 1022
Geostationary Operational Environmental
Satellite (GOES), 1022
Geostationary satellites, 1022
GLEAMviz, 1113
Global Change Information System
(GCIS), 430
Goddard Space Flight Center (GSFC), 962
Google Earth, 1333
Government off-the-shelf (GOTS), 959,
962, 970
Government Performance Reporting Act
(GPRA), 1031
Government Reference Architecture (GRA),
747, 748, 1042
Gradle build scripts, 948
Grand Canyon, 660
Graphical user interface (GUI), 476, 683,
692–693
Greedy policy, 1168
Guidance, Navigation and Controls (GN&C),
959, 975

H
Hardware-in-the-Loop (HIL), 966, 968
testbeds, 678
Hardware-software (HW-SW) integration, 383,
661
activity parameters, 520
allocation relationships, 512
applications, 504, 514
capturing data, properties, 509, 510
communication protocols, 522
1553 Controller interface, 518, 519
data elements, 502
data intensive applications, 509
digital interfaces, 503
electrical characteristics, 516
engineering development process, 504
execution platforms, 504
functional allocation, 506–508
interfaces, 503
internal interfaces, 515
Item Flows, 511, 512
MBSE approaches, 503
modelling approaches, 506
modelling environment, 502
object flows, 521
physical elements, 502
power-on activity, 514
power supply, 513

1553_Protocol Block Definition Diagram,
517
software command, 502
system software level, 506
typing ports, 510, 511
UML, 504
wireless connector, 503
Hardware systems, 657
HAZOP analysis, 1056
Heterogeneous agents, 1162
Hierarchical nesting of requirements, 362
Hierarchy of reuse practices, 792
High fidelity numerical methods, 640
High level architecture (HLA), 427
High Level Architecture (HLA)—Object
Model Template (OMT), 427
High-level conceptual ontology, 806
High-Level Discrete (HLD), 514–516
High-level operational processes, 904, 905
Homonyms, 934
Hopf rule, 1289
House-of-Quality (HoQ), 899, 909
Houston's concept, 372
Hover Test Vehicle (HTV), 963, 964
HSUV security issues, 1063–1071
Human-centered design (HCD), 472
Human-centered modeling, 493
Human-computer cooperation, 682
Human decision-making, 237
Human factors, 332
Human Factors and Ergonomics (HFE), 475
Human-in-the-loop simulation (HITLS), 472,
476, 491, 493
Human-machine interfaces, 1086
Human-machine system, 7, 472
Human performance models, 446, 466, 467
Humans-hardware-software model, 667
Human systems integration (HSI)
activity, 474
automation, 485, 486
cognitive-physical structure-function
system, 480
complexity, 481
declarative knowledge, 473
design flexibility, 477, 478
digital twins (DT), 478–480, 493, 494
emergence, 482, 483
emergent functions, 482, 483
evolution, 473, 475–477
experience, 484, 485
experience feedback, 484
functional Context-Resource
hyperspace, 482

infrastructure, 481
maturity, 482, 483
people and social groups, 480
problem-solving, 486
procedural knowledge, 473
procedures, 486
PRODEC, 473
recursive definition, 481
regulatory-based requirements, 484
resource management, 477
separability, 482, 483
sociotechnical system, 472
structures, 482, 483
system knowledge, 477
systems-engineering framework, 473
task, 474
ultra-safe industries, 484
Human system integration (HSI) ontology
behavior, 450–452
development process, 447
export task allocation, 455
factors, 447
human agent, 449, 450
human agent functions, 459
human functions, 459
IMPRINT model, 459, 461
mechanism, 447
model based environment, 455
modeling tools, 447
parametrics, 452
requirements, 448
stakeholders, 447
structural diagrams, 452, 453
SysML, 447
system diagrams, 455
top level, 447
workflow, 455
Hybrid-electric propulsion systems, 622, 623
Hybrid evolution, 995
Hypertext transfer protocol (HTTP)
requests, 98

I
Identity, 1265
Image Analyst, 456, 457
IMCE vocabularies, 949–950
Immersive virtualization labs/facilities, 1328
Immersive virtual reality, 1330
Import/export, 826, 827
Imputation, 614
Incompatibility constraints, 610
Incremental development, 284

Independent Technical Analysis (ITAs), 872
Independent Technical Assessments
(ITAs), 872
Industrial Internet of Things (IIoT), 637
Industry 4.0, 634, 1239–1240, 1326, 1334,
1335
digital engineering, 657, 658
digital revolution, 656
digital transformation, 656
MBSE, 656
OPM cyber-physical system applications,
668–671
OPM ISO 19450, 663–667
Industry transformation, 1335
Infinite-horizon discounted model, 1169
Information fusion system (IFS), 1005
Inheritance, 787, 792, 797, 799, 804, 805
InPort set, 1214, 1215
Institutional stakeholders, 312
Instrument catalog development, 1038
Instruments, 665
Integrated concept
abstraction, 301
complexity, 302
decomposition, 301
interactions, 301
operands, 302
potential vertical breakdowns, 302
Integrated Concept Engineering
Framework, 919
Integrated-Conceptual-Design-Method (ICDM)
framework, 898–903, 908–910,
912, 914
Integrated Definition Methods (IDEF)
community, 445
Integrated model-based engineering, 121
Integrated modeling environment, 259
Integrated Systems Engineering and Pipelines
of Processes in Object-Oriented
Architectures (ISE&PPOOA), 55, 80
Integrated Test and Operations System
(ITOS), 971
Integrated trade-off analytics framework, 135
Integrated vehicle health management
(IVHM), 638
Integrating heterogenous models, 418
bridging mechanisms, 431–433
causes of ontological differences, 422–424
LCIM, 424–426
ontologies and models, 419–420
preserving model correspondence, 420–422
semantic ontologies, 430–431
standards and tools, 426–428

1348   Index

Integration, test, and evaluation (IT&E), 880, 881
Intellectual property (IP), 651
Intelligent Digital Twin, 644
Intended users, 282
Interface Block, 365, 502, 511, 515, 517, 518, 522
Interface Control Documents (ICD), 503, 513, 968, 978–979
Internal Block Diagram (IBD), 501, 502, 506, 507, 510–512, 515, 558, 560, 562, 809
Internal knowledge, 390
International Council of Systems Engineering (INCOSE), 333, 443, 506, 552
International Council on Systems Engineering (INCOSE) Model-Based Capabilities (MBCA) Matrix, 659
International Resource Identifier (IRI), 938
International Space Station (ISS), 1319
Internet of Things (IoT), 634, 637, 656, 1240–1241
Interoperability, 816, 913
  import/export, 826, 827
  linked data, 827
  link management, 828, 829
  proprietary adapter, 828
  services, 830
Interpersonal communication, 817
Isolated system, 481
Isomorphism, 1272
Item flows, 511, 512
Iterated dynamics, 1278
Iteration, 209
Iterative mapping, 604
Iterative system specification (ISS), 1191
Ivory tower, 338

J

Japanese Meteorological Agency (JMA), 1022
Jet Propulsion Laboratory (JPL), 918, 1137–1138, 1163
Joint Polar Satellite System (JPSS), 1022
JPL's Deep Space Network (DSN), 1181

K

Karban's example, 112
Kepler Space Telescope, 786
Key performance parameters (KPP), 215
Killer trades, 215
Kirchoff's laws, 1290
Knowledge base, 27
Knowledge-based and probabilistic methods, 539

L

Labeling, 1264
Lack of stakeholders' engagement, 915
LADEE Spacecraft Observatory, 959
Lane Keep Assist (LKA) function, 402
Language, 926–927
Level Control System (LCS), 561
Levels of Conceptual Interoperability Model (LCIM), 424–426
Life-cycle cost estimate (LCCE), 869
Life cycle cost model, 121
Lifecycle modeling language (LML), 352
Linearized models, 8
Local time of ascending node (LTAN), 1035
Logical Architecture, 386, 388, 392, 393, 395–398, 410
Logical engineering, 819
Logical process, 797–799, 801, 802, 804
Loose metamodel, 22
Low Earth Orbit (LEO), 788, 1022
Lowest technical definition, 395
Lunar Atmosphere Dust Environment Explorer (LADEE), 956, 959, 962
  ambiguous ICDs, 978–979
  Command and Telemetry Dictionary, 970–972
  commanding, modes and transitions, 975
  ConOps-driven testing methodology, 972–974
  development cycle, 964–966
  emergent behavior, 979–980
  fault management, 976
  final load, 982–983
  formal inspection, 977
  GN&C, 975
  GOTS components, 970
  model-based unit testing, 969–970
  prototyping, 963–964
  reaction wheels, 979
  requirements, 966–967
  simulation hardware, 968–969
  Star Tracker system, 979
  system reboot, 980–982
  test infrastructure and traceability, 969
  testing cadence, 976
  training and mission operations, 977–978
Lunar C3I, 1001
Lunar Laser Communications Demonstration (LLCD), 957, 980, 981, 983
Lunar Module's Descent Propulsion system, 786
Lunar orbit rendezvous (LOR), 620

Index 1349

Lunar Reconisance Orbiter Camera (LROC), 984
Lunar rover robotic arm system, 807
Luxury of numbering games, 1311

M

Machine learning, 487, 639, 641, 647, 1174
Maintenance cycles, 638
Manifestation layer, 819, 820
Manufacturing anomalies, 638
Manufacturing Assurance core MA Process, 880
Mapping, 396
Market economics model, 1089, 1091–1093
Market Share (MS), 1096
Markov Decision Process (MDP), 1166–1169
Markov Property, 1167
Markov Reward Process (MRP), 1168
Mars Cube One (MarCO) CubeSats, 787
Mars Exploration Rovers, 807
Mars InSight lander mission, 787
Mars Science Laboratory (*Curiosity*), 787, 807
Mass-damper system, 597
Materiel solution analysis, 731–732, 743–746
Mathematical Function metamodel, 1203
MATLAB, 835, 874
Maximum expected value (MEV), 1139, 1142
MBSE environment, 816
    graphical elements, 821
    interoperability, 816
    logical model, 821
    mental/manifestation layers, 818
    text document, 818
    tool/repository, 816, 817
MBSE methods for reuse
    4A Method, 796
    capabilities, 794
    context, 798
    design reuse logical process, 797
    hierarchy, 794
    high-level conceptual ontology, 806
    implementation, 802
    lunar rover robotic arm system, 807
    methodology dashboard, 808
    robotic arms on the MoI and RCM, 810
    space industry (*see* Space industry design reuse)
    SysML, 801, 802
    SysML custom tables, 811
    SysML Reuse profile, 803
    technical inheritance, 799, 800
    value assessment process, 800
MBSE models, 636
MBSE patterns, 151

MBSEsec methodology, 1059–1061, 1064–1067
MBSE testbed, 636, 638
Mealy machine, 1284, 1285
Mechatronics, 397
Mediation, 487
Mental grand canyon
    brainstorm, 292
    stakeholder, 292
    system architecture decisions, 293
Mental layer, 819, 821
Merge nodes, 566
Merging-sorting-pruning process, 910
Metamodel, 269, 270, 1137, 1199, 1201–1212
Metamodeling, 930
Metamodels
    constructs, 27
    definition, 20
    loose, 22
    ontologies, model and, 21
    strict, 22
Methodical Approach to Executable Integrative Modeling (MAXIM), 667, 668
Methodology, 48
    cost and time savings, 340
    interfaces, 340
    interoperability, 341
    problem analysis, 339
    requirements, 342
    stakeholders, 340
    verification and validation, 345, 346
Methodology-as-a-Model approach, 901
Mission assurance (MA), 864
    audits, 882
    benefits, 868
    challenges, 865, 866
    Core MA Process, 889
    design assurance, 885–887
    digital format, 864
    digital threads, 866
    FireSat model, 883
    IT&E assurance, 887, 888
    MBSE modelling techniques, 868
    model artifacts, 866
    NASA OMA's MBMA vision, 867
    processes, 863
    program assurance, 883, 884
    reviews, 882
    terminology hierarchy, 864
Mission Assurance Guide (MAG), 862, 875
Mission Assurance Plan (MAP), 863, 865
Mission & System Specification and Design Process, 883
Mission-enemy-troops-terrain-time available-civilian (METT-TC), 701

Mission of Interest (MoI), 793
Mission Operation Systems (MOS), 977
Mission success, 862, 864, 867
Model-based approach, 258, 340, 1086, 1088, 1089
Model-based concept representation
  OPM modeling language, 296
  solution-specific environment, 299
  stakeholders domain (D1), 297, 298
  system architecture framework, 296
Model-based-conceptual-design (MBCD). 291, 743
  data management, 913
  designing alternative solutions, 910
  implementation issues, 900
  lack of "stopping criteria" for modeling, 914
  lack of common taxonomy, 916
  objectives, 897
  operational aspects, 904, 907
  organizational support for, 913
  performance-based specification, 909
  requirements' uncertainty, 916
  stakeholders' needs, 904, 907
Model-Based-Conceptual-Design Working group (MBCD WG), 897
Model based conceptual framework, 594
Model-based constructs, 1174
Model-Based Cyber-Physical Systems Engineering (MBCPSE), 381
Model-based error detection, 1175
Model-based health evaluations, 1174–1176
Model-Based Human Systems Integration (MB-HSI), 472
Model-based integrated decision support tool flow chart, 137
Model-Based Mission Assurance (MBMA), 867
Model-Based Off-Nominal State Identification and Detection (MONSID), 1175
Model-based reliability, 553–555
Model-based requirements
  elicitation, derivation, and trade-off analysis, 369–372
  property-model methodology, 360
  requirement elements, 353
  system models as requirements, 353–355
  theoretical framework, 351–352
  Wymorian requirements modeling, 356
Model Based Safety Analysis
  advantages, 552, 570, 571
  aileron command, 575–577, 579–582
  Airbus A320, 572, 573
  aircraft, 572

aircraft manufacturers, 551
application, 572
attributes, 550
back-up hydraulic actuator, 574
challenges, 553
civil transport aircraft, 550
disadvantages, 552
lack of traceability, 552
limitations, 552
nominal behavior, 574
off-nominal behavior, 574
practice approach, 563
research community, 553
safety, 551
safety analyses, 550
summary table, 564
system development process, 552
systems reliability, 551
traceability, 551
Model-Based System Architecting (MBSA)
  cognitive process, 305
  conceptualizations and decisions
    (see Decision making)
  functionality, 305
  holistic system, 292
  inherent capability, 324
  language model, 292
  SIMILAR process, 292
Model-based systems engineering (MBSE), 4, 5, 7, 10, 88, 120, 198–199, 258, 259, 274, 276, 280, 290, 333, 350, 352, 353, 370, 410, 418, 442, 472, 501, 529–530, 552, 591, 595, 634, 646, 656, 676–678, 702, 712, 726–728, 739, 740, 749, 865, 988, 1108, 1137
  acquisition life cycle, 748–749
  advancing simulation and virtualization, 1328–1330
  AI-driven systems, 6
  approaches, 1315
  artificial intelligence/machine learning, 1241–1244
  biases in human decision making, 1237
  CAD/CAE, 1316–1317
  capabilities, trade-off analyses, 123–130
  challenges, 6, 144, 593
  closure, 773
  communication tools, 774
  complexity, 6–7
  concepts from other disciplines useful for, 1248
  concurrent design facilities, 1314–1316
  contracting and procurement, 746–748

Index 1351

crowdsourcing, 1244
dashboard tool, 681
and decision analysis, 1233–1238
definition, 4
deployment, 765
description, 754
design, operation and digital ecosystems, 103–113
determining the deliverables, 768
development of, 765
and digital engineering, 1238–1240
digital enterprise, 1333–1334
and digital thread, 1239
and digital twins, 1239, 1325–1328
distributed hardware environment, 685–687
DoD Digital Engineering Strategy, 1322–1323
economic aspects, 334
effort, 768
for enterprise gap analysis, 739–743
and entertainment arts, 1245–1247
execution, 771–772
exemplar repositories, packages, libraries, 684
expectations, 334
experimentation setup, 535–538
experimentation support, 688–689
extensible architecture, 682
formal representation, 5
graphical user interface, 683
hardware and connectors, 684
human-computer cooperation, 682
illustrative example, 774–778
implementation, 334, 538
implications for curriculum, 1248–1249
Industry 4.0, 1239–1240
initiation and scope definition, 766–767
institutional barriers, 1238
Internet of Things, 1240–1241
landscape, 1324
logical architecture, 684–685
materiel solution analysis, 743–746
and MBE, 1322
MBSE2 approach, 109–111
measurement, 769
measurement tools, 774
methodology, 533–534
model and scenario refinement, 687–688
modeling architecture, 595
modeling methods, 683
modelling requirements, decision making in system design, 132–134
models needed for, 1320–1322

multi-perspective visualization, 691–692
optimization, control and learning algorithms, 683
plan management, 769
predefined scenarios, 681
preferences, values and utility, 1233–1234
preliminary experiments, 698–699
process, 534–535
process planning, 767
project handovers, 1319–1320
project management, 754–763
project management tools, 773
project planning and tracking tools, 774
prototype testbed implementation, 687–698
quality requirements, 769
quantitative analysis, 540–545
rapid scenario authoring, 689–690
resource allocation, 768
review and evaluation, 772–773
risk analysis, 10
risk management, 768
risk management tools, 774
ROI, 10–11
scenario elements, 682
schedule, 768
SE issues, 1318
simulation platforms, 684
smart dashboard, 692–697
and social networks, 1244–1245
software programming environment, 683
stakeholder, 334
systematic approach, 764
system modeling, 682
system modeling and verification, 687
testbed benefits, 685
testbed concept, 680
testbed ontology, 678–680
testbed repository, 688
three-phase approach, 764
UAV design case study, 131–132
uncertainty, 1235
updating belief, 1236
valuing information and experimentation, 1236
virtual world for SE, 1325
Model-based systems engineering (MBSE) methodology, 48
adoption and tailoring, 74–78
evaluation, 73–74
Model-Based System, Safety, and Security (MB3SE), 1057–1059
ModelCenter™, 135, 428
Model centric engineering (MCE), 114

Model Characterization Pattern, 177
Model development, 264, 265
  agile development, 273
  architecture deployment, 271
  architecture utilization, 273, 277
  data collection, 270, 272
  model creation, 274
  step 1 (problem framing), 266
  step 2 (metamodel development), 269
  step 3 (data collection), 270
  step 4 (model creation), 271
Model health & complexity management
    project, 113
Modelica, 927
  Language™, 427
  simulation model, 603
Modeling, 897, 898, 900–903, 907–910,
    913–916, 919
  languages, 262–264, 269, 270, 273
  patterns, 264, 270
  profiles, 263, 270, 274
  templates, 264, 270, 279
Modeling and Analysis of Real-Time and
    Embedded Systems (MARTE) profile,
    504
Modeling and simulation progression, 1192
  DEVS formalism (see Discrete event system
    specification (DEVS))
  framework, 1194
  types of models in, 1193
Modeling & simulation (M&S) technology, 531
Modeling languages, 23, 342
  mapping, 343
  Object Process Methodology (OPM), 342
  stereotypes, 344
  SysML, 343
  traceability, 344
Modeling methodology, 262
Modeling scenarios, 1140
Modeling tools selection, 344, 345
Modeling Triads, 332
Modeling waste heat, 1140–1141
Model integration, 1088, 1089
Model interoperability, 913
Model island, 337
Model management system (MMS), 98, 830
Modular common bus (MCB) spacecraft, 957
Modular interface, 616
MOHICAN, 491
MoI Description stereotype, 804
MoI-RCE Requirement Pairs, 809
MoI-RCE requirements, 809
MoI Required Interface stereotype, 805

MoI Technical Need, 798–801, 804, 805
Molecular dynamics (MD), 434, 435
Monad, 1283
Monte Carlo dynamics, 1283
Monte Carlo simulation (MCS), 462, 1039,
    1190
Morphological matrix, 599–601
Morphological Table, 900
MSL "pencils down" design, 808
Multi-agent system (MAS)
  additional resources, 1177
  agent health, 1178
  Bayesian Belief Networks, 1164–1166
  centralized/distributed architecture, 1180
  centralized trust management, 1178
  conventional error detection, 1173, 1174
  dependable system, 1155
  disruptive events, 1154, 1155
  distributed architecture, 1178
  distributed trust management, 1179
  factors, 1155, 1156
  form of trust, 1154
  hardcore, 1179, 1180
  health evaluation, 1174
  MDP, 1166–1169
  ML-health evaluations, 1176, 1177
  model-based health evaluations, 1174–1176
  network health policy, 1180
  non-compliant health, 1180
  ontologies, 1157, 1159, 1160
  peer's subjective trust, 1155
  POMDP, 1169, 1170
  POMDP trust evaluation, 1170–1172
  practical and theoretical evaluation
    methods, 1154
  problem statement, 1154, 1155
  psychologists, 1155
  self-health evaluations, 1179
  software component, 1154
  transaction malfunctions, 1180
  trust and reputation study, 1182–1184
  trust architectures, 1178
  trust evaluation, 1160, 1161, 1163, 1164
Multi-attribute decision-making (MADM)
    techniques, 89
Multi-attribute tradespace exploration (MATE),
    314
Multi-attribute utility theory (MAUT), 1030
Multidisciplinary Design Optimization (MDO),
    596, 606, 618, 745
Multi-mission Modular Spacecraft (MMS), 788
Multimodality sensors, 651
Multi-model approach, 1115–1123

Multi-model based decision support
  benefits, 1109
  candidate factors stakeholders, 1116–1119
  conceptual model-based, 1122
  costs, risks, and benefits, 1107
  COVID-19
  COVID-19 pandemic, 1106
  decision-making infrastructures, 1108
  disease model, 1125
  feedback loops, 1107
  healthcare workers, 1124
  initial question set, 1114–1115
  interdependencies, 1107
  medical models, 1126
  ontologies, 1110–1111
  ontology fragment consistent, 1120
  pandemic decision support framework,
    1109–1110
  pandemic interventions, 1114
  pandemic planning, 1108
  population and workforce model, 1126
  protection model, 1126
  public policy, 1115
  sociotechnical systems, 1107
  vulnerability Model, 1126
Multi-Objective Evolutionary Algorithms
    (MOEA), 616
Multi-objective genetic algorithms, 601
Multi-perspective visualization, 651
Multiple objective decision analysis (MODA),
    130, 137
  custom, 137–140
Multi-scale modeling, 418, 432, 433
Multi-source/multi-sensor information, 648
Multistream variant management research, 838

N
NASA Apollo Mission Design, 618, 619
NASA/JPL Mars Rovers, 1317
NASA Mars rovers, 807
NASA Procedural Requirements (NPR), 961
NASA's Space Shuttle program, 786
NASA Systems Engineering Handbook, 787
National Aeronautics and Space Administration
    (NASA), 774, 867
National Oceans and Atmosphere
    Administration (NOAA), 265
  analysis, 1026
  application of MBSE, 1026
  architectural decision, 1029
  business model, 1039
  climate and science researchers, 1032

concept development, 1019
constellation architectures, 1040
conventional models, 1026
cost, 1018
decision analysis framework, 1032
design vectors, 1039
development phases, 1021
development strategy, 1019
DoD Architecture Framework, 1029
efficient frontier plot, 1040, 1041
follow-on activities, 1041
functional model, 1029
hierarchy of decisions, 1024, 1025
implementations, 1018
ISO 42010 framework, 1042
JPSS programs, 1023
LEO satellite, 1023
metrics, 1032
modeling processes, 1037–1039
models, 1021
models *vs.* design vectors, 1029, 1030
model types, 1026
optimism *vs.* pessimism, 1028
performance and value modeling, 1030
performance attributes, 1033
physical designs, 1018
physical development, 1018
policy, 1027, 1028
problem space, 1019
radical cases, 1039
requirements, 1027
satellite data, 1031
satellite development and manufacturing
    timelines, 1024
satellite lifetimes, 1027
satellite products, 1032
satellites, 1024
satellite sensors, 1032
solution space, 1019
stakeholder assessment, 1029
42010 standard, 1028
strategic objectives, 1033
support decision making, 1039
SysML language, 1020
system development, 1018, 1020
system development activities, 1020
system development and production, 1019
system-of-interest, 1030
transition, 1018
value model, 1029, 1030
variance analysis, 1040
weather satellite program, 1023
weather satellites perform, 1022

National Weather Service (NWS), 1022, 1031
NATO architecture framework (NAF), 259, 260
Natural isomorphism, 1274
Natural transformation, 1273
Net Option Value (NOV), 1090
Net present values (NPVs), 1090, 1094
Network category, 1281
Network time protocol (NTP), 686
Neural network-based supervised machine
    learning algorithm, 644
NOAA Observing System Architecture
    (NOSA), 265, 268, 269, 271, 272
Nodes, 389
Nominal behavior, 553
Nominal Group Technique (NGT), 900
Non-determinism, 1280–1283
Non-dominated Sorting Genetic Algorithm
    2 *NSGA-II*, 606
Nonelectronic Parts Reliability Data (NPRD),
    551
Nonlinearities, 8
Non-propagating battery design, 711
Non-real-time simulation, 647
North Atlantic Treaty Organization (NATO),
    445
NSOSA design vector models
    alternative set, 1034
    constellation configuration, 1034–1036
    cost-performance, 1036
    human intervention, 1037
    launch policy, 1035
    modeling loop, 1035
    satellite configurations, 1034
    solar observation instruments, 1037
    U.S. legacy program, 1036
Numerical weather prediction (NWP), 1022,
    1031
NumPy, 684
Nunn-McCurdy breach on, 649

O

Object Management Group (OMG), 71, 504,
    818
Object Modeling group (OMG), 506, 901
Object-Oriented Programming (OOP), 617, 803
Object-oriented software development, 835
Object-Oriented Systems Engineering Method
    (OOSEM), 51–53, 79, 386
Object Process Diagrams (OPDs), 664
Object Process Language (OPL), 664
Object-Process Methodology (OPM), 53–54,
    79, 293, 1261

Observing System Simulation Experiment
    (OSSE), 1032
Offline system architecting process, 305, 325
Older adults, 1083
OMG Unified Architecture Framework (UAF),
    262
Onboard flight software (OFSW), 961, 962, 967
Ontological metamodeling
    abstraction, 927–929
    automation, 929
    combining modeling languages, 930–933
    instantiation relationship, 933
    language, 926–927
    linguistic metamodeling paradigm, 933
    modeling, 929
    natural language, 934
    OWL 2, 933
    OWL 2 DL, 934
    semantics, 930–933
    SWRL rules, 934
Ontological Modeling Language (OML)
    Bikeshed, 945
    description, 941
    description bundle, 942
    description formalisms, 939
    load, 947
    merge, 945–946
    openCAESAR, 944
    OWL 2 DL, 940
    OWL Reason, 946–947
    OWL syntax, 946
    SPARQL, 947
    textual syntax, 943
    vocabulary, 941
    vocabulary bundle, 941
    workflows, 942–943
Ontological reasoning, 937–939
Ontology(ies), 24, 164, 419, 420, 678–680,
    1110–1111
    Bikeshed, 945
    DL, 934
    IRI, 938
    linguistic, 934
    model, metamodel and, 21
    pizza, 29
    problem-solving, 926
    process for developing, 36, 37
    satellite parts, 41
    structure and ownership, 34, 35
    tools for development and reasoning, 40
    uses, 24
Ontology for digital twin-enabled model,
    535–536

Index

Ontology Web Language (OWL), 1111
    reasoner, 1148
    representation, 1147
OntoUML, 937
OPCloud, 667
Open Biological and Biomedical Ontologies
        (OBO), 1111
OpenCAE, 105
OpenCAE DECO, 105
openCAESAR
    analysis services, 944
    Eclipse-based workbench, 943
    Github repositories, 948
    ontology-based system modeling, 943
    Semantic Web community, 943
    tools, 943
OpenMBEE project, 98, 1314
OpenMETA, 427
Open Model-Based Engineering Environment
        (OpenMBEE), 830
Open Services for Lifecycle Collaboration
        (OSLC), 828, 913, 934
Open-source libraries, 640
Open system, 351
Operads, 1293
Operand, 665
Operating principle, 391, 395, 403–405
    degradation, 405
    diagnostics, 405
    functional misbehavior, 404
    safety, 405
    security, 405
Operational Concept Description (OCD), 292
Operational concept (OPSCON), 455
Operational environment data, 639
Operational-level issues, 740
Operational life, 647
Operational performers, 872
Operational readiness assurance (ORA), 881
Operational readiness training (ORT), 978
Operations, Maintenance, and Sustainment core
        MA process, 882
Optimization, 615, 616
Option-decisions (OPT), 612
Option-pricing theory, 1095
Organization
    business case, 335
    compromise, 332
    culture and general resistance, 333
    distribution, 335
    document-based approach, 333
    infrastructure, 346, 347
    interaction, 333

    learning curve, 338
    methodology, 332
    responsibility, 336
Organizational hierarchy, 737
Original equipment manufacturer (OEM),
        1096, 1097
Overall Evaluation Criteria (OEC), 601
Overarching process, 235–236
    confirmation bias, 237–239
    framing, 240–242
    human decision-making on, 237
    severity amplifiers, 239–240
    uncertainty, 244
Overmodeling, 337
OWL 2 DL, 932–934, 936, 938–940, 943
OWL SHACL, 947
OWL SPARQL, 947
OWL Start/Stop Fuseki, 947–948

P
Pandas, 684
Pandemic decision support, 1109, 1122, 1129
Pandemic response process management
        system, 1123
Parallel resistance, 1291
Parameterization construct, 358
Parametric estimating, 847
Parental communication, 394, 395
Partially observable Markov decision process
        (POMDP), 8, 687, 1169, 1170
Partitions, 903
Pattern(s), 151, 706, 719
    matching, 599
    mining, 707, 709–719
Pattern-based systems engineering (PBSE), 707
Pedestrian traffic information, 650
Performance metrics, 610, 615
Performance models, 199
Performance Work Statement (PWS), 733, 747
Perform Test, 880
Permutation-decisions, 612, 613
Perseverance, 787
Personalization, 240
Personalized medicine, 9
Physical architectures, 1308
Physical article reuse, 791
Physical environment, 635
Physical system, 638
Physical twin, 636, 644
Physics-based approach, 640
Physics-based modeling, 640
Picturing quantum processes, 1293

Planned reuse, 792
"Plug-and-play" fashion, 788
Polar Orbiting Environmental Satellites (POES), 1022
Policy iteration (PI), 1169
Policy makers, 1113
Polymorphism, 319
Polynomial regressions, 97
Ports, 609
Post-AoA, 747
Power analysis workflow
  example, 1149
  HTML documents, 1148
  openCAESAR tools, 1147
  OWL representation, 1147
  report views, 1148
  SPARQL query, 1147
Power Equipment List (PEL), 1141–1143
Power-On activity diagram, 520
Power Supply Card, 515, 518
Power system architectures, 603
Powertrains, 395
Pragmatic interoperability, 425, 436
Predictive lifecycle models, 767
Pre-Digital Twin, 642, 643
Price-to-Win (PTW), 848
Primitive concepts, 25
Prioritize, 209, 222, 225, 226, 233, 235, 243, 247, 248, 251
Probabilistic risk assessment (PRA), 931
Problem framing, 266, 275
  advance planning, 280
  afterwards, 283
  agile development of models using sprints, 273
  approach, 276, 277, 280
  architecture frameworks, 259–262
  homework assignment, 281
  incremental development, 284
  intended users and uses of the models, 277
  model-based approach, 258
  model development, 264
  model development method evolution, 274
  modeling methodology, 262, 263
  NOSA model, 266
  prototyping, 284
  re-framing the problem, 284
  six step method for model development, 265
  sponsor participation, 281
  step 1.2 (model scope & context), 278
  step 1.3 (information & data needs), 279
  step 1.4 (model views & products), 279
  storyboarding, 281
Problem space
  definition, 351
  of functions, 351
  of outcomes, 351
Problem statement, 340
Process, 49
  assets library, 246
  functor, 1280
  interface, 1277
Process Integration and Design Optimization (PIDO), 596
Processor-in-the-Loop (PIL), 966, 968
PRODEC method
  BPMN-CPSFA PRODEC process, 491
  cognitive science, 488
  collaboration, 491
  computer science, 488
  contexts, 490
  declarative knowledge, 488–490
  human-centered design, 488
  hybrid Human/AI systems, 492
  machine learning, 492
  procedural knowledge, 488–490
  task analyses, 491
  uncertainty management, 492
  virtual assistant, 492
Production system reference modeling, 113
Product lifecycle management (PLM), 9, 530, 734
Product line engineering (PLE), 164, 838
Profit Margin (PM), 1096
Program Assurance Core MA Process
  activities, 868
  Capabilities and Operational Activities, 870
  capabilities and strategic roadmap, 870
  document-based acquisitions, 869
  MBSE tools, 871
  models, 871
  traceability, 869, 870
  UAF consistency, 871
Program management overlap (PMO), 1309–1310
Programmatic decisions, 307
Program of Record 2025 (POR2025), 1028
Project-level issues, 741
Project management, MBSE, 754
  characteristics, 760–763
  description, 755
  importance of, 759–760
  methodology, 755
  model-based approach, 755
  non-trivial and challenging, 757–759
Project Management Institute (PMI), 755

Index 1357

close, 756
execution, 756
initiation, 755
performance and monitoring, 756
planning, 756
Project performance, 334
Property-model methodology, 360
Prospect theory, 237
Protégé language, 29
Prototyping, 284
Proxy element, 823, 824
Proxy Ports, 517
Pugh matrix-based process, 900
Purpose-driven methodology, 333
pyCycle, 617
Python, 835

Q
Quadcopter control panel, 693
Quadcopter indicator panel, 693
Quality Adjusted Life Years (QALYs), 1091
Quality Function Deployment (QFD), 899, 909, 919, 1097
Quality issues, 865
Quantitative evaluation, 591
Quasi-isolated system, 481
Query-driven modeling, 333
Query reformulation, 209

R
Reachability matrix, 563
Read entity B list (RBL), 825
Realization, 393
    models, 408
Real Time and Embedded Systems (RTES), 504
Real-Time Operating System (RTOS), 968
Recovery-dependent reuse, 791
Recurrent convolutional neural network (RCNN) model, 537, 699
Redundant hydraulic actuator, 579
Reference elements, 579
Reference system, 420
Re-framing the problem, 284
Reinforcement learning, 539, 641, 646
    approaches, 1176
Reliability, availability, maintainability, and safety (RAMS) analysis, 877
Reliability analysis, 562, 563
Reliability Block Diagram (RBD), 562, 569
Reliability Configuration Model (RCM), 558
Reliability values, 562

Remote physical systems visualization, 650
Request for information (RFI), 848
Request for Proposal (RFP), 733, 748, 848, 872
Request for Tender (RFT), 733, 748
Requirements Analysis and Validation Core
    MA Process
    document-based SE approaches, 872, 874, 875
    elements, 873
    parametric diagrams, 874
    requirements traceability mechanisms, 873
    techniques, 872
    verification method field, 875
Requirements discovery process, 208–209
    chocolate chip cookie making system, 211–213
    functional requirements, 212
    nonfunctional requirements, 213
    test plan, 214
    use case template, 209–211
Requirements engineering, 353, 369, 372
Requirements Engineering and Management
    for software-intensive Embedded
    Systems (REMsES), 386
Requirements Interchange Format (ReqIF), 818, 873
Resilient design, 252
Resistance, 336
Resistor networks, 1293
Resource-level issues, 741
Return-on-investment (ROI), 10, 11, 650, 915, 1098
Reusable Asset Archive, 804
Reusable Asset Record, 804, 805
Reusable prototype, 643
Reusable resources, 839
Reusable work efforts, 789
Reuse, 384, 385, 395, 399, 401, 404, 406, 409
    Development for Reuse (DFR), 841
    Development with Reuse (DWR), 839
    GRF (see Generalized Reuse Framework (GRF))
    in research and literature, 838
    reusable resources, 839
Reuse Candidate Elements (RCE's), 799
Reuse Candidate Element stereotype, 805
Reuse Candidate Mission (RCM), 800
Reuse of system definition, 842
Reuse of system design, 843
Reuse of system implementation, 844, 845
Reuse of validated system component, 845
Reuse Requirement stereotype, 805
Reuse taxonomy, 790, 791

Rework Action types, 809
Rework Set stereotype, 805
Risk analysis, 207
    chocolate chip cookie making system, 220–222
    process, 220, 224
Risk assessment, 1312
Risk-based validation, 10
Risk estimation, 1176
Risk management (RM), 862, 1312–1313
Risk matrix, 1313
Risk register, 221, 226
Robotic arm system, 807
Robust MBSE environments, 322
Rough order of magnitude (ROM), 848
RS422_Electrical_Characteristics, 510

S

Safety and Mission Assurance (SMA), 867
Safety assessment, 553–555
Satellite Electronics Box (SEB), 513–515, 517, 520, 522
Satisfiability analysis, 939
Scaled Agile Framework (SAFe), 388, 864
Scoring functions, 226, 228
Second generation product line engineering, 838
Security assurance, 1049
Security Context Element, 1159
Security domain model, 1064
Security policy, 1049
Security requirement, 1064
SEIR model, 1125
Self-agents, 1176
Self-driving vehicles, 641
Self-healing mechanisms, 638
Semantic(s), 930–933
    domain, 19
    interoperability, 168, 425, 428, 430
    link, 393
    linkage, 384, 394
    ontologies, 430–431
Semantic Application Language (SADL), 937
Semantic modeling
    electrical power, 1136
    model-based approach, 1136
    problem statement, 1136–1137
Sensing Fire, 905
Sensitivity analysis, 204–206, 233, 1309
Sensor attribute set, 671
Sensory disabilities, 1083
September 2020 ransomware attack, 1046

Sequence diagrams (SD), 365, 560, 909
Service-level issues, 741
Service-oriented architecture (SOA), 835
SESAR (Single European Sky ATM Research), 483
Set-based design, 122, 132, 1097
Set Theory Utility package, 1203–1204
Severity amplifiers, 239–240
Shared SA (SSA), 480
SIMILAR process, 229–230
    assess performance, 234
    integration, 233
    investigate alternatives, 232
    model the system, 232–233
    problem statement, 230–232
    re-evaluation, 234
    system launch, 233–234
Simulation, 9, 531, 539, 543, 647, 676–681, 683–687, 691, 692, 694, 695, 702, 703, 1112–1114
Simulation-based analysis, 639
Simulation-based modeling, 651
Simulation Program with Integrated Circuit Emphasis (SPICE), 927
Simulink®, 427, 874
Simulink Interface Layer (SIL), 966
SIR Model, 1111–1112
Situation awareness (SA), 480
Situation calculus, 1158
Six Sigma frameworks, 755
Size drivers, 848
Smart dashboard, 692–697
    prototype, 694
Smart functions, 650
S*Metamodel, 156
S*Models, 162
Social inclusion, 1083, 1084
Social Security Administration (SSA), 1091
Social Security Disability Insurance (SSDI) Program, 1091
Social Security Supplemental Security Income (SSI) Program, 1091
Society of Automotive Engineers (SAE), 504
Sociopolitical complexity, 6
Sociotechnical SoS infrastructure, 488
Software as a service (SAAS), 1092
Software development environment, 290
Software engineering, 960, 977
Software error detection, 1174
Software realization, 408
Software systems, 382, 657
Solutioneering, 900, 914
Solution-neutral environment (SNE), 298

Index 1359

Solution-neutral functions (SNPs), 314
Solution-neutral operand (SNO), 298
Solution-neutral process (SNP), 298
Solution space, 351
Solution-specific architecture decisions
    generic form, 323
    morphism, 321
    operands, 322
    problem-orientation, 321
    solution-neutral functions, 322
Solution-specific environment
    conceptual design, 301
    generalization-specialization
        relation, 300
    OPM, 300
    principal solutions, 301
Solution-specific operands (SSOs), 299
Source lines of code (SLOC), 838
Source Selection Award, 872
Source selection process, 733
Spacecraft, 42, 1136
Spacecraft power management
    Earth. Clipper, 1139
    electrical power, 1136
    remote sensing instruments, 1138
    semantic web technologies, 1136
    specification and design, 1137
Space Flight Operations Facility (SFOF), 1181
Space Habitat design, 789, 790
Space industry design reuse
    design knowledge reuse, 786, 788, 789
    hardware reuse, 785, 786
    planned and unplanned reuse, 792, 793
    reuse taxonomy, 790, 792
    structured and unstructured methods, 793
    work effort reuse, 789, 790
Space Mission Analysis and Design (SMAD)
    textbook, 904
Space Platform Requirements Working Group
    (SPRWG), 1033
Space Systems Working Group (SSWG), 897
S*Patterns, 162
Specific forms (SFs), 301
Sponsor participation, 281
Stakeholder, 198, 208, 209, 231, 243, 333
    decomposition, 310
    traceability, 312
    value, 122
Star Tracker system, 979
State Machine Diagrams, 559, 560, 579
Statement of Work (SOW), 733, 747
State-of-the-Art, 763–765
State transition matrix, 1167

Steep learning curve, 334
Storyboarding, 281
Strategic Fit (SF), 1098
Strategic-level issues, 740
Strict metamodel, 22
String diagrams, 1276–1280
Structural models, 199
Structural SysML model, 541
Structured and unstructured methods, 793
STS-46 (Atlantis), 786
STS-75 (Columbia), 786
SUAVE, 617
Sufficiency, 327
Sunk cost bias, 1237
Supervised learning, 641
Supervision, 487
Susceptibility, infection, recovery
    (SIR), 1124
Synchronous Dynamic Random-Access
    Memory (SDRAM), 980
Syndeia, 935
Synergy between MBSE and digital twin
    braking systems, 639
    ethics and data privacy, 641
    high-cost systems, 640
    machine learning and AI, 641
    modeling technologies, 640
    system life cycle, 639
    virtual system model, 639
Syntactic interoperability, 424, 426, 427, 435,
    437
SysML Activity Diagrams, 556, 565
SysML-based MBCD interpretation, 900
SysML Block Definition Diagrams, 556
SysML Internal Block Diagrams, 556, 563,
    569, 572
SysML4Modelica, 596, 597
SysML profile, 795, 797, 801–803, 811
SysML Requirement Diagrams, 556
SysML Reuse profile, 803
SysML-Sec methodology, 1057
SysML Sequence Diagrams, 556, 567
SysML State Machine Diagrams, 568
SysML Use Case Diagrams, 556
SysMLv2, 936
SYSMOD infrastructure process, 347
SYSMOD Methodology Adoption Process
    (SMAP), 338
System architecting, 731, 743, 747, 1308–1309
System architecture(s), 386, 388–390,
    392–396, 399, 402, 405, 410–412, 590,
    592, 602
    models, 530

Index

System architecture decisions
  agent models, 446
  approach changes, 467
  architecture workload comparison, 465
  automated assist functions, 464
  automation, 446
  automation function, 466
  behavioral allocation analysis, 456
  cognitive models, 443, 445
  communication systems, 467
  design phases, 444
  engineering process, 458
  formal models, 444
  functional allocation, 446
  geospatial UAV system model, 456
  heuristic-driven process, 444
  human agents, 465, 466
  human behavior, 445
  human centered model-based
    systems, 460
  human characteristics, 444
  human functions, 460, 461
  human-machine interfaces, 442
  human performance, 467
  human-system collaborations, 443
  human-system interaction, 445, 452
  human system interface, 443
  human-to-human task distribution, 465
  image analyst, 460, 466
  IMPRINT VACP Scale Values, 462
  inside-out/outside-in approach, 443
  integral system element, 444
  intelligence sources, 458
  lexical semantics, 442
  mental models, 442
  mission constraints, 467
  model, 443
  multiple perspectives, 444
  perform geospatial intelligence, 457
  process comprises, 452
  semantics, 446
  sensitivity analysis, 466
  stakeholders, 442, 443
  supervisory role, 466
  syntax, 446
  SysML diagrams, 445
  system operation, 445
  tasks, 460
  types of functions, 446
  UAV Image Analyst, 466
  user interfaces, 445
  workload analysis, 461
Systematic design space exploration, 591

Systematic exploration, 591
System attribute, 839–846, 852
System conceptualization, 649
System developing process (SD1), 53
System diagram (SD), 665
System-driven digital enterprise, 1331
System dynamics, 1113, 1119, 1124, 1125,
    1129
System effectiveness model, 122
System elements, 394
System exploit full motion video imagery task
    network, 459
System knowledge, 477
System launch, 234
System life cycle processes, 385
System maintenance, 649
System Modeling Language (SysML), 262,
    264, 274, 275, 442, 504, 505, 554, 901,
    906, 960
System models, 1321
System-of-interest (SOI), 352, 898, 904, 907
System-of-systems (SoS), 33, 481, 603, 636,
    988, 989, 1108
  abstraction phase, 998, 999
  artificial intelligence, 1010
  definition phase, 996, 997
  exploratory evolution, 995
  fully specified evolution, 994
  hybrid evolution, 995
  implementation phase, 1000
  implications, 990
  modeling, 993
  organizational Structure, 996
  preliminary phase, 994
  research vignettes, 1001, 1002
  risk detection, 992
  stakeholder, 991
  SySML language, 1003
System on a Chip (SoC) applications, 503
System performance, 908
  model, 122
System requirements review (SRR), 111
Systems engineering, 4, 245, 291, 492, 635,
    726, 784, 785, 789, 792–794, 801, 812,
    834, 836, 838, 839, 845, 848, 853–855,
    926, 1136, 1260, 1261
  assumptions, 1309
  circular economy, 1305
  end-to-end approach, 1306–1308
  Engineering simplicity, 1310–1311
  failures, 1311–1312
  framework, 862
  historical perspective, 1304

MBSE (*see* Model Based Systems
 Engineering)
and program management overlap,
 1309–1310
risk management, 1312–1313
sensitivity analysis, 1309
system architecting, 1308–1309
system boundaries, 1306
"Vee" life cycle development, 727
Systems engineering and technical analysis
 (SETA), 843
Systems Engineering Research Center
 (SERC), 485
Systems modeling environment
 (SME), 346
Systems Modeling Language (SysML),
 88, 291, 352, 383, 501, 503, 599, 740,
 743, 818, 901, 903, 904, 906,
 908–913, 915, 916, 1190
 AHP conceptualization in, 91
 CRUD, 824
 data elements, 501
 design, operation and digital ecosystems,
  103–113
 developing systems, 502
 generic decision concepts in, 90
 integrates disciplines, 502
 management tool, 821
 multi-fidelity multi-phase tool integration,
  100–103
 physical elements, 501
 single repository, 821
 surrogate model, 96
 tools, 90, 92
Systems Modeling Toolbox (SYSMOD),
 79, 81–82
 actor maintenance use case, 67
 actor operator use case, 67
 FFDS problem statement, 64–65
 FFDS system use case, 67
 logical architecture, 68–73
 logical grouping of, 62
 methods, 58–61
 problem solving process, 63–64
 relationships, 58
 system use cases, 65–69
System-software engineering gap
 cyber-physical system, 660
 development process, 660
 digital and non-digital components,
  660
 digital transformation, 660
 physical system, 660

product development processes, 660
streamlining, 659
traditional tools, 660
UML/SysML dominance, 661, 662
UML/SysML software/system, 662, 663
Systems operational dependency analysis
 (SODA), 1004
System Status Evaluation (SSE), 1177
Systems Test and Operation Language (STOL),
 971

T
Tabu-search algorithm, 601
Tactics, techniques, and procedures (TTP), 749
Tagging model elements, 320
Taxonomy, 791
Team-X and Team-A, 918
Technical Architecture, 388, 392, 393, 396,
 397, 399, 405, 407, 411, 412
Technical elements, 392, 396, 405
Technical inheritance process, 799
Technical interoperability, 424
Technical performance measures (TPMs), 233
Technical requirements document (TRD), 733,
 747
Technical solution, 405
Technological space, 933
Technology, organization and people (TOP),
 486–488
Technology Readiness Level (TRL), 787
Terms of Reference (TOR), 1025
Ternary criterion, 310
Testbed ontology, 678–680
Testbed repository, 688
Test-driven development (TDD), 336
Test Exit Reviews (TERs), 881
Test plan, 214
Test Readiness Reviews (TRRs), 881
Tethered Space Satellite, 786
The Europa Clipper project, 1138–1140
Theia web browser-based workbench, 945
Theory of codesign model system, 1293
Theory of structured cospans, 1293
The Program of Record in 2025 (POR2025),
 1023
Thermal Management Systems (TMS), 603
ThermalSensor, 907
Tight binding (TB), 434, 435
Time-to-critical-effect (TTCE), 1177
Tool adapters, 948
Topological data analysis, 1293
Torque adjustment, 392

Traceability diagrams, 906
Traceability matrices, 906
Trade-off analysis, 597
  custom MODA models, 137–140
  framework for integrated, 135
  MBSE capabilities to enable, 123–130
Tradeoff study, 214–216
  chocolate chip cookie acquisition system, 216–220
  requirements, risk processes and, 222–224
  with tradeoff matrices, 232
Tradespace, 122, 130, 132–134, 142
  exploration, 307, 316
Trajectory control module, 643
Transactional databases, 1123
Transformees, 665
Traveling Road Show (TRS), 968
True model-based requirements (TMBR), 364
Trust Attitude Object, 1159
Trusted Model Repository Pattern (TMR), 170
Trustee, 1158
Trust evaluation, 1160, 1161, 1163, 1164
Trust in belief, 1158
Trust in performance, 1158
Trustor, 1158

U
UAF Operational Connectivity Diagram, 871
UAF Strategic Roadmap, 870
UAV Mission Control Station, 457
Uncertainty, 197
  contributors to, 237
  human decision making, 244
  overarching process, 244
  sources, 215, 222
  system for handling, 244–249
Unidirectional relationships, 822, 823
Unified Architecture Framework (UAF), 262, 869, 873, 1055
  grid, 262
  metamodel, 269–271, 274
  modeling language, 263, 269
  UPDM, 263, 274
  views, 258, 262, 263
Unified Architecture Framework (UAF) Modeling Language (UAFML), 740, 743
Unified Architecture Method (UAM), 260
Unified Foundational Ontology (UFO), 937
Unified Modeling Language (UML), 383, 444, 504, 554, 802

Unified Profile for DODAF and MODAF (UPDM), 263
"Unique Heritage" reuse, 792
University Affiliated Research Center (UARC), 485
Unmanned aerial vehicles (UAV), 6, 456, 532
Unplanned reuse, 792, 793
Unsupervised learning, 641
US Census Bureau, 1082
US Centers for Disease Control & Prevention, 1082
US Department of Defense (DoD), 456, 658, 659
U.S. National Academy of Sciences, 551
Use case, 231, 901
  analysis, 341
  models, 1089
Utility, 240

V
Vaccination, 1127
Validated for Reuse, 845
Value iteration (VI), 1169
Value model, 1026
Value Model Development, 1038
Variant modeling with SysML, 82
Vehicle-integrated solution, 323
Vehicle Interface Library, 644
Very High-Speed Integrated Circuit Hardware Description Language (VHDL), 927
Vicinity Screening Subsystem, 670
Virtual Assistant, 492
Virtual environment, 636
Virtual HCD (VHCD), 472, 473, 494
Virtual prototypes, 473, 495, 528, 638, 642
Virtual representation, 642
Virtual Satellite approach, 790
Virtual simulations, 538
Virtual system models, 636, 643, 650
Visualization, 403
V-model, 384, 387
Vocabulary(ies), 937–939, 941
  bundle, 941

W
Wearable Coach^{TM}, 1080, 1081, 1084, 1085
Web ontology language (OWL), 30, 430, 1158
Weight of importance, 216–218, 226
White box view, 403
Witches, 238
Work Breakdown Structure (WBS), 800

Index

Workload analysis, 452, 460, 461, 463, 466
Work Models that Compute (WMC), 446
Workspace Reach analysis, 812
Workstation Simulation Models (WSIM), 966
WorldFrame, 644
Wymore's scoring functions, 240
Wymorian process, 198
Wymorian requirements modeling, 356

X
XML Metadata Interchange (XMI), 818

Z
Zig-zagging, 593